T0174168

Size, Structure, and the Changing Face of American Agriculture

Size, Structure, and the Changing Face of American Agriculture

EDITED BY

Arne Hallam

Routledge
Taylor & Francis Group

LONDON AND NEW YORK

First published 1993 by Westview Press

Published 2019 by Routledge
52 Vanderbilt Avenue, New York, NY 10017
2 Park Square, Milton Park, Abingdon, Oxon OX14 4RN

Routledge is an imprint of the Taylor & Francis Group, an informa business

Copyright © 1993 Taylor & Francis

Library of Congress Cataloging-in-Publication Data
Size, structure, and the changing face of American agriculture / edited by Arne Hallam.
    p. cm.
    Includes bibliographical references.
    1. Farms, Size of—United States. 2. Agricultural industrie—
United States. 3. Land use, Rural—United States. 4. United
States—Rural conditions. 5. Rural families—United States.
6. Agriculture—Economic aspects—United States. 7. Agriculture and
state—United States. I. Hallam, Arne.
HD1470.5.U6S59 1993
338.1'6—dc20   93-29046
                                                                        CIP

ISBN 13: 978-0-367-28732-0 (hbk)

Contents

Preface

This book's purpose is to provide a comprehensive discussion of concepts, methods of analysis, and empirical results related to the structure of agriculture, with particular emphasis on the United States. The focus is on changes in structure in the twentieth century and projected changes in the first decades of the next century. Particular attention is given to the dramatic changes in structure that have occurred in the last forty years and how the integration of agriculture both nationally and globally will affect structure in the near future.

The book addresses such questions as: Are profitable small family farms a thing of the past? Will agriculture become as industrialized as the rest of the economy? Is there a place for large mechanized farms in a society that values the environment? Can medium-size farms obtain the resources necessary to make a profit? Will rural communities survive as the number of farms falls? Does farm size affect the welfare of consumers or the quantity and quality of the food supply? Does technological change tend to favor particular types of farms? What is the role of government in encouraging or discouraging particular agricultural structures? Although the book does not provide definite answers to these questions, it provides a rich and deep framework and significant empirical data for analyzing these issues in a scientific manner.

The book is intended for several audiences. It is a source of information for undergraduate and graduate students studying agriculture, economics, rural sociology, or the political economy of rural areas. It also provides an excellent reference for professional economists and other researchers interested in the issues related to size, scale, and scope with an empirical emphasis on agriculture. It contains extensive references to the literature. The chapters on methods and data are intended for professional audiences, but most other chapters are written with a general audience in mind. Thus the bulk of the book is intended for anyone with an interest in agricultural issues, particularly those related to farm size, competition, and the future of agriculture.

Arne Hallam
Iowa State University
Tall Grass Prairie

Acknowledgments

This book is a product of North Central Regional Committee NC-181 entitled "Determinants of Farm Size and Structure in North Central Areas of the United States," which was initiated in 1986. Cooperating agencies were the agricultural experiment stations of Arkansas, Idaho, Illinois, Iowa, Kansas, Michigan, Minnesota, Nebraska, New York, North Dakota, Ohio, South Dakota, and Wisconsin and the Economic Research Service of the U.S. Department of Agriculture. The Farm Foundation provided financial as well as intellectual support for various meetings of this group. The idea for a book as a product of the committee came at the prodding of the administrative advisor, Robert Jolly, Iowa State University, and the public virtue of Lindon Robison, Michigan State University, who as committee chair felt obliged to satisfy the project's stated objectives. The committee, through annual meetings, identified topics to be addressed and selected tentative authors. Initial drafts were presented at committee meetings. These presentations were formally critiqued and informally discussed at the same meetings. Second drafts of the chapters were sent to outside reviewers. Some of these reviewers were authors of other chapters in the book, but most were outside professionals. Thus special thanks are due to Phil Garcia and Steve Sonka, University of Illinois; Sherrill Nott, Craig Harris, LuAnne Lohr, and Steve Hansen, Michigan State University; David Harrington, Economic Research Service, USDA; Robert Burton, University of Kansas; Loren Tauer, Cornell University; and Philip Paarlberg, Purdue University. Following this outside review, some chapters were dropped or significantly revised and a third set of drafts prepared. The final drafts were then edited for consistency.

A task of this magnitude is not without its pitfalls and headaches. Bruce Bullock, who replaced Robert Jolly as administrative advisor in 1989-90, was instrumental in securing funds for the publication of the book and running interference for the editor. The Department of Economics at Iowa State University contributed substantial intellectual and in-kind support throughout the project. The Economics Department at Brigham Young University provided a comfortable atmosphere during the sabbatical leave when most of the editing took place. Donna Otto of

Iowa State University donated two plus years of her time and probably more of her patience and sanity in preparing the manuscript. She endured several changes in the book's format, tardy authors, unintelligible writing in the margins, and constant interruptions to produce a manuscript precise, accurate, and easy to read. My sincere appreciation goes out to her for her outstanding attention to duty and responsibility. The final version is her product more than that of anyone else.

Special appreciation also is due to my wife, Susan, and my four patient, but ever-wondering-when-this-would-be-done, children (Ezra, Elizabeth, Justus, and Emmeline), who had to hear about all the problems with the manuscript when no one else would listen. They deserve high praise for patience with a project that conferred no direct benefits but was important to their father. Thank you.

A.H.

About the Contributors

Mary C. Ahearn is an agricultural economist with the Economic Research Service, USDA.

Jay Dee Atwood is an agricultural economist at the National Headquarters of the Soil Conservation Service, Division of Strategic Planning and Policy Analysis.

Joseph A. Atwood is an associate professor of economics at Montana State University.

Peter J. Barry is a professor of agricultural economics at the University of Illinois, Urbana-Champaign.

Marvin T. Batte is an associate professor of agricultural economics and rural sociology at Ohio State University.

Philip A. Bergen is an economist with the Agricultural Development Branch of Agriculture Canada.

Boris E. Bravo-Ureta is an associate professor of agricultural economics at the University of Connecticut.

Thomas A. Carlin is deputy director of the Agriculture and Rural Economy Division, Economic Research Service, USDA.

George L. Casler is a professor of agricultural economics, New York State College of Agriculture and Life Sciences, Cornell University.

Virginia L. Clark is dean and professor, College of Home Economics, South Dakota State University.

Steven R. Denault is a business analyst with Country Company Insurance, Bloomington, Illinois, and part-time instructor of economics at Hartland Community College, Normal, Illinois.

Ed Dickson is an economist with the Ontario Ministry of Agriculture and Food, Ontario, Canada.

Hisham El-Osta is an agricultural economist with the Agriculture and Rural Economy Division, Economic Research Service, USDA.

Glenn Fox is associate professor of agricultural economics at the University of Guelph, Ontario, Canada.

Cole R. Gustafson is an associate professor of agricultural economics at North Dakota State University, Fargo.

Arne Hallam is an associate professor of economics at Iowa State University.

Glenn A. Helmers is a professor of agricultural economics at the University of Nebraska.

Robert H. Hornbaker is associate professor of agricultural economics at the University of Illinois.

Michael A. Hudson is Bruce F. Failing, Sr., Professor of Personal Enterprise and Co-Director of Entrepreneurship and Personal Enterprise, Cornell University.

Larry Janssen is a professor of economics at South Dakota State University.

James D. Johnson is chief of the Farm Sector Financial Analysis Branch, Agriculture and Rural Economy Division, Economic Research Service, USDA.

Roger Johnson is a professor of agricultural economics at North Dakota State University.

Michael A. Mazzocco is assistant professor of agricultural economics at the University of Illinois.

Kent D. Olson is an associate professor of agricultural and applied economics at the University of Minnesota.

Wayne D. Rasmussen is an agricultural historian who was with the USDA for fifty years.

Lindon J. Robison is a professor of agricultural economics at Michigan State University.

William E. Saupe is a professor of agricultural economics at the University of Wisconsin-Madison/Extension.

Bruce J. Sherrick is assistant professor of agricultural finance at the University of Illinois.

Vincent H. Smith is an assistant professor of economics at Montana State University.

B. F. Stanton is a professor of agricultural economics at Cornell University.

Ron G. Stover is professor of rural sociology at South Dakota State University.

Luther G. Tweeten is Anderson Professor of Agricultural Marketing, Policy, and Trade, Department of Agricultural Economics and Rural Sociology, The Ohio State University.

Myles J. Watts is a professor of economics at Montana State University.

Gerald W. Whittaker is an agricultural economist with the Economic Research Service, USDA.

1

The Importance of Size and Structure in U.S. Agriculture

Arne Hallam

Agriculture in the United States has changed dramatically in the 20th century. The technology, organization, and structure of agriculture are dynamic, and future changes may dwarf past ones. The evolution of agriculture will have important impacts on farmers and society at large. The purpose of this book is to explain past changes in structure, identify factors leading to structural changes, analyze the effects of structural change on the well being of society, and suggest where structural change will lead in the future.

Notable characteristics of the United States economy over the past 150 years are significant output growth, rapid industrialization, and the shrinking of the agricultural sector in terms of contribution to GNP, of farmers, and of number of farms. Gross national product rose 40 fold in real terms over period 1869-1990.[1] Gross private domestic product originating in the farm sector fell from 35 percent in the period 1869-1879 to 10.3 percent in 1929, and to 5.8 percent by 1960.[2] Data based on current methods for computing national income show a drop in the farm sector share of GDP from 7.7 percent in 1929 to 1.9 percent in 1990.[3] Farm employment, which dropped from 13.5 to 9.9 million in the period 1910-1950, fell to 2.9 million by 1989.[4] The number of farms fell from a historic high of 6.4 million in the first two decades of this century to 2.1 million in the 1987 Census of Agriculture. This period was also characterized by significant increases in agricultural productivity. The index of farm output per man-hour increased from 8.22 in 1910 to 18 in

1948 to 158 in 1989 using 1977=100 as a base.[5] Much of this increase was due to substitution of capital for labor. Even accounting for changes in the mix of inputs, productivity increases have been substantial. Recent work by USDA on total factor productivity (total output per unit of total production input) shows an increase from 64 in 1948 to 92 in 1989 using 1982=100 as a base. Thus, productivity has increased by more than one-half in the post war years alone.[6]

Important changes in the way farm products are produced and the communities where production takes place have accompanied these changes in the number of people on farms and agriculture's contribution to national income. The United States economy has gone from one based heavily on subsistence, and production from many small family farms, to an industrial and service based economy with agricultural production taking place on fewer and larger farms. These changes in the structure of agriculture and the general economy have had profound impacts on aggregate and individual welfare. National income per capita, total food production and income and wealth per commercial farm have all increased dramatically. Yet, many farmers still receive production subsidies to raise income, numerous rural areas are impoverished, and hunger in the United States stubbornly persists. Several questions arise. How have changes in agriculture in this century affected the welfare of producers and consumers? How stable is the current agricultural sector? What are the factors that lead to changes in agricultural structure and its relationship to other sectors? Will future changes in agriculture have a large impact on the U.S. economy? Will changing attitudes about the environment, rural living, community, and sustainable economic policies dramatically affect the agriculture of the future? And how does a particular agricultural structure affect the well-being of society? Responses to these questions form the basis for this study on size and structure in American agriculture.

Agricultural Structure and Economic Welfare

There are many reasons for interest in the structure of agriculture. Some are economic and others relate to social and political issues. The most compelling motivation for the study of industry structure is its potential impact on socio-economic welfare. The way in which agriculture is organized has the potential to affect productivity, long run growth, and the stability of food supplies to name just a few elements of socio-economic well-being. Dramatic increases in the productivity of agriculture have influenced the well-being of the average U.S. citizen.

Many productivity changes are directly related to structure. Structural change in the future has the possibility of affecting national economic welfare both directly through market power and indirectly through its effect on the economic environment.

The Direct Effects of Firm Size and Structure

Economic theory implies that in competitive industries with constant or increasing costs all firms will earn zero economic profits and products will be produced at minimum average cost. In such industries prices will reflect costs of production and consumer welfare will be maximized for a given distribution of resources. If firms have declining costs, there will be a tendency for firms to grow and competition may break down, leading to monopolistic pricing. In such cases consumer welfare may be less than under competitive pricing. Agriculture is traditionally viewed as one of the bastions of perfect competition in the present day economy. Thus some view the movement towards larger firms as the precursor for a breakdown of competition and eventual monopolization of the food industry. While there are few signs of such a happening in the near future, the potential for such changes creates an interest in measuring size economies and other forces leading to fewer, larger firms.

The structure of farms can also affect individual profitability as the economy moves toward (but never reaches) long run equilibrium. During periods of adjustment to optimal size and technology, individual firms have widely differing rates of profit. While these differences may be eliminated in the long run, they have significant wealth effects in the shorter run. As will be discussed below, if early adopters of a new technology receive extra profits, which can then be invested in additional land or equipment, the distribution of current returns affects future firm size and numbers.

The rising size of a farm necessary to provide a reasonable level of family income is a direct link between farm structure and producer welfare. As optimal capital-labor ratios change over time, the fixed investment necessary to complement and reward the individual farmer may increase, thus leading to larger farms, and forcing those with smaller holdings to expand beyond a single family operation, or seek off-farm employment.

Economies of size can affect international competitiveness and changes in the terms of trade. In the less than perfectly competitive real world, economies of size in one country may be exploited to maximize domestic welfare; while economies of size in another country may be stifled and an industry protected to attain other social goals.

Indirect Effects of Structure on Society's Well Being

Non-competitive pricing and slow adjustment to equilibrium affect economic welfare directly. Agricultural structure also affects societal welfare in a number of indirect and important ways. These include technology, food quantity and safety, the environment, and interactions with rural communities.

Technological Change. The size and profitability of firms in an industry may affect the rate of technological progress in the industry and vice versa. For example, new technologies may favor large firms as opposed to small firms. Alternatively, larger firms may be more inclined to adopt new technology since they can apply it to a bigger resource base or spread its cost over more units of output. An industry characterized by competition may be more likely to adopt cost-reducing technologies. But an industry with some profit cushion may be more willing to risk capital on research and development.

Stability of Food Supply. An industry with numerous producers may be more likely to supply food in both good and bad economic times. If an industry is dominated by a few firms, profit opportunities in non-food crops or other sectors, poor management, attempts to manipulate the market, organized labor unrest, or bad weather in concentrated production regions could lead to sudden and unpredictable changes in food supplies. Given the importance of food for survival, concentration in the food industry may be viewed as being more serious than in other sectors. On the other hand, if there are significant economies of size in food production, then an industry characterized by many small and high-cost firms may not be the best way to obtain a stable and plentiful food supply.

Safety of Food and Food Products. Large or small firms struggling to survive in a competitive environment may take shortcuts in regards to food safety. While large firms in a less than perfectly competitive environment with a branded product may have more to lose by cutting back on quality or safety, smaller firms in a more competitive setting may be more restrained by the rigors of a market disciplined by sheer firm numbers. A distribution system under fairly centralized control may be better able to ensure product safety through the various stages of the production and marketing process. Specialized, expert workers such as those often hired by large firms may be better able to provide quality control, as opposed to workers for small firms who are spread too thin. On the other hand, workers who share in firm ownership may have greater incentive to follow accepted food safety procedures.

Structure and the Environment. There is significant debate about the effects of farm structure on the environment. Do large firms tend to ignore environmental impacts or are they better able to follow accepted procedures due to superior management and access to newer technologies. Are there diseconomies of size in disposal of wastes? Were the family farms of the last century more concerned with soil conservation than the large corporate farms of today. Solid answers to these questions have not been provided by either research or experience. Such questions are of priority interest and are dealt with in some detail in Chapter 21.

Rural Communities and Rural Life. The decline in the number of farmers has been associated with a decline in the population of many rural communities. Can such smaller communities attain sufficient economies in the provision of public services such as schools, roads and government to remain viable? Do larger farms tend to lead to poorer rural communities through the commodification of the labor force or do they provide more jobs through value added and articulation with input suppliers? These issues are highlighted in Chapter 20.

Working Conditions and Individual Freedom. Small and medium size farms tend to be family owned and operated with little hired labor. Many larger farms are also family owned and operated. As farms become larger and hire more labor do working conditions tend to deteriorate? Does the repetitive nature of labor tasks on large farms lead to a less enjoyable life style than the variety inherent in owner-operated agriculture. Are hired workers given the same chances to enjoy the "agrarian lifestyle" as owners, or do large farms improve quality of life for disadvantaged workers by providing jobs unavailable elsewhere?

Economic Policy and the Political Process. Operators of large farms may be more motivated and better able to lobby for economic support than operators of smaller farms. Payments from government commodity programs tend to be disproportionately concentrated on larger farms. Larger farms may be better able to obtain assistance from the state and local extension and related public programs. But programs to improve the efficiency of agriculture may be more conveniently delivered to a few large farms than to many small ones.

The Intrinsic Value of Certain Agricultural Structures

While many arguments about farm structure turn on economic well-being, various popular arguments are more related to the intrinsic values attributed to small family farms and notions about "utopian" rural communities. While such arguments may not have the "rigor" of traditional economic analysis, they are important politically and socially.

Many authors are convinced that the current economic system is destructive to both the global system and economic community in a holistic as opposed to a narrow profit or efficiency motivated paradigm (Daly and Cobb 1989). For such groups, arguments based on traditionally measured economic welfare miss most of the key issues.

Farm Fundamentalism and the Inherent Goodness of Family Farms. One of the earliest proponents of agricultural fundamentalism in America was Thomas Jefferson. In 1788 he wrote:

> Those who labor in the earth are the *chosen people of God*, if ever he had a chosen people. . . . Corruption of morals . . . is the mark set on those who, not looking up to heaven to their own soil and industry as do the husbandman for their subsistence, depend for it on the casualties and caprice of consumers.

William Jennings Bryan (1896) was a strong proponent of the magic of agriculture. In classic lines he cried: "Burn down your cities and leave our farms, and your cities will spring up again as if by magic; but destroy our farms and the grass will grow in the streets of every city in the country." In an article in *Cosmopolitan* in 1904 he further extolled the virtues of the farm life as "an *independent* way of living", requiring "*less capital* to begin work", as "*healthful*", and as teaching "*the true basis of rewards*". He also argued that "farm life cultivates hospitality and generosity, and, without entirely removing temptation, gives parental influence a chance to strengthen the child before the seeds of disobedience are planted by evil associations".

More recent writers have echoed a similar tune. Horace Hamilton (1946) summarized the views of many on the contrast between family and large scale commercial farms at a conference in 1946. He characterized the family farm as producing "men of strong character and moral consciousness", as opposed to the labor forces on commercial farms who frequented "pool rooms, honky-tonks, cheap picture shows . . .", and had ". . . flashy, back-slapping personalities."

Similar sentiments were echoed during periods of financial stress in the 1980s by groups as diverse as the American Agriculture Movement, The Center for Rural Affairs (Strange 1988), and the Catholic Church. Some economists (Tweeten 1989) have discounted these arguments but they still have immense public sway.

Small Is Beautiful. The popular best seller in 1973 by E.F. Schumacher entitled *Small is Beautiful* spawned a movement towards smaller and more worker-oriented firms in all sectors of the economy including farming. Some have argued that work is a more meaningful experience when carried out in small, more decentralized firms. While economists might argue that workers should make these preferences known in their wage offers and the jobs they seek, others counter that the current industrial system allows no such freedom of choice.

The Amenity Value of "Small" Farms and "Pastoral" Rural Areas. While perhaps not a rallying cry of agricultural groups, many arguments about size in agriculture implicitly attribute a value to the sight of energetic farmers scratching out a livelihood from the fertile yet hostile environment, and a countryside covered with rolling pastures, clear streams, and the "good" shepherd quietly tending his flocks. Many current urban dwellers attach some value to rural areas as places to visit, enjoy nature, and engage in recreation. Some of these activities are viewed as being more enjoyable in a countryside dotted with small well kept farms, as opposed to large smelly feedlots or poultry factories.

Whatever the reasons, many in America today feel that the family farm should be preserved (Jordan and Tweeten 1987). Many of the current debates about structure relate to the family farm issue and how it will fare over the next century.

Other Interest in Agricultural Structure

Anticipating the changes in agricultural structure is of interest in and of itself. Over time, industries adapt to meet changes in technology, consumer preferences, and world conditions. The ability to predict and understand such changes reduces uncertainty for individual firms, mitigates stress for consumers, allows policy makers to design programs that direct change towards desired ends, and helps investors to make wiser decisions on resource allocation. Observing current economies of size, as one of many factors that affect the future path of an industry or sector, may help predict the changing structure of U.S. agriculture.

Agricultural structure is an important issue for a variety of reasons. How that structure has evolved and will continue to evolve over time constitutes a major purpose of this book.

Purpose and Plan of the Book

The objective of this book is to provide information on the changing structure of U.S. agriculture, increase understanding of the factors leading to structural change, and render insight into the potential future structure of agriculture. The book accomplishes these objectives by setting forth six major purposes. The purposes of the book are: to define agricultural structure, to discuss the structure of agriculture and how it has changed, to identify factors leading to structural change, to assess how agricultural change affects economic welfare, to analyze the relationship between agricultural structure and economic policy, and to inform concerning the techniques and data used to study agricultural structure. The sections of the book address different parts of the structure issue; each section accomplishing a portion of each purpose. Chapters 1 through 6 describe structure, chapters 7 through 11 discuss methods and data, chapters 12 through 19 are concerned with the determinants of size and structural change, while chapters 20 through 22 examine the implications of structural change for agriculture and other sectors of the economy.

What Is Meant by Agricultural Structure

Structure means different things to different groups. At least some of the debate about agricultural structure is rooted in different understandings of what constitutes structure and how to measure it. The book makes a serious attempt to provide a new and comprehensive definition of agricultural structure such that the key forces and actors are clearly identified. The book also presents an approach for categorizing the numerous elements of structure in an organized fashion. Chapter 2 discusses these issues and provides a framework for the rest of the book.

The Structure of American Agriculture and How It Has Changed

American agriculture has changed dramatically since the revolution. The book documents changes in agriculture with special emphasis on the 20th century. This history gives insights into the forces that have changed forever the face of the countryside. The book also describes the current structure of American agriculture in detail. The material gathered here provides a comprehensive source of information on the structure of American agriculture. The book also examines the stability of the current structure and its likelihood to endure.

Factors that Bring About Change in Agricultural Structure

A major purposes of the book is to discover factors which have the potential to significantly influence structure. Factors within agriculture impact structure, but some of the most important factors may be outside of agriculture altogether. For example, the financial health and stability of agricultural lenders may influence the growth of the sector, but such effects may be dominated by the nation's overall credit market. The book will attempt to determine which factors have been important in the past.

The book will present theoretical and empirical arguments about the factors that have the greatest potential for impact, and will attempt to identify which factors will be most important in the future.

How Does Agricultural Structure Affect Economic Welfare?

This introductory chapter has discussed some of the ways in which agricultural structure contributes to both sectoral and general welfare. The book will flesh out these arguments and provide insights into just how important (or insignificant) agricultural structure is to the rest of the economy. Factors that may lead to major income distribution changes in agriculture may have little impact outside the sector. Alternatively, major income transfers to agriculture may do little to affect the structure of the sector.

The Relationship Between Agricultural Structure and Economic Policy

Many agricultural policies are enacted with the stated purpose of influencing agricultural structure. For example, the "family farm" has been the focus of a variety of policies ranging from the Homestead Act to the formation of the Farmers Home Administration and commodity programs. Can policy actually affect structure? Have past policies addressed towards structure had the desired impact, or any impact at all? Agricultural structure also affects policy. The lobbying power of agriculture has declined over time as the congressional farm bloc has lost size. Can large farms continue to tilt the economic scale in their direction? Do organized farm groups have an appreciable impact on policy? Will antitrust policy become a major factor in the food and agriculture sector of the future?

Methods and Data Useful in Characterizing Size and Structure

A variety of analytical methods have been used to measure agricultural structure. These methods range from simple surveys and frequency tabulations to complicated nonlinear statistical models. There is significant debate about the results of these studies and frequent criticism of the data and methods used. The book will provide a comprehensive review of methods used in studying economies of size and scale, ways of analyzing profitability, and techniques for describing industry structure. Data for analyzing agricultural structure comes from a variety of sources and using a variety of collection techniques. The book provides a thorough discussion of the data used to study size and structure issues, and some of the problems in using these data.

Organization of the Book

The Importance of Size and Structure Issues in American Agriculture. Chapters 1 through 6 present an overview of farm size and structure and detail the past and current situation. Chapter 2 discusses the concept of farm structure, considers alternative definitions, proposes a framework for analyzing farm size and structure issues, and mentions some important areas of disagreement about measuring farm size and structure. Chapters 3 and 4 discuss U.S. agricultural structure in an historical context and bring the reader up to date on current agricultural structure. Chapter 4 also discusses some alternative ways to characterize structure such as by sales, acreage, and tenure. Chapter 5 is a detailed study of current agricultural structure in the North Central region. The chapter gives more detail than is possible at the aggregate level about one important agricultural area. Chapter 6 is an empirical study documenting cost size relationships for corn, soybeans, and wheat. Using data from the Farm Cost and Returns Survey, the authors examine the cost of production for farms of various size classes.

Methods and Data Pertaining to Size and Structural Change in Agriculture. Chapters 7 through 11 of the book review methods and data used in studying size and structure issues. Chapter 7 is a review of methods used to measure economies of size and scale. The chapter is self-contained and provides a extensive discussion of techniques that have been used to measure economies of size, scale and scope. Chapter 8 is the empirical complement of chapter 7 and reviews studies of size and scale in agriculture. The chapter attempts to develop a consensus about economies of size in U.S. agriculture. Chapter 9 focuses on the question of why some farms are more successful than others, and on how management may affect firm viability and sustainability. Chapters 10 and 11 provide descriptions of data used for size and structure studies as collected at the state and national levels. The chapters pull together information not before available in one consistent source.

Factors Influencing Structural Change in Agriculture. The bulk of the book is comprised of new studies that highlight key factors influencing agricultural structure. The factors were identified in part by members of NC-181, a regional project dealing with size and structure in U.S. agriculture. Authors were chosen based on their expertise in the various areas identified. Chapter 12 discusses how technology has and will continue to impact the structure of American agriculture. Chapter 13 analyses the impact of government commodity programs on farm numbers while chapter 14 analyses the effects of tax policy on farm structure. Given the often ascribed structural intent of government policy, these chapters evaluate the effectiveness or impotence of such policy. Chapter 15 presents arguments about the effects of agricultural finance on the structure of the sector and the implications of changes in financing arrangements on future structure. Chapter 16 investigates the changing structure of the agribusiness sector and its implications for production agriculture. Chapter 17 is an examination of the effects of risk and uncertainty on the optimal structure of farms, while Chapter 18 explores changing tenure patterns in agriculture and how they affect other aspects of structure. Chapter 19 discusses how family organization and style of decision making affect the structure of farm businesses and how they are run.

Impacts of Structural Change in U.S. Agriculture. The final chapters look at the broader implications of agricultural structure. Chapter 20 examines the relationship between agricultural structure and the health of rural communities. The effects of rural communities on agriculture and its well-being are also discussed. Chapter 21 addresses issues related to agriculture and the environment. The effects of farm size and structure on the environment are analyzed. The ways in which agricultural structure may influence environmental regulations are also discussed. Chapter 22 projects future farm size and structure in the United States.

What the Book Is Not

This book is broadly based and includes input from economists, individuals in management science, sociologists, and political scientists. The book addresses issues from a general social science perspective, but emphasizes models developed by economists. The book has no political agenda and any biases are those of the individual authors. Individual authors have approached the issues from different backgrounds and perspectives, and thus conclusions may not always agree between chapters. The authors have tried to follow a common framework and a use a common set of empirical observations. The approach of the book is to present the facts and then interpret them in light of historical experience and our understanding of human behavior. Almost all the chapters were presented orally in meetings attended by other authors. The very lively and sometimes heated discussions at these meetings has had a significant impact on the final content of chapters. All chapters were also reviewed by individuals knowledgeable in the respective areas so as to validate approaches and conclusions.

The book does not cover all the important issues related to agricultural structure. While the chapter on rural communities examines some of the external aspects of structure, the book does not address many important issues related to the overall economic system, freedom, worker satisfaction, and quality of life. The book is noticeably silent on topics related to economic development and international trade. The book also has relatively little to say about the impact of the general economy on agricultural structure, except in a rather static sense. One must draw the line somewhere, however, and the topics selected here will provide sufficient discussion material for all but the most ravenous of consumers. For those interested in further material, the proceedings of NC-181 edited by Robison (1988) and Hallam (1989, 1990) are a good place to start.

Notes

1. Data on GNP are taken from Romer (1986) and *Economic Report of the President* (1991).

2. Data on gross domestic product originating in the farm and non-farm sectors are taken from the Department of Commerce publication *Historical Statistics of the United States* (1960, 1975). These data are inexact but give the flavor of aggregate changes.

3. See *Economic Report of the President* (1991).

4. See *Economic Report of the President* (1991), and *Historical Statistics of the United States* (1975).

5. Data were taken from *Production and Efficiency Statistics* (1989), ECIFS 9-4 and additional unpublished data from the same source.

6. USDA reports Tornqvist output, input, and productivity indices in the publication *Production and Efficiency Statistics* (1989), ECIFS 9-4.

References

Bryan, William Jennings. 1896. *The First Battle: A Story of the Campaign of 1896.* Pp. 199-206. Chicago: W. B. Conkey Company.

Bryan, William Jennings. "Farming as an Occupation." *Cosmopolitan* 34 (January 1904): 369-371 as compiled in *William Jennings Bryan: Selections*, Ray Ginger, ed. New York: Bobbs-Merrill, 1967.

Daly, Herman E., and John B. Cobb. 1989. *For the Common Good: Redirecting the Economy Toward Community, the Environment and a Sustainable Future.* Boston: Beacon Press.

Economic Report of the President. 1991. Washington, DC: GPO.

Hallam, A. 1989. *Determinants of Farm Size and Structure*. Proceedings of the program sponsored by the NC-181 Committee on Determinants of Farm Size and Structure in North Central Areas of the United States, held in January, Tucson, AZ. Ames, IA: Department of Economics, Iowa State University.

_____ . 1990. *Determinants of Farm Size and Structure*. Proceedings of the program sponsored by the NC-181 Committee on Determinants of Farm Size and Structure in North Central Areas of the United States, held in January 1990, Albuquerque, NM. Ames, IA: Department of Economics, Iowa State University.

Hamilton, Horace. 1946. "Social Implications of the Family Farmer," in J. Ackerman and M. Harris, eds., *Family Farm Policy: Proceedings of a Conference on Family Farm Policy*. Chicago: University of Chicago Press.

Jefferson, Thomas. 1788. *Notes on the State of Virginia*. Philadelphia: Prichard and Hall.

Jordan, Brenda, and Luther Tweeten. 1987. "Public Perceptions of Farm Problems." Research Report No. P-894. Stillwater: Agricultural Experiment Station, Oklahoma State University.

Robison, L. 1988. *Determinants of Farm Size and Structure*. Proceedings of the program sponsored by NC-181 Committee on Determinants of Farm Size and Structure in North Central Areas of the United States, held January 16, 18, and 19, 1988, San Antonio, TX. Michigan Agricultural Experiment Station, Journal Article No. 12899.

Romer, Christina. 1986. "The Prewar Business Cycle Reconsidered: New Estimates of Gross National Product, 1869-1918." *NBER* Working Paper 1969.

Schumacher, Ernst Friedrich. 1973. *Small is Beautiful: A Study of Economics as if People Mattered*. London: Blond and Briggs.

Strange, Marty. 1988. *Family Farming*. Lincoln: University of Nebraska Press.

Tweeten, Luther. 1989. *Farm Policy Analysis*. Boulder: Westview Press.

United States Department of Agriculture. *Agricultural Statistics*, various issues. Washington, DC: GPO.

United States Department of Agriculture. Economic Research Service. *Production and Efficiency Statistics*. 1989. ECIFS 9-4. Washington, DC: GPO, April 1991.

United States Department of Commerce. Bureau of the Census. *Census of Agriculture*. Washington, DC: GPO, various issues.

United States Department of Commerce. Bureau of the Census. *Historical Statistics of the United States, Colonial Times to 1957*. Washington, DC: GPO, 1960.

United States Department of Commerce. Bureau of the Census. *Historical Statistics of the United States, Colonial Times to 1970*. Washington, DC: GPO, 1975.

2

Farm Structure:
Concept and Definition

B. F. Stanton

Farm structure, like a number of terms which are widely used by the general public, has no single, widely accepted definition. The word *structure* suggests a framework around which a more complete entity or whole is constructed. Perhaps *framework* is the ideal word or concept to have in mind as we examine alternative ways in which "farm structure" has been defined and used. The basic objective of studying *farm structure* is to understand more fully how and why the sector of the U.S. economy that produces agricultural products is changing, and what such change may mean in the future. Farm structure has become one of the central concerns of those interested in agricultural policy in the second half of the twentieth century. The meaning of this concept and reasons for interest in it are central to this book.

This chapter is designed to help the reader grasp the concept of farm structure and more fully appreciate the discussion surrounding this concept. In the first section, the concept or definition of structure is explored and developed. Since the measurement of farm size is a critical component of structure, the second section describes the different methods of classifying farms by size along with the problems, caveats, and pitfalls of each method.

Definitions of Farm Structure

One way to understand how the term *farm structure* is used and understood is to examine a few definitions and statements that have been published by recognized authorities in the field. These quotations provide additional insight into the problems and concerns that are commonly addressed under the heading *farm structure*. The first citation is an excerpt from a standard dictionary. This is followed in turn by direct quotes from three relatively recent and important sources. "Structure. . . . The arrangement or interrelation of all the parts of a whole; manner of organization or construction: as they studied the structure of the atom, the structure of society . . ." – *Webster's New World Dictionary of the American Language,* World Publishing Company, 1966, p. 1447.

The *structure of agriculture* refers to the number and size of farms; ownership and control of resources; and the managerial, technological and capital requirements of farming. The issues of the structure of agriculture are illustrated by such questions as:

- Will the family farm survive?
- Do farm programs help or injure the chances of family farm survival?
- Who controls production and marketing decisions at the farm level?
- What is the balance of market power among input suppliers, farmers, and marketing firms?
- Will U.S. agriculture eventually become industrialized and controlled by large agribusiness corporations?
- What type of agriculture is wanted in America?

(From Knutson, Penn and Boehm, *Agricultural and Food Policy,* 2nd edition, Prentice Hall, 1990, Chapter 11, p. 270.)

Structure is not an easy concept to define. It involves the following components:

- Organization of resources into farming units.
- Size, management, and operation of these units.
- Form of business organization, whether a sole proprietor or several individuals in a partnership or corporation.
- The degree of freedom to make business decisions, and the degree of risks borne by the operator.
- Manner in which the firm procures its inputs and markets its products.
- Extent of ownership and control of the resources that comprise the farming unit.
- Ease of entry into farming as an occupation.
- Manner of asset transfer to succeeding generations.
- Restrictions on land use; immediate 'sovereignty' versus stewardship for future generations.

16

The term *family farm structure*, although loosely and imprecisely used, often means a relatively large number of modest-sized farms, each operated by a family unit, perhaps employing some nonfamily labor, but with the husbandry and management decisions by the operator and family and the inputs purchased from and products sold in open, easily accessible, competitive markets. Obviously, a wide range of structural configurations would fit within this definition.

(Penn, J. B., "The Structure of Agriculture: An Overview of the Issue," *Structural Issues of American Agriculture*, ESCS, USDA, Agr. Econ. Rpt. 438, November 1979, p. 5.)

Farm Structure refers to farm size and numbers, tenure patterns, legal organization (sole proprietorship, partnership or corporation), the market arrangements under which farmers buy and sell, and the institutional arrangements (including, of course, the public sector) influencing the farming industry . . .

The four components of the definition of the family farm given in the introduction provide a framework to examine the prospective future structure of farming, which is moving toward:

1. fewer farms and less production under family sole proprietorships and more under nonfamily-type corporations, partnerships and conglomerate firms. Vertical coordination in the form of production contracts and vertical integration is on the rise and extending to additional commodities such as swine. Farm operators increasingly must share control of their decisions with the government, landlords, banks and others;
2. more farms and production under arrangements whereby the operator furnishes less than one-half the labor;
3. more part-time operations in which the operator and family receive more income from off-farm than from farm sources;
4. an industry in which farms of all types have gradually if erratically improved their rate of returns on resources and incomes to parity with those in the nonfarm sector in normal years . . .

(From Tweeten, Luther, *Causes and Consequences of Structural Change in the Farming Industry*, NPA Food and Agriculture Committee, FPA Rpt. 207, P. 54, 1984.)

Common Elements in These Definitions

These quotations have a number of elements in common which are fundamental to an understanding of structural issues in a policy context.

It is important to recognize first that all farm structure issues are concerned simultaneously with: (1) farms and farm businesses, (2) farm households, and (3) agricultural resources. There are some important reasons for examining each of these aspects of farm structure separately because the forces of change touch each of them somewhat differently.

1. *Farm Businesses* -- At the center of any study or discussion of farm structure are a set of business entities with balance sheets, profit and loss statements, commonly referred to as farm businesses. These operations combine the services of land, labor, capital, and management in production and sustain profits and losses like any other business. They are somewhat unusual in the sense that land is commonly such an important component in the mix of resources used in production. Yet even here there is wide variation among them from intensive greenhouse operations at one extreme to cow/calf operations making use of large areas of open range and pasture for every animal unit. The business as a productive enterprise contributing value-added to the national economy is basic to this concept.

2. *Farm Households* -- It is important to recognize and differentiate decisions and actions taken by the farm household from decisions and actions taken by the farm business. Increasingly, such decisions are separate, even though interrelated. In terms of priority, it is the welfare of the members of the household that usually takes precedence in decision-making if there is a conflict between the household and the business. Following the same logic, decisions about off-farm and on-farm employment by family members is taken in the context of what is best for the farm household in total, not necessarily what is best for the farm business. Understanding the nature of change in farm structure requires conscious recognition of the role of the household in actions of family members, especially as more and more members of farm households hold off-farm jobs.

3. *Agricultural Resources* -- Fundamental to the organization of agricultural businesses is the resource base associated with farming in all its many dimensions. For most farms, land is a necessary and key component. For livestock rearing enterprises, it may encompass large areas of land with substantial amounts of capital invested per worker in the business. On crop farms, whether they produce extensive crops like hay and grain, or intensive crops like vegetables and fruit, the quality of the land resource and the associated climate and location with respect to markets has much to do with production alternatives and the

way farming is organized. Land, labor, capital, and management collectively are the fundamentals around which farm structure is constructed and the basic mechanisms through which structural change occurs is carried out. Thus, the quality and quantity of these resources available in each region are one of the key determinants of farm structure as it evolves, as well as to the alternatives open to farm businesses and farm households.

Public interest in and concern for the protection of the environment has increased in recent decades. Greater priority is being given to the ways in which agricultural resources are used, and the incentives provided to insure that such use is wise with appropriate concern for future generations. This increased focus in public policy may well influence structural change, particularly in the ways environmental constraints affect farm size and costs associated with new requirements.

Efforts to understand farm structure and analyze changes observed over time must recognize the separate ways in which: (1) the nature of the agricultural resources, (2) the requirements of the farm household, and (3) the needs of the farm business itself influence the processes of change. The old idea, that farming is a way of life and that all the members of the farm household focus on the farm business for life, no longer holds for much of American agriculture. Farms are operated as businesses or for other reasons such as recreation or a place to live in the country. Interests of household members may dominate farm decisions. So may environmental considerations imposed outside the farm or household. The resource base and its possibilities continue to influence what the business can and cannot do.

Major Issues and Concerns About Farm Structure

Changes in the structure of agriculture are often discussed in the context of such things as the size distribution of farms, the degree to which production has been vertically integrated in a given industry, the importance of tenancy, and the freedom of operators to make key production decisions. All of this requires that the farm sector be considered in the larger context of the food and fiber industries of which farming is one part. One can think of the larger industry as made up of five major parts: (1) input supply, (2) farming, (3) processing and manufacture, (4) retailing, and (5) food away from home. Farming is increasingly dependent on many sources for the inputs they combine with their land, labor and capital to produce crop and livestock products. They market their products through a large and complex sector of processing and manufacture. Structural change within farming is clearly influenced by events and change in the rest of the food and fiber industry.

Some of the most important questions concerning farm structure must be considered in this larger industry context. Among those which are most widely discussed are:

1. *Changing Distributions in an Industry Context* -- Nearly all discussions of farm structure include an examination of changes in numbers and sizes of farms. It is changes in cropland per farm or capital investment per farm or gross sales per business unit that are used in describing structural change or in addressing major concerns about structure. It is from aggregate distributions that one can learn what proportion of industry sales is provided by the 10,000 largest farms or what has happened to the numbers of small, or mid-sized farms, however these are defined. No single frequency distribution can effectively describe farm structure. Hence, distributions of farms by sales class, land area, labor force, acres of key crops, or numbers of livestock are all used in examining structure and change through time. What happens to these distributions remains a focus of public interest and debate.

   The ways in which distributions of numbers of farms by size are organized provide evidence for alternative explanations of the declines in numbers observed over various time spans. It is particularly important to recognize that $100,000 of sales in 1969 is roughly equivalent to $200,000 of sales in 1979 because of the inflation which occurred and the doubling of prices in that decade. Changes in size distributions based on gross sales data can be easily misinterpreted. As economies of size concentrate more of total farm output in the hands of larger producers, efforts to look at percentages of total product sold by size class may be important to recognize. Particular concerns about the effects of changing size distributions on access to markets and the implications of smaller numbers of large producers for rural communities emerge.

2. *Production Decisions and Organization* -- Associated with the industrialization of agriculture has come market demands for consistent supplies of uniform products in relatively large quantities. Input suppliers on one side of farm producers, and processors and marketing firms on the other side have become fewer in number, with the power to impose substantial requirements on the production and marketing decisions of producers.

Complete independence for the producer in deciding what to plant or produce and when and how much to sell has become less and less possible for an increasing number of relatively perishable products.

The organization of production on the farm must increasingly fit into an industry context. The more perishable the product, the more important that it move to market in an efficient and well established manner. For those farmers who wish to control both production and marketing decisions, they must establish a special marketing niche in localized markets, utilize roadside markets or pick-your-own operations, or some other retailing relationship. Alternatively, they will contract or build a relationship with an integrated production and marketing system or become the integrator. The more storable the product, in general, the less pressure to be directly tied to some formal arrangements for buying inputs and selling products. Increasingly, farm businesses examine such options or participate in cooperative or corporate systems that allow efficiency in buying, selling or obtaining services.

3. *Resource Ownership and Control* -- Fixed investment requirements per worker in commercial agriculture are high by comparison with nearly any other modern industry. Real estate requirements explain an important part of this; capital for livestock and machinery are another major component. Historically, individual proprietorships have been the dominant form of business organization in farming. Thus, access to resources for farm production and control over them has been a major concern in the evolution of farm structure in nearly all industrialized societies.

Historically, control over the use of land and tenancy issues have been major concerns of people in both developed and less developed societies. In the United States, tenancy and share cropping were major policy issues in the 1920s and 1930s. By the 1950s, most farmers in the United States owned at least part of the land they farmed and rented some land in addition. In the later years of this century, part owners (as contrasted with full owners) have become the dominant group in commercial agriculture. The Jeffersonian notion of a nation of independent freeholders operating their homesteads with family labor does not describe any segment of American agriculture today even though this myth lives on in the minds of many Americans.

As less and less land and the associated capital is owned by the farmers who operate it, the way in which returns to these resources is distributed will be of increasing interest to owners, operators, and society at large. The potential for concentration of power in the hands of a few key decision makers continues to undergird the discussion of farm structure issues. In individual communities, there are the beginnings of such control or positions of power, even though at regional or national levels, such concentration is small compared to that in most other industries. The broiler industry, fed cattle, and some processed vegetables provide examples of situations where public concerns are already evident (Reimund, Martin, and Moore 1981).

Nearly all of the commonly raised issues with respect to changes in farm structure are related in some degree to the three general issues just reviewed. There is substantial overlap in such discussions. Data on changing numbers of farms are usually cited to support arguments about elements of monopoly power in the hands of producers or integrators. Substantial space is devoted to many of these issues in succeeding chapters of this book with substantial citations to methodology and research experience in each area.

Study and Analysis of Farm Structure Issues

The selected definitions of farm structure and the preceding paragraphs call attention to common elements and issues associated with discussion about farm structure. They are intended to provide a frame of reference for succeeding chapters in this book. Questions about structure and size distributions, including methods of analyzing them, revolve around farms, households and agricultural resources. Increasingly, the focus of attention is on public policy issues. The effects of government programs on the size distribution of farms; the ability of farms operated primarily with family labor to compete with larger farms; the degree of control by major processors over all phases of production in some sectors of agriculture; the effect of farm organization on the environment and the rural community -- these are the kinds of issues which generate interest in understanding more fully what is happening to farm structure and the forces which bring about change. No single size distribution or method of analysis can effectively characterize the complexity of farm structure. A sense of the larger context in which structural change will continue to occur is essential.

The word, size, is used throughout the discussion of farm structure. In fact, size is such an important part of the structure discussion that the phrase "size and structure" is used widely instead of just the word, structure. Even though size does have this important role, there is not a universally accepted procedure for measuring size. In the next section, alternative methods of measuring size are presented and discussed.

Efforts to Classify Farms and Measure Farm Size

Substantial efforts have been made by various units of government, both in the United States and other developed countries in the world to develop systematic ways to classify farms into meaningful groups to aid in discussing public policy issues and to describe more accurately changes as they occur in the structure of agriculture. Most of these efforts use multiple criteria to classify farms into meaningful groups. An internationally accepted classification system has not emerged. The major descriptive measures center on: (1) economic size in terms of output, and (2) amount of labor used in production, and (3) the share of household income provided by the farming enterprise. Measures using physical quantities, such as acreage or numbers of animals, have not been used widely in the U.S. because of the problems of aggregation across different types of farming. It is quite common to talk about farm size in terms of tillable acres on crop farms, especially when the type of farm is held constant. In Western Europe, areas of cropland expressed in hectares is the most commonly used measure of farm size despite the problems of comparability for extensive and intensive production systems.

Census Classification into Economic Classes

Under the leadership of Ray Hurley at the Bureau of the Census and with the encouragement of the Bureau of Agricultural Economics, USDA, as well as his Census Advisory Committee, an Economic Classification of Farms was developed for the 1950 Census of Agriculture. All farms were first divided into two groups: "commercial" and "other" (Table 2.1). The *commercial* farms were further divided into six classes on the basis of the value of farm products sold. The *other* farms were subdivided into three groups with the general titles of "part-time," "residential," and "abnormal." The major criteria used in classification were value of farm products sold, days of work off-farm by the operator, and income of family members from off-farm sources.

In many respects, this system divided farms into three major categories: *full-time, part-time* and *residential*. The subdivision for economic Class VI differs from part-time only on the reported number of days of work off the farm and income from off-farm sources. If one were to assume that most of the 717,201 farms in economic Class VI were, in

TABLE 2.1 Distribution of Farms by Economic Class, 1950

Class	Criteria Used: Value of Farm Products Sold	Criteria Used: Other	Number of Farms
Commercial:			
I	$25,000 and over	None	103,231
II	10,000 - 24,999	None	381,151
III	5,000 - 9,999	None	721,211
IV	2,500 - 4,999	None	882,302
V	1,200 - 2,499	None	901,316
VI	250 - 1,199	Less than 100 days of work off farm by operator; income of family members from off-farm sources less than value of farm products sold.	717,201
			3,706,412
Other:			
Part-time	$250 - 1,199	100 days or more of off-farm work by operator; income of family members from off-farm sources greater than value of farm products sold.	639,230
Residential	Less than $250	None	1,029,392
Abnormal	Not a criterion	Institutional farms, experimental farms, grazing associations, etc.	4,215
			1,672,838
Total number			5,379,250

Source: U.S. Census of Agriculture, Volume II, 1950, pp. 1109-10.

fact, individuals who necessarily were getting more than half their net income from off-farm sources, they could well be counted with the part-time units. Thus, 56 percent of the total, just under three million, could be considered *full-time* farms; 25 percent were *part-time* or close to that designation; and 19 percent were *residential*.

Hurley continued to experiment with Economic Classes during the next two decades, adjusting the six commercial categories to reflect both changes in prices and technology (Table 2.2). Most of the sales class intervals doubled between 1950 and 1969 even though the Producer Price Index for farm products and processed foods and feeds had only increased from 93.9 to 108.0 over those 20 years. (The increase in prices was not as important as the rapid adoption of new technology and its impact on output per unit of labor). The "other" categories now included *part-time* and *part retirement* with the use of an age criterion as well as days of work off the farm.

In 1974, the economic classes were dropped and have not reappeared in subsequent Census publications. No doubt the tremendous changes in prices and technology for agriculture between 1969 and 1974 were part of the reason. While there were obvious problems in establishing meaningful criteria in which to group farms by size, the lack of such classes has left interpretation of these distributions to each reader, often unskilled in thinking about the many different forces which shifted farms from one sales class to another. One consequence of dropping the official economic classes is the conclusion by some that the great restructuring of American agriculture, which occurred between 1950 and 1969, is continuing at the same rates in the 1970s, 1980s and 1990s even though the evidence provided by the Census suggests quite strongly to the contrary.

In the 20 years between 1950 and 1969, the number of farms in the United States were cut in half. They declined from 5.4 million to 2.7 million. In the span between 1969 and the 1987 Census, numbers fell much less dramatically to 2.1 million. Moreover, the change in the definition of a farm in 1974 which raised the requirement for minimum agricultural sales from $250 to $1,000 accounted for half of the loss in numbers including a necessary adjustment for inflation. In this period, the more important issues were not the loss of farm numbers but the growing importance of the 50,000 largest farms in terms of the growing proportion of total output they provided. (Additional discussion of these issues is provided in Chapters 4 and 22.)

The European Community

Given the number of problems just demonstrated in using value of farm products sold to define farm size when making comparisons over

TABLE 2.2 Distribution of Farms by Economic Class, 1969

Class	Criteria Used: Value of Farm Products Sold	Other	Number of Farms
Commercial:			
1	$40,000 and over	None	221,690
2	20,000 - 39,999	None	330,992
3	10,000 - 19,999	None	395,472
4	5,000 - 9,999	None	390,425
5	2,500 - 4,999	Less than $2,500 sales if normally would have had sales in excess of $2,500 (crop failure, new farms, large inventories).	395,104
6	50 - 2,499	Operator under 65 years of age and did not work off-farm more than 100 days.	192,564
Part-time	50 - 2,499	Operator under 65 years, worked off-farm more than 100 days.	574,546
Part retirement	50 - 2,499	Operator who is over 65 years of age.	227,346
Abnormal	Not a criterion	Institutional, experimental and research farms, and Indian reservations.	2,111
Total number			2,730,250

Source: U.S. Census of Agriculture, 1969, Volume II, Chapter 7, p. 7.

time, some other alternatives have been developed around the world. The European Community now uses a system of economic size classes denominated in *European Size Units* (ESU). There are nine size classes; the smallest is Class I with less than 2 ESU; the largest includes farms with 100 ESU or more.

A European Size Unit is defined in terms of the European Currency Unit (ECU). One ESU is equal to 1000 ECU's of Standard Gross Margin. Gross Margin is the difference between gross receipts and variable costs per unit of production, usually land area or animal unit. Standardized Gross Margin (SGM) is calculated in each of the 12 countries of the EC for every productive agricultural enterprise annually. These values are standardized using ECU's for the 1980 reference period. Thus, if one hectare of wheat has an average gross margin of 120 ECU's in France in 1988 and the index of prices is 150 on the 1980 base, the SGM will be 80 ECU's per hectare using the 1980 reference period. Put another way, if prices increased 50 percent between 1980 and 1988, one ESU = 1500 ECU in 1988 prices.

The ESU and the nine economic size classes have worked well for the Europeans. Both the Farm Accountancy Data Network used throughout the EC, and the Community Surveys of Agricultural Holdings, similar to our Census, use these classifications. Standard Gross Margin (SGM) has the additional advantage of being an approximation of Value Added which makes comparisons of size across enterprises much more appropriate than gross sales. All that is needed to calculate ESU for a farm is the number of hectares of each crop and the number of animal units. SGM is provided for each enterprise by individual national governments, usually the Ministry of Agriculture based on local farm accounting data.

Japanese Classification of Farms

Japanese statistics on agriculture and farming are well established with a long history of rather complete national and local records by prefecture dating back to the nineteenth century (Arayama). A standard economic classification of farms has been in place since the 1950s. The definition of a farm is based on a minimum level of sales of agricultural products; in 1965, it was Y30,000 and in 1985 it was Y100,000 or the rough equivalent of U.S. $750 (MAFF, Japan).

Farms are divided into three major categories:

1. *Full-time* -- At least one person who works 250 days or more on the farm; family income comes primarily from farming; accounted for 14 percent of farms in 1984-85.

2. *Part-time, Type I* – Farm income is larger than off-farm income; farm operations require less than 250 days per year; accounted for 18 percent of farms in 1984-85.

3. *Part-time, Type II* – Farm income is less than off-farm income; farm operations account for less than 250 days per year; accounted for 68 percent of farms in 1984-85.

Two other economic classifications have been developed as well. The first is *core farm*; it includes all farms with males aged 16-59 years old who are engaged in farming for more than 150 days a year. This classification accounted for 20 percent of the farms and 46 percent of the arable land in 1984-85. The second is designated as *viable farm*. These farms have an average agricultural income per full-time worker that at least matches the average income of non-farm employees in neighboring urban areas.

A Labor-Based Classification

Much of the technology applied in agricultural production has sought to increase labor productivity. Labor is a key input around which production is organized. It can be a common denominator across all types of production and is an input which can be measured in physical units on a consistent basis over time. Thus, it has many of the key elements which might be used in an economic classification system for U.S. farming. A labor-based classification system might include the following general categories:

1. *Full-time, Large*. Establishment where agricultural production and marketing is the primary occupation of the operator (manager), and where 60 months or more of operator, family, regular hired or day labor are employed.

2. *Full-time, Family*. Establishment where agricultural production and marketing is the primary occupation of the operator (manager), and where from 10 to 60 months of operator, family, regular hired or day labor are employed.

3. *Part-time*. Establishment where agricultural production is an important contributor to family income and where from 2 to 10 months of operator, family or day labor in total is required in business operations.

4. *Residential*. Establishment where agricultural production occurs but is not an important contributor to family income; less than 2 months of total labor are required under average conditions to carry out agricultural operations.

This classification system uses some of the original descriptive terms from Hurley's economic classification system for the 1950 Census. It provides four major categories within which size groups based on value of production or value added could be constructed as well. If the basic classes were used regularly, it would help to identify more clearly the major groups of farms within agriculture and help to reduce confusion about the number of farms affected by different types of public policy. Such a system would require that more information be obtained systematically by the Census and the National Agricultural Statistics Service, USDA, about labor provided by family members in agricultural operations. Essentially, no other new information is required.

An alternative approach for a labor-based classification system is to use standardized labor requirements for each of the productive enterprises on a farm and determine size of operations in this manner after determining acres of crops and numbers of livestock. Activities of direct marketing, farm processing and similar activities would then have to be counted in days of labor used in the business. Estimates of labor requirements by enterprise have not been updated in most of the United States for many years. Advances in technology have been substantial so that any such effort would also require indicators of the technology used for each enterprise.

One of the reasons for examining different ways to classify farms into logical economic categories is to improve the public's ability to understand changes in farm structure over time. Substantial difficulties arise when comparing farm numbers by sales classes over a span of years because inflation and technology can affect the results so strongly (Young, et al. 1988). Building a consistent series based on physical characteristics and primary sources of family income could reduce misunderstanding of how many farmers there are by category and what is happening to them.

Concluding Comment

Farm Structure results from the interplay of many economic, social and political forces. We can be sure that this structure will continue to change through time as individual households take action in their own best interests. The adoption of new technology and opportunities for employment outside agriculture were major forces in cutting farm numbers in the United States in half between 1950 and 1970. Such an absolute loss in numbers cannot occur again. Yet, the potential for striking change in individual sectors of production agriculture remains. Public interest in this process is likely to continue as 10,000 to 15,000 of our largest farms provide successively larger and larger shares of national agricultural output in the years ahead.

References

Arayama, Yuko. 1989. "Out-Migration from Agriculture and Time Allocation Within Farm Households." *Agricultural and Governments in an Interdependent World*. Pp. 63-73. United Kingdom: Dartmouth Publishing Company.

Hurley, Ray. 1965. "Problems Relating to Criteria for Classification of Farms." *Journal of Farm Economics* 47: 1565-1571.

Kislev, Yoav, and W. Peterson. 1986. "Economies of Scale in Agriculture: A Survey of the Evidence." *World Bank Report* No. DRD 203, 23 pp.

Knutson, Ronald D., J. B. Penn, and W. T. Boehm. 1990. *Agricultural and Food Policy*, 2d ed., Chapter 11, Prentice-Hall.

McDonald, Thomas, and George Coffman. 1980. "Fewer, Larger U.S. Farms by Year 2000—and Some Consequences." 20 pp. ESS, USDA, Agr. Info. Bull. 439.

Ministry of Agriculture, Forestry and Fisheries (Japan). 1980. *Census of Agriculture*.

Raup, Philip M. 1985. "Structural Change in Agriculture in the United States." 34 pp. Univ. of Minnesota, Dept. of Agr. & App. Econ. Staff Paper P85-41.

Reimund, Donn A., J. R. Martin, and C. V. Moore. 1981. "Structural Change in Agriculture: The Experience for Broilers, Fed Cattle, and Processing Vegetables." Tech. Bull. 1648, ESS, USDA.

Reithmuller, Paul. 1988. *Japanese Agricultural Policies, A Time of Change*. 359 pp. Australian Bureau of Agricultural and Resource Economics, Policy Monograph No. 3.

Ruttan, Vernon W. 1988. "Scale, Size, Technology and Structure: A Personal Perspective." 18 pp. Univ. of Minnesota, Dept. of Agr. & App. Econ. Staff Paper P88-1.

Sumner, Daniel A. 1986. "Farm Programs and Structural Issues," in B. L. Gardner, ed., *U. S. Agricultural Policy: The 1985 Farm Legislation*. Pp. 283-328. American Enterprise Institute.

Tweeten, Luther. 1984. *Causes and Consequences of Structural Change in the Farming Industry*. 72 pp. National Planning Assn., NPA Report 207.

USDA, ESCS. 1979. *Structural Issues of American Agriculture*. 305 pp. Washington, Agr. Econ. Rpt. 438.

Young, Douglas L., et al. 1988. "Farm Size Classifications and Economies of Size: Some Empirical Issues." *Determinants of Farm Size and Structure*. Pp. 115-128. Michigan Agr. Exp. Sta. Journal.

3

The Structure of Agriculture
in an Historical Context[1]

Wayne D. Rasmussen and B. F. Stanton

The American people have been concerned about structural issues from the beginnings of our country. Not surprisingly, that interest continues today. Most of the new settlers when they arrived on the eastern shores of the North American continent had a primary concern for their food supply and their ability to produce it with the resources at their disposal. Decisions had to be made about the rights to use land and its ownership; a government had to be established to protect these rights and to provide for the common defense; an exchange system and markets were needed for this new agrarian economy to grow. In a very real sense, there were already structural issues of importance that had to be decided. The organization of production and control of agricultural resources differed substantially up and down the Atlantic coast.

A number of different systems for the conduct of agriculture and commerce had been brought to the New World. Farms, households and agricultural resources were the basic fabrics out of which the cloth of these different structures were woven. But they differed in the ways that production systems were organized, in the degree to which ownership and control was concentrated in the hands of the few or the many, and the ways in which returns from production were distributed. A plantation system along with small holder production grew up in the South. Large tracts of land were granted to individual owners. Production of export crops fostered a labor intensive system of agricultural output. The opportunity to obtain slave labor was exploited by the leaders in agricultural communities. An agrarian economy built around large production units evolved.

In New England with its hardwood forests and rocky soils, it was difficult to produce an agricultural surplus. Small farms remained the rule. Once cleared, the land was best suited for growing grass and a few cereals. Livestock production was important with small flocks of sheep and dairy animals. Cottage industries flourished. Water power was harnessed. The bounty of the forest was turned into lumber, furniture and ships. Mills and small industry grew with farming. Farms, households and agricultural resources were organized and developed into a different structure from the one that evolved in the South.

In the middle Atlantic States, the patterns of agricultural growth and development were somewhat independent and unique, but the agricultural structure which evolved was more akin to what had emerged in New England than in the plantation South. There were large land grants made to the patroon of New Amsterdam that the English left undisturbed when they assumed control in New York. Nevertheless, small farms, whether rented or owned, were the most common way of obtaining agricultural products, even though plantations were established in some cases on what we now know as the Delmarva peninsula. Small business and manufacture developed side by side with farming. Trade up and down the Atlantic Coast as well as with Europe was well established by the middle of the eighteenth century.

Structural Issues at the Time of the Revolution

One can argue that the American Revolution was fought by farmers and their leaders in part because of the issues associated with farm structure. A number of the first English colonies on the Atlantic seaboard were settled under what was called the quit-rent system. That is, a farmer could get title to his land subject to a perpetual small fee, usually to be paid to an absentee landlord in England. Many farmers saw no reason why their hard-earned money should go to a person who had done nothing for them and had title to the land only because of a King's whim. Farmers were further antagonized by the British government in 1763 when it forbade settlement west of the Alleghenies, and sent an army to drive settlers already west of the mountains back across. Farmers felt, quite rightly, that a major purpose of the ordinance was to protect the interests of British land speculators. Many farmers, including the large tobacco planters in the South, also resented the efforts of the British to control and tax the trade in colonial farm products. These limitations on land titles, settlement, and marketing of farm products, all elements of farm structure, were major causes of the Revolution.

Once the war was won, the new states did away with quit-rents, the Constitution forbade taxes on exports, and a system was devised to promote the settlement of western lands. The land system, largely devised by Thomas Jefferson, provided for the sale of western lands with fee simple titles and for bringing the western territories into the union as states equal in every way to the original thirteen. Jefferson called for a free, independent farm population, extolling its virtues as the foundation of the nation.

Jefferson articulated much of the agrarianism which has influenced farm policy throughout our history. Moral corruption, according to Jefferson, "is the mark set on those, who, not looking up to heaven, to their own soil and industry, as does the husbandman, for their subsistence, depend for it on the casualties and caprice of customers." Jefferson's ideal farmer provided for his family from his own land by his own efforts. He carried on a self-sufficient agriculture, buying and selling as little as possible. He did not rent his land but owned it in fee simple. He did his own work. As an independent, self-supporting member of society, he was his own boss, responsible for his own managerial decisions.

As a practical matter, the structure of American agriculture never followed the Jeffersonian model. Jefferson himself operated a commercial plantation with slave labor, producing crops for market, and importing goods from England. Every farmer from the earliest settlements produced at least some goods for market since true self-sufficiency was impossible. Even though many farmers acquired title to land under the ordinances Jefferson sponsored, both investors and speculators were buying land and leasing it to farmers. A few of these early large landed estates, regularly leased to tenants, are still found scattered across the nation. From the beginning, there were important differences in the ways businesses were organized, the ways in which households contributed to both to output and business decisions, and the ownership and control of agricultural resources -- key components of farm structure issues as they developed over time (Chapter 2).

Farm Structure and the Civil War

Unrest over farm structure helped establish the new nation, and continuing problems in this area threatened the very existence of the United States. The industrializing North and the plantation South were fixed upon divergent courses, with the divergence centering upon differing farm structures. The North's farms were operated by family labor, the South's plantations by slave labor. The South's dependence upon the export of one major crop opened it up to economic exploitation by northern industrial interests, primarily through the tariff. The Civil War, resulting from these conflicts, brought major changes in farm structure in both the North and the South.

The war led many Northern farmers to increase production by increasing the size of farms and by replacing hand labor by horse-drawn machinery. The result was that these farmers, after the war, were caught up in commercial agricultural production in order to pay for the land and machinery. Southern farmers, virtually without economic resources, turned to an economic system known as share-cropping, damaging to the cropper, the landowner, and the land itself.

In 1862, after the South withdrew from the Union, Congress passed four laws vitally affecting the structure of American agriculture. All four sought to encourage family farms. The Homestead Act offered 160 acres of Federal land without charge to a person willing to improve the land and live on it for five years. The Department of Agriculture was established to develop useful information and carry it to the farmers. The land-grant colleges, established under the Morrill Land Grant College Act, were to educate young people in agriculture so that they could apply the newest knowledge of improved farming methods to their land. Finally, a transcontinental railroad was financed at least in part to provide better access to lands opened under the Homestead Act, and to give farmers a way of getting their products to market.

The Opening of the West and Experiments in Structure

The increased commercialization of agriculture resulting from the Civil War and the opening of Western lands led to large increases in production and to recurring market surpluses. As a result, farmers organized national associations, including the National Grange and the Farmers Alliances, and turned to cooperatives, to try to gain what they regarded as a fairer share of national income. Under these conditions, individual farmers or their organizations were unable to gain enough market power to bargain effectively with the railroads over transportation rates or the grain merchants over the prices for their crops. They turned to politics and the populist movement for support. But most of these efforts were not particularly successful.

Meanwhile, the West saw important, large-scale experiments with corporate operation and foreign ownership of farms and livestock ranches. Large, mechanized bonanza wheat farms in the Red River Valley grew out of the depression of 1873, while foreign-owned cattle ranches were developed by eastern speculators, who convinced foreign investors that they could get rich by buying cattle, grazing them without cost on Federal land, and selling the resulting two-year olds. The wheat farms seemed to offer an opportunity to apply the most modern technology to production and to develop economies of scale, but they failed. A couple of poor crop years, fluctuations in the price of wheat, increasing costs of farm machinery, and the problem of hiring workers who could and would operate the machinery inevitably led to the failures.

A similar fate overtook the large scale cattle ranches. Animal predators and disease took a toll, while rustlers were not uncommon. However, the extremes of climate caused the greatest losses. Most foreign investors lost everything after two very dry summers, and the terrible winter of 1886-87.

The First Half of the Twentieth Century and Government Intervention

By the turn of the century, however, farmers were entering one of the "golden ages" of American agriculture. Total production was not increasing as rapidly as demand, so the two reached a certain equilibrium. Nevertheless, there were still problems. In 1908, President Theodore Roosevelt appointed a Country Life Commission, the first national effort to examine the structure of farming and of farm life. The Commission, which stated that in "independent and strong citizenship, the agricultural people constitute the very foundation of our efficiency," reported that agriculture was generally prosperous, but that country life was deficient. The Commission made many recommendations which were eventually carried out, such as the establishment of land banks and of a nationwide agricultural extension service. However, many of the problems it discussed are still with the nation 80 years later.

The Great Depression and the New Deal of the 1930s had a major impact upon farm structure -- an impact that is still continuing. In the 1920s, farm management specialists and other agricultural economists had encouraged farmers to analyze their costs and returns and to reorganize their farms to get the greatest returns. Helpful as this was to individual farmers, it could not overcome the impacts of the depression. Farm legislation was a first order of business for President Franklin D. Roosevelt and Secretary of Agriculture Henry A. Wallace. The Agricultural Adjustment Act of 1933 was signed on May 12. It provided for supporting the prices of major farm products for farmers who agreed to reduce their production. Laws, with the purpose of supporting farm prices and relating production to demand, have been in effect from then to the present and have had a major impact on farm structure.

Other New Deal legislation, too, has had a continuing impact. Laws establishing the Farm Credit Administration, the Soil Conservation Service, the Farmers Home Administration, and the Rural Electrification Administration are still important to farm structure.

Change After World War II

The greatest changes in farm structure the nation has seen, however, came during and just after World War II. The second American agricultural revolution, marked by the completion of the transition from animal to tractor power and the application of systems analysis to farming began during the war and came to full fruition in the 1950s. In addition to moving from animal to tractor power, this agricultural revolution included the use of hybrid seeds, the adoption of strains of livestock that could make the most efficient use of feed, the careful use of fertilizers, irrigation when and where necessary, and the use of agricultural chemicals when this promoted more efficient production. Many of these changes called for increased capital in farming and for adjustments in size of farms, usually upward, to make the best use of the technology. Unforseen at the time was the fact that with the unprecedented increase in productivity, the nation would need fewer farmers and fewer farms.

During the 1950s and 1960s, a number of proposals and studies were made with the idea of strengthening the family farm (the term was still being used by the Department and by agricultural economists), developing a more rational price support system, and ending rural poverty. Little of substance resulted, although the passage of the Agricultural Trade Development and Public Assistance Act in 1954, known as Public Law 480, provided a base for exporting large quantities of our surplus commodities to nations in need. This law, considerably modified, is still in effect and is still used.

In the same year, on January 11, 1954, President Dwight D. Eisenhower asked that particular attention be given to problems peculiar to farmers with low incomes. The first result was a 44-page report entitled, *Development of Agriculture's Human Resources*. Implicit in the report was the idea that too many resources were devoted to farming, and that financial resources, land, and people should be moved out of agriculture. The general approaches recommended for attacking the problems were to increase productivity in agriculture, improve prospects in part-time farming and nonfarm jobs, increase opportunities for training, and utilize surplus labor in decentralized defense industries. Pilot projects were established, some of which were helpful to particular individuals and communities, but the problem remained. During the 1960s and 1970s, legislation was passed, funds were appropriated, and projects were undertaken, generally under the heading of rural development. However, the nation has not yet made a firm commitment to rural development and there would seem to be little prospect of such commitment in the immediate future. The major and most effective actions appear to have been taken by the State Extension Services, and such programs will likely continue. But the structural changes, particularly for rural households, brought about by rural development have not been substantial.

Both economists and politicians have called for important changes in price support and adjustment legislation to secure desired changes in structure. Virtually nothing has been done in spite of proposals, committee reports, hearings, seminars, and on and on. Little can be expected in the next round of price support legislation. However, two changes have had some effect -- limitations on payments to be made to one farm and, more important, a step toward requiring participation in solid conservation programs in order to receive support payments.

In recent years, policy makers have stressed the importance of exports and export subsidies to bring about a reasonable balance between production and demand. Substantial amounts of money have been spent on increasing exports, but with somewhat unclear results. We have seen exports of meat blocked supposedly because our producers almost universally bring up our beef cattle on growth hormones. We have seen what appears to be a reasonable proposal to gradually cut out all production and export subsidies in order to promote a freer world trade in farm commodities sidelined by the European Community. Perhaps it is time to try and understand why many other governments are unwilling to accept our proposals. One of the major reasons may be that important differences in farm structure can be recognized. Public policies have been put in place to either maintain the status quo or to encourage change. There are important ways in which business organization, household decision making and resource management as practiced in the United States account for some of the differences observed in farm structure here and in other developed countries.

Structural Issues in Other Developed Countries

Farm structure has evolved over time in different general patterns in Europe, North America, Australia and New Zealand. Clearly, the amounts of natural resources available relative to numbers of people, the form of government, and past experiences of war and its consequences have all influenced structural developments. It is important for Americans to recognize that history has had its strong hand in the ways different societies have responded to the industrialization of agriculture in the twentieth century.

Such problems as the number of large versus small farms remain common to all these locations. So do such issues as corporate ownership and operation versus the maintenance of many small farms across the landscape. Policy to maintain a sizeable population on the land rather than encouraging farm people to leave for more productive employment in other industries is a matter of debate. So is the degree to which private initiative versus government intervention determines net farm income. Even the question of ensuring that an important part of a nation's food supply can be produced within its borders can be at issue -- something which remains difficult for most Americans to accept and recognize, but which remains very real in Western Europe and Japan. In fact, after World War I, the determination of most of the European nations to build up enough food production to maintain themselves in time of war brought a virtual end to international trade in food products.

The Common Agricultural Policy of the European Community recently published a statement entitled, "The New Agricultural Structures Policy," which addresses many of these concerns. According to a statement in the introduction:

> Structural measures, aimed at improving the efficiency of farming and living, working and production conditions in agriculture particularly on those smaller holdings most seriously affected by natural or socio-economic deficiencies, are therefore a prerequisite to the attainment of the objectives of the CAP.

The measures were summarized and then developed somewhat more fully in respect to their application to particular situations. The main features included the following. Investment should be made to reduce production costs except where this would lead to the increased production of surplus products. Farm management services should be made available. Special aid should be given to the transfer of holdings to young farmers. Less-favored areas should receive compensatory allowances and increased aid for crafts and tourist activities – similar to some rural development policies in the United States. Forestry should be developed as an alternative to the production of surplus commodities. There must be greater emphasis on education and training. Payments to aid farmers in carrying out conservation and environmental protection programs were authorized. Finally, the use of new technology, the production of new products, and the development of new market outlets were to be encouraged to improve the added value of agricultural production and enable farmers to share in the benefits.

This list is not unlike one which might be drawn up by a group of American agricultural economists. One point, which is implicit but not forthrightly stated in the Common Agricultural Policy report, was made by a French agricultural economist recently: "We must pay whatever subsidies are necessary to keep our people on the land because we have no place else to put them."

In other parts of the world, situations differ. In some less developed nations, large blocks of land are held by the very wealthy, with the poor being landless. In some, land holdings are too small to allow efficient production. In some well-to-do nations, people are kept on the land for defense, to avoid overcrowding in the cities, or for other reasons.

Clearly, farm structure and the public policies that influence structural change differ among the several countries of the world as well as from that in the United States. Major forces such as the ratio of land to people, the stability of government and its associated institutions, and past experience with war and its consequences, have had major influences on what has evolved. Many of the same issues are important in each of these societies. The responses have been different because priorities are different. Hopefully, we can learn from the experience in other nations as well as our own as efforts are made to influence changes in structure over time.

Summary

Obviously, the structure of farming has a major impact on the well-being of any nation. It is hard to quantify this because our concepts of structure undergo change; we are not always sure of what we want or expect to occur. For example, Jefferson believed that ownership of the land by those who farmed it was an essential element of a sound democracy and a stable government. Yet today, except for those who believe racial minorities who work the fields in some parts of the nation should own the land they work, few regard land ownership as a problem. Farm land is generally available for purchase or rent. If rented, the tenant usually makes the management decisions regarding the operation of the farm.

There are other questions related to land. In times of farm prosperity, good farmland may be scarce and valued at prices which make it almost impossible for newcomers, particularly young people, to enter farming, either as owners or renters. Another problem about which many good city folk wring their hands is the continuing growth of very large farms and their domination of the market. In fact, this is a key subject in almost any study of farm structure.

Small and part-time farms, usually with most of the income from outside farming, is another subject in which many people are interested and one that we can no longer easily dismiss. Such farms must be fitted into any model or program concerned with farm structure.

Studies of farm structure face a number of problems, of which definition is only the first. It is necessary to determine what our present farm structure is and then decide if it should be or could be changed. Is our present farm structure providing the American people with what they want?

One of the difficulties of dealing with structure is its almost fluid nature. It is subject to pressure from many directions. For example, changes in technology have always brought changes in structure, and there is every reason to believe this will continue as computers and biotechnology bring sweeping new efficiencies, as tractors and hybrid corn did in another generation. Agricultural policy, including both price supports and tax measures, has been charged with bringing about a greater concentration of land ownership and the development of very large farms operated by fewer and fewer people. Most Americans would agree with Jefferson that a strong farming sector helps maintain political stability, especially when the people farming the land own it and make the decisions. In recent years though, there have been charges that societal concern for open space, conservation of natural resources, and the environment are threatening the independence and the livelihood of farmers and ranchers. This is but one example of the many and conflicting influences on agricultural structure.

American agriculture is a much smaller part of the U.S. economy than it was 200 years ago, yet it remains important both politically and economically. Farming undergirds the food industry which has grown increasingly international in its scope. Specialization of function continues. Farmers are less and less self sufficient; more and more of the inputs of production are purchased, either from other farmers or from a vast array of specialized suppliers. Those who call themselves farmers include a substantial range of people from young to old, poor to rich, fully dependent on their own labor to dependent on hiring large numbers of workers.

History suggests that the structure of agriculture will continue to be dynamic. In an open society, many forces will interact to bring about change. Farm households will continue to be the crucial decision makers in allocating their own labor and the resources they control, some of which will go to farming, some to other enterprise. More and more of total agricultural production may well come from a smaller and smaller proportion of what are officially defined as farms. Agricultural resources may become yet a smaller part of the nation's total land, labor and capital, or on the other hand, the race between population and food supply in the world may put new demands on the nation's farmers and require additional resources for agricultural production.

History tells us that structural change will continue. No doubt, some in society will view future change with alarm as they have in the past. And government and society will try to respond to influence the process. While many of the changes in previous decades have been painful for individual families and communities, a largely strong and productive agriculture survives. Perhaps most important of all will be a society, free, confident, and open enough, to allow change to continue.

Note

1. There is significant literature on the history of U.S. agriculture. A selected bibliography is included at the end of the chapter.

Bibliography

Benedict, Murray R. 1953. *Farm Policies of the United States, 1790-1950*. New York: The Twentieth Century Fund.

Cochrane, Willard W. 1979. *The Development of American Agriculture, A Historical Analyses*. Minneapolis: University of Minnesota Press.

Danbom, David B. 1979. *The Resisted Revolution: Urban America and the Industrialization of Agriculture, 1990-1930*. Ames, IA: Iowa State University Press.

Drache, Hiram M. 1964. *The Day of the Bonanza: A History of Bonanza Farming in the Red River Valley of the North*. Fargo, ND: North Dakota Institute for Regional Studies.

Dyson, Lowell K. 1986. *Farmers' Organizations*. Westport, CT: Greenwood Press.

Edwards, Everett E. 1943. *Jefferson and Agriculture*. P. 23. USDA Agricultural History Series No. 7, Washington, DC.

Ellsworth, Clayton S. 1960. "Theodore Roosevelt's Country Life Commission." *Agricultural History* 34: 155-172.

European Community Commission. 1985. "The New Agricultural Structures Policy." *Green Europe*, No. 211, p. 1.

Gates, Paul W. 1965. *Agriculture and the Civil War*. New York: Alfred A. Knopf.

Loehr, Rodney C. 1952. "Self-Sufficiency on the Farm." *Agricultural History* 26: 37-41.

Mahé, L. P., and K. J. Munk. 1988. "The EC Grain Price Policy at the Core of the CAP." Paper at European Agricultural Association's 16th Seminar, Bonn, Germany.

Rasmussen, Wayne D. 1965. "The Civil War: A Catalyst of Agricultural Revolution." *Agricultural History* 39: 187-195.

_____ . 1982. "The Mechanization of Agriculture." *Scientific American* 247: 77-89.

_____ . 1983. "The New Deal Farm Programs: What They Were and Why They Survived." *American Journal of Agricultural Economics* 65: 1158-1162.

_____ . 1987. "Public Experimentation and Innovation: An Effective Past but Uncertain Future." *American Journal of Agricultural Economics* 69: 890-899.

_____ . 1989. *Taking the University to the People: Seventy-Five Years of Cooperative Extension.* Ames, IA: Iowa State University Press.

Report of the Commission on Country Life. 1908. Reprint. P. 17. Chapel Hill, NC: University of North Carolina. 1944.

Saloutos, Theodore. 1982. *The American Farmer and the New Deal.* Ames, IA: Iowa State University Press.

Schlebecker, John T. 1963. *Cattle Raising on the Plains, 1900-1961.* Lincoln, NE: University of Nebraska Press.

Socolofsky, Homer E. 1979. *Landlord William Scully.* Lawrence, KS: Regents Press of Kansas.

U.S. Congress, 96th, 2nd Session, Senate, Committee on Agriculture, Nutrition, and Forestry. 1980. *Farm Structure: A Historical Perspective on Changes in the Number and Size of Farms.* Washington, DC.

U.S. Congress, Office of Technology Assessment. *Technology, Public Policy, and the Changing Structure of American Agriculture: A Special Report for the 1985 Farm Bill.* OTA-F-272. Washington, DC.

U.S. Department of Agriculture. 1981. *A Time to Choose: Summary Report on the Structure of Agriculture.* Washington, DC.

U.S. Department of Agriculture, Economics, Statistics, and Cooperatives Service. 1979. *Structure Issues of American Agriculture.* Ag. Econ. Report 438, Washington, DC.

4

Changes in Farm Size and Structure in American Agriculture in the Twentieth Century

B. F. Stanton

The dawn of the twentieth century heralded great change in American agriculture. Homesteaders had staked their claims to most of the productive land in the West. Mechanization had begun to save labor in a range of applications. Animal and human power, which had felled the forests, broken the prairie sod, and established farming across the continent, was to be aided and then replaced in large measure by mechanical power and the magic of electricity.

A nation of small farmers and tradesmen would rapidly become an industrial and service economy. Like most other sectors, agriculture would become industrialized; farm labor would move off the land to a myriad of new occupations; often these transitions would be painful and disruptive. Yet, the same hardy spirit which had carved out farms and ranches across the hills and plains would sustain another transformation: the consolidation of land, labor and capital into a new agriculture where science and machines would allow one worker to do what many had been required to do in previous generations.

The process of structural change in agriculture during the twentieth century in the United States has not been easy. From the "Golden Age of Agriculture" before World War I to the depths of the Great Depression in a span of less than 20 years was traumatic for everyone; especially those who had to leave the land when there were no jobs and no places to start again. With economic expansion in the 1940s and continued growth in the postwar years, the great exodus out of agriculture between

1950 and 1970 was much less painful, but no less dramatic. Farm numbers fell in those years at the greatest rate in the century. Industrialization and the adoption of mechanical and electrical power was in full swing. Capital was substituted for labor across the land. A healthy economy absorbed displaced workers from the farm sector with substantial success. Yet, rural poverty and the "people left behind" remained no less a continuing problem, touched but not emancipated by the programs of "the Great Society."

Land in Farms and Farm Numbers

The story of change in American agriculture is documented effectively in Census statistics starting in 1850. The early Census counts chart the sweep of settlers out of the East and Midwest into new lands as they opened. The land in farms doubled between 1850 and 1890 (Table 4.1). The largest addition to land in farms in any decade occurred between 1890 and 1900 when more than 215 million acres, over one-fifth of total land in farms today, was added to the national total (Figure 4.1).

Cropland harvested[1] was recorded in each Census period starting in 1880. Here, too, the greatest addition to the total cropland occurred between 1890 and 1900. Further additions to the cropland base occurred in each succeeding decade until 1930. From this base there were important fluctuations in the next 50 years and some shifts between regions, but the national totals remained relatively steady. Government programs, the weather, and economic conditions influenced acres planted and harvested from year to year.

Land in farms reached a high point in the 1950s. In each succeeding five year period the total has fallen modestly so that in 1987 land in farms returned to about the same total as in 1925. It is important in looking at subsequent statistics on changes in farm numbers and size distributions to keep the land area used for farming in perspective. Land was being added to the agricultural base until 1950. The cropland total shifted out of some of the less productive areas in the Eastern United States to the West between 1930 and 1950, one of the results of animal power being replaced by tractor power. The cropland base in total remained close to 400 million acres throughout all of that period.

The number of farms grew steadily from 1850 to 1910. Between 1910 and 1935, farm numbers remained relatively constant between 6 and 6.5 million as modest amounts of land were added to the total and mechanization became more important. The fall in farm numbers which started in the decade of the 20s was slowed by the depression of the 30s. Once World War II was over the great decline in numbers, held back by the depression and the war, began in earnest.

TABLE 4.1   Farm Numbers and Land in Farms, 1850-1987

Year	Number of Farms (millions)	Land in Farms (million acres)	Average Farm Size (acres)
1850	1.4	294	203
1860	2.0	407	199
1870	2.7	408	153
1880	4.0	536	138
1890	4.6	623	137
1900	5.7	839	146
1910	6.4	879	138
1920	6.4	956	148
1930	6.3	987	157
1940	6.1	1061	174
1950	5.4	1161	216
1954	4.8	1158	242
1959	3.7	1124	303
1964	3.2	1110	352
1969	2.7	1063	389
1974	2.3	1017	440
1978	2.3	1015	449
1982	2.2	987	440
1987	2.1	964	462

Source: Census Data, United States, 1850-1987.

Farm numbers decreased by more than 1.6 million in the decade of the 1950s. Undoubtedly, part of this decrease resulted from the adoption of new technology that would have occurred earlier but for the war and the lack of tractors and associated machinery. Consolidation of small units, particularly in the states east of the Mississippi, was common. Off-farm opportunities for employment were good and commuting to jobs from rural locations became possible as a network of all weather roads was extended.

The rapid consolidation of farms into larger units and the decrease in farm numbers continued in the decade of the 1960s. In a span of 20

FIGURE 4.1 Land in Farms and Cropland Harvested (U.S. Census, United States, 1880-1987)

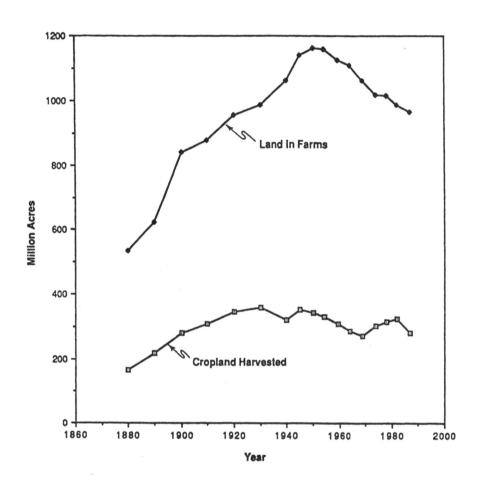

years, farm numbers were cut in half with little fanfare. The great readjustment resulting from the introduction of tractor power and electrical energy, accompanied by the adoption of many technological developments in the plant and animal sciences, brought about striking advances in agricultural productivity. Excess production capacity was a continuing problem throughout these decades as government programs to limit acreages planted to *basic* crops and a system of price supports became institutionalized.

The decade of the 1970s brought a modest reduction in farm numbers, less than 500,000, compared to the two immediately preceding decades. Shortfalls of food and feed grains in other parts of the world led to rapid increases in farm prices in the early 1970s. Agricultural land prices rose more rapidly than the rate of inflation and a boom mentality led to rapid expansions on an important number of farms with large increases in debt.

After the boom of the 1970s came the inevitable readjustments in land prices and the debt-led reorganizations and liquidations of the early 1980s. Farm numbers in total decreased but modestly. The loss of 2.65 million farms between 1950 and 1970 could never be experienced again, even though the readjustments of the late 1970s and early 1980s caught much more public attention and debate. Structural change was still an issue; but much more nearly in terms of the proportions of total agricultural output that would be produced by different economic classes of farms, than in declines in farm numbers as such.

While there are many things wrong with trying to describe American agriculture in terms of the average number of acres per farm, because of the vast differences between intensive and extensive forms of production, the statistics in Table 4.1 help to tell something about the nature of change. Average farm size fell in successive decades of the 19th century in a time when human labor and animal power were the primary sources of energy for agriculture. Average farm size began to increase after 1920 as tractor power began increasingly to replace horses. The great leaps forward occurred between 1950 and 1969, at the same time as farm numbers were cut in half, another indication that this was the period of greatest structural change in U.S. agriculture. Average farm size remained surprisingly constant between 1974 and 1987.

The Measurement and Identification of Structural Change

The study of structural change in agriculture has been approached in many different ways. In Chapter 2, it was suggested that farm structure be considered with respect to changes in: (1) distributions, such as farm size, measured in terms of land area, output, and other resources used; (2) production decisions and organization; and (3) resource ownership and control of enterprise. Most of the historical data are most readily organized in the form of size distributions. Even here it is difficult to insure that comparability of definitions and data sources has been established. Substantial reliance is placed on data from the U.S. Census of Agriculture which was collected every ten years at the same time as the Census of Population until 1920. A Census of Agriculture was obtained every five years from 1925 onward to 1974. Intervals of four years were followed to 1982 to coordinate the Census of Agriculture with the Census of Manufactures. The apparent inconsistencies in the length of periods reflect these changes during the century.

Less numerical data are available to examine changes in production decisions and methods of organization. These will be considered using available census data and other sources. Changes in resource ownership and control are considered in greater detail in Chapter 16. Discussion of issues such as tenancy and the importance of rented land will be considered here.

No single measure of structure can reflect the many facets of change associated with the technological revolution that is still in progress and had its roots in the 19th century. Because so much of this change has occurred in the years since World War II, the process is even more difficult to place into an historical context. The various ways of looking at size distributions remain as the most important evidence to evaluate.

Definition of a Farm

The official definition of a farm has changed eight times since the first definition was provided for the Census of 1850. All of the definitions required that agricultural operations involving crops and/or livestock be conducted and operated as a single unit under the direction of one management (individual, partnership, or corporation).

From the beginning there was a requirement that there be some minimum level of sales, $100 in both 1850 and 1860. No minimum acreage was required initially; from 1870-1890 a minimum of 3 acres was needed unless total sales exceeded $500 when this requirement was waived. In 1900, a new condition was added: the full-time services of at least one person. This requirement continued until 1925 when it was dropped and operators reported in four categories about days worked off the farm. Additional changes were made in definitions relating to minimum acres or minimum sales of agricultural products.

The definition in place for the Censuses of 1974, 1978, 1982, 1987, and 1992 is the official one used for all government statistics. "Any place from which $1000 or more of agricultural products were sold or normally would have been sold during the census year." The acreage requirements used in 1959-69 were dropped and the minimum sales requirement increased. In all of these definitions, the minimum requirement to qualify as a farm unit was small enough to insure that nearly any unit that could be thought of as a farming operation was included. From the beginning, many small, part-time operations were included in the farm count (Appendix B).

Size Distributions in Acres of Land

Acres of land in farms have been recorded in each of the census years (Table 4.2). It provides a general indicator of change through time in the size distribution of farms. Clearly all acres are not the same. In a composite picture of farms across the country, it does indicate, however, something about the way in which the basic land resource used in operations changed with technology and economic conditions.

Perhaps the most striking thing about comparing the size distributions from 1900 through 1940 is their similarity. Farm numbers increased in each decade to 1920 and then fell slightly in 1930 and again in 1940. But the patterns remained relatively constant. About the same proportions remained in each of the classes. The proportion of the total that were under 50 acres in size actually increased slightly between 1900 and 1940. Not surprisingly, the proportion of farms over 260 acres increased from 9.2 percent in 1900 to 11.9 percent in 1940. The stability of the distributions over these 40 years is the most noteworthy thing to recognize.

In contrast, there were marked changes in the decades following 1940 (Table 4.3). This is the period when the great reductions in farm numbers occurred. There were 2.286 million farms with less than 50 acres in the 1940 Census and only 0.596 million in 1987. As a proportion of the total, the number of farms with less than 50 acres also declined from 37.5 percent in 1940 to 28.5 percent in 1987.

The shrink in numbers for farms with 50-279 acres was equally impressive between 1940 and 1987. Most of the drop in numbers occurred between 1950 and 1969 but the largest proportional shift occurred between 1969 and 1987 with only 40.0 percent of all farms remaining in the medium size category of 50-279 acres (Table 4.4).

The changes from decade to decade in the categories of large farms in Table 4.3 is of special interest. The total number of farms with 260

TABLE 4.2   Size Distribution (in acres) of U.S. Farms, 1900-1940

Size Category	Thousands of Farms				
(in acres)	1900	1910	1920	1930	1940
Small:					
Under 10	267	335	289	359	506
10-49	1664	1918	2010	2000	1780
Medium:					
50-99	1366	1438	1475	1375	1291
100-174	1422	1516	1450	1343	1310*
175-259	490	534	531	521	486*
Large:					
260-499	378	444	476	451	459
500-999	103	125	150	160	164
1000 and over	47	50	67	81	101
Total	5737	6362	6448	6289	6097

*The Census classes were 100-179 and 180-259 in 1940.

Source: Census Data, United States, 1900-1940.

acres or more increased from 724,000 in 1940 to 781,000 in 1950 and 808,000 in 1969. The total dropped back to 786,000 in 1969 but fell by more than 100,000 units by 1987 to 655,000. Most of the *full-time,* commercial units in the 1980s fall in this general size category. Much of the shrink in numbers occurred in the 260-499 acre category, especially between 1969 and 1987.

An overview of the shifts in farm numbers grouped into three somewhat arbitrary size categories is presented in Table 4.4. The farms with less than 50 acres were more than one-third of the total until after 1950. Even in the 1980s they included more than one-fourth of the total. The medium size category of 50-259 acres decreased in relative importance in nearly all of the decades but remained the largest category. A large part of these units are *part-time* units in the 1980s but in the 1950s and earlier included many *full-time* farms[2].

The impact of the adoption of new technology, mechanical power, and other labor saving devices is particularly evident in the increased proportion of total farms in the "large" category that occurred between 1950 and 1969. In 1969, 13.4 percent of total farms had 500 or more acres; in 1987, it had grown to 17.7 percent.

TABLE 4.3   Size Distribution (in acres) of U.S. Farms, 1940-1987

Size Category	Thousands of Farms					
(in acres)	1940	1950	1959	1969	1978	1987
Small:						
Under 10	506	485	244	162	151	183
10-49	1780	1478	813	473	392	413
Medium:						
50-99	1291	1048	658	460	356	311
100-179	1310	1103	773	542	403	334
180-259	486	487	415	307	234	192
Large:						
260-499	459	478	472	419	348	286
500-999	164	182	200	216	213	200
1000 and over	101	121	136	151	161	169
Total	6097	5382	3711	2730	2258	2088

Source: Census Data, United States, 1940-1987.

TABLE 4.4   Size Class of U.S. Farms as a Percent of Total, 1990-1987

	Size Class (as a percent of total)		
Year	Small Under 50 Acres	Medium 50-259 Acres	Large 260 Acres and Over
1900	33.7	57.1	9.2
1910	35.4	54.8	9.7
1920	35.7	53.6	10.7
1930	37.5	51.5	11.0
1940	37.5	50.6	11.9
1950	36.5	49.0	14.5
1959	28.5	49.7	21.8
1969	23.2	48.0	28.8
1978	24.0	44.0	32.0
1987	28.5	40.0	31.5

When land in farms is aggregated for each of the acreage classes so that total land in farms by size class can be considered in each of the Census years, the continuing shift of agricultural land into larger operating units is seen more clearly (Table 4.5). In 1910, over 53 percent of the farm land was in units of less than 260 acres; more than one-third of the land was in units of 50-179 acres. By 1930, a modest shift to larger units was evident. Land in farms of 260 acres or more had increased by 8.1 percent.

Between 1930 and 1950, an important shift of land from farms with less than 260 acres to larger units had already occurred. A combination of consolidation of small farms into larger units and renting of part of the land farmed was in process. The period between 1950 and 1969, when half of the farms dropped out of the statistics, is when the two largest size categories increased at the expense of the other four. Farms with 260-499 acres continued to be an important category in 1969, but now 68.3 percent of all the farm land was in operating units of 500 acres or more. Again, it is important to remember that in many cases only part of the land farmed was owned by the operators.

The changes between 1969 and 1987 were the least dramatic of any of the comparisons. The same direction of change held true with more and more of the total agricultural land operated in units of 500 acres or more. By 1987, 87.5 percent of the land was in farms with 260 or more acres. The proportion of total agricultural land farmed in units of 1000 acres or more has increased steadily across the twentieth century to 62.4 percent in 1987. Farms with 5000 acres or more accounted for 33.7 percent of all land in farms in 1987.

With more than 169,000 operating units farming 1000 acres or more in 1987, concentration is far from a major problem, when compared with most businesses or industries. It is also easy to forecast that more of the total farm land can be expected to be included in operating units of 1000 acres or more in each of the remaining Census years in this century. It is also likely that the number of farms in this category will increase as more of those in the 500-999 acre category seek to enlarge their operations by bidding away land now operated in some of the smaller sized farms.

Size Distributions by Gross Sales

One of the most common methods of measuring size of business, regardless of the type of industry, is to look at output in terms of gross sales. This is an internationally accepted way of comparing firms both within and between industries. It has been widely used in the United

TABLE 4.5 Percent of Total Farm Land by Size Category, 1910-1987

Size Category (in acres)	Percent of Farm Land in Size Category				
	1910	1930	1950	1969	1987
Small:					
Under 50	6.2	5.7	3.6	1.3	1.2
Medium:					
50-179	35.1	28.3	19.4	10.2	7.0
180-259	12.0	11.2	9.1	6.2	4.3
Subtotal (Under 260)	53.3	45.2	32.1	17.7	12.5
Large:					
260-499	18.2	15.8	14.4	14.0	10.7
500-999	9.5	11.0	10.9	13.9	14.4
1000 and over	19.0	28.0	42.6	54.4	62.4
Subtotal (over 260)	46.7	54.8	67.9	82.3	87.5
Acres of land in farms, United States, millions	879	987	1161	1063	964

Source: Census Data, United States, 1910-1987.

States in looking at distributions of farms particularly in the second half of the century.

One of the major disadvantages in using gross sales in discussing changes in farm size or structure is the difficulty of adjusting for the effect of changes in prices in these distributions when comparisons are made across time periods. A farm that sold $25,000 of farm products in 1950 is far different from one that had sales of $25,000 in 1969 or 1987. Moreover, there is more than price level changes involved in seeking comparability. Changes in technical efficiency have occurred which affect the prices of both outputs and inputs. Capital has been substituted for labor so that a farm requiring one or two full-time workers in 1940 is substantially different from one using one or two full-time workers in the 1980s.

The following list summarizes some of the commonly recognized problems with using sales as a measure of farm size in any given year:

1. Effects of changing price levels are not easily accounted for in comparisons between years.
2. Changes in crop or livestock inventories are not considered. Sales from two years or only part of a year may be included.
3. Government payments are not included as a source of income as in the case of the Census in 1987.
4. Crop failures or livestock losses understate the size of a business when there are relatively few sales, but many acres, workers or expenses may be involved.

Despite these well-recognized problems, gross sales persists as the most commonly used method of describing farm size and presenting size distributions.

The dimensions of the problems of making comparisons across time are suggested by the data in Table 4.6 taken from the Censuses of 1969-1987. Farm numbers declined only slightly during this period. There was one change in the definition of a farm when the minimum level of sales to qualify as a farm was increased from $250 to $1000 in 1974. The price level essentially doubled between 1969 and 1978; it increased by about 15 percent between 1978 and 1982 and then decreased by 5.3 percent between 1982 and 1987.

It is quite easy to see how individuals could look at these unadjusted data and see a substantial shift to "larger" farms especially between 1969 and 1978. It implies major structural changes in a very short time period. But the changes in prices obscures logical comparisons across time.

One way to approximate "true" distributions of farm numbers on a common base of prices is shown in Table 4.7. Because agricultural prices

TABLE 4.6   Distribution of Farm Numbers by Sales Class, 1969, 1978, 1987

| | Census Year | | |
Description	1969	1978	1987
Producer Price Index, Farm Products (1967=100)	109.1	212.5	230.1
Prices Received by Farmers (1977=100)	59	115	126
Value of Farm Products Sold:		*Number of Farms*	
$500,000 or more	4,079	17,973	32,023
200,000 - 499,999	12,608	62,645	61,148
100,000 - 199,999	35,308	141,050	202,550
40,000 - 99,999	169,695	360,093	287,587
	(221,690)	(581,761)	(583,308)
$ 20,000 - 39,999	330,992	299,175	225,671
10,000 - 19,999	395,472	299,215	250,594
5,000 - 9,999	390,425	314,088	274,972
2,500 - 4,999	395,104	300,699	262,918
Under $2,500	994,456	460,535	490,296
Abnormal	2,111	2,302	–
Total	2,730,250	2,257,775	2,087,759

Source: Census of Agriculture and Statistical Abstract of the United States, 1969, 1978, 1987).

essentially doubled between the census years of 1969 and 1978, the 390,425 farms listed in Table 4.6 in the sales class, $5,000-9,999, were advanced in Table 4.7 into the sales class, $10,000-19,999. Thus, by a similar process of approximation, the 1969 distribution of farm numbers was put on a 1978 base in Table 4.7 to allow direct comparisons.

Something more than a change in prices was also at work, even though much of the change in the unadjusted distributions can be attributed to price inflation. This is especially evident if comparisons are made between numbers of farms in the four largest size classes.

TABLE 4.7 Comparison of Farm Numbers by Adjusted Sales Class, 1969 and 1978

Description	1969 Census Data Distributed on 1978 Base*	1978 Census
Producer Price Index, Farm Products (1967=100)	109.1	212.5
Index of Prices Received by Farmers (1977=100)	59	115
Value of Farm Products Sold:	*Number of Farms*	
Full-time:		
$500,000 or more	11,535	17,973
200,000 - 499,999	40,460	62,645
100,000 - 199,999	103,990	141,050
40,000 - 99,999	396,697	360,093
Subtotal	552,682	581,761
Part-time:		
$ 20,000 - 39,999	395,472	299,175
10,000 - 19,999	390,425	299,215
Subtotal	785,897	598,390
Primary residential:		
$ 5,000 - 9,999	357,922	314,088
2,500 - 4,999	339,444	300,699
1,000 - 2,499	346,732	460,535
Subtotal	1,044,098	1,075,322
Abnormal	2,111	2,302
Total	2,384,788†	2,257,775

*Adjusted Census distributions from Table 4.6.
†Reduced from 2,730,250 to account for all farms with sales of $500 or less in 1969 (definition change).
Source: United States Census Data, 1969 and 1978.

The loss in farm numbers between 1969 and 1978 came in part from the units with sales of less than $20,000. One source of the loss in numbers of small farms between 1969 and 1978 was the change in the definition of a farm, when the minimum sales requirement was increased from $250 to $1000. This accounted for about 350,000 of the drop in numbers.

The important conclusion is that farm size, measured in terms of gross sales, is affected by more than the rate of inflation, especially among the larger units. If prices doubled between 1969 and 1978 and everything else remained the same, then a farm with $50,000 of sales in 1969 would be equivalent to one with $100,000 of sales in 1978. But the application of output increasing, new technology, such as machinery, that allowed one man to handle more acres of crops in a timely fashion, have also affected the results. Greater productivity must be considered as well. Many of the shifts observed among the four largest size classes in Table 4.7 are partly a result of this increased productivity.

The Effect of Changes in Productivity

Evidence of the combined effects of price and technology on data for farms of relatively constant size in terms of labor, cropland, and management is suggested by averages taken from the Illinois Farm Business Record Summaries (Table 4.8). This source includes records from a large number of continuing farmers over a long span of years using the same summary procedures and full inventory adjustments annually. Groups of farms with essentially the same resource base are averaged.

Between 1960 and 1975, these Northern Illinois grain farms used about the same amount of cropland annually but the average months of labor used per farm decreased from 20 to 14 months. Specialization in production of corn and soybeans increased. Cash receipts were clearly influenced by yields and prices. A comparison of the averages for 1970 and 1975 reflects both of these effects.

Between 1975 and 1980 the analysts summarizing records for grain farms in Northern Illinois broadened the acreage base from 340-499 to 340-799 acres. The average amount of labor used per farm, however, remained nearly the same and by 1985 the average used for the larger land base was 14 months, the same as ten years earlier. If one simply looks at average cash receipts on these farms across this span of years, one sees substantial growth in size. From 1960 to 1985, cash receipts increased 5.6 times. Corn prices were 2.5 times higher; corn yields were up by 180 percent. Less labor harvested more land and much more product.

TABLE 4.8 Farm Business Summary Averages (Northern Illinois Grain Farms, 340-499 Acres, Soils Rated 76-100)

Characteristic	Record Summary Averages for:					
	1960	1970	1975	1980*	1985*
Number of farms	122	408	235	534	487
Months of labor	20	15	14	16	14
Acres of tillable land	384	395	405	534	487
% land in corn and soybeans	77	87	95	98	93
Yield of corn/bushel	92	93	146	100	166
Price received, corn/bushel	$1.04	$1.18	$2.78	$2.64	$2.57
Capital investment	$223,600	$342,600	$691,300	$2,020,000	$1,309,900
Cash receipts	$ 33,089	$ 48,707	$113,267	$ 178,315	$ 186,031

*In 1980 and 1985, the acre interval was 340-799.

Sources: Summaries of Illinois Farm Business Records.

TABLE 4.9   Land in Farms by Tenure (Census Data, United States, 1900-1987)

| Year | Tenure Class | | | |
	Full Owners	Part Owners	Tenants	Total
	Million Acres			
1900	519	125	195	839
1910	519	133	227	879
1920	515	176	265	956
1930	435	246	306	987
1935	452	266	337	1055
1940	449	300	312	1061
1945	519	371	252	1142
1950	526	423	212	1161
1954	495	470	193	1158
1959	459	498	167	1124
1964	432	533	145	1110
1969	375	550	138	1063
1974	360	535	122	1017
1978	332	561	122	1015
1982	342	531	114	987
1987	318	519	127	964

This brief examination of farm records helps to demonstrate why gross sales or cash receipts, even when corrected for changes in prices, do not capture the full nature of structural change in agriculture as effectively as they might. Particularly in the years since 1940, one worker has been able to handle more units of livestock and more units of cropland with the aid of substantial investments of additional capital.

Resource Ownership and Control

Farm Numbers and Land Use by Tenure Class

The Census has classified farms throughout the twentieth century into three important tenure classifications: full owners, part owners, and tenants (Figure 4.2). The basic definitions are implied by the titles. *Full owners* operate only land they own. *Part owners* operate land they own

FIGURE 4.2   Land Tenure Patterns (United States, 1900-1987)

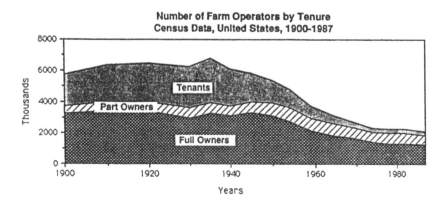

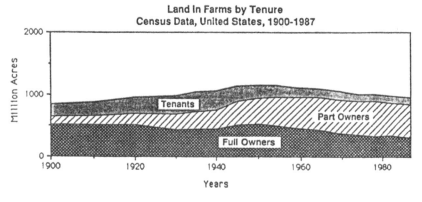

and as well as land they rent from others or work on shares for others. *Tenants* operate only land they rent from others or work on shares for others.

Tenancy was an important issue of public policy in the years before World War II. The number of tenant farmers grew in each decade until the mid 1930s when the count reached more than 2.8 million. An important part of this number were sharecroppers, often on relatively small holdings; many of these were located in the Southeast. The decline in tenant operated farms began before the end of the 1930s. Between 1935 and 1950 over 1.4 million tenant operated farms had dropped from the count. At the same time, full-owner farms held steady at more than 3.1 million and part owner farms increased from 689,000 to 825,000 (Appendix, Table 4.12).

Tenant operated units became a smaller and smaller part of the total number of farms between 1950 and 1974. Since 1974, tenant farms have accounted for 11 to 13 percent of the total number. In 1987, tenant farms were no longer located primarily in the Southeast. The only states with 10,000 or more such farms were California, Illinois, Iowa, Kansas, Minnesota, Nebraska, and Texas.

The relative importance of tenancy is indicated by the proportion of all land in farms operated by tenants in different Census periods. Land operated by tenants increased steadily from 1900 to 1935. At its peak, one-third of the total was tenant operated (Table 4.9).

The steady decline of full tenancy starting in 1940 and continuing into the 1980s reflects an important structural change in American agriculture. Part ownership has become the dominant form of farm operations. A farmer owns part of the land he operates and rents the rest. The rented land may be one parcel of cropland or some pasture; it may also be 80 or 90 percent of the land he farms. The urge to own all the land one operates has been replaced by a desire to bring together a large enough resource base to make an effective business. Renting part of the land is now a natural part of much of the commercial sector in American agriculture. Since 1969, part owners have operated more than half of America's farm land and the trend continues.

Full owners were the dominant tenure class in the first half of the century both in numbers and land operated. The relative decline in importance of full ownership since 1950 does not make this an unimportant group. It is still the largest in terms of numbers including many small, part-time and residential farms. Most farmers want to own their land; for many, however, the most efficient way to expand operations is to rent rather than buy additional cropland. The social status of a renter or tenant has changed during the course of the century. Renting is seen as part of successful operations. Tenancy is not generally viewed as an important social problem. It is simply a component of the way in which commercial agriculture is organized and operated. Landlords provide an important part of the capital to both tenants and part owners in a capital intensive industry.

Is there a relationship between tenure status and size? As suggested in the preceding table, part owners have become more and more important in the commercial sector over time. In 1987, almost 60 percent of all farms were operated by full owners but 85 percent of these farms had sales of less than $40,000. In contrast, part owner farms made up 29 percent of the total numbers with more than 50 percent having sales of $40,000 or more. Tenant farms which made up only 11.5 percent of the numbers were intermediate in their size distributions compared to the other two categories (Table 4.10).

If one examines the approximately 296,000 farms with sales of $100,000 or more in 1987, there were important numbers in each of the three categories. The number of large, full owner farms is substantial. They make up 37 percent of the farms with sales of $500,000 or more. Part owners are the group that have taken on increasing importance in the last two decades and are generally thought to continue to grow in importance relative to the other two.

Business Organization

Most farms have been operated as sole proprietorships throughout this country's history. That pattern continues (Table 4.11). A written, formal business organization is not required. The business as such is not separated from the other personal assets of the operator. Formal partnerships have become more common as more than one individual has been involved in the ownership and operation of the business. Most of these written agreements have been established in the second half of the twentieth century. Commonly they are between a father and son or two brothers; nevertheless limited partnerships among unrelated individuals are occurring with increasing frequency as the size of businesses grow.

Corporate organization has become more common as well (Table 4.11). The first efforts to obtain systematic information on types of business organization was made in the Census for 1969. At that stage, family-held corporations were not distinguished from others. Only the number of stockholders was reported. More and more farm businesses were incorporated between 1969 and 1978. Family-held corporations with 10 or less stockholders predominated. Data on four types of corporations were collected.

TABLE 4.10 Average Size and Tenure Status (Census Data, United States, 1987)

Description	Full Owners	Part Owners	Tenants	All Farms
Number of farms	1,239,000	609,000	240,000	2,088,000
Land area, million acres	317.8	519.8	126.9	964.5
Average acreage	257	854	528	462
Average sales per farm	$42,200	$105,500	$81,300	$65,200
Number of farms with sales of $100,000 or more	84,815	168,575	42,331	295,721
Percent of all farms with sales of $100,000 or more	28.7	57.0	14.3	100.0

TABLE 4.11 Type of Business Organization (Census Data, United States, 1969, 1978, 1987)

Type of Business Organization	1969	1978	1987
	Number of Farms		
Sole proprietorship (individual or family)	2,477,031	1,965,860	1,809,324
Partnership	221,535	232,538	199,559
Corporation	21,513	50,231	66,969
All other	10,070	9,146	11,907
Total	2,730,250	2,257,775	2,087,759
Type of Corporations:			
Family-held:			
10 or less stockholders	19,716*	43,138	59,599
10 or more stockholders	1,797*	1,275	1,172
Other than family-held:			
10 or less stockholders		4,688	5,379
10 or more stockholders		1,130	819
Total	21,513	50,231	66,969
Percent of Total Value of Sales:	*Percent*		
Sole proprietorship	67.8	61.6	56.3
Partnership	17.4	16.1	17.1
Corporation	14.2	21.7	25.6
Other	0.6	0.6	1.0
Total	100.0	100.0	100.0

*Not classified in 1969 except by number of shareholders.

A comparison of the data for 1978 and 1987 suggests some interesting trends. The number of partnerships has decreased in both absolute and percentage terms. The number of corporations has increased; they were 3.2 percent of the total in 1987 compared with 2.2 percent in 1978. Family-held corporations with 10 or less stockholders are the category with the greatest growth. The other than family-held corporations with less than 10 stockholders also grew. It seems likely that most of the larger farm businesses will incorporate or establish written partnership agreements in the years ahead.

Corporations were particularly important in 1987 in a few sectors of production agriculture. For cattle and calves, they accounted for 46 percent of the sales from all types of farms; cattle and calves represented 35 percent of total sales by all farm corporations. Corporations represented 71 percent of all nursery and greenhouse sales; they also had 46 percent of all vegetable sales nationally, 43 percent of fruit, berry and nut sales; and 42 percent of other crop sales such as potatoes and sugar. Of the 11,093 farms selling $1,000,000 or more of products in 1987, 5,165 were corporations of which 4,177 were family-held. As suggested by the data in Table 4.11, the corporate form of organization is used increasingly but sole proprietorships remain the dominant form of organization both in numbers and percent of total sales.

Summary and Conclusions

Farm Numbers

The 20th century has been a time of great change in the structure of agriculture in the United States. Farm numbers increased steadily throughout the 19th century as did land in farms. In 1900, there were 5.7 million farms and 839 million acres in farms. Farm numbers continued to increase until there were more than 6.4 million units in 1935. Farm numbers held at more than 6.0 million until after 1940 and U.S. entry into World War II. After this, with good job opportunities available, farm numbers declined rapidly, especially between 1950 and 1969 when farm numbers were cut in half. After 1969, the drop in numbers has continued but at a much slower rate. In 1987, there were 2.1 million farms.

Land in Farms

Land in farms continued to increase in each decade during the first half of the century. The peak in land in farms at 1,161 million acres occurred in 1950. In subsequent years, farm land has slowly been converted to forest, recreational uses, and urban and suburban development. Land in farms in 1987 had declined to 964 million acres, a drop of 17 percent in four decades.

Technology

Farming at the turn of the 20th century was powered by horses, mules and human labor. The mechanical revolution in agriculture had started; machines were used to harvest many important crops; the first agricultural experiment stations and colleges had been started. Applications of science and technology to solve agricultural problems and reduce human toil and drudgery had just begun to make their mark.

Between 1900 and 1940, there was only modest structural change. Farm size changed little; tractor power began to replace horses; the agricultural depression of the 1920s followed by the general depression of the 1930s slowed the adoption of new technology developed to improve agricultural productivity.

The sweeping structural changes between 1950 and 1969 were foreshadowed by developments within agriculture during World War II and the immediate postwar years. People were uprooted from their old patterns of life by the War. New skills were learned and new jobs were made available. Electricity and all weather roads made life in the country and commuting to industrial jobs a fine alternative. Applying the new agricultural technology developed over the previous 30 years now became possible.

Size Distributions

Whether measured in terms of acres of land in farms or in value of sales per farm, adjusted for price changes, the great changes in farm size occurred between 1950 and 1969. Farm numbers were cut in half. Labor productivity increased dramatically; excess capacity in agriculture became a chronic problem. A shift away from general crop and livestock farms to specialization in one or two enterprises became the general rule.

Resource Ownership and Tenure

Full owners were the dominant tenure class in the first half of the century both in terms of numbers and land operated. Tenancy and sharecropping were important issues of public policy between 1920 and 1940 when about one-third of the land in farms were tenant operated. By 1960, the availability of off-farm jobs and the advent of mechanized agriculture had reduced tenancy dramatically. Part owner farms have become the dominant tenure class in the second half of the decade.

These operators own part of the land they farm and rent the rest. A large share of the full owners in the 1980s operated small farms with sales of less than $40,000 annually.

Business Organization

Sole proprietorships are the dominant form of business organization in farming with 87 percent of the total number in 1987. Partnerships account for a little less than 10 percent. Family corporations with 10 or less stockholders are of growing importance. In terms of total sales, corporations and partnerships accounted for 42.7 percent of the total in 1987, while sole proprietorships have decreased in relative importance between 1969 and 1987.

Structure at the Beginning of the 1990s

About 50 percent of all units defined as farms in the 1980s sold less than $10,000 of farm products. Most of these can be thought of as places to live in rural areas where farming provides a small fraction of family income for the operators. Farms with sales between $10,000 and $40,000 accounted for another 22 percent of farm numbers and 10 percent of products sold. These are largely part-time units where family income in most cases comes from off-farm as well.

At the beginning of the 1990s, 600,000 farms with $40,000 or more of sales, accounted for 90 percent of agricultural products sold. These are the units where family income, in most cases, comes from farming. The 30,000 largest farms accounted for 35-40 percent of total sales and have increased in importance during the 1980s. Structural change occurring in the last decade is expected to continue. (A set of projections are provided in Chapter 22.)

A Final Note

The current policy debate about farm structure in part relates to how rapidly the largest farm units will come to dominate production and marketing of key commodities within commercial agriculture. It is important to remember that the competitive structure of agriculture, characterized by many thousands of farms, stands in stark contrast to most industries in the United States, including those that sell inputs to farmers on one side and those that buy farm products on the other. Structural change, so important in farming, is still modest when compared to the changes in farm machinery, meat packing, or the grain trade.

Appendix A

TABLE 4.12 Tenure of Farm Operators, 1900-1982

| Year | Tenure Class | | | Total |
	Full Owners	Part Owners	Tenants	
	Thousands			
1900	3261	451	2025	5737
1910	3413	594	2355	6362
1920	3435	558	2455	6448
1930	2968	657	2664	6289
1935	3258	689	2865	6812
1940	3121	615	2361	6097
1945	3340	661	1858	5859
1950	3113	825	1444	5382
1954	2757	857	1168	4782
1959	2140	811	760	3711
1964	1836	782	540	3158
1969	1706	671	353	2730
1974	1424	628	262	2314
1978	1298	681	279	2258
1982	1326	656	259	2241
1987	1239	609	240	2088

Source: Census Data, United States, 1900-1982.

Appendix B

The Farm Definition

When the first census of agriculture was conducted in 1840, there was no official attempt to define what exactly constituted a farm. The first census definition, for 1850, was simple; any place that had $100 or more in total agricultural products sales value was a farm. Since that time, acreage and dollar values of sales limits have been added, changed, or removed, but the requirements that the land be involved in, or connected with, agricultural "operations," and that it be under the day-to-day control of a single management (individual, partnership, corporation, etc.) have been retained.

The most important requirement is, of course, the connection with agricultural operations, which -- again for Census purposes -- are the production of livestock, poultry, and animal specialties and their products, and/or crops, including fruits, greenhouse, and nursery products. The land involved in these operations need not be contiguous to comprise a single farm, it must only be operated as a single unit." (For an exception to this general rule, see the section on the definition used in 1950-1954 censuses.)

The changes in the various criteria used for the definition of a farm are outlined below, by census:

1. *1850-1860.* No acreage requirement, but a minimum of $100 in total sales value of agricultural products.
2. *1870-1890.* A minimum of 3 acres was needed for a tract to qualify as a farm. Places with less than 3 acres were considered farms if they had a minimum of $500 in agricultural product sales.
3. *1900.* The acreage and minimum sales requirements were removed, and cranberry marshes, greenhouses, and city dairies were included, provided they required the full-time services of at least one person.
4. *1910-1920.* A minimum of 3 acres, with $250 or more in total value of sales, unless the individual operation required the full-time services of at least one person.
5. *1925-1945.* The requirement for continuous services by at least one person was dropped for the 1925 and following censuses; otherwise the definition used in the 1910-1920 censuses was unchanged.
6. *1950-1954.* The acreage qualification was retained, but places of less than 3 acres were counted as farms if they had $150 or more in total sales value of agricultural products during the census year. Places that would normally have had at least $150 in sales, or that had begun operating as a farm for the first time in 1954, were also counted as farms. If a place had sharecroppers or other tenants, the land assigned to each was treated as a separate farm, even though the landlord handled the entire holding as a single unit. Land retained and worked by the landlord was considered a separate farm.
7. *1959-1974.* Any place with 10 acres or more, and with $50 or more in agricultural products sales, or any place with less than 10 acres, but with at least $250 in total sales qualified. If sales were not reported, or if the reported sales figures were obviously incorrect, average prices were applied to reported estimates of harvests and livestock produced to arrive at estimated sales values.
8. *1978-1987.* The minimum acreage requirement was dropped. Any place that had, or normally would have had, $1,000 or more in total

agricultural products sales during the census year was counted as a farm.

Source: *1982 Census of Agriculture*, AC82-SS-4, Volume 2 Subject Series, Part 4, History, U.S. Department of Commerce, Bureau of the Census, p. 72.

Notes

1. Cropland harvested in acres indicates the land area from which crops like corn, wheat, hay, fruits and vegetables were harvested. It includes orchards, vineyards and land used for perennial crops. It does not include permanent pasture or land not tilled on a regular basis. Land in farms includes all the land area owned and operated as farm units. This large land area includes forest, open rangeland and permanent pasture, as well as cropland. Large areas of the forested East and the dry lands of the West are parts of farms even though cropland on these units involves small percentages of the total.

2. Widely accepted definitions of *full-time* and *part-time* farms have not been established. *Full-time* units would indicate that at least one worker equivalent is employed in farming and that 50 percent or more of family income is provided in most years from farming. *Part-time* farms include all those with less than one worker equivalent involved in farming and where family income commonly comes primarily from off-farm sources.

References

Ahearn, Mary. 1986. *Financial Well Being of Farm Operators and Their Households*. ERS, USDA, Agricultural Economics Report No. 563.

Bureau of the Census. *Censuses of Agriculture*. Vol. 2, General Reports, Washington, DC, Department of Commerce, 1930, 1945, 1950, 1959, 1969, 1978, 1982.

Carlin, Thomas, and John Crecink. 1979. "Small Farm Definition and Public Policy." *American Journal of Agricultural Economics* 61: 933-939.

Congressional Budget Office. 1978. "Public Policy and the Changing Structure of American Agriculture."

Daugherty, Arthur, and Robert Otte. 1983. "Farmland Ownership in the United States." ERS Staff Report AGES 830311.

Economics, Statistics, and Cooperatives Service, U.S.D.A. 1979. "Structure Issues of American Agriculture." Agricultural Economics Report No. 438.

Edwards, Clark, M. G. Smith, and R. N. Peterson. 1985. "The Changing Distribution of Farms by Size: A Markov Analysis." *Agricultural Economics Research* 37: 4.

Harrington, David, and Thomas Carlin. 1987. "The U.S. Farm Sector: How Is It Weathering the 1980s?" ERS, USDA, Agricultural Information Bulletin No. 506.

Heady, Earl. 1983. "Economic Policies and Variables Relating to the Structure of Agriculture." Department of Economics, Iowa State University.

Lin, William, George Coffman, and J. B. Penn. 1980. "U.S. Farm Numbers, Sizes, and Related Structural Dimensions: Projections to the Year 2000." ESCS, USDA, Technical Bulletin No. 1625.

Miller, Tom, Gordon Rodewald, and R. McElroy. 1981. "Economies of Size in U.S. Field Crop Farming." ESCS, USDA, Agricultural Economics Report No. 472.

Nikolitch, Radoje. 1972. "Family Size Farms in U.S. Agriculture." ERS, USDA, Agricultural Economics Report No. 499.

Office of Technology Assessment, The Congress. 1986. *Technology, Public Policy and the Changing Structure of American Agriculture.* Washington, DC.

Raup, Philip M. "Some Questions of Value and Scale in American Agriculture." *American Journal of Agricultural Economics,* pp. 303-308.

Reimund, Donn A., J. Rod Martin, and C. V. Moore. 1981. "Structural Change in Agriculture: The Experience for Broilers, Fed Cattle and Processing Vegetables." ESS, USDA, Technical Bulletin No. 1648.

Reimund, Donn A., T. A. Stucker, and N. L. Brooks. 1987. "Large-Scale Farms in Perspective." ERS, USDA, Agricultural Information Bulletin No. 505.

Robison, Lindon J., ed. 1988. "Determinants of Farm Size and Structure." Michigan Agricultural Experiment Station Journal Article 12899.

Schertz, Lyle, et al. 1979. "Another Revolution in U.S. Farming?" ESCS, USDA, Agricultural Economics Report No. 441, p. 445.

Stanton, B. F. 1988. "Changes in Farm Structure: The United States and New York, 1930-1982." Cornell University Agricultural Economics Staff Paper No. 84-23.

Stanton, B. F. 1988. "Classification and Integration of Sector Accounts in Agriculture." Cornell University Agricultural Economics Staff Paper No. 88-8.

Sumner, Daniel A. 1985. "Farm Programs and Structural Issues," in B. L. Gardner, ed., *U.S. Agricultural Policy: The 1985 Farm Legislation.* American Enterprise Institute, Washington, DC, pp. 283-328.

Tweeten, Luther. 1984. "Causes and Consequences of Structural Change in the Farming Industry." National Planning Association, NPA Report No. 207.

United States Senate, Committee on Agriculture, Nutrition and Forestry. "Farm Structure, A Historical Perspective on Changes in the Number and Size of Farms." 379 pp.

5

Recent Changes in Size and Structure of Agriculture: A Study of Selected States in the North Central Region

Robert H. Hornbaker and Steven R. Denault

The changing structure of American agriculture has long been an issue discussed by agricultural policy analysts. Structural changes affect opportunities for beginning farmers, the viability of many existing farms, sociological aspects of the rural community, and the cost and availability of food to consumers.

A major change in farm structure began with the end of World War II as many new technologies were used in agriculture. Farmers moved from horsepower to tractor power, and used more hybrid seeds, fertilizers, chemicals, irrigation and more productive livestock. It was expected that there would be fewer farms after the war, but farmers expanded their operations to make the best use of new technologies and farm numbers decreased even more than expected.

In 1979, Secretary of Agriculture Bob Bergland renewed interest in the structure subject by holding nationwide hearings. The event prompted additional studies of the factors affecting the structure of agriculture and the type of structure that might exist in the future. There was support from across the nation for the "family farm."

According to Stanton (1989), structural change during the 1970s and 1980s was still an important issue, but concern was for the proportions of agricultural output that would be produced by different economic classes of farms rather than the decline in farm numbers. Lin, Coffman and Penn illustrate these ideas with the following statistics. In 1969, 80 percent of the total agricultural output was produced by approximately

655,000 farms (the largest 24 percent). By 1974, 493,000 farms (the largest 20 percent) produced the same amount. They estimated that by the year 2000, the 50,000 largest U.S. farms would produce two-thirds of all agricultural output with half of the total agricultural output coming from the largest 1 percent of the farms.

To examine the factors affecting the structure of agriculture, Baab (1979) identified five dimensions of structure. These dimensions are: number and size of farms, specialization, owner control, entry barriers and socioeconomic characteristics. Stanton (1989) cites two variables commonly used in studying agricultural structure changes: 1) status of the operator, measured in terms of tenancy (e.g., full-owner, part-owner or tenant) or type of business organization (e.g., family farm or sole proprietor, partnership, or corporation) and 2) farm size distribution measured in terms of acreage or output. Both distribution measures have limitations, but the acreage criterion still indicates the direction of agricultural structure and the sales measurement is largely accepted for the sake of convenience. The U.S. Census of Agricultural provides farm sales figures for the entire U.S. In addition, sales of agricultural output reflects the combined productivity of all factors of production used in farming.

Farm size is commonly measured in acreage and/or value of sales. This dimension can be used to describe the proportion and distribution of all farms within a geographic region. A number of factors influencing farm size are: technology, economies of size, variations in input prices, variations in commodity prices, risk and expectations, price-cost margin, capital requirements, goals of the farmer, managerial ability, foreign demand and exchange rates, and government fiscal and monetary policy.

In this chapter, changes in farm size and structure from 1978 to 1987 are examined for the following eight states in the North Central region: North Dakota, Kansas, Iowa, Illinois, Wisconsin, Michigan, Indiana and Ohio. Distributions of farms by size and structure categories are obtained from the 1978, 1982, and 1987 U. S. Census of Agriculture. In addition, U.S. Census of Agriculture data are obtained for selected counties in Illinois and compared to characteristics of individual farm records.

U.S. Census of Agriculture Data

Three definitions of farm size are chosen for examining the size distributions of farms in the selected states. The size definitions include gross value of sales, acreage and market value of machinery and equipment. The U.S. Census categories are summarized in five classes for each size definition as depicted in Table 5.1. The gross value of sales

TABLE 5.1 Size Categories and Number of Farms in Eight Selected States in the North Central Region[a]

	Gross Farm Sales Categories ($ 1,000)				
Sales	S1 1 - 10	S2 10 - 20	S3 20 - 100	S4 100 - 250	S5 250+
1978	297.1	113.8	146.5	61.6	13.3
1982	253.7	89.7	136.0	84.9	25.1
1987	237.3	81.0	114.1	77.8	24.1

	Farm Acreage Categories				
Acres	A1 1 - 100	A2 100 - 220	A3 220 - 500	A4 500 - 1000	A5 1000+
1978	213.3	165.2	167.4	74.6	38.3
1982	221.7	145.1	147.7	73.2	42.0
1987	198.0	127.8	130.8	70.7	45.8

	Machinery and Equipment Investment Categories ($ 1,000)				
Investment	M1 1 - 20	M2 20 - 50	M3 50 - 100	M4 100 - 200	M5 200+
1978	282.8	170.9	120.7	67.7	17.2
1982	245.2	153.5	118.3	80.0	30.7
1987	222.2	148.1	106.3	69.0	26.4

[a]Number of farms are in 1000s.

includes the gross values of all crops, animal and poultry and specialty products sold.

The number of farms in each size category is also shown in Table 5.1. Based on the sales definition of farm size, the absolute number of farms decreased in each category from 1982 to 1987. However, between 1978 and 1987, the percentage of farms in S2 and S3 decreased, while categories S4 and S5 increased. Likewise, using the acreage definition, the percentage of farms in the largest two categories has increased, but the absolute numbers have only increased for farms larger than 1000 acres. Within the North Central region, the proportions of the smallest farms, based on sales and acreage, has not decreased, while the absolute number has declined.

The only absolute increases in farm numbers, as classified by machinery investment, occurred in the largest two categories between 1978 and 1982. The only proportional increase from 1982 to 1987 was in M2. Moreover, agriculture's structure does not appear to be moving toward the "bimodal" distribution of farms that was anticipated (McDonald et al. 1980). Tweeten (1988) adds that the distribution of farms is skewed and has led to a dual agriculture--a few large farms that account for most of the agricultural output and part-time small operations that account for most of the farms. The size distributions here indicate that while the proportion of very small part-time farms remains constant, there exists a natural growth in the distributions from small to medium to larger farms.

Inappropriate interpretations of changes in farm size and structure may occur if policy makers rely solely on aggregate information. Many of the important changes within regions, states and counties are lost in the national average statistics. To examine changes at the state level, number of farms and size statistics for the eight states (North Dakota, Kansas Iowa, Illinois, Wisconsin, Michigan, Indiana, and Ohio) are reported in Table 5.2. One of the most important disclosures from the state statistics in Table 5.2 is the relative change in the farm number decline from the first to second period and the variation in those changes among states. In general the average annual rate of decline in farm numbers was approximately 0.6 percent more during the second period (1.74 percent per year) for a total 8.70 percent decline over the five year period. All states except North Dakota and Kansas experienced farm number declines exceeding 8.5 percent from 1982 to 1987. Michigan's numbers declined by 12.77 percent during the second period and only 2.92 percent from 1978 to 1982. South Dakota was the only state experiencing more rapid decline during the 1978 to 1982 than during the 1982 to 1987 period (9.7 percent and 3.1 percent respectively).

TABLE 5.2   Total Number and Average Size of Farms for Eight Selected States

	1978	1982	1987
North Dakota			
number	40,357	36,431	35,289
average acres	1,033	1,104	1,143
Kansas			
number	74,171	73,315	68,579
average acres	640	642	680
Iowa			
number	121,339	115,413	105,180
average acres	274	211	229
Illinois			
number	104,690	98,483	88,786
average acres	282	292	321
Wisconsin			
number	86,505	82,199	75,131
average acres	206	210	221
Michigan			
number	60,426	58,661	51,172
average acres	183	187	202
Indiana			
number	82,483	77,180	70,506
average acres	204	211	229
Ohio			
number	89,131	86,934	79,277
average acres	177	177	189

Table 5.3 includes the percentages of farms raising major grain and livestock products. These statistics indicate the percentage of farm operators who reported raising corn, soybeans, wheat, cattle, dairy cows, beef cows and/or hogs. Information on the business organization and tenure of farms in the eight states is reported in Table 5.4. Farms are classified as *full owners* if the operator farms only land he/she owns, *part*

TABLE 5.3 Percentage of Farms in Selected North Central States Producing Grain and Livestock

	Grain			Livestock			
	Corn	Wheat	Soybeans	Cattle	Dairy	Beef	Hogs
North Dakota							
1978		85		51	12	43	9
1982		80		51	10	43	7
1987		80		49	8	41	7
Kansas							
1978	14	62	22	54	3	42	9
1982	11	67	23	50	2	37	5
1987	13	56	28	47	1	34	5
Iowa							
1978	83	1	65	57	10	37	47
1982	80	2	63	54	9	36	40
1987	79	1	65	47	7	31	35
Illinois							
1978	75	21	70	44	7	30	27
1982	75	29	68	42	6	29	19
1987	75	24	69	38	5	26	19
Wisconsin							
1978	65		5	74	53	18	18
1982	64		7	76	54	18	15
1987	65		7	71	50	14	12

Michigan							
1978	56	23	23	42	16	18	14
1982	52	26	25	44	16	18	13
1987	49	20	25	38	13	16	11
Indiana							
1978	68	24	56	47	9	30	27
1982	65	31	55	48	9	30	23
1983	64	26	57	43	7	27	21
Ohio							
1978	62	36	50	48	14	27	20
1982	61	38	45	49	13	27	16
1987	58	33	46	44	12	24	14

TABLE 5.4   Ownership and Tenure of Farms in the Eight North Central States

| | Percentage of | | | | | |
| | Farms | | | Acres | | |
	1978	1982	1987	1978	1982	1987
North Dakota						
Full Owner	34	34	32	23	24	19
Part Owner	49	49	50	65	66	67
Tenant	7	17	18	12	11	13
Individual	88	87	88	79	78	78
Partnership	12	12	10	15	14	13
Family Corp.	0	0	1	1	1	1
Kansas						
Full Owner	41	43	44	19	22	19
Part Owner	43	41	41	67	65	66
Tenant	17	16	16	14	13	15
Individual	88	88	88	80	78	78
Partnership	10	9	9	14	13	13
Family Corp.	2	2	3	5	7	7
Iowa						
Full Owner	46	46	46	29	29	25
Part Owner	32	33	33	50	52	55
Tenant	22	21	21	21	19	20
Individual	86	85	85	80	78	79
Partnership	12	11	11	14	13	12
Family Corp.	2	3	4	5	7	7
Illinois						
Full Owner	42	44	44	21	22	19
Part Owner	36	36	37	57	59	60
Tenant	22	20	19	22	20	20
Individual	86	85	85	79	78	78
Partnership	13	12	12	17	16	16
Family Corp.	1	2	2	3	4	5

(continued)

TABLE 5.4 (*continued*)

| | Percentage of | | | | | |
| | Farms | | | Acres | | |
	1978	*1982*	*1987*	*1978*	*1982*	*1987*
Wisconsin						
Full Owner	62	59	58	47	43	41
Part Owner	31	32	33	46	49	51
Tenant	7	9	8	7	7	7
Individual	88	86	86	81	77	77
Partnership	10	11	11	13	15	15
Family Corp.	2	2	3	5	7	8
Michigan						
Full Owner	61	60	61	39	36	34
Part Owner	32	33	33	56	59	61
Tenant	6	6	6	6	5	5
Individual	89	89	88	80	77	76
Partnership	9	9	9	16	17	18
Family Corp.	1	1	2	3	4	5
Indiana						
Full Owner	57	57	58	31	30	28
Part Owner	31	32	31	.57	59	60
Tenant	12	11	11	13	11	12
Individual	86	85	85	77	75	75
Partnership	12	12	11	17	16	15
Family Corp.	2	3	4	6	8	9
Ohio						
Full Owner	59	60	59	37	36	34
Part Owner	29	29	29	51	53	55
Tenant	12	11	11	12	11	12
Individual	86	86	86	79	77	77
Partnership	12	12	11	17	17	17
Family Corp.	1	2	2	3	4	5

owners if the farm includes owned land and rented land and *tenants* if all land is rented or the operator works on shares.

Three types of business organizations are reported in Table 5.4. *Individual or family operations* are defined as those organizations controlled and operated by an individual. *Partnerships* are only those operations in which an agreement to the shares of contribution and income has been made by two or more partners. *Family corporations* are those corporations in which more than 50 percent of the stock is owned by related persons. Any other business forms constitute the residual percentage.

The following sections include a discussion of the size distributions of farms in each state and type of farm information. Farms are discussed from west to east beginning with North Dakota. Distributions of farms by size categories are depicted in Figures 5.1 to 5.8.

North Dakota

As shown in Figure 5.1, 36 percent of the farms in North Dakota in 1978 had gross sales between $1,000 and $10,000. By 1987, only 30 percent of the farms remained in this smallest sales category. The proportion of farms declined in both S1 and S2 from 1978 to 1982. The increases in sales categories also took place between 1978 and 1982. The $100,000 to $250,000 class increased from 7 percent to 14 percent, and the $250,000 and above class from 1 percent to 3 percent. Very little change occurred in the sales distribution between the 1982 and 1987 census.

The acreage distribution of farms in North Dakota is uncharacteristic of farms within the North Central region of the U.S. Around 40 percent of the farms in the state are larger than 1000 acres. This category of farms increased from 38 percent in 1978 to 43 percent by 1987. The only other acreage class of farms to increase in size during the nine-year period was A1 (100 to 220 acres).

North Dakota also has a larger percentage of farms with higher levels of investment in machinery and equipment than most other states in the region. More than 50 percent of the farms have at least $50,000 invested in machinery and equipment, and about 30 percent exceed an investment level of $100,000.

As indicated in Table 5.3, about 80 percent of the farms in North Dakota produce wheat and one-half feed cattle. Most of the farms are individual operations (84 percent in 1987) while 14 percent are partnerships. Many of the partnerships include family-operated partnerships. More than 30 percent of the farms are fully owned by the operator and about 17 percent are tenant farms. The 50 percent part-owned farms operate about two-thirds of the acreage.

FIGURE 5.1 Distribution of farms in North Dakota by gross value of sales, acreage, and value of machinery and equipment size categories.

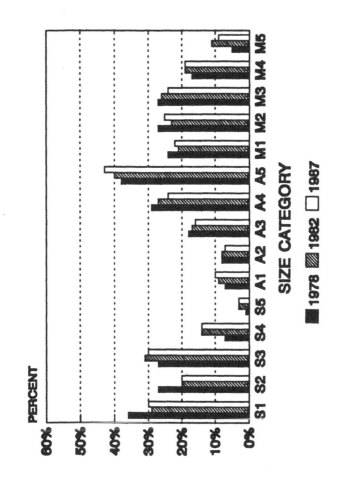

Kansas

The distribution of farms in Kansas (Figure 5.2) is characterized by approximately 45 percent of the farms in the smallest sales category ($1,000 - $10,000). In 1987, 15, 18, 10 and 4 percent of the farms were in the next four sales categories, respectively. Only the smallest sales category exhibited an increase in the proportion of farms from 1982 to 1987. The largest two sales size groups increased in 1982 but remained constant between the 1982 and 1987 census.

Kansas farms are more evenly distributed within the acreage size classes than farms in most of the other states. Each acreage category includes approximately 20 percent of the farms. Only farms smaller than 100 acres and larger than 1000 acres have increased in proportion since 1978. The proportion of small farms (S1) increased from 21 percent to 23 percent between 1978 and 1982, while the percentage of the largest farms grew from 18 percent to 21 percent between 1978 and 1987. Reductions in the proportion of farms occurred in the three middle size classes during the nine-year period.

As with the sales classification of farms, there is a very high percentage of farms in the smallest machinery and equipment investment category. The proportion of farms in the two smallest categories (M1 and M2) decreased 2 to 3 percent, while the percentages in the three largest categories increased 2 to 3 percent from 1978 to 1987.

The predominate crop grown on Kansas farms is wheat, while the number of farms growing soybeans has increased from 22 percent to 28 percent since 1978. The number of farms which produce livestock has declined. In 1978, 54 percent fed cattle, down to 47 percent in 1987. Likewise, beef cow farms have diminished from 42 percent to 34 percent. The percentage of fully owned farms increased in Kansas from 41 percent to 44 percent, but the acreage they control was 19 percent of the total in 1987. The business organization of the farms has changed little, with 88 percent organized as individual units farming 78 percent of the land. Approximately 9 percent are partnerships and 3 percent family corporations.

Iowa

In 1978, 31 percent of the farms in Iowa had gross value of sales between $20,000 and $100,000 (Figure 5.3). By 1987, only 25 percent were in the S3 category and 30 percent had sales between $1,000 and $10,000. The two largest sales categories increased in proportion from 1978 to 1982, but changed little in 1987.

FIGURE 5.2 Distribution of farms in Kansas by gross value of sales, acreage, and value of machinery and equipment size categories.

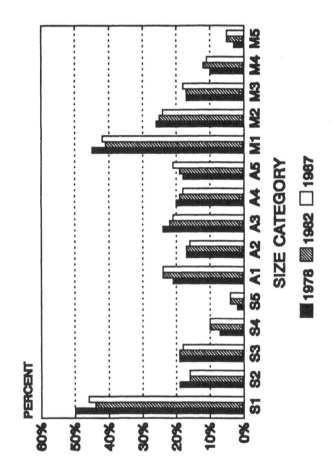

84

FIGURE 5.3   Distribution of farms in Iowa by gross value of sales, acreage, and
value of machinery and equipment size categories.

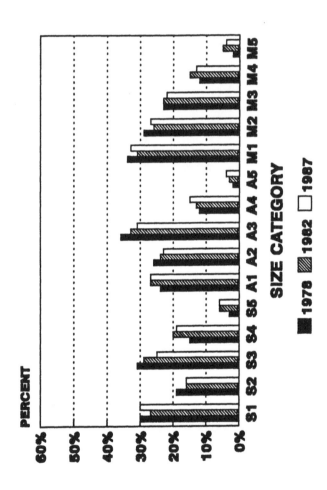

The largest acreage category of farms continues to be A3 (220 - 500 acres), which included 31 percent of the farms in 1987. Categories A4 and A5 increased in size, while A2 and A3 decreased from 1976 to 1987. Between 1982 and 1987, all but the smallest two classes of farms, by value of machinery and equipment, decreased in size. In these latter five years, there emerges a decreased investment in machinery and equipment.

Corn and soybean farms predominate in Iowa. The proportion of farms producing livestock has decreased significantly since 1978. The tenure and business organization of farms in Iowa have remained fairly constant over the nine-year period. Of all Iowa farms, 45 percent are fully owned by the operator and about 85 percent are operated as sole individual businesses. However, one-third of the part owner farms have increased their share of land from 50 percent to 55 percent.

Illinois

Figure 5.4 indicates that the distribution of farms in Illinois is similar to Iowa's, but with a larger percentage of farms in the smallest classes. Like most other states, the S1 and S2 categories increased in size from 1982 to 1987, while categories S3 to S5 decreased. In 1987, 37 percent of the farms had gross sales less than $10,000, and for 4 percent of the farms, gross sales exceeded $250,000.

The distribution of farms by acreage category, much like sales, includes a large proportion of the smallest farms (~32 percent) and a small but increasing proportion of large farms. As in Iowa, the two largest categories are increasing in size and categories A2 and A3 are decreasing.

The proportion of farms in M2 increased by 4 percent and M4 decreased by 4 percent from 1982 to 1987. All other machinery investment categories remained constant in the two most recent years. Illinois farms remain primarily corn and soybean production units, with about 25 percent of the farms also producing wheat. More and more, Illinois farms are also specializing in grain production, reducing cattle feeding 6 percent and hog operations 8 percent. Of all farms, 44 percent are classified as fully owned and operate 19 percent of the acreage. The 37 percent part owner farms have increased their share to 60 percent of the land.

Wisconsin

Wisconsin is an example of how changes in number of farms within gross sales categories can be very misleading. The census data in Figure 5.5 show large reductions in the two smallest farm classes (S1 and S2)

FIGURE 5.4   Distribution of farms in Illinois by gross value of sales, acreage, and
value of machinery and equipment size categories.

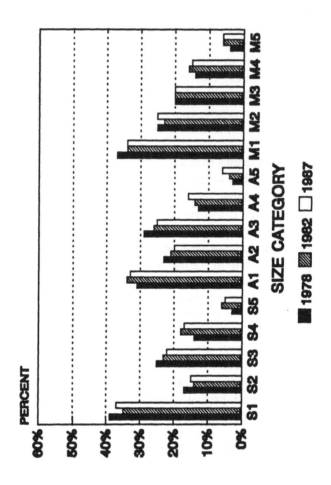

FIGURE 5.5   Distribution of farms in Wisconsin by gross value of sales, acreage, and value of machinery and equipment size categories.

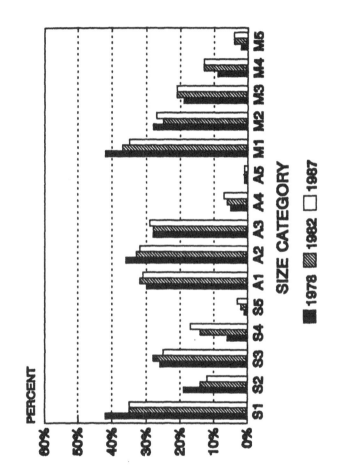

and a large increase in S4. However, the acreage distribution indicates a decline in A2 and only modest increases in A3 and A4 from 1978 to 1982. Growth in Wisconsin has occurred in the 220 to 1000-acre range, but not above 1000 acres. The machinery and equipment investment categories indicate increased proportions in the largest three categories from 1978 to 1982, but no change in these classes between the 1982 and 1987 census. In these last two census counts, M1 decreased and M2 increased by approximately 2 percent.

Of the three major crops listed in Table 5.3, corn is produced on 65 percent of the Wisconsin farms and soybeans on only 7 percent. The proportion of farms producing livestock in Wisconsin is the highest of any of the eight North Central states. Although the percentages have declined 71 percent of the farms fed cattle and 50 percent produced milk in 1987. Wisconsin also has a high but declining ownership rate. Of all farms, 62 percent were fully owned in 1978, 59 percent in 1982 and 58 percent in 1987. The growth in Wisconsin is also in the part owner farms with an increase in acreage from 46 percent to 51 percent.

Michigan

Michigan has the highest concentration of farms in the smallest categories S1, A1 and M1 of any of the eight states (Figure 5.6). Of all farms, 53 percent had gross sales less than $10,000 in 1987 and just fewer than 50 percent were less than 100 acres. In 1978, 58 percent of the census farms had gross sales between $1,000 and $10,000. Around 20 percent of the farms were in the $10,000 to $100,000 categories and the $250,000 plus category increased from 2 percent to 4 percent. Only the largest category by sales and acreage has increased in each of the last two census counts. About 50 percent of the farms invested less than $20,000 in machinery. The M1 category decreased 8 percent during the nine years, whereas M2 through M5 increased.

In 1987, about half of the Michigan farms produced corn, 20 percent produced wheat and 25 percent produced soybeans. Michigan has the highest full ownership and the lowest tenant rates at 61 percent and 6 percent respectively in 1987. However, in 1987 61 percent of the land was controlled by part owners.

Indiana

Numbers in Indiana are also highly concentrated in the smallest farm categories (Figure 5.7). Moreover, there was practically no change in farm size, by sales classification, from 1982 to 1987. By acreage, farms between 220 to 500 acres decreased 2 percent and farms greater than 1000

FIGURE 5.6   Distribution of farms in Michigan by gross value of sales, acreage, and value of machinery and equipment size categories.

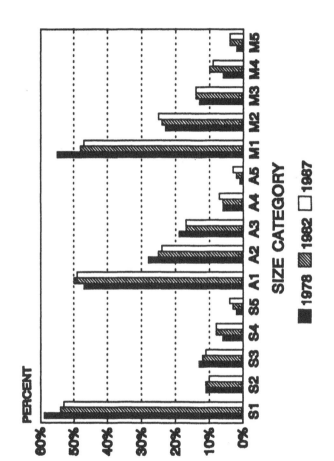

FIGURE 5.7   Distribution of farms in Indiana by gross value of sales, acreage, and
value of machinery and equipment size categories.

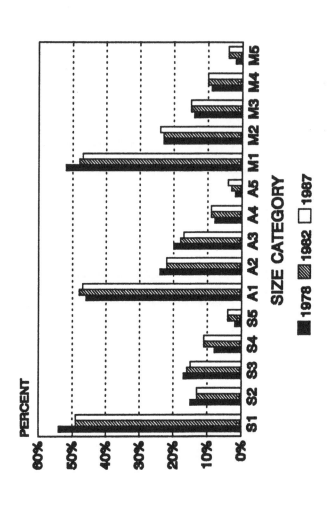

acres increased by 2 percent during the nine-year period. The machinery investment categories also exhibited very little change from 1982 to 1987.

The number of farms growing corn declined 4 percent, while a slightly larger percentage are producing wheat and soybeans. As in the rest of the North Central region, the number of livestock operations has declined.

Ohio

Fifty-seven percent of the farms in Ohio had gross sales less than $10,000 and only 2 percent were more than $250,000 in 1978 (Figure 5.8). As of 1987, 54 percent were in S1 and 4 percent in S5. As noted in most of the other states in the North Central region, S3 was the second largest sales category, but S3 has declined each year while S4 and S5 increased. In the acreage classifications, A1 and A5 increased in proportion, from 1976 to 1982, while A2 and A3 decreased. From 1982 to 1987 practically no change took place in the acreage classifications of farms. The 500 to 1000-acre category remained constant over the nine-year period. In Ohio, 53 percent of the farms had investment in machinery less than $20,000 in 1978 and 47 percent in 1987. The second category (M2) increased from 24 percent in 1978 and 1982 to 27 percent in 1987. The two largest categories increased in 1982 but remained fairly constant from 1982 to 1987.

Ohio has seen a 4 percent reduction in farms growing corn and soybeans and a 3 percent decline in those producing wheat. The livestock farm percentages have likewise declined. Ohio also has the lowest proportion of part owner farms (29 percent) in the eight states.

Illinois Agriculture Census and FBFM Comparison

Tables 5.5 - 5.7 depict the distribution of farms by sales, acreage and investment in machinery and equipment for selected counties in Illinois. One county was selected from each of the nine crop reporting districts in Illinois. For each of nine counties, data from the Census of Agriculture are compared to records from the Farm Business Farm Management Associations. Four of the nine counties and totals for the state of Illinois are presented in Tables 5.5 - 5.7. These data are used to compare the structure of the FBFM sample farms to those in the U.S. Census of Agriculture.

The FBFM farms were selected from over 7,500 which participate in the FBFM Associations within Illinois. The majority of these farmers are

FIGURE 5.8   Distribution of farms in Ohio by gross value of sales, acreage, and value of machinery and equipment size categories.

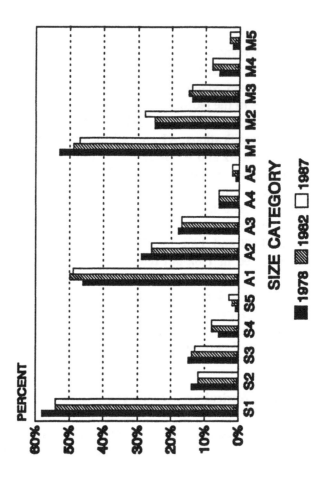

TABLE 5.5 Distribution of Farms by Sales Census of Agriculture, Illinois (Percent of Total Farms)

	S1 1 - 20	S2 20 - 40	S3 40 - 100	S4 100 - 250	S5 250+
Whiteside County					
U.S. Census					
1978	27	18	29	20	4
1982	23	14	27	25	9
1987	25	17	24	20	9
FBFM					
1982	0	0	28	49	23
1987	0	2	25	39	35
1987*	0	2	30	44	25
Iroquois County					
U.S. Census					
1978	21	20	35	19	3
1982	18	16	30	28	7
1987	22	16	31	24	5
FBFM					
1982	0	0	28	60	13
1987	0	2	22	56	20
1987*	0	5	29	54	12
McDonough County					
U.S. Census					
1978	36	17	28	15	2
1982	31	16	24	21	6
1987	35	16	23	17	4
FBFM					
1982	0	0	20	55	25
1987	0	0	13	40	47
1987*	0	0	20	40	40
Randolph County					
U.S. Census					
1978	55	19	16	7	0
1982	50	13	19	9	2
1987	53	15	14	10	2
FBFM					
1982	0	4	25	67	4
1987	0	0	20	60	20
1987*	0	4	24	64	8

*Values deflated to 1982 nominal dollars.

TABLE 5.6 Distribution of Farms by Acreage, Illinois (Percent of Total Farms)

	A1 1-100	A2 100-220	A3 220-500	A4 500-1000	A5 1000+
Whiteside County					
U.S Census					
1978	27	29	31	10	3
1982*	28	24	32	13	4
1987*	29	23	29	16	4
FBFM					
1982*	0	9	58	30	4
1987*	0	14	51	28	7
Iroquois County					
U.S. Census					
1978	18	22	39	18	3
1982*	21	19	35	20	5
1987*	20	18	32	24	6
FBFM					
1982*	0	0	23	63	15
1987*	0	5	15	61	20
McDonough County					
U.S. Census					
1978	29	20	33	17	2
1982*	29	20	31	16	4
1987*	30	18	29	17	7
FBFM					
1982*	0	0	45	50	5
1987*	0	0	27	33	40
Randolph County					
U.S. Census					
1978	36	28	24	9	2
1982	40	25	22	10	3
1987*	38	24	23	11	5
FBFM					
1982*	0	8	46	33	13
1987*	0	8	44	28	20

*Values deflated to 1982 nominal dollars.

TABLE 5.7 Distribution of Farms by Market Value of Machinery and Equipment, Illinois (Percent of Total Farms)

	M1 1-20	M2 20-50	M3 50-100	M4 100-200	M5 200+
Whiteside County					
U.S. Census					
1978	34	28	24	11	3
1982	28	26	24	17	5
1987	30	18	26	18	5
FBFM					
1982	13	34	38	16	0
1987	49	39	11	2	0
1987*	51	40	7	2	0
Iroquois County					
U.S. Census					
1978	20	29	27	18	6
1982	22	19	25	26	8
1987	25	22	23	24	6
FBFM					
1982	13	30	43	15	0
1987	32	54	15	0	0
1987*	49	42	10	0	0
McDonough County					
U.S. Census					
1978	34	30	17	16	3
1982	31	21	19	18	10
1987	31	27	20	19	3
FBFM					
1982	20	45	20	15	0
1987	47	27	27	0	0
1987*	53	20	27	0	0
Randolph County					
U.S. Census					
1978	54	23	9	11	4
1982	41	25	15	14	4
1987	45	20	16	14	4
FBFM					
1982	8	25	54	13	0
1987	56	44	0	0	0
1987*	60	40	0	0	0

*Values deflated to 1982 nominal dollars.

full time operators and are thus likely to be, on average, larger than the average Census of Agriculture farms within Illinois. The Census of Agriculture includes any operation with more than $1,000 in sales annually. The FBFM sample was also chosen from those farms which had certified records from 1982 to 1987. Therefore, the sample consists of the same farms in 1987 as in 1982.

Distributions by Sales and Acreage

Tables 5.5 and 5.6 indicate that almost all FBFM farms are found in the largest three size categories. In all four counties the proportion of farms in A4 decreases while the proportion in A5 increases. In contrast the Census data show both categories A4 (500 - 1000 acres) and A5 (1000+ acres) increasing in proportion from 1978 to 1987. However, the number of FBFM farms does not change from 1982 to 1987, while entrants and exits can be observed in the Census pool.

For the sales categories, the Census data show increases in the proportion of farms, between 1982-87, for only the smallest sales categories, $1,000 to $20,000 and $20,000 to $40,000. A similar trend was noted earlier in North Dakota, Kansas, Iowa, and Indiana. The increase in only small farms could be explained by one of two reasons: 1) the net number of small part-time operations remained constant while the net exits to farming occurred in the larger full-time operations, or 2) exits from farming occurred somewhat equally across size categories, but new entrants have been at the small part-time level.

The nominal FBFM data show increasing proportion of large farms. However, the change is primarily due to inflation. Except for McDonough county, the deflated (by GNP) sales distributions in 1987 are almost identical to those in 1982. This indicates that the Census data actually understate the decline in proportion of farms in at least the two largest sales categories between 1982 and 1987 ($100,000 to $250,000 and $250,000+).

Distributions by Value of Machinery and Equipment

The distributions of farms by value of machinery and equipment investment (Table 5.7) are quite different from those for sales and acreage. In this case, approximately 85 percent of the FBFM farms are found in the smallest two categories (less than $50,000) while around 65 percent of the Census farms are identified in M1 and M2. Part of the discrepancy may be different accounting of machinery and equipment value.

It is apparent from both sources of data and the lack of growth in the largest categories for any of the other states, that investment in new machinery and equipment has not kept pace with that in the 70s or current depreciation on this equipment. Except for McDonough county, proportion of farms has decreased in the larger categories and increased in the smaller.

Table 5.8 presents average yield, fertilizer expense and chemical expense for Illinois Census of Agriculture and FBFM records. It is clear from these data that average yields are higher for the FBFM sample than the Census population at large. However, it is not possible to test the significance of these differences since the variation in yield is not available for the Census data. It is also clear from this table that per acre fertilizer and chemical expenses have declined in both nominal and real terms since 1982. This fact is also substantiated by other studies.

Tables 5.9, 5.10 and 5.11 depict other descriptive information about size and structure of farms which is not readily available in the Census of Agriculture data. These data are again presented by the three size measures used in the earlier part of the chapter.

In Table 5.9, the smallest size category, A1, is not depicted because, as indicated in Table 5.5, none of the FBFM farms are smaller than 100 acres and very few are included in the 100 to 220 acre category. It should be apparent from this table that it would be difficult to exist as a full-time grain farm with an acreage less than 220. Farm living expense averaged approximately $28,000 between 1987 and 1990 (University of Illinois). Only farms of at least the size of the largest three categories generate enough before tax net farm income to cover the average family living expense.

It is also worth recognizing, the change in tenure from small to larger farming operations. The percentage ownership declines by approximately 10 percent from category A2 to A5. Corn yield increases from category A2 to A3 but is moderately constant across categories A3 to A5. However, soybean yield does not vary by a significant amount between four size categories.

Three significant factors are noted when farm size is characterized by investment in machinery and equipment. First, the proportion of land owned by the operator is higher with larger investment. Although this trend is not as pronounced in 1987 as in 1982. In 1982, 36 percent of the land was owned by operators in the largest category and only 20 percent in the $1,000 to $20,000 category. The 1987 adjusted numbers are 35 percent and 30 percent for the largest and smallest categories respectively. The second factor is that yield is reasonably constant across categories. Actually, the highest corn and soybean yields are noted in the middle categories M2 and M3. This and the first factor are likely explained, in part, by a larger percentage of livestock production for the farms in M4 versus M2 and M3. A higher proportion of livestock production may also constitute a larger proportion of corn acreage. Therefore, explaining the lower yields and the third factor which is higher pesticide cost per acre. Also noted is the large reduction in interest expense which

TABLE 5.8 Average Annual Yields, Fertilizer, and Chemical Expenses (Census of Agriculture and Illinois Farm Business Farm Management Association)

	Yields			Fertilizer	Ag. Chem.
	Corn	Soybeans	Wheat		
Whiteside County					
U.S. Census					
1978	109	34	33	38.88	15.79
1982	123	40	33	37.79	18.26
1987	123	43	41	29.76	17.66
FBFM					
1982	145	44	49	36.99	16.87
1987	133	48	45	23.94	12.24
Iroquois County					
U.S. Census					
1978	108	35	36	30.75	11.96
1982	132	41	47	32.41	14.60
1987	133	40	68	23.76	12.80
FBFM					
1982	146	46	57	26.71	11.60
1987	137	43	78	17.71	9.31

McDonough County				
U.S. Census				
1978	104	37	33	27.53	12.86
1982	133	41	35	27.07	14.25
1987	121	36	52	20.75	14.76
FBFM				
1982	134	43	38	22.52	13.20
1987	115	35	56	17.66	9.43
Randolph County				
U.S. Census				
1978	90	29	36	21.06	9.37
1982	98	33	42	25.38	11.15
1987	102	32	45	19.24	10.26
FBFM				
1982	103	35	45	26.04	11.83
1987	105	35	51	21.52	9.06

TABLE 5.9 Annual FBFM Income, Production, and Structure Information Classified by Acreage

	A2 100-220	A3 220-500	A4 500-1000	A5 1000+
Average Acres				
1982	173	365	699	1,297
1987	165	356	703	1,426
Net Farm Income				
1982	26,146	21,194	23,815	32,106
1987	15,969	31,104	45,888	74,008
1987*	13,568	26,427	38,988	62,878
Total Capital Investment				
1982	242,927	413,681	752,123	1,010,444
1987	159,156	257,503	409,692	690,500
1987*	135,222	218,779	348,082	586,661
Percent Owned Land				
1982	37	30	32	24
1987	38	33	29	27
Avg. Corn Yield				
1982	133	138	139	140
1987	121	139	139	135

Avg. Soybean Yield				
1982	44	44	44	45
1987	41	45	43	42
Fertilizer Expense per crop acre				
1982	43.48	30.81	28.99	27.19
1987*	27.58	26.46	23.94	24.92
1987	23.43	22.48	20.34	21.18
Pesticide Expense per crop acre				
1982	16.66	12.92	14.27	13.07
1987*	13.90	11.73	13.15	12.86
1987	11.81	9.97	11.17	10.93
Interest Expense per crop acre				
1982	60.60	31.71	37.34	37.01
1987*	19.71	26.41	25.27	21.31
1987	16.75	22.44	21.47	18.11

*Values deflated to 1982 nominal dollars.

TABLE 5.10 Annual FBFM Income, Production, and Structure Information Classified by Market Value of Machinery and Equipment

	M1 1-20	M2 20-50	M3 50-100	M4 100-200
Average Acres				
1982	385	453	617	1,016
1987	469	716	952	1,158
1987*	488	724	1,068	1,158
Net Farm Income				
1982	25,482	20,644	25,847	21,808
1987	31,840	43,435	70,162	77,542
1987*	27,614	39,975	60,764	65,881
Total Capital Investment				
1982	278,765	436,541	703,542	1,100,891
1987*	248,543	400,963	656,029	911,406
1987*	217,582	378,533	555,990	774,347
Percent Owned Land				
1982	20	32	32	36
1987*	31	31	31	35
1987	30	33	27	35
Avg. Corn Yield				
1982	136	137	138	142
1987*	134	139	149	131
1987	135	140	148	131

Avg. Soybean Yield				
1982	42	45	44	45
1987	43	43	47	43
1987*	43	44	47	43
Fertilizer Expense per crop acre				
1982	27.38	29.60	30.43	33.86
1987	23.69	25.28	30.08	21.11
1987*	20.23	22.44	24.12	17.93
Pesticide Expense per crop acre				
1982	12.28	12.89	14.47	14.87
1987	11.54	13.39	13.66	15.31
1987*	9.59	12.30	10.63	13.02
Interest Expense per crop acre				
1982	19.62	34.34	37.31	42.74
1987	23.03	24.65	24.77	25.22
1987*	18.56	22.65	20.14	21.43

*Values deflated to 1982 nominal dollars.

TABLE 5.11 Annual Income, Production, and Structure Information Classified by Sales

	S3 40 - 100	S4 100 - 250	S5 250+
Average Acres			
1982	412	582	878
1987	382	598	984
1987*	451	631	1,013
Net Farm Income			
1982	9,905	22,668	50,076
1987	20,030	34,723	78,717
1987*	18,758	32,159	75,186
Total Capital Investment			
1982	296,997	590,978	1,200,395
1987	158,701	315,371	694,481
1987*	150,248	309,180	626,552
Percent Owned Land			
1982	23	33	41
1987	23	31	35
1987*	26	32	36
Avg. Corn Yield			
1982	134	138	143
1987	132	140	139
1987*	134	139	140
Avg. Soybean Yield			
1982	43	44	48
1987	43	43	44
1987*	43	43	45
Fertilizer Expense per crop acre			
1982	26.26	30.18	38.17
1987	20.62	24.95	29.86
1987*	18.42	21.70	25.58
Pesticide Expense per crop acre			
1982	11.39	13.68	18.15
1987	9.65	12.01	16.43
1987*	8.91	10.35	14.71
Interest Expense per crop acre			
1982	19.96	33.13	53.43
1987	12.43	22.73	29.99
1987*	9.47	21.52	26.23

*Values deflated to 1982 nominal dollars.

occurred from 1982 to 1987. Interest expenses is now quite constant across size categories.

Characterizing farm size by sales, rather than acreage or investment, presents a slightly different picture. First, net farm income rises much more directly with sales than by acreage or investment. Also, ownership increases more rapidly. Both of these factors are, again, likely due to an increase in the proportion of livestock in the larger size category. Expenses, in all categories (fertilizer, pesticide and interest) are significantly higher in the $250,000+ category.

Conclusions

The size of farms and structure of agriculture varies somewhat from state-to-state. North Dakota has the largest farms, as classified by acreage and Ohio the smallest. The acreage distribution of farms in North Dakota is skewed to the left while the distribution of farms based on value of machinery and investment has changed toward a uniform distribution. In contrast, all of the size distributions in Ohio are skewed to the right and the number of small part-time farms has changed little while the largest farms have grown in number at a very modest rate.

All of the states have exhibited a disinvestment in livestock diversification as the proportion of farms which produce livestock has declined in each year. Increases in acres farmed by part owner operators have increase significantly in all the states except Kansas and North Dakota. The Farm Business Farm Management data indicate approximately 30 percent of the land owned by the operator, although the largest farm (classified by sales) own 35 to 40 percent but receive a higher portion of sales and income from livestock. The FBFM data also suggest that a viable full time operation needs more than 400 acres, $100,000 in sales and $300,000 invested in machinery and equipment to cover normal family living expense.

The distribution depicted in the chapter do not indicate that agriculture's structure in the North Central region is becoming bimodal. Certainly the distribution of output is becoming more skewed in that the largest farms are producing a larger percentage of that output. However, that would be true for any industry where the number of plants is decreasing but the largest plants are growing in size. In general, the proportion of small farms (classified by sales) increases from west to east within the North Central region. Small farms in Ohio and Michigan are much different than those in Kansas and North Dakota with fewer off-farm opportunities in the western states.

References

Baab, E. M. 1979. "Some Causes of Structural Change in U.S. Agriculture." *Structure Issues of American Agriculture*. USDA, ESCS Agricultural Economic Report 438, pp. 51-59.

Lin, William, George Coffman, and J. B. Penn. 1980. *U.S. Farm Numbers, Sizes, and Related Structural Dimensions: Projections to Year 2000*. National Economics Division; Economics, Statistics, and Cooperatives Services; U.S. Department of Agriculture, Technical Bulletin No. 1625.

McDonald, Thomas, and George Coffman. 1980. "Fewer, Larger U.S. Farms by Year 2000 - and Some Consequences." National Economics Division, Economics and Statistics Service, U.S. Department of Agriculture, Agriculture Information Bulletin No. 439.

Stanton, B. F. 1989. "Changes in Farm Size and Structure in American Agriculture in the Twentieth Century." Proceedings of the NC-181 Committee on Determinants of Farm Size and Structure in North Central Areas of the United States, January 7-10, Tucson, AZ, pp. 11-46.

Tweeten, Luther. 1988. "World Trade, Exchange Rates, and Comparative Advantage: Farm Size and Structure Implications." Proceedings of the NC-181 Committee on Determinants of Farm Size and Structure in North Central Areas of the United States, January 16-19, San Antonio, TX, pp. 165-176.

U. S. Census of Agriculture, 1978, 1982, 1987, U.S. Department of Commerce, Bureau of the Census.

University of Illinois. 1990. *Summary of Illinois Farm Business Records*. Circular 1316, College of Agriculture Cooperative Extension Service.

6

The Production Cost-Size Relationship: Measurement Issues and Estimates for Three Major Crops

Mary C. Ahearn, Gerald W. Whittaker, and Hisham El-Osta

Introduction

It is a commonly held belief that economies of size exist in U.S. agricultural production and that these economies have been a significant factor, perhaps the most significant factor, in explaining our current agricultural structure. More specifically, the view is that the most economically efficient size of farms will prosper and other farms will tend to exit or gravitate to that farm size. Underlying production technologies drive the relationship (Heady 1952). This chapter will address the issue of costs of production with an emphasis on measurement issues and the variation in costs by farm size and other indicators of structure. Economic costs per bushel for corn, soybeans, and wheat will be examined.

We commonly consider the causal relationship between costs and structure to be one-way—costs affect structure, but it is important to realize that it may be a two-way relationship since factors other than production technology affect structure. That is, factors which affect structure, other than production costs, can inhibit farms from moving to economically efficient sizes. Structure then affects costs of production. For example, the current structure in agriculture has evolved from one which was significantly affected by the early settlement laws. The purpose of many of these laws was to specify a farm size. Other institutional constraints are in place today, such as inheritance laws or the acreage restrictions in reclamation polices affecting Western water rights. Both the past and the current laws have had an impact on the structure in agriculture as we know it today (Bills and Dideriksen 1980; Opie 1987). This two-way relationship will not be discussed further in this chapter, but the recognition that structure can affect costs should be borne in mind.

Empirical Evidence and Measurement Issues of Cost-Size Relationships

Excellent reviews of the empirical evidence on cost-size relationships are available elsewhere, including in this volume (see Chapter 8 of this volume, and Hallam). Also available are reviews of the approaches to measuring cost-size relationships (see Chapter 7 of this volume, and Garcia and Sonka 1983). Some of these analytical approaches are based on economic theory, for example, an attempt is made to measure the long run average cost curve or the production frontier. Others are strictly empirical measurements of observed production cost-size relationships where technology varies across farms. The approaches are generally categorized as: classification analysis, synthetic approaches (generally using linear programming), parametric approaches (based on profit functions, cost functions, and production functions), and nonparametric approaches (such as nonparametric frontier production functions). Although the approaches vary widely, after more than two decades of research since Madden's notable article on economies of size in agriculture, most have concluded that an L-shaped relationship exists between costs and level of output and that the relationship exists as a result of economies of size. More recently, Hall and LeVeen (1978) suggested that the general shape is still L-shaped since many of the studies were completed, but the production level at which the curve begins to flatten out may now be at a lower level.

An Alternative Explanation

Seckler and Young (1978) take exception to the commonly-held interpretation of the cost-size relationships measured empirically in the literature. They argue that there is another possible explanation for the L-shaped relationship found between average costs and production level and the increase in farm sizes over time other than existence of economies of size. The other explanation relates to the existence of superior managerial abilities and the desire to increase net income or wealth. Superior managers, as indicated by profitability, will have the ability and incentives to expand and leave the inferior, high-cost producers in the smaller sizes of farms.

Furthermore, Seckler and Young (1978) believe too much weight is generally given to the small, higher cost farms in the conventional analysis. By way of example, Seckler and Young (1978) observe that the L-shaped size-cost curve presented in Carter and Dean (1961) was heavily influenced by the seven smallest and inefficient farm observations below approximately 180 acres in size. A cost-returns ratio for only the farms of 180 acres and above was very close to one. In contrast, the Carter and Dean (1961) analysis which included the seven smaller observations found that economies of size, as measured by the cost-returns ratio, did not cease to exist until 640 acres. They then go on to argue that the high-cost farms at the lower extreme of the size distribution "should not be permitted to have such enormous impact on the analysis." A similar effect of small farms (less than $100,000 in value of production) on a size-cost curve has been shown using averages for grouped data from USDA's Farm Costs and Returns Survey for U.S. farms which specialized in dairy and wheat in 1985 (Ahearn, Dubman, and Hanson 1987a, 1987b).

We are generally sympathetic to Seckler and Young's (1978) concern about the effect and the corresponding interpretation of small farms on the conventional cost-size relationship found. In Figure 6.1, we have shown a scatterplot of the costs of production by farm size for the 1,222 observations from the 1987 Corn Version of the USDA's Farm Costs and Returns Survey. There are three noteworthy points to make regarding this scatterplot: (1) A frontier of this relationship would yield a right angle, that is, low cost producers are found at all sizes of farms. (2) More variation in costs exists for the small farms. (3) When a standard power function was fitted to the data a L-shape relationship occurred. However, when a translog function (which allowed for the existence of diseconomies of size at large farm sizes) was fitted to the data a slight U-shape relationship occurred.

Measurement Conventions

Regardless of the analytical technique used to view the cost-size relationship, certain conceptual and measurement issues must be addressed regarding how to handle the raw data like that presented in Figure 6.1. Commonly practiced conventions for handling data may ensure the conventional findings or at least color the results. We will discuss five issues which must be addressed in an empirical analysis of the cost-size relationship. These relate to the use of whole farm vs. individual commodity costs data, the specification of farm size, the use

FIGURE 6.1   Scatterplot of Average Cost per Bushel for Corn Versus
Gross Income

Dollars/bushel

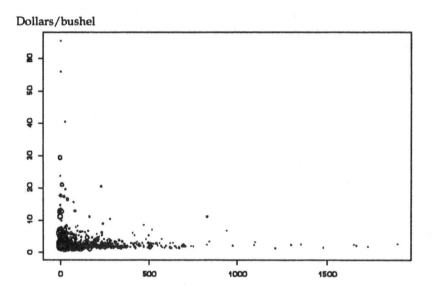

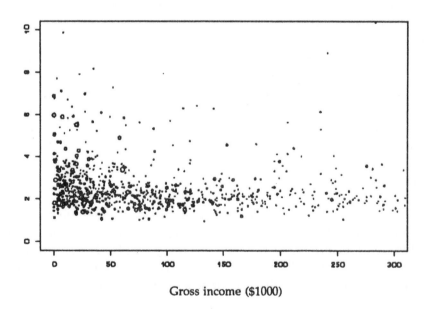

Gross income ($1000)

of individual vs. grouped data, pecuniary economies, and the treatment of unpaid, owned resources.

Whole Farm Versus Individual Commodity Costs. Economists face a real dilemma in analyzing production costs of individual commodities. First, critical market and policy information are available on a per-unit-of-output basis, such as market price or target price for wheat. Those exogenous pieces of information are some of the major factors guiding an operator's decision of how to allocate inputs. Ex post, they are critical to determining the profitability of a farm firm. Therefore, it is natural for users of agricultural data to demand cost estimates and analysis, such as the relationships between costs and production levels, on a per-unit-of-output basis, for comparison to output prices or support levels. However, there are two critical drawbacks to examining costs per unit of output from a theoretical basis. First, calculation of these costs assumes that all inputs are allocable to final products, that is, that production is separable. Very little empirical evidence exists on this subject, but what is available suggests that nonjoint production can generally not be supported (Just, Zilberman, and Hochman 1983). The second, and related, theoretical difficulty with examining costs per unit of output is that farm operator's are generally not optimizing over a single output. Rather, their objective function refers to all enterprises of the farm firm and, less recognized though no less important, farm household unit as a whole (Huffman 1988; Singh, Squire, and Strauss 1986). The majority of U.S. farms are organized in the traditional mode of a single farm operator or a single farm operator household associated with a closely-held farm business making allocation decisions regarding the resources they control in order to maximize the welfare of the household. The relevant unit is the farm firm-household unit. For this reason, the lowest cost producer, of say corn, will not necessarily be the most economically efficient farm firm-household unit and not even the most economically efficient farm firm. Moreover, with respect to growth in farm size, and hence structural change in agriculture, the critical factor is how profitable the farm firm is relative to the off-farm activities of the farm firm-household unit.

Almost all studies of cost-size relationships for U.S. agriculture focus on a particular commodity sector, as we do later in this chapter, rather than cost-size relationships for the whole sector. For one reason, this helps to control for some exogenous factors, such as differing demand and supply conditions, among commodities. Two approaches to the data are possible for specifying costs of producing an individual commodity--one approach is a specialized, whole farm approach and the other is limited to costs associated with the production of the individual commodity in relation to physical output quantities. The results from the two approaches to measurement will not necessarily be consistent.[1] The answer to the question about which is the preferred approach is intimately tied to the question of nonjointness in production.

The first approach relies on specifying costs for farms that specialize in the commodity of interest. Obviously, this approach necessitates that output of the farm be specified as a value of production rather than in physical terms, such as bushels, which cannot be aggregated across commodities. This negates the usefulness of the analysis for comparing commodity costs to relevant output prices. However, the major problem with this approach is that the results are sensitive to how a specialized farm is specified and the effect of the varying mix of other enterprises on the costs and returns of the specialized farm. Often analysts confine the coverage to a particular production region to minimize this effect, thereby limiting the scope of the results. There is no single accepted procedure on how to define a specialized farm. Two common approaches are to define a specialized farm as one which produces any of the commodity of interest and the other is to define a specialized farm according to the Standard Industrial Classification (SIC) system. The SIC system, used by the Census of Agriculture, generally classifies farms based on which commodity accounts for at least 50 percent of sales.

The second data approach for examining the cost-size relationship is to include only costs associated with the commodity of interest and to measure a cost per unit of physical output. The advantages to this approach are that costs can be compared to output prices and the results for a particular commodity are allegedly not biased by the effects of the differing commodity mixes being produced across farms. The disadvantage to this approach is that, since relatively few farms produce only one output, assumptions must be made about how the joint inputs, such as machinery, are shared in production. If production is separable, the allocation of inputs is a discovery issue. But as mentioned above, production may not be separable and so the allocation of inputs may be arbitrary and misleading. Nevertheless, standard commodity costs and return estimates, or budgets as they are sometimes called, are routinely produced by USDA and land grant economists, although usually not by size of farm.

Specification of Size. Size can be specific to the individual commodity enterprise or it can relate to the whole farming operation. Again, if production is not separable, realized economies of size may be related to the whole farm and not specific to the commodity enterprise. Input and output related measures of size are common for specifying size of a commodity enterprise. For example, planted acres of the crop, or herd size, or quantity or value of output are all common measures of size. The most common measures of whole farm size relate to output of the farm, however, the Census of Agriculture does provide grouped data by farm size measured in acres.

If the size is directly a function of this year's output, whether commodity-specific or whole farm, stochastic factors such as weather, disease, or pestilence can both affect farm size and costs per unit of output. For example, a normally large farm, in terms of production, which experienced severely reduced production levels due to weather would be classified as a small farm under most farm size classification systems and would have very high production costs per unit of output. In agriculture there will always be some farms in any production period who experienced an unusually low production. The range of farms viewed as small will contain the "transitory small," high-cost farms who would normally be classified as large farms along with the small farms that are normally classified as small. The effect of these farms who, for a single disastrous year, are classified as small farms with high costs is to inflate the average production costs in the size category. This problem was first described by Stigler (1952) in reference to using OLS to estimate cost curves based on actual yields and is known and widely recognized as the regression fallacy, but it is generally ignored in empirical applications. Of course, the effect of expected vs. realized output can go in the opposite direction, but then the effect is constrained by physical limits. For example, it is possible for a farm which typically has high production levels to have a total failure, but a farm which typically has low production levels can only expect to increase production a relatively modest amount due to the usual resource constraints. For this reason, one would expect to find a greater divergence between expected and realized output for farms with low realized output than for farms with large realized output. This is consistent with the results found by Whittaker, Ahearn, and El-Osta (1990) for a 1987 sample of cash-grain farms in the Midwest.

Another problem associated with the conventional measure of farm size based on production levels is due to the nature of government programs, particularly those for which the terms vary annually and so farm operators may regularly carry costly excess production capacity. However, in this case the solution is somewhat more tractable—the corresponding value of the programs can be included with the production. Unfortunately, the alternative measures of size based on inputs rather than outputs, such as value of assets (which are affected by nonfarm influences) or acres of farmland (of widely varying quality) are even more problematic.

Individual Versus Grouped Data. The issues associated with size measures based on production are relevant for cost-size relationships based on both grouped and individual data. The third empirical issue commonly encountered in studies of cost-size relationships is an issue only for grouped data. Namely, the arbitrariness of the size groupings has an effect on the relationships found. The effect of outliers can be accentuated or downplayed based on how the size groupings are defined. Once the groupings are defined, all groups will carry the same weight when, for example, a cost-size curve is plotted with the grouped data-- whether there are 30 or 300 farms in the group.

Pecuniary Economies of Size. The fourth measurement issue deals with pecuniary economies. It is commonly recognized that more research is needed on pecuniary economies of size and that what little evidence there is suggests that they do exist (Krenz, Heid, and Sitler 1974; Smith, Knutson, and Richardson 1986). The reason the evidence is sparse is because data are generally collected the way farm operators usually keep their records, in terms of expenditures and receipts, rather than prices and quantities. On the output side, although most operators are readily able to recall their yields and most common output price, most data collection efforts have not included these items. Given the data constraints, we are usually forced to ignore pecuniary economies when interpreting the data.

Opportunity Costs of Owned Inputs. The final empirical issue in measuring costs we wish to highlight has to do with how opportunity costs for owned inputs are valued. This includes the opportunity cost for land, nonland capital, unpaid operator and family labor and management. In the most common types of income statements in our agricultural data system, these costs are not included in calculating the costs and returns of the whole farms. For example, this is why Hall and LeVeen (1978) had to impute some of these costs when using Census of Agriculture data to analyze the cost-size relationships with that data set.

Costs are more frequently imputed for owned inputs when costs of producing individual commodities are measured. In some ways, the most problematic of these costs is unpaid labor and management. For land it is common to impute a cost based on rental market data for comparable land in the area. Alternatively, for both land and nonland capital, many researchers ask an owner-operator to provide an estimate of the current market valuation of the capital stock through the data collection process. The issue then is one of specifying the appropriate discount rate. (The only problem with this approach for estimating the costs of agricultural production is if the value of the capital reflects the use value for non-agricultural activities.) In the case of labor and management, however, since no slave market for farm operators exists, one cannot ask operators to value themselves or other unpaid laborers. More realistically, researchers do not even generally ask operators to value their or others' labor and management services in terms of what their wages or salaries might be in the next best market alternative. Instead, the usual practice is to ask the operator for a count of the hours involved or estimate the hours involved based on known machinery use and impute a cost based on the state or U.S. average wage rates for hired farm workers or hired farm managers. The quality of an hour of work time varies considerably across farms and likely varies by farm size, making this approach seriously flawed. It especially becomes a concern for economies of size work when it is recognized that results about the existence of economies of size are sensitive to this imputation as was found for U.S. dairy farms (Ahearn, Dubman, and Hanson 1987a) or for Australian wheat and sheep operations (Vlastuin, Lawrence and Quiggin 1982). However, no better alternative approach for estimating costs has been presented, short of ignoring all unpaid labor and management costs. Miller, (1983) in fact, suggests omitting operator costs from cost studies used for policy purposes.[2]

The Cost-Size Relationship: Corn, Soybeans, and Wheat

In this section we will present data on production costs for three major crops: corn, soybeans, and wheat. We will use the so-called classification analysis approach to examine the cost-size relationship. This approach simply measures average production costs for farms classified by size of operation and does not purport to measure a production frontier or the long run average cost curve. With respect to the five measurement elements discussed above: (1) costs for single commodity enterprises will be examined; (2) farm size will be measured in two ways--by acres of the commodity planted and by economic class of the whole farm; (3) the data are for individual farming operations and commodities, rather than grouped data; (4) data limitations preclude us from separating out the pecuniary effects; and (5) cost estimates will be presented both including imputations for owned inputs and excluding imputations.

As mentioned, we will examine the average costs for farms classified by size using two different measures of farm size. The first measure relates to the size of the whole farming operation, the economic class measure, which is based on the sum of sales, value of product removed, government payments, and other farm-related income. Four economic classes of farms have been defined: less than $40,000, $40,000 to $99,999, $100,000 to $249,999, and $250,000 or more. The second measure relates only to the corn enterprise and is measured by acres planted to corn. Four acreage classes are defined: less than 25 acres, 25 to 99 acres, 100 to 499 acres, and 500 acres or more.

Besides examining the costs of production by size, we will also examine the costs by region and by indicators of structure other than size, such as tenure arrangements of the farm, major occupation of the operator, and off-farm income of the household. Because the data presented are from a sample, rather than a census, we have conducted pairwise t-tests of significance between groups for the economic costs of production per bushel of output.

Data Source

The uniqueness of these cost estimates is due to the high quality and representativeness of the underlying data. The data are from various versions of USDA's annual Farm Costs and Returns Survey (FCRS). The FCRS data are the first data available which are statistically representative of the cost of producing selected commodities in the U.S. The FCRS has been conducted annually by the Economic Research Service (ERS) and the National Agricultural Statistics Service (NASS) beginning with the 1985 survey for calendar year 1984. Each year there are multiple versions of the FCRS: an in-depth, whole-farm version, and commodity cost-of-production versions, and more recently, a version aimed at collecting input prices and quantities separately. All versions have questions about whole-farm expenses and income in common; each cost-of-production version gathers detailed information about input use, cultural practices, and production costs of a particular commodity. Because of survey costs, USDA cannot undertake detailed surveys of every commodity in every year; the FCRS covers each commodity about every four years. The FCRS has replaced the old Cost of Production (COP) surveys which were jointly conducted by ERS and NASS. Unlike the FCRS, the COP surveys were not probability-based surveys and, hence, were not statistically representative of costs of production.

The FCRS is a multiframe stratified survey. The sample is drawn randomly from stratified list and area frames. For a particular cost-of-production version the list frame is made up of farms known to have previously produced the commodity. The list frame is stratified by size. The area frame is made up of land segments and is stratified by land use. Each farm sampled represents a number of similar farms, the particular number being the survey expansion factor, which is the inverse of the probability of the sampled farm being selected. Data are expanded by the survey expansion factors to produce estimates for the population of all farms producing the commodity. (See Chapter 11 in this volume for more information on the FCRS.) Because of the complex sample design, the usual statistics to test for significance can not readily and accurately be computed without taking into consideration the survey design

(Cochran 1977). Fuller et al. (1986) have developed special software to correctly account for the variance of estimates for a limited array of statistics when the sample has a complex design. Our pairwise t-tests for statistical differences in means between groups reported in this chapter have accounted for the complex sample design.

The corn data in this analysis are from the 1987 FCRS and the soybean and wheat are from the 1986 FCRS. Although the surveys are designed to be statistically representative, they are known to undercount farms, especially small farms. The corn survey represented 80 percent of U.S. corn production and 83 percent of corn acreage. The corn survey represented 482,500 farms, which is 77 percent of those reported in the 1987 Census of Agriculture. The soybean survey represented 71 percent of the U.S. soybean production of 1986 and 68 percent of the soybean acreage. For soybean-producing farms, the representation of the survey was 267,669, compared to the 441,899 farms a year later as reported in the 1987 Census of Agriculture. The wheat survey represented 82 percent of U.S. wheat production and 74 percent of wheat acreage. The survey represented 280,846 wheat farms in 1986, compared to the 1987 Census number of 352,237.

About the Estimation System

The cost and return estimates in this chapter generally follow the USDA procedures (USDA). Those procedures have been developed over time with input from the National Agricultural Standards Cost of Production Review Board which was established under the 1981 Farm Bill. The cost estimates presented here are not identical to the official estimates published annually (USDA). They differ because the official estimates are calculated at the state level and the estimates presented here are based on the newly developed farm level model for estimating costs (Glaze 1988).[3] The differences are minor and are mainly a result of the differing levels of aggregation. A major advantage to the farm level estimates are that they allow us to analyze the distribution of costs and returns in a variety of ways, for example, by size of farm.

Four measures of production costs are reported in this article: variable cash expenses, fixed cash expenses, capital replacement, and economic costs. Capital replacement is an economic depreciation--the portion of the value of machinery and equipment that is used up in the production of a specific commodity valued at current market values. Economic costs are designed to account for the value of all inputs in production, whether owned, rented, or financed.

The estimates include the costs incurred by both the operation and the landlord, they do not include marketing costs, and the *direct* effects of government programs are excluded. The combined operation-landlord account means that estimates of cash expenses do not include an expense for cash and share rent expenses paid by the farm operation to the landlord. This is because what is a rental expense to the farm business is exactly canceled as an income to the landlord. Of course, the economic costs do include an imputation for all land used in production. Estimates of cash expenses include an interest expense because generally the interest is paid to those other than the combined operation-landlord entity.

The treatment of government programs is problematic. It is not possible to eliminate the pervasive, indirect effects of the programs on the agricultural sector, e.g., the effects of the programs on market prices. The direct effects on costs and returns, such as the deficiency payments received and the costs of complying with the program, are eliminated in the standard USDA estimates because of the interest in providing policy makers information on production costs and returns in the absence of programs. Historically, USDA has not collected the data necessary to compute costs and returns including all of the effects of income support programs until the 1988 rice version of the FCRS. Therefore, data do not exist to allow us to include the direct effects of the programs for the 1986 and 1987 data discussed in this chapter.[4] It is not sufficient to estimate the effect on returns given known support levels without actual farm-level data, because producers incur expenses to participate in the program. No acceptable basis exists for imputing the relevant costs. For example, since a participating producer is required to set-aside or conserve a portion of his or her acreage that would have been planted to a particular crop, opportunity costs are incurred to set-aside that acre. Participants may also be required to incur direct costs by maintaining a cover crop or controlling weeds on set-aside acreage. In the case of rice for 1988, when the direct effects of government programs on costs were included, the greatest effect was on the cost for land (Salassi, et al. 1990). This would likely be the case for all of the program crops. The reason land costs are so affected is because land costs for share rented land under the standard USDA procedure are based on valuing the production at the market price, rather than a target price. If the direct effects of the program were to be included on costs, not only would the cost of compliance be included, but production under share rental arrangements would be valued at a higher rate. Since yields would be constant under the two approaches for measuring costs, including and excluding the direct effects of the programs, the effect of the increased costs would be greatest for those groups where yields are higher. For example, under the system which includes the cost effects of the government programs, cost advantages per bushel will be lessened, or even disappear, for high-yielding, low-cost regions. Of course, the increased returns from including the deficiency payments will more than compensate for these increased costs, in terms of the net effect.

Cost estimates for individual inputs are calculated in one of four general ways, either by valuing quantities of inputs at the most appropriate price available, direct costing, allocating a whole farm cost based on an acceptable allocation scheme, or through the use of a budget generator. Which approach is most suitable depends in large part on the operator's ability to provide the required information. For additional detail about the methods and the data sources see USDA, 1990.

The standard and preferred approach is to collect the quantities of inputs from the operator, where possible. Quantities are generally valued at the state average price, but increasingly the data are being collected on prices actually paid in order to retain the farm-level variability. For example, the 1989 Wheat version of the FCRS collected prices and quantities of fertilizers and seed. The quantity-valuing approach is used to estimate the costs of seed, fertilizer, lime and gypsum, hired labor, unpaid labor, and return to land. Under the direct costing approach, the operator specifies the expenditure on an input for the commodity of interest. The direct cost approach is used for chemicals, custom operations, purchased irrigation water, and technical services, such as soil testing. Fixed expenses are generally those whole farm expenses which are not allocable to an individual commodity and so are allocated to individual commodities based on a commodity's share in the total value of production.

In order to calculate the economic depreciation of farm capital, fuel expenses, repairs, and drying, a budget generator is used. The budget generator uses actual, detailed farm-level data on the machinery complement used in the production of the commodity and the hours of use or miles driven. This information is used in conjunction with known current market information about the value of the capital and secondary information on the average technical efficiency of machinery.

Corn

Corn is the most valuable U.S. crop. Only beef cattle and hay are produced on more U.S. farms than is corn. During the 1980s, normal U.S. production of corn has been about 7 billion bushels per year with 20 percent or more being exported. In 1987, the average cash (variable plus fixed) cost of producing a bushel of corn was $1.40, but 25 percent of producers had cash costs of $1.13 or less.

Costs of Production and Characteristics by Size. Table 6.1 presents the costs and characteristics of corn farms by economic class of farm.

TABLE 6.1 U.S. Corn Costs and Farm and Operator Characteristics by Economic Class, 1987

Item	Economic Class				All Farms
	Less than $40,000	$40,000-$99,999	$100,000-$249,999	$250,000 or More	
Number of farms	206,164	123,713	105,833	46,790	482,500
Share of all farms, %	43	26	22	10	100
Share of corn production, %	11	21	33	35	100
Corn yields, bu./acre	96	115	123	126	118
Acres planted to corn, acres/farm	33	85	145	344	101
Variable cash costs:					
Dollar per acre	102.98	104.84	115.68	132.77	117.24
Dollar per bu.	1.07	0.91	0.94	1.06	0.99
Fixed cash costs:					
Dollar per acre	48.46	46.11	46.86	49.05	47.65
Dollar per bu.	0.51	0.40	0.38	0.39	0.40
Capital replacement:					
Dollar per acre	17.12	18.09	18.27	18.93	18.29
Dollar per bu.	0.18	0.16	0.15	0.15	0.15
Economic costs:					
Dollar per acre	242.55	243.07	255.75	264.13	253.96
Dollar per bu.	2.53	2.12	2.08	2.10	2.15

Farm finances:					
Net cash income, $	2,258	21,813	46,488	93,148	25,787
Government payment, $	3,149	10,657	18,579	47,281	12,738
Assets, $	189,065	298,783	522,289	1,102,539	378,870
Debt, $	17,500	63,731	107,641	254,950	72,152
Debt/asset ratio, %	9	21	21	23	19
Operator characteristics:					
Farming as a major occupation, %	64	89	96	97	80
Older than 65 years, %	27	11	14	3	18
Household off-farm income, $	21,343	13,865	12,506	17,133	17,079

Source: 1987 Corn Version, Farm Costs and Return Survey, USDA.

Over 40 percent of all corn-producing farms were in the smallest economic class, but they accounted for just over 10 percent of corn production. On the other hand, 10 percent of the largest farms accounted for 35 percent of the corn production. Variable cash costs per bushel were significantly higher for the smallest economic class of farms and the largest economic class of farms than for the two mid-sized classes. However, when all costs are included (variable and fixed cash costs, capital replacement, and opportunity costs of owned inputs), there was only a statistical difference between the smallest economic classes and the other three. The average economic cost per bushel of corn for corn producers in the smallest economic class was $2.53, compared to $2.10 for the largest economic class of producers, $2.08 for the $100,000 to $249,999 class, and $2.12 for the $40,000 to $99,999 class. None of the pairwise t-statistics were statistically significant for the average economic costs of any of the other pairs of larger size classes.

The FCRS data show that 26 percent of corn-producing farms have an acreage of less than 25 acres (Table 6.2). Only 3 percent are in the largest category of 500 acres or more of corn acreage. However, that largest 3 percent of farms accounted for more than 20 percent of the corn produced in 1987. As corn acreage increased across the groups, the average yield increased from 91 to 128 bushels per acre. Both cash and economic costs per acre also increased. However, the higher level of costs was more than compensated for by higher yields, so that the largest farms realized an economic cost per bushel of $2.07, compared to the average of all farms of $2.15. The group of smallest corn producers averaged $2.64 in economic costs per bushel, fully 28 percent greater than the group of largest corn producers. The largest corn operations of 500 acres or more and the 100-499 group had almost identical costs per bushel for the most inclusive definition of costs, economic costs. The 100-499 acre corn operations averaged slightly lower cash expenses per bushel due to their lower variable cash expenses. The difference in both variable cash and economic costs were statistically significant between the smallest acreage class of farms and the other larger acreage classes. For total economic costs, there were also statistical differences found between the farms with 25 to 99 acres and the two larger acreage classes.

The major variable expenses are for seed, fertilizer, chemicals, and repairs. On a per-acre basis, all of these input costs increased as the size of the corn operation increased. On a per-bushel basis these costs were very similar across size classes. Hired labor charges per bushel were higher on the large farms, but paid and unpaid labor combined were less. Land charges per acre were significantly greater on the larger farms. This is in part because of the higher yields associated with the larger corn operations and the correspondingly higher rent paid as a share of production per acre.

TABLE 6.2 U.S. Corn Costs and Farm and Operator Characteristics by Corn Acreage Class, 1987

Item	Corn Acreage Class				
	Less than 25	25-99	100-499	500 or More	All Farms
Number of farms	124,683	207,256	135,244	15,317	482,500
Share of all farms, %	26	43	28	3	100
Share of corn production, %	2	20	54	23	100
Corn yields, bu./acre	91	106	121	128	118
Acres planted to corn, acres/farm	12	53	190	675	101
Variable cash costs:					
Dollar per acre	108.09	110.56	115.11	130.97	117.24
Dollar per bu.	1.19	1.04	0.95	1.03	0.99
Fixed cash costs:					
Dollar per acre	57.87	46.73	46.88	49.05	47.65
Dollar per bu.	0.64	0.44	0.39	0.38	0.40
Capital replacement:					
Dollar per acre	14.94	17.08	18.71	19.03	18.29
Dollar per bu.	0.16	0.16	0.15	0.15	0.15
Economic costs:					
Dollar per acre	239.01	243.11	255.37	264.20	253.96
Dollar per bu.	2.64	2.29	2.10	2.07	2.15

(continued)

TABLE 6.2 (continued)

Item	Corn Acreage Class				All Farms
	Less than 25	25-99	100-499	500 or More	
Farm finances:					
Net cash income, $	6,290	19,267	43,936	112,471	25,787
Government payment, $	1,660	7,113	24,758	72,905	12,738
Assets, $	190,369	316,430	552,210	1,227,643	378,870
Debt, $	22,040	53,572	120,686	302,944	72,152
Debt/asset ratio, %	11	17	22	25	19
Operator characteristics:					
Farming as a major occupation, %	61	83	92	99	80
Older than 65 years, %	25	20	9	5	17
Household off-farm income, $	21,685	16,404	13,413	21,091	17,079

Source: 1987 Corn Version, Farm Costs and Return Survey, USDA.

Costs of Production and Characteristics by Corn Region. Five major corn regions are defined in this analysis: the Northeast (Connecticut, Delaware, Maine, Maryland, New York, Pennsylvania, and Vermont), Southeast (Alabama, Kentucky, Louisiana, North Carolina, South Carolina, Tennessee, Virginia, and Kentucky), Corn Belt (Illinois, Indiana, Iowa, Michigan, Minnesota, Missouri, Ohio, and Wisconsin), Northern Plains (Colorado, Kansas, Nebraska, North Dakota, and South Dakota), and the West (California, Texas, and Washington). During the survey year, 1987, corn yields in all regions were record highs or near-record highs, and this is reflected in the relatively low costs per bushel.

Two-thirds of all corn farms are in the Corn Belt, and the Corn Belt accounts for about three-quarters of all corn production (Table 6.3). Corn producers in this region are more specialized in corn and generally have the highest yield per acre. Corn farms in the Corn Belt tend to be somewhat smaller than those in the Northern Plains and West, but larger than those in the East both in terms of average corn acreage and economic class of the whole farm. Average economic costs per bushel were the lowest in the Northern Plains at $2.01 and in the Corn Belt at $2.08--there was not a statistical difference between the two regions. Although the yields are lower in the Northern Plains, their significantly lower per-acre costs keep their per-bushel costs as low as the Corn Belt's. The Southeast has historically had the highest costs per bushel of all the major corn regions--in 1987 costs were $3.08 per bushel. This is clearly due to their lower yields--the per-acre costs of corn production in the Southeast are lower than the other regions. The economic cost per bushel in the Northeast was $2.55 and $2.62 in the West; these were not statistically different from each other, although their costs per acre were statistically different. Both yields and costs in the West were high in large part as a result of irrigation.

Corn Costs of Production and Characteristics by Non-Size Structural Indicators. Costs of production vary by other indicators of structure, besides size of farm, although indicators such as tenure and household characteristics are often related to farm size. Farm size is generally related to the major occupation of the farm operator and the proportions of income earned on and off the farm, the amount of production specialization, the legal form of organization, and the age of the operator. For example, one-quarter of operators with corn acreage of 25 acres or less were 65 years old or more. They were less likely to have farming as a major occupation and their farm incomes were low and off-farm incomes were high. Their assets, debt, and net worth were all low relative to farms with larger corn acreage.

Table 6.4 presents the average variable cash cost and the average economic cost per bushel of corn by selected indicators of structure. By

TABLE 6.3 Corn Characteristics, by Regions, 1987

Item	Northeast	Southeast	Corn Belt	Northern Plains	West
Share of U.S. corn farm, %	8	14	66	10	2
Share of U.S. production, %	3	6	74	15	2
Corn yields, Bu./acre	92	77	125	115	120
Corn cropland, Acres/farm	40	60	108	154	118
Economic cost, Dol./acre	235.99	238.20	260.19	232.58	314.53
Economic cost, Dol./bu.	2.55	3.08	2.08	2.01	2.62
Share of group's farm in economic class:					
Less than $40,000, %	42	66	39	24	27
$40,000-$99,999, %	25	15	28	25	19
$100,000-$249,999, %	20	10	23	36	27
$250,000 or more, %	13	9	10	15	27
Farm finances:					
Net cash income, $	27,296	17,825	11,917	30,062	44,231
Government payments, $	2,071	6,228	8,340	21,976	39,207
Assets, $	473,775	297,078	284,772	460,631	648,475
Debt, $	61,941	54,670	42,733	91,669	129,317
Debt/asset ratio, %	13	18	15	20	20
Operator characteristics:					
Farming as major occupation, %	79	66	82	96	95
Older than 65 years, %	18	17	19	26	6
Household off-farm income, $	23,940	21,609	16,352	10,122	13,913

Source: Ahearn, M., et al., 1990a.

TABLE 6.4 Variable Cash and Economic Costs of Producing Corn per Bushel by Other Structural Indicators, 1987

Indicator	Average Variable Cash Costs	Average Economic Costs
	Dollars per Bushel	
Tenure		
Own all acres	.98	2.06
Rent-in some acres		
Some share, no cash	.91	2.20
Some cash, no share	1.13	2.24
Both cash and share	.96	2.07
Major Occupation		
Farming	.98	2.13
Other	1.09	2.41
Operator Age		
Less than 35	.93	2.05
35 - 49	.99	2.13
50 - 64	1.01	2.18
65 or more	1.02	2.24
Major income source of household		
Farming greater than off-farm	.98	2.11
Off-farm greater than farming	1.06	2.33

Source: 1987 Corn Version, Farm Costs and Returns Survey, USDA.

tenure arrangements, the lowest cost producers were those who owned all of the land they planted to corn or used a combination of owned, cash-rented, and share-rented land. There were no significant differences in the cash variable costs or economic costs by whether the major occupation of the farm operator was farming or not and by whether or not the farm operator worked at least 2,000 hours per year on his or her operation. However, there were significant differences in costs based on whether the farm operator household was most dependent on farm or off-farm income. Farm operator households more dependent on their farm for their income had economic costs of $2.11 per bushel, compared to $2.33 for households more dependent on their off-farm sources. This significant difference based on major income source of the household and lack of significance for the variables based on the operator's work time alone, indicates the importance of examining household linkages to the farm business in order to understand farm firm behavior. Significantly lower economic costs, but not variable cash costs, were found for the younger operators (less than 35 years old) compared to those operators in the age groups above 50 years old.

Soybeans

Soybeans are the second most valuable crop in the U.S. The U.S. is the world's largest producer and exports more soybeans than any other crop. In 1986, the year in which the soybean data were collected, the Southeast and Delta experienced a drought and unusually low yields. The average cash cost of producing a bushel of soybeans in 1986 was $2.46, but 25 percent of producers had cash costs of $1.88 or less.

Soybean Costs of Production and Characteristics by Size. Similar to corn, a large share of soybean-producing farms, 37 percent, are in the smallest economic class, but account for a disproportionately small share of production, 13 percent (Table 6.5). There were no statistically significant differences by economic class for variable cash costs only. The smallest farms did have statistically significant larger economic costs per bushel than the other sizes of farms. The major factor for their high costs is due to their relatively low yields of 29 bushels per acre, rather than differing uses of inputs per acre. Costs of production were also significantly higher between the farms in the $40,000 to $99,999 economic class and the largest farms. There were no significant differences in costs between the two largest groups of farms and the two mid-sized groups of farms.

When size is measured in terms of soybean acreage class, although the average economic costs per bushel appear to be higher across the acreage classes, the variance is so great within the acreage classes that the differences are not statistically different, with one exception (Table 6.6). Differences in average economic costs are only statistically significant between farms in the two mid-sized groups, those with 25 to 99 acres in soybeans and those with 100 to 499 acres.

Soybean Costs of Production and Characteristics by Soybean Region. Table 6.7 presents costs and characteristics of soybean producing farms by region. The North Central region (Illinois, Indiana, Iowa, Minnesota, Missouri, Nebraska, and Ohio) dominates in U.S. soybean production. The other soybean regions are the Southeast (Alabama, Georgia, Kentucky, North Carolina, and Tennessee) and the Delta (Arkansas, Louisiana, and Mississippi). The North Central region accounts for over 80 percent of the production. Yields are considerably higher in the North Central region, as well, and although per-acre costs are higher in the

Table 6.5 U.S. Soybean Costs and Farm and Operator Characteristics by Economic Class, 1986

Item	Economic Class				All Farms
	Less than $40,000	$40,000-$99,999	$100,000-$249,999	$250,000 or More	
Number of farms	99,953	68,931	77,288	21,497	267,669
Share of all farms, %	37	26	29	8	100
Share of soybean production, %	13	25	41	22	100
Soybean yields, bu./acre	29	35	34	37	34
Acres planted to soybean, acres/farm	61	142	215	391	153
Variable cash costs:					
Dollar per acre	44.85	50.22	45.83	52.13	48.03
Dollar per bu.	1.55	1.43	1.34	1.40	1.40
Fixed cash costs:					
Dollar per acre	37.82	32.35	38.04	36.88	36.04
Dollar per bu.	1.31	0.92	1.11	0.99	1.06
Capital replacement:					
Dollar per acre	11.18	10.47	10.78	10.85	10.78
Dollar per bu.	0.39	0.30	0.31	0.29	0.31

(continued)

TABLE 6.5 (continued)

Item	Economic Class				All Farms
	Less than $40,000	$40,000-$99,999	$100,000-$249,999	$250,000 or More	
Economic costs:					
Dollar per acre	159.33	162.83	150.17	154.31	155.42
Dollar per bu.	5.50	4.63	4.38	4.14	4.53
Farm finances:					
Net cash income, $	2,153	16,409	25,750	119,825	22,088
Government payment, $	1,864	10,730	19,464	34,964	11,887
Assets, $	155,596	208,744	504,019	916,693	331,014
Debt, $	19,870	87,206	173,033	317,824	105,365
Debt/asset ratio, %	13	42	34	35	32
Operator characteristics:					
Farming as a major occupation, %	64	86	95	95	81
Older than 65 years, %	18	10	6	3	11
Household off-farm income, $	18,066	16,585	40,795	14,134	23,932

Source: 1987 Soybean Version, Farm Costs and Return Survey, USDA.

TABLE 6.6 U.S. Soybean Costs and Farm and Operator Characteristics by Soybean Acreage Class, 1987

	Acreage Class				
Item	Less than 25	25-99	100-499	500 or More	All Farms
Number of farms	39,105	108,905	106,369	13,290	267,669
Share of all farms, %	15	41	40	5	100
Share of soybean production, %	1	16	59	23	100
Soybean yields, bu./acre	37	36	37	29	34
Acres planted to soybean, acres/farm	13	59	213	855	153
Variable cash costs:					
Dollar per acre	62.97	50.17	46.30	49.61	48.03
Dollar per bu.	1.70	1.40	1.26	1.73	1.40
Fixed cash costs:					
Dollar per acre	52.80	40.36	37.95	30.37	36.40
Dollar per bu.	1.42	1.13	1.03	1.06	1.06
Capital replacement:					
Dollar per acre	12.74	10.32	10.90	10.71	10.78
Dollar per bu.	0.34	0.29	0.30	0.37	0.31

(continued)

TABLE 6.6 (continued)

Item	Acreage Class				All Farms
	Less than 25	25-99	100-499	500 or More	
Economic costs:					
Dollar per acre	200.40	176.46	159.38	133.73	155.42
Dollar per bu.	5.40	4.93	4.34	4.68	4.53
Farm finances:					
Net cash income, $	8,174	14,654	34,073	28,031	22,088
Government payment, $	4,648	8,311	15,475	33,777	11,887
Assets, $	163,776	285,197	383,583	777,796	331,014
Debt, $	43,360	68,447	147,745	251,129	105,365
Debt/asset ratio, %	26	24	39	32	32
Operator characteristics:					
Farming as a major occupation, %	54	81	89	97	81
Older than 65 years, %	10	18	5	3	11
Household off-farm income, $	21,022	24,658	25,302	15,576	23,932

Source: 1987 Soybean Version, Farm Costs and Returns Survey, USDA.

TABLE 6.7 Soybean Characteristics, by Region, 1986

Item	North Central	Southeast	Delta
Share of U.S. soybean farms, %	83	11	6
Share of U.S. production, %	83	8	9
Soybean yields, bu./acre	39	23	19
Soybean cropland, acres/farm	132	163	437
Economic cost, dol./acre	169.50	129.91	112.44
Economic cost, dol.bu.	4.27	5.59	5.94
Share of group's farm in economic class:	35	59	30
Less than $40,000, %	27	12	35
$40,000-$99,999, %	30	19	26
$100,000-$249,999, %	8	10	9
$250,000 or more, %			
Farm finances:			
Net cash income, $	25,027	6,193	10,561
Government payments, $	12,184	6,866	17,388
Assets, $	329,203	356,885	306,743
Debt, $	104,080	89,390	155,160
Debt/asset ratio, %	32	25	51
Operator characteristics:			
Farming as major occupation, %	80	82	91
Older than 65 years, %	10	21	2
Household off-farm income, $	25,392	18,337	13,698

Source: Ahearn, M., et al., 1990a.

North Central region, the higher yields makes this region the lowest-cost producing region on a per-bushel basis. This is similar to the case of corn, where the major low cost region, the Corn Belt, had the highest per-acre costs but also the highest yields.

Certainly a major reason for the higher costs in the Southeast and the Delta in 1986 lies in the drought of that year, but even in "normal" times, the North Central region has lower costs per bushel of soybeans. The Southeast had significantly higher per-acre costs of production than did the Delta and somewhat higher yields. Although the cost estimates per bushel differ between the two regions, this difference is not significant. Since at least 1975, the per-bushel economic costs in the Delta and

134

Southeast have been very similar, with the Southeast usually showing somewhat higher costs. The Southeast is characterized by a large share of small farms. Almost 60 percent of the soybean-producing farms in the Southeast were in the economic class of $40,000 or less.

Soybean Costs of Production and Characteristics by Non-Size Structural Indicators. There were no statistically significant differences in cash variable costs for the structural characteristics presented in Table 6.8. There were differences for economic costs per bushel for some tenure arrangements and by the major income source of the household. The relatively high production costs for operations which had some cash rental arrangements but no share rental arrangements ($5.09 per bushel) were significantly higher than those operations which had only share rental arrangements and those which had a mix of share and cash rental arrangements. Again, we found that there were statistically lower production costs for those producers whose households were most dependent on their farm for their major income source, $4.43, compared to the $4.89 per bushel for those more dependent on their off-farm income sources.

Wheat

More than half of the U.S. wheat is exported, accounting for 30-40 percent of the world trade during the 1980s. Compared to most other agricultural commodities, wheat is not a homogenous commodity. There are five major classes of wheat, each with different final uses which affect demand on the international market. About half of the wheat production in the U.S. is hard red winter wheat. In 1986, fewer acres were planted to wheat than in previous years and production was down due to drought conditions in the Southern Plains and the North Central regions. The average cash cost of producing a bushel of wheat was $2.01, but 25 percent of producers had cash costs of $1.58 or less.

Wheat Costs of Production and Characteristics by Size. As was the case for corn and soybeans, the largest economic class in terms of farm numbers is the $40,000 or less class, but these farms account for a rather small share of wheat production (12 percent in 1986). Compared to corn and soybean producers, wheat producers and their households earned less cash income from both farm and off-farm sources. The estimates of per-bushel costs of producing wheat are included in Table 6.9. Cost levels appear to differ, but these differences are not significantly different either for variable cash expenses or economic costs per bushel due to the large within-class variation.

There were ten percent of all wheat producing farms in the largest acreage class in 1986 and they accounted for nearly half of all production

TABLE 6.8 Variable Cash and Economic Costs of Producing Soybeans per Bushel by Other Structural Indicators, 1986

Indicator	Average Variable Cash Costs	Average Economic Costs
	Dollars per Bushel	
Tenure		
Own all acres	1.45	4.71
Rent-in some acres		
Some share, no cash	1.22	4.41
Some cash, no share	1.71	5.09
Both cash and share	1.37	4.32
Major Occupation		
Farming	1.40	4.51
Other	1.41	4.70
Operator Age		
Less than 35	1.38	4.38
35-49	1.44	4.54
50-64	1.33	4.52
65 or more	1.48	4.85
Major income source of household	1.39	4.43
Farming greater than off-farm	1.43	4.89
Off-farm greater than farming		

Source: 1986 Soybean Version, Farm Costs and Returns Survey, USDA.

(Table 6.10). This is much different than for corn and soybeans where the largest acreage class accounted for less than one-quarter of the production. These large farms average nearly 1,000 acres of planted wheat. There were no statistical differences between any of the pairs of variable cash costs. There were statistical differences between the economic cost per bushel for the smallest acreage class of $4.92 and all of the other three classes. In addition, the $3.78 average economic cost per bushel for the farms in the acreage class of 25 to 99 was statistically different from the economic cost of the largest acreage class. No differences were found between the two largest classes.

TABLE 6.9 U.S. Wheat Costs and Farm and Operator Characteristics by Economic Class, 1986

Item	Economic Class				All Farms
	Less than $40,000	$40,000–$99,999	$100,000–$249,999	$250,000 or More	
Number of farms	106,991	84,835	65,696	23,323	280,846
Share of all farms, %	38	30	23	8	100
Share of wheat production, %	12	24	39	25	100
Wheat yields, bu./acre	27	30	34	37	33
Acres planted to wheat, acres/farm	70	162	310	502	190
Variable cash costs:					
Dollar per acre	36.70	38.34	44.63	55.13	44.19
Dollar per bu.	1.34	1.26	1.32	1.49	1.35
Fixed cash costs:					
Dollar per acre	19.19	21.24	21.30	23.83	21.54
Dollar per bu.	0.70	0.70	0.63	0.65	0.66
Capital replacement:					
Dollar per acre	11.36	10.85	11.27	10.70	11.05
Dollar per bu.	0.41	0.36	0.33	0.29	0.34
Economic costs:					
Dollar per acre	110.42	111.82	112.36	123.91	114.48
Dollar per bu.	4.03	3.68	3.33	3.35	3.50

Farm finances:

Net cash income, $	-1,161	15,830	33,704	-23,105	10,305
Government payment, $	2,961	13,200	25,109	40,537	14,356
Assets, $	165,531	311,695	541,495	1,335,664	394,806
Debt, $	19,694	86,432	144,962	389,673	99,882
Debt/asset ratio, %	12	28	27	29	25

Operator characteristics:

Farming as a major occupation, %	71	90	96	88	84
Older than 65 years, %	35	9	6	5	18
Household off-farm income, $	19,099	13,579	9,991	16,543	15,089

Source: 1986 Wheat Version, Farms Costs and Returns Survey, USDA.

TABLE 6.10 U.S. Wheat Costs and Farm and Operator Characteristics by Wheat Acreage Class, 1986

Item	Wheat Acreage Class				All Farms
	Less than 25	25-29	100-499	500 or More	
Number of farms	71,236	93,072	88,807	27,731	280,846
Share of all farms, %	25	33	32	10	100
Share of wheat production, %	2	10	39	49	100
Wheat yields, bu./acre	34	35	33	32	33
Acres planted to wheat, acres/farm	13	54	238	946	190
Variable cash costs:					
Dollar per acre	55.00	50.48	44.35	42.47	44.19
Dollar per bu.	1.63	1.46	1.36	1.31	1.35
Fixed cash costs:					
Dollar per acre	28.68	25.13	21.33	20.77	21.54
Dollar per bu.	0.85	0.73	0.65	0.64	0.66
Capital replacement:					
Dollar per acre	11.79	10.07	11.32	10.99	11.05
Dollar per bu.	0.35	0.29	0.35	0.34	0.34
Economic costs:					
Dollar per acre	166.30	130.50	117.36	107.23	114.48
Dollar per bu.	4.92	3.78	3.60	3.31	3.50

Farm finances:					
Net cash income, $	9,928	14,286	22,203	-40,191	10,305
Government payment, $	4,948	7,598	18,813	46,927	14,356
Assets, $	259,606	300,229	428,840	950,537	394,806
Debts, $	47,892	77,668	127,393	219,893	99,882
Debt/asset ratio, %	18	26	30	23	25
Operator characteristics:					
Farming as a major occupation, %	80	75	93	93	94
Older than 65 years, %	30	17	12	9	18
Household off-farm income, $	16,150	15,171	15,484	10,821	15,089

Source: 1986 Wheat Version, Farm Costs and Returns Survey, USDA.

Costs per bushel of two major inputs, seed and fertilizer, decreased as wheat acreage increases. This is in contrast to the relationship found for the other two major crops, corn and soybean. Land costs per bushel also declined as farms increase in wheat acreage. The only major costs which did increase as wheat acres increased were interest expenses and repairs.

Wheat Costs of Production and Characteristics by Wheat Region. There are six production regions specified, due to the versatility of the various classes of wheat and the differing associated cultural practices. The following regions are defined: North Central (Illinois, Indiana, Missouri, and Ohio), Northern Plains (Minnesota, Montana, North Dakota, and South Dakota), Central Plains (Colorado, Kansas, Nebraska, and Wyoming), Southern Plains (New Mexico, Texas, and Oklahoma), Northwest (Idaho, Oregon, and Washington), and the Southeast (Alabama, Arkansas, Georgia, Kentucky, Louisiana, Mississippi, North Carolina, South Carolina, Tennessee, and Virginia). States in the Northeast and Southwest are not included as separate regions because of small sample sizes, but they are included in the U.S. totals. The cost comparison by region for wheat is interesting in that only one pair of regions had a statistically different average economic cost per bushel. All other pairwise comparisons were not statistically different. However, on a per-acre basis, almost all pairwise comparisons were different. The exceptions are noted below.

The Central Plains is the largest wheat-producing region in the U.S. and in 1986, as in many other years, it was also the lowest cost region (Table 6.11). Economic costs per bushel were $3.29, however, this cost estimate was not significantly different from the other regions' costs, except for the North Central region. The three Plains regions together account for over 70 percent of U.S. production. All three of them, had statistically similar and relatively low costs per planted acre and similar yields, although yields in the Southern Plains were down from their usual levels in 1986 due to drought conditions. Wheat farms in the Southern Plains are unusual in that 43 percent of them were in the smallest economic class but all wheat farms combined had an average planted wheat acreage of a relatively large 321 acres.

The Northwest also had relatively low economic production costs per bushel in 1986 of $3.49. The uniqueness of wheat production in the Northwest is due to its greater use of irrigation compared to most other regions and, for dryland wheat, its greater use of fallowing. As a result, yields and per-acre costs are up from the U.S. average. The North Central region accounted for one-third of all wheat-producing farms in 1986, but only nine percent of production due to the greater proportion of smaller wheat farms in this region. About 45 percent of all wheat

TABLE 6.11 Wheat Characteristics, by Region, 1986[a]

Item	North Central	Northern Plains	Central Plains	Southern Plains	North-west	South-east
Share of U.S. wheat farms, %	32	18	19	10	6	10
Share of U.S. production, %	9	31	26	14	13	4
Wheat yields, bu./acre	34	31	31	27	48	34
Wheat cropland, acres/farm	51	322	280	321	295	68
Economic cost, dol./acre	132.79	108.23	101.76	98.22	168.80	125.18
Economic cost, dol./bu.	3.86	3.48	3.29	3.61	3.49	3.74
Share of group's farm in economic class:						
Less than $40,000, %	45	28	33	43	20	45
$40,000-$99,999, %	33	37	36	19	20	23
$100,000-$249,999, %	14	25	24	34	44	25
$250,000 or more	8	10	7	4	16	7
Farm finances:						
Net cash income, $	10,077	30,225	-19,964	8,811	31,833	13,897
Government payments, $	7,462	21,788	17,468	17,824	26,129	8,727
Assets, $	339,726	454,504	376,930	409,000	590,572	306,937
Debt, $	76,829	139,119	77,678	117,844	168,349	76,181
Debt/asset ratio, %	23	31	21	29	29	25

(continued)

TABLE 6.11 (continued)

Item	North Central	Northern Plains	Central Plains	Southern Plains	North-west	South-east
Operator characteristics:						
Farming as major occupation, %	75	96	75	88	92	90
Older than 65 years, %	19	11	14	22	17	22
Household farm income, %	15,560	9,407	18,343	26,166	10,538	12,586

[a]Due to sample size restrictions data for the Southwest and Northeast are not provided separately. They are included in the U.S. totals, however.

Source: Ahearn, M., et al., 1990a.

farms in the North Central were in the smallest economic class and the average wheat acreage there was only 51 acres. The yields of 34 bushels per acre were close to the national average, although down for the region due to drought conditions, and economic costs per bushel were $3.86, somewhat above the U.S. average. Per-bushel economic costs were $3.74 in the Southeast. Like the North Central region, wheat-producing farms in the Southeast tend to be smaller than other wheat regions.

Wheat Costs of Production and Characteristics by Non-Size Structural Indicators. Significant differences in average economic costs per bushel of wheat were found for several indicators of structure, other than size (Table 6.12). By tenure arrangements, full owners had significantly higher costs than operators who rented in some of their land. With respect to how a farm firm-household allocates its resources, wheat operations who had an operator whose major occupation was farming had significantly lower production costs and one where farming was the major source of income had significantly lower costs. Finally, operators 65 years or older had significantly higher costs than all other age groups.

Review and Conclusions

Costs of producing corn, soybean, and wheat vary significantly in the U.S. A major factor affecting the variation is the differences in yields. Yield differences are most obvious by region. Size of farm and the factors associated with being a small farm are also related to costs of production. Size of farm was measured in two ways: by economic class of the whole farm and by acres planted of the commodity. Since most farms with small planted acreage of the respective crops were also small farms as measured by the economic class criteria, the results were similar by the two size classification schemes.

Although size was found to be a major factor in differentiating cost levels, the threshold under which size was important was rather small. Corn farms in the economic classes of $40,000-$249,999 were found to have similar average economic costs as the largest group of producers. For soybeans, there were no statistical differences between the two largest size groups over $100,000. When size was measured by economic class, no significant differences were found by size of farm for wheat production. When size was measured by acreage class for wheat, however, the small farms did have significantly higher production costs than the larger farms, but no cost differences were found for the two largest acreage classes above 100 acres.

It is important to note that, although the higher costs are occurring on very small farms, these small farms represent large proportions of

TABLE 6.12   Variable Cash and Economic Costs of Producing Wheat per Bushel by Other Structural Indicators, 1986

Indicator	Average Variable Cash Costs	Average Economic Cost
	Dollars per Bushel	
Tenure		
Own all acres	1.74	4.41
Rent-in some acres		
Some share, no cash	1.20	3.30
Some cash, no share	1.38	3.52
Both cash and share	1.32	3.32
Major Occupation		
Farming	1.30	3.43
Other	2.09	4.49
Operator Age		
Less than 35	1.34	3.44
35-49	1.35	3.39
50-64	1.30	3.47
65 or more	1.61	4.27
Major income source of household		
Farming greater than off-farm	1.31	3.41
Off-farm greater than farming	1.59	4.05

Source: 1986 Wheat Version, Farm Costs and Returns Survey, USDA.

producers. For example, 43 percent of corn producers are in the economic class of $40,000 or less. Small farms heavily weight most average statistics in the agricultural data system; averages for all farms generally differ greatly from averages for what are generally considered to be the major producers in the sector. Data users need to use average statistics of all farms in the U.S. with this awareness. The purpose may dictate that it is most appropriate that the small farms be eliminated from the field of observation because their viability is much less a function of the prices and policies affecting the larger producers.

Our classification analysis basically supports the established L-shape relationship between production costs and size (see Figures 6.2 and 6.3). These curves should not be viewed as long run average cost curves. We also offer the caution that the concentration of producers at small size

FIGURE 6.2 Economic Costs for U.S. Corn, Soybean, and Wheat by Gross Income
(Source: FCRS, 1986 and 1987.)

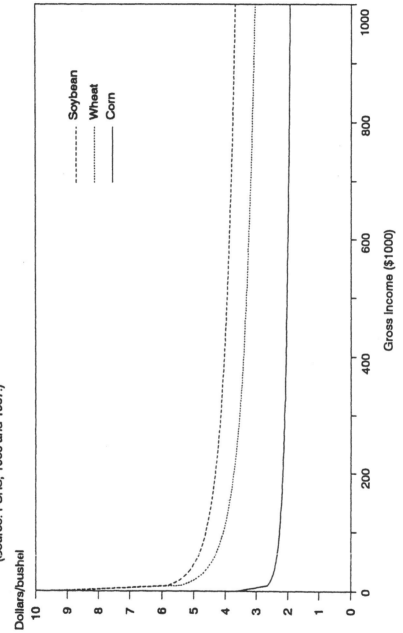

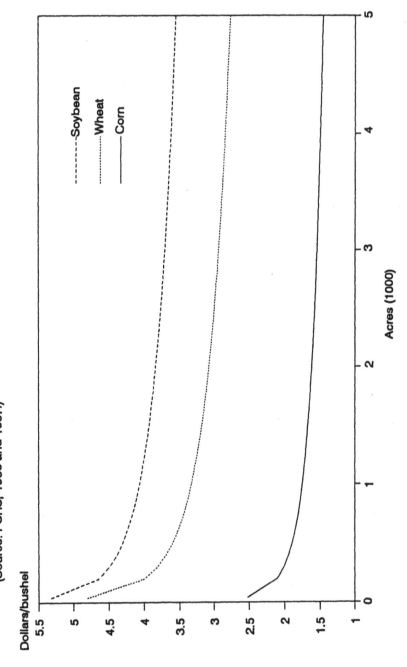

FIGURE 6.3 Economic Costs for U.S. Corn, Soybean, and Wheat by Size of Acreage (Source: FCRS, 1986 and 1987.)

levels and the high cost of production of some of these producers may give the impression that more economies of size exist than actually do when standard cost-size relationships are plotted graphically.

The other structural characteristics that were related to cost levels were the major source of income of the household, farm vs. off-farm, and in the case of wheat, the major occupation of the operator. These factors are also associated with farm size. Small farms in the U.S., in general, rely more on their off-farm income than their farm income. These results reemphasize the importance of jointly accounting for both the farm business and the farm household well-being when analyzing the agricultural sector.

Notes

1. To test this, Ahearn, Whittaker, and Glaze (1990) compared the technical efficiency by farm size for a sample of Midwestern corn-producing farms using only the inputs associated with the corn enterprise to the results of measuring efficiency in the multi-output case for the whole corn-producing farm. Their results indicated that for that sample of farms, the two approaches yielded similar conclusions about technical efficiency by farm size.

2. There may even be a problem with the costs of hired labor in that many operators pay family members, not based on the value of their contributions to labor and management, but as an income tax avoidance strategy.

3. The authors are indebted to the many statisticians and economists who have contributed to the development of the data bases. In particular, Dargan Glaze (1988), with assistance from Mir Ali, is responsible for the development of the farm-level cost of production model for 1986 wheat and soybeans and 1987 corn.

4. The expectations are that all future cost of production versions will include the necessary survey questions in order to estimate costs and returns including the effects of the programs. More recent data for wheat have been collected for the 1989 calendar year and include these program-related data, but they are not yet available.

References

Ahearn, Mary, Mir Ali, Robert Dismukes, Hisham El-Osta, Dargan Glaze, Ken Mathews, Bill McBride, Robert Pelly, and Mike Salassi. 1990a. "How Production Costs Vary." AIB 599, Econ. Res. Serv., USDA.

Ahearn, Mary, Robert Dubman, and Gregory Hanson. 1987a. "Financial Performance of Specialized Dairy Farms." AIB-519, Econ. Res. Serv., USDA.

———. 1987b. "Financial Performance of Specialized Wheat Farms." AIB-528, Econ. Res. Serv., USDA.

148

Ahearn, Mary, Gerald Whittaker, and Dargan Glaze. 1990b. "Cost Distribution and Efficiency of Corn Production," in *Proceedings of the 181 Meetings on Determinants of Size and Structure in North Central Areas and the United States,* Albuquerque, NM, January 6-8, 1990, edited by Arne Hallam. Ames, IA: Dept. of Economics, Iowa State University.

Bills, Nelson L., and Raymond I. Dideriksen. 1980. "Land and Water Base: Historical Overview and Current Inventory," in *Farm Structure: A Historical Perspective on Changes in the Number and Size of Farms,* Committee on Agriculture, Nutrition, and Forestry, United States Senate, 96th Congress, 2d Session. Washington, DC: U.S. Government Printing Office.

Carter, H. O., and G. W. Dean. 1961. "Cost-Size Relationships for Cash Crop Farms in a Highly Commercialized Agriculture." *Journal of Farm Economics* 43(2): 264-277.

Cochran, W. G. 1977. *Sampling Techniques.* New York: John Wiley and Sons.

Fuller, Wayne A., William Kennedy, Daniel Schnell, Gary Sullivan, and Heon Jin Park. 1986. *PC-CARP.* Statistical Laboratory, Iowa State University, Ames, IA.

Garcia, Philip, and S. T. Sonka. 1984. "Methodological Issues in Assessing Economies of Size: Selected Positive Analytic Approaches," in *Economies of Size Studies.* A collection of papers presented August 3-4, 1983, Purdue University, West Lafayette, Ind. Ames: Center for Agricultural and Rural Development.

Glaze, Dargan. 1988. "A New Approach to Estimating COP Budgets." *Agricultural Income and Finance.* USDA, ERS. AFO-29.

Hall, Bruce F., and E. Philip LeVeen. 1978. "Farm Size and Economic Efficiency: The Case of California." *American Journal of Agricultural Economics* 60(4): 589-600.

Hallam, Arne. 1991. "Economies of Size and Scale in Agriculture: An Interpretative Review of Empirical Measurement." *Review of Agricultural Economics* 13: 155-172.

Heady, E. O. 1952. *Economics of Agricultural Production and Resource Use.* New York: Prentice-Hall.

Huffman, Wallace E. 1988. "Production, Consumption, and Labor Supply Decisions of Farm Households: A Review of the Evidence for North America." Staff Paper No. 183, Department of Economics, Iowa State University.

Just, Richard E., David Zilberman, and Eithan Hochman. 1983. "Estimation of Multicrop Production Functions." *American Journal of Agricultural Economics* 65(4): 770-780.

Krenz, Ronald D., Walter G. Heid, Jr., and Harry Sitler. 1974. "Economies of Large Wheat Farms in the Great Plains." AER-264, Econ. Res. Serv., U.S. Dept. Agr.

Madden, J. Patrick. 1967. "Economies of Size in Farming." U.S. Department of Agriculture, Econ. Res. Serv., AER No. 107.

Miller, Thomas A. 1984. Conceptual Issues in Economies of Size Research: How Assumptions Drive Away Research," in *Economies of Size Studies.* A collection of papers presented August 3-4, 1983, Purdue University, West Lafayette, Ind. Ames: Center for Agricultural and Rural Development.

Opie, John. 1987. *The Law of the Land: Two Hundred Years of American Farmland Policy.* Lincoln: The University of Nebraska Press.

Salassi, Michael, Mary Ahearn, Mir Ali, and Robert Dismukes. 1990. "Effects of Government Programs on Rice Production Costs and Returns, 1988." Econ. Res. Serv., USDA, AIB 597.

Seckler, David, and Robert A. Young. 1978. "Economic and Policy Implications of the 160-Acre Limitation in Federal Reclamation Law." *American Journal of Agricultural Economics* 60(4): 575-588.

Singh, Inderjit, Lyn Squire, and John Strauss. 1986. *Agricultural Household Models: Extensions, Applications, and Policy.* Baltimore: The Johns Hopkins University Press.

Smith, Edward G., Ronald D. Knutson, and James W. Richardson. 1986. "Input and Marketing Economies: Impact on Structural Change in Cotton Farming on the Texas High Plains." *American Journal of Agricultural Economics* 65(3): 716-720.

Stigler, G. J. 1952. *The Theory of Price.* New York: Macmillan.

U.S. Department of Agriculture. 1990. *Economic Indicators of the Farm Sector: Costs of Production--Major Field Crops, 1988.* Econ. Res. Serv., ECIFS 8-4.

Vlastuin, Chris, Denis Lawrence, and John Quiggin. 1982. "Size Economies in Australian Agriculture." *Review of Marketing and Agricultural Economics* 50(1).

Whittaker, Gerald W., Mary C. Ahearn, and Hisham El-Osta. 1990. "Planned vs. Realized Output of Farmers and the Implications for Cost-Size Relationships." Unpublished manuscript.

7

Economies of Size: Theory, Measurement, and Related Issues

Arne Hallam

While economies of size are often invoked as a factor explaining everything from changes in the structure of the meat packing industry, to mergers in the banking industry, to the demise of the family farm, the appropriate ways to analyze and measure this phenomenon remain a subject of much disagreement.

As discussed in Chapter 1, the presence or absence of economies of size has a variety of implications for the structure and growth pattern of an industry. Three broad areas of concern seem particularly relevant. These are competition and welfare, preservation of a particular structure (the family farm), and forecasting future change.

The first issue deals with the interaction of size economies, perfect competition, and welfare economics. If the long-run cost curve for a firm is downward sloping in the relevant region, then marginal cost will lie below it and competitive marginal cost pricing will not prevail. This could lead to monopolization of the industry and the associated problems with obtaining a competitive and welfare maximizing equilibrium. While societal well-being can be maximized in the presence of monopoly using a variety of instruments, the conventional wisdom in economics seems to prefer the welfare implications of many firms. Economies of size are related to international competitiveness and changes in the terms of trade. In the less than perfectly competitive world, economies of size and associated market power in one country may be exploited to maximize domestic welfare; while potential economies of size in another country may be stifled and an industry protected to attain specific social goals. Lumpiness in the acquisition and installation of fixed plant, or demand uncertainty may tend to create short-run economies of size that encourage the growth of larger firms than would be justified by societal cost minimization.

While most economists tend to be concerned about size economies only as they affect efficiency and the sustainability of competitive equilibrium, much research in agricultural economics has focused on firm "size" as an independent issue. This interest is based on the normative desirability of the "family farm." If it can be shown that there are no economies of size for firms larger than the "typical" family farm, then policies which protect this entity are justified from a societal point of view. Given the current structure of most of production agriculture (broiler production is a probable exception), it seems reasonable to postulate that almost all interest in size economies is related to this normative issue and not to maintaining competition in primary agricultural markets. In fact, much of the research undertaken seems deeply devoted to the task of demonstrating "L" shaped average cost curves (Madden 1967; Miller, et al. 1981; Castle 1989). Research has also focused on the sustainability of the family farm over time and whether new technologies (Kalter and Tauer 1987) will remain neutral in the battle for agricultural resources.

Concern with sustainability of the family farm also relates to the third reason for interest in economies of size. Over time industries adapt to meet changes in technology, consumer preferences, and world conditions. Being able to predict and understand such changes reduces uncertainty for the firm, mitigates stress for the average consumer, helps the politician anticipate changes in support and lobbying forces, and provides recreation for demographers, sociologists, and science fiction writers. By understanding why and which firms grow and prosper, investors can make wiser decisions on how to allocate societal resources in an uncertain world. While economies of size are only one of many factors that affect the growth path of industry, whether important or not, they are almost always cited.

Given the interest in economies of size, their definition and measurement is important. Because economies of size measures may be used for distinctly different purposes, clarity on definition and appropriate measurement is crucial. Alternative theoretical measures of size economies result from different assumptions on technology, length of run, and residual resource claimants. While positive economic analyses often focus on the identification of size economies, the results of such positive efforts are often the primary input into unapologetically normative proscriptions both at the firm and industry level. On the other hand, conditionally normative or synthetic studies are often used to demonstrate or measure tendencies toward economies of size. The compatibility between theory and the measurement methods used is a prime consideration in evaluating the worth of empirical work and a focus of this chapter.

The next section of the chapter addresses the definition of economies of size and the relationship between economies of size, scale, and scope. Particular attention is devoted to the multiproduct firm and issues relating to firm homogeneity in empirical work. Extensions to the basic static concepts are also discussed. A short section on the market implications of size economies is followed by a discussion of alternative ways to measure economies of size, scale, and scope using a variety of methods and data sources. The conclusion outlines areas of clear direction, areas of continued debate, and guidelines for future research.

Economies of Size: Some Tentative Definitions

Single Product Firms Under Certainty

Before defining economies of size, it is useful to define the related concept of returns to scale. Assume the production function is given by

$$y = f(x_1, x_2, ... x_n) = f(x) \qquad (1)$$

where y is output and x is the vector of inputs $x_1...x_n$. Let f have the usual neoclassical properties of quasiconcavity, strict positive monotonicity, and twice continuous differentiability in x. Then the function is said to exhibit nonincreasing returns to scale if for all $x \, \varepsilon \, R_+^n$, $\lambda \geq 1$, and $0 < \mu \geq 1$

$$f(\lambda x) \leq \lambda f(x) \text{ and } \mu f(x) \leq f(\mu x)$$

Thus the function increases less than proportionately as x increases along a ray and decreases more than proportionately as x decreases. In a similar fashion nondecreasing returns to scale exist if for all $x \, \varepsilon \, R_+^n$, $\lambda \geq 1$, and $0 < \mu \geq 1$

$$f(\lambda x) \geq \lambda f(x) \text{ and } \mu f(x) \geq f(\mu x)$$

The function exhibits constant returns to scale if for all $x \, \varepsilon \, R_+^n$ and $\theta > 0$

$$f(\theta x) = \theta f(x)$$

This global definition of returns to scale is often supplemented by a local one that has a specific numerical magnitude. The elasticity of scale (Ferguson 1981) is implicitly defined by

$$\varepsilon = \frac{\partial \ln f(\lambda x)}{\partial \ln \lambda} \quad | \quad \lambda = 1 \tag{2}$$

This simply explains how output changes as inputs are changed in fixed proportions (along a ray through the origin). Intuitively, this measures how changes in inputs are scaled into output changes. The elasticity of scale can also be represented as

$$\varepsilon = \sum \frac{\partial f}{\partial x_i} \frac{x_i}{y} \tag{3}$$

Thus elasticity of scale is the sum of the output elasticities for each input. If ε is less than one, then the technology is said to exhibit decreasing returns to scale; if it is equal to one, then the technology exhibits constant returns to scale; and if ε is greater than one, the technology exhibits increasing returns to scale. While alternative levels of ε have important implications for productivity analysis and the neoclassical theory of distribution, it is non-proportionate changes in inputs (those changes not along a ray) that are relevant for most economic questions and economies of size in particular.

Associated with a production function is its dual cost function. Given a neoclassical production function the cost function is defined as

$$c(w,y) = \min_{x}(wx: f(x) = y) \tag{4}$$

The first order conditions for a minimum imply that

$$w_i = \lambda \frac{\partial f}{\partial x_i} \tag{5}$$

where λ is the Lagrange multiplier associated the output constraint. Shephard's lemma implies that the derivative of the cost function with respect to input prices yields Hicksian input demands, while the derivative with respect to output yields the multiplier λ or marginal cost.

$$\frac{\partial c(w,y)}{\partial w_i} = x_i(w,y)$$

$$\frac{\partial c(w,y)}{\partial y} = \lambda(w,y)$$

(6)

Given these relationships, the elasticity of scale can be represented in terms of the parameters of the cost function. Specifically

$$\varepsilon = \Sigma \frac{\partial f}{\partial x_i} \frac{x_i}{y}$$

$$= \frac{\Sigma w_i x_i}{\lambda y}$$

$$= \frac{\Sigma w_i x_i}{y \frac{\partial c(w,y)}{\partial y}}$$

(7)

$$= \frac{AC}{MC}$$

$$= \frac{1}{\frac{\partial \log c(w,y)}{\partial \log y}}$$

This cost function measure of elasticity of scale will be equivalent to the production function measure at appropriately defined points. Specifically, for any point $(x^0, y^0 = f(x^0))$, pick the vector w^0 such that

$$x^0 = \underset{x}{argmin} \, [wx \mid f(x) = y^0]$$

Then the cost function $C(w,y)$ evaluated at (w^0, y^0) implies the point $(x^0, y^0 = f(x^0))$ and the measures will be the same. Thus at all points along the expansion path the cost and production function measures will be equal since the prices will support the given input-output combination. As one moves away from a given point, the measures will differ if the direction of movement is not the same. For example, moving along a ray to a new point (x^1, y^1) and to a new cost minimizing point $C(w^1, y^1)$ will not yield the same point in input-output space unless x^1 is cost minimizing for w^1. And at different points, returns to scale are not necessarily the same. However, at points where again input prices support (x,y) in the sense of x being cost minimizing for y, the measures are equal. Is this sense, the statement that returns to scale measured using the production and cost functions are only equal along the expansion path is a misnomer since the measures are equal at all consistent points.

Returns to size is best defined using the cost indirect production function (Shephard 1974) given by

$$IP(\frac{w}{c^0}) = \max_y[y \mid C(w,y) \leq c^0] = \max_y[y \mid C(\frac{w}{c^0},y) \leq 1] \qquad (8)$$

This gives the maximum production for a given expenditure level. Returns to size considers how $IP(w/c^0)$ changes as the fixed expenditure level c^0 changes when input prices are held fixed. In a manner analogous to returns to scale, if maximal output increases proportionately more than cost (expenditure level), the technology exhibits increasing returns to size. If the maximal output increases proportionately less than cost, the technology exhibits decreasing returns to size and if they increase proportionately the same, the technology exhibits constant returns to size. Returns to size is usually measured using the reciprocal of the cost elasticity with respect to output as defined by

$$\gamma = \frac{1}{\left(\dfrac{\partial c(y,w)}{\partial y}\right)\left(\dfrac{y}{c}\right)} = \frac{c(y,w)}{\left(\dfrac{\partial c}{\partial y}\right)y} = \frac{AC}{MC} \qquad (9)$$

Then by definition, the elasticity of size is the ratio of average to marginal cost. If the elasticity of size (cost elasticity) is greater (less) than one, then the firm exhibits increasing returns to size and the average cost curve is downward sloping. If the average cost curve is upward sloping, then the size elasticity is less than one and the size elasticity is equal to one when average cost is at its minimum (Hanoch 1975).

The elasticity of size and the elasticity of scale are, of course, related to each other and are equal at all points along the expansion path. Unfortunately, for most technologies the expansion path is not along a ray through the origin. The only technologies for which this is the case are homothetic technologies (Sandler and Swimmer 1978). A homothetic technology can be written as $y = F(\phi(x))$ where ϕ is linearly homogeneous and F is a monotonically increasing function. This can also be written $h(y) = \phi(x)$, where $h(y) = F^{-1}$. For such technologies, the cost function can be written as

$$c(y,w) = h(y) \; c(w) \qquad \qquad (10)$$

where h(y) is a representation of the production technology. To see the equivalence of scale elasticity and size elasticity in the case of homotheticity, find the scale elasticity for the production function by implicitly differentiating $h(y) = \phi(x)$ as follows

$$h'(y)\frac{\partial y}{\partial x_i} = \frac{\partial \phi}{\partial x_i}$$

$$\Rightarrow \frac{\partial y}{\partial x_i} = \frac{\left(\dfrac{\partial \phi}{\partial x_i}\right)}{h'(y)} \qquad \qquad (11)$$

By substituting $\partial\phi/\partial x_i/h'(y)$ for $\partial y/\partial x_i$ $(\partial f/\partial x_i)$ in the expression for ε and using Euler's theorem the following is obtained:

$$\varepsilon = \frac{h(y)}{h'(y)y} \qquad \qquad (12)$$

The cost elasticity is $(\partial C/\partial y)(y/C)$ which for equation (10) gives $(h'(y)y)/h(y)$ and so the scale elasticity is the reciprocal of the cost elasticity.

Size and scale elasticities vary with output levels unless the production function is homothetic or ray homothetic (Färe et al., 1985). Hanoch specifically showed that the change in the elasticity of scale with respect to output holding x in fixed proportions is different from the change holding w constant and allowing x to adjust in an optimal fashion as discussed previously. He also demonstrated that average costs are U-shaped if and only if $\varepsilon(x) = 1$ for some $y = f(x)$, and $\varepsilon = 1$ implies $(\partial \varepsilon/\partial y)$ I w < 0. Intuitively this means that the scale elasticity decreases through 1 along the expansion path. He further showed that simply requiring a negative derivative of the scale elasticity with respect to output is not sufficient for "U" shaped cost curves.

Since economies of size are directly related to the elasticity of cost with respect to output, the cost function seems a natural candidate for the estimation of economies of size. While more will be said about this in a later section, it is well to point out that the profit function can also be used for the same purpose in cases where it is defined. For a profit function, returns to scale are given by

$$\varepsilon = \Sigma \frac{\partial y}{\partial x_i} \frac{x_i}{y} = \Sigma \frac{w_i}{p} \frac{\dfrac{-\partial \pi}{w_i}}{\dfrac{\partial \pi}{p}} \qquad (13)$$

$$= \frac{\Sigma w_i x_i}{p \ y} = \frac{cost}{revenue}$$

This follows directly from the definition of ε and the use of Hotelling's lemma to define x and y. Of course, this is just the ratio of cost to revenue and will be less than or equal to one for profit maximizing firms. This implies that no increasing returns to scale should be identified by using the profit function. As discussed by Stefanou and Madden (1988), while the profit function may be useful in analyzing the structure of production, it provides no new information as to size economies. This point is also made in Chambers (1984) who presents a fuller discussion of some of these issues.

Multiproduct Firms

Returns to size and scale are less clear in the case of multi-product firms. Consider the production possibility set of the multi-product firm:

$$T = \{(x,y) \mid y \text{ can be produced by } x\} \qquad (14)$$

where y and x are vectors of outputs and inputs, respectively. This technology can also be represented in terms of various functions. Consider first the output distance function (Shephard 1970). It is defined as

$$D_0(y,x) = \min_{\gamma} \ [\gamma > 0: (x, \frac{y}{\gamma}) \varepsilon T] \qquad (15)$$

Intuitively, the output distance function tells us how much we can expand a given output vector y and still be able to produce it with a fixed input vector x. If D_0 is less than one, the vector x will produce y, possibly with some to spare. Define the input distance function (Shephard 1953) as

$$D_i(y,x) = \max_{\lambda} \ [\lambda > 0: (\frac{x}{\lambda}, y) \varepsilon T] \qquad (16)$$

The input distance function tells us how much we can shrink a given input vector x and still be able to produce a fixed output vector. If D_i is greater than one, the vector x will produce y, possibly with some to spare. In order to be able to use calculus and to provide results that look very similar to the single output case some authors (McFadden 1978, Baumol, Panzar and Willig 1982) assume conditions necessary to define the transformation function $t(x,y)$. The transformation function is implicitly defined by

$$t(x,y) > 0 \quad if \ (x,y) \ \varepsilon \ T \tag{17}$$

While there may be no easy way to construct $t(x,y)$ for some technologies T, appropriate conditions on T insure its existence. Furthermore, assume that $t(x,y)$ is continuously differentiable in x and continuously differentiable in y_i, for $y_i > 0$, at points (x,y), where x is cost efficient for y. This implies differentiability at all points except perhaps where an output is increased from zero, possibly resulting from fixed set-up costs.

A input based measure of returns to scale for the production possibility set is defined by

$$\varepsilon_m = sup\{r \mid there \ exists \ a \ \delta>1 \ such \ that(\lambda x, \lambda^r y) \ \varepsilon \ T \ for \ 1 \leq \lambda \leq \delta\} \tag{18}$$

This gives the maximum proportional growth rate of outputs along their ray, as all inputs are expanded proportionally (Baumol, Panzar, and Willig 1982). This measure of returns to scale in terms of the input distance function is

$$\varepsilon_m(x,y) = \sup_r \ \{r>0: D_i(\lambda x,\lambda^r y) \geq 1\} \tag{19}$$

The measure explains how much y can be proportionately expanded for a given proportionate increase in x. This is the measure most commonly reported in the literature. This can also be expressed in terms of the gradient of D_i as

$$\varepsilon_m(x,y) = \frac{-D_i(x, \ y)}{y' \nabla_y D_i(x,y)} \tag{20}$$

This can be proven by showing the right hand side of 20 is greater than or equal to ε_m but that it cannot exceed it. Define x^* as the value of x for which $D(x^*,y) = 1$. In other words, x^* is the level of x which is just able to produce y. Now define the set R as

$$R = [r > 0 : D_i(\lambda x^*, \lambda^r y) \geq 1, \lambda \geq 1) \tag{21}$$

Now consider an element r of R and define the function H as

$$H(\lambda) = D_i(\lambda x^*, \lambda^r y) \tag{22}$$

Now consider the derivative of H with respect to λ evaluated at $\lambda = 1$

$$
\begin{aligned}
\frac{\partial H(\lambda)}{\partial \lambda} &= \sum_i x_i^* \frac{\partial D_i(\lambda x^*, \lambda^r y)}{\partial (\lambda x_i)} + r\lambda^{r-1} \sum_j y_j \frac{\partial D_j(\lambda x^*, \lambda^r y)}{\partial (\lambda y_j)} \\
&= \sum_i x_i^* \frac{\partial D_i(x^*, y)}{\partial x_i} + r \sum_j y_j \frac{\partial D_j(x^*, y)}{\partial y_j} \\
&= D_i(x^* y) + r\, y' \nabla_y D(x^* y)
\end{aligned}
\tag{23}
$$

For all elements r of R, $D_i(\lambda x^*, \lambda^r y) \geq 1$. Therefore the function $H(\lambda)$ achieves its lower bound at $\lambda = 1$. Since $H(\lambda = 1) = 1$, then $H'(\lambda \geq 1) \geq 0$ for all r in R. This then gives us an upper bound on r. Specifically this implies

$$\frac{\partial H(\lambda)}{\partial \lambda} \geq 0$$

$$\Rightarrow D_i(x^*, y) + r\, y' \nabla_y D(x^*, y) \geq 0$$

$$\Rightarrow \frac{-D_i(x^*, y)}{y' \nabla_y D(x^*, y)} \geq r \tag{24}$$

$$\Rightarrow \frac{-D_i(x^*, y)}{y' \nabla_y D(x^*, y)} \geq \varepsilon_m(x, y) = \sup_r [r > 0 : D_i(\lambda x, \lambda^r y) \geq 1)$$

To complete the proof we need to show that the leftmost expression in 24 cannot be larger than ε_m. Denote this expression as $\tilde{\varepsilon}$ and consider that there is space between it and ε_m. Label a point between $\tilde{\varepsilon}$ and ε_m r^*, such that $\tilde{\varepsilon} > r^* > \varepsilon_m$ with r^* not contained in R. Now define a function $G(\lambda)$ as follows

$$G(\lambda) = D_i(\lambda x^*, \lambda^{r^*} y) \tag{25}$$

Clearly, $G(1) = 1$. Consider now the possibility that $G'(1) > 0$. If $G(1) = 1$ and $G'(1) > 0$, then there exists a number $\delta > 1$ such that $G(\lambda) \geq 1$ for $1 \leq \lambda \leq \delta$. In other words, by the continuity of G, there must be a point λ between 1 and δ such that $G(\lambda)$ is between $G(1)$ and $G(\delta)$. But if this is the case, r^* is an element of R, contrary to the assumption. Therefore $G'(1) \leq 0$. How calculate $G'(1)$ as follows

$$\frac{\partial G(\lambda)}{\partial \lambda} \leq 0$$

$$\Rightarrow D_i(x^*, y) + r^* y' \nabla_y D(x^*, y) \leq 0$$

$$\Rightarrow \frac{-D_i(x^*, y)}{y' \nabla_y D(x^*, y)} \leq r^* \tag{26}$$

$$\Rightarrow \tilde{\varepsilon} \leq r^*$$

This contradicts the assumption that $\tilde{\varepsilon} > r^*$ and thus implies that $\tilde{\varepsilon} = \varepsilon$.

An equally intuitive measure of returns to scale in terms of the output distance function is

$$\eta_m(y, x) = \inf_r [r > 0: D_0(\lambda y, \lambda^r x) \leq 1] \tag{27}$$

The measure explains how much x must be proportionately increased in order to still be able to produce a proportionate increase in y. The input based measure of elasticity of scale can also be defined in terms of the parameters of the transformation function using an argument similar to that above as

$$\varepsilon_m = \frac{-\sum_{i=1}^{n} x_i \frac{\partial t}{\partial x_i}}{\sum_{j=1}^{m} y_j \frac{\partial t}{\partial y_j}} \tag{28}$$

where the derivatives are evaluated at $t[x^*(y, w), y]$.

All these measures are based on the proportionate increase in all outputs from a proportionate increase in all inputs. Nonproportionate increases in y cannot be measured using any of these definitions since an increase in inputs allows many tradeoffs between output levels, none of which is inherently more interesting than any other.

In the single product case, returns to scale is derived from the cost function as the ratio of cost to marginal cost times output (AC/MC). In the multiproduct case, returns to scale is equal to the ratio of total cost to the inner product of the marginal cost vector and the output vector. This can be shown using the duality between the input distance and cost functions. The multiproduct cost function is defined as

$$C(w,y) = \min_{x}(wx: D_i(y,x) \geq 1) \; w>0 \tag{29}$$

If the functions are sufficiently differentiable, this problem can be solved using standard Lagrangian techniques.

$$L(w,y) = \min_{x}(wx - \mu(D_i(y,x) - 1)) \tag{30}$$

The first order conditions are as follows

$$w - \mu \nabla_x D_i(y,x) = 0$$
$$-D_i(y,x) + 1 = 0 \tag{31}$$

Since the constraint is satisfied at the optimum, the resulting cost function can be written

$$C(w,y) = wx(w,y) - \mu(w,y)[D_i(y, x(w,y)) - 1] \tag{32}$$

Differentiating this with respect to w and using the first order conditions will give Shephard's lemma

$$\nabla_w C(w,y) = x(w,y) + w \nabla_w x(w,y) - \mu(w,y)[\nabla_x D_i(y, x(w,y))\nabla_w x(w,y)]$$

$$- \nabla_w \mu(w,y)[D_i(y, x(w,y)) - 1]$$

$$= x(w,y) + \nabla_w x(w,y)[w - \mu(w,y)\nabla_x D_i(y, x(w,y))] + 0 \tag{33}$$

$$= x(w,y)$$

Differentiating this with respect to y gives the following

$$\nabla_y C(w,y) = w\nabla_y x(w,y) - \mu(w,y)[\nabla_y D_i(y,x(w,y))$$

$$+ \nabla_x D_i(y,x(w,y))\nabla_y x(w,y)] - \nabla_y \mu(w,y)[D_i(y,x(w,y)) - 1)]$$

$$= \nabla_y x(w,y)[w - \nabla_x D_i(y,x(w,y))] - \mu(w,y)[\nabla_y D_i(y,x(w,y))] + 0 \tag{34}$$

$$= -\mu(w,y)\ \nabla_y D_i(y,x(w,y))$$

This can be interpreted more meaningfully by considering the optimal multiplier μ. At the optimum, the cost function is equal to the value of the multiplier. This can be shown by forming the inner product of the first order conditions in 24 with the optimal x vector and then solving for μ.

$$x'(w,y)w - \mu x'(w,y)\nabla_x D_i(y,x) = 0$$

$$\Rightarrow \mu = \frac{x'(w,y)w}{x'(w,y)\nabla_x D_i(y,x)} \tag{35}$$

$$= \frac{c(w,y)}{\nabla_x' D_i(y,x)\ x(w,y)}$$

Since the distance function is homogeneous of degree one in x, the denominator is equal to one by Euler's theorem and the result follows. Now combine this information with 27 to obtain

$$\mu = c(w,y)$$

$$\Rightarrow \nabla_y C(w,y) = -\mu(w,y)\ \nabla_y D_i(y,x(w,y)) \tag{36}$$

$$= -c(w,y)\ \nabla_y D_i(y,x(w,y))$$

Multiply both sides by y and then divide the right hand side by $D_i(x,y)$ = 1 to obtain

$$\nabla_y c(w,y) = -c(w,y)\ \nabla_y D_i(y,x(w,y))$$

$$\Rightarrow \frac{y\ \nabla_y c(w,y)}{c(w,y)} = -\frac{y\ \nabla_y D_i(y,x(w,y))}{D_i(y,x(w,y))} = \frac{1}{\varepsilon_m} \tag{37}$$

$$\Rightarrow \frac{c(w,y)}{y\ \nabla_y c(w,y)} = \varepsilon_m$$

The marginal costs associated with each output holding the others fixed are weighted by the associated output levels. This measure can also be written

$$\Psi_m = \frac{C(y,w)}{\sum\limits_{j=1}^{m} \frac{\partial C}{\partial y_i} y_i} = \left(\frac{\partial \ln C\ (\lambda y, w)}{\partial \ln y} \right)^{-1} \Big|\ \lambda = 1 \qquad (38)$$

It is the reciprocal of the cost function elasticity as outputs are increased along a ray. The two measures are equal along the expansion path and so will be equivalent global measures only for homothetic technologies. Färe, Grosskopf, and Lovell (1986) have shown that the two primal measures based on the distance function are equivalent. They also show that convexity and efficiency in production (production along an isoquant) gives equivalence between the primal and dual measures.

In order to provide additional intuition for the measurement of multiproduct returns to scale using the cost function some additional concepts must be introduced. For a single product firm, average cost is the ratio of total cost to output. For the multiproduct firm, there are several outputs and no natural divisor exists. A composite good defined over the outputs y could be used to measure average costs. Average costs at different levels of this composite good measure average costs for the technology along a ray in output space since proportionate increases in output lead to similar changes in the composite good. Consider then a particular output level y^0 which is designated as the unit bundle for all output vectors with the same proportions. Average costs along a ray through y^0 are called ray average costs and are defined as

$$RAC = \frac{C(\lambda y^0, w)}{\lambda} \qquad (39)$$

The factor λ is the units of y^0 in the bundle $y = \lambda y^0$. The question then is which bundle to assign the unit value along each ray. The convention (Baumol, Panzar and Willig 1982) is to use the bundle whose components sum to one. This sum is known as the L^1 norm (since y^i is always positive) on R^m if there are m outputs. An alternative would be use the standard L^2 (Euclidean) norm given by $(\sum\limits_{i=1}^{m} y_i^2)^{\frac{1}{2}}$ and set it equal to one. Denote the norm of y by $|y|$. Then let any y along the ray through y^0 be given by $y = \lambda y^0 = |y|\ y^0$, since for $y = y^0$, $|y| = 1$. Then ray average costs are given by

$$RAC = \frac{C(\lambda y^0, w)}{\lambda} = \frac{C(||y|| y^0, w)}{||y||}$$

$$= \frac{C(y, w)}{||y||}$$

$$= \frac{C(y, w)}{\displaystyle\sum_{i=1}^{m} y_i} \tag{40}$$

if we use the sum as the norm. The use of the sum of the outputs as a divisor may smack of adding apples and oranges, but the weighting is truly arbitrary since all changes are measured along a ray. The elasticity of ray average cost is given by

$$e = \frac{\partial RAC(\lambda y)}{\partial \lambda} \frac{\lambda}{RAC(\lambda y)} \tag{41}$$

The elasticity of ray average cost is related to scale economies and it can be shown that (Baumol, Panzar, and Willig 1982, p. 51)

$$\psi_m = \frac{1}{1+e} \tag{42}$$

Thus returns to scale will be greater (less) than one as the elasticity of ray average cost is less (greater) than zero. Constant returns to scale are equivalent to a zero elasticity of ray average cost.

In the single product case, returns to size, defined as the reciprocal of the cost function elasticity, measured changes in costs as inputs expand in a cost minimizing non-proportional manner with increases in output. No similar measure exists in the case of the multiproduct firm since there is not a natural way to expand multiple outputs other than proportionately, and this gives multiproduct returns to scale as discussed above. Of course, profit maximizing output levels could be used but this would give a measure that depends on things other than technology.

Empirical studies often use alternative denominators in measuring average costs for the multiproduct firm in order to avoid this proportionate expansion. A common choice is revenue so that analysis is based on the changes in cost revenue ratios as output expands. While this may be a useful measure of firm profitability, it does not measure a technical relationship since responses to price effects determine the levels of output evaluated. In particular, average cost defined in this way would be shaped differently depending on the prices of outputs (Chambers and Hallam 1992).

While overall returns to scale provide information on economies as the firm expands, information about changes in costs as individual products expand is also useful. To analyze such changes the concept of product specific returns to scale is used. This is related to the incremental cost of producing the good. Consider a firm producing m outputs and denote this set of outputs by M. The incremental cost of producing the ith good in quantity y_i is given by

$$IC_i(y) = C(y) - C(y_{M-i})$$

(43)

and the average incremental cost is given by:

$$AIC_i = \frac{IC_i(y)}{y_i}$$

(44)

The notation y_{M-i} denotes the vector y with a zero in the y_i place and $C(y_{M-i}) = C(y_1, y_2, ...y_{i-1}, 0,...,y_m)$ if it exists. Product specific returns to scale are then defined as the ratio of incremental cost to the product of marginal cost and output y_i i.e.,

$$\psi_i = \frac{IC_i}{y_i \frac{\partial C}{\partial y_i}}$$

(45)

When ψ_i is greater than one, there are increasing product specific returns to scale, etc. In a similar fashion, returns to scale for a group of products can be defined by evaluating the incremental cost of increasing these products from zero to some level holding all other products constant, i.e., for a set $L \subseteq M$

$$\psi_L = \frac{IC_L(y)}{\sum_{j \varepsilon L} \frac{\partial C}{\partial y_j} y_j}$$

(46)

If the subscript L denotes a subset of the outputs and M denotes all outputs, then

$$\psi_m = \frac{\alpha_L \, \psi_L + \left(1-\alpha_L\right) \psi_{M-L}}{\dfrac{\left(IC_L + IC_{M-L}\right)}{C}} \qquad \text{where}$$

(47)

$$\alpha_L = \frac{\displaystyle\sum_{j\varepsilon L} y_j \, \frac{\partial C}{\partial y_j}}{\displaystyle\sum_{j\varepsilon M} y_j \, \frac{\partial C}{\partial y_j}}$$

Note that if the denominator in the last expression were unity, then overall economies of scale would be the weighted average of those of the groups. If the production process is nonjoint, the denominator will be one; otherwise, overall economies of scale depend on interactions between the outputs.

While economies of scale or size explain cost changes that occur as output expands, these may also be changes in cost due to the product mix chosen. If there are cost advantages from the production of several products simultaneously as contrasted with their production in separate firms or processes, then economies of scope are said to occur. A formal definition requires the specification of several subsets of M, the set of all products. Let S be a subset of M and let P be a non-trivial partition of S with an element of the partition denoted L_i. The union of the L_i is equal to the set S and the sets L_i and L_j are nonintersecting for $i \neq j$. Then economies of scope are said to exist if

$$\sum_{i=1}^{k} C\left(y_{L_i}\right) > C(y_S)$$

(48)

For example, if M = {1,2} and P = [{1},{2}], then economies of scope exist when:

$$C(y_1, y_2) < C(0, y_2) + C(y_1, 0)$$

(49)

The degree of scope economies provides a measure of this cost savings and is given by:

$$SC_L (y) = \frac{\left[C(y_L) + C(y_{M-L}) - C(y)\right]}{C(y)}$$

(50)

Overall returns to scale are related to returns to scope and product specific returns to scale through the identity:

$$\psi_m = \frac{\alpha_L \ \psi_L + \left(1-\alpha_l\right) \ \psi_{M-L}}{1 - SC_L} \tag{51}$$

There are two major sources of economies of scope. The most obvious is joint production which results from the presence of public inputs. Public inputs are those which can be used by one production process without reducing the amount available for other processes. An example might be the use of ponds for fish culture and providing watering for grazing animals. While exact public inputs are rare, quasi-public inputs are very common. Quasi-public inputs are those which can be shared by two production processes without complete congestion. The use of the same planting equipment for corn and soybeans is a typical example since the demands do not typically occur at the same time. Many allocated fixed inputs in agriculture may lead to economies of scope. In agribusiness firms, economies of scope may exist in sales and advertising since sales efforts can promote more than one product or line.

Economies of scale and scope are often related to the concept of subadditivity of the cost function. A cost function is said to be strictly subadditive at y if for any and all outputs y^1, y^2, $...y^k$ (where the superscript denotes different vectors and not elements of a vector) with $y^j \neq y$, $j = 1, 2, ..., k$ such that $\sum_{j=1}^{k} y^j = y$ and $C(w,y) < \sum_{j=1}^{k} C(w,y^j)$. This condition is usually the one used to define a natural monopoly. While this seems closely related to the idea of scope economies, Baumol, Panzar and Willig (1982) have shown that multiproduct economies of scale and scope do not necessarily imply strict subadditivity and so a definitive test for natural monopoly remains elusive.

Other Economies of Operation

In addition to economies of scale, size, and scope, other types of economies may give firms advantages in costs or marketing. Pecuniary economies are those economies that result from lower costs of inputs or reduced costs of marketing due to size of operation. Volume discounts on purchased inputs are perhaps the most common form of pecuniary economy for farm firms (Smith et al. 1981). Economies can also result from economies of size or scale in the selling function apart from any production economies. Besides pecuniary economies that can occur without significant market power firms may also obtain pecuniary economies due to market power. These types of economies, while important in some sectors or crops, are beyond the scope of this chapter.

Dynamic Production Processes

The discussion thus far has concentrated on a static analysis of firms facing certain prices. The addition of dynamic production processes or uncertainty further complicates the analysis. Standard analysis of the firm's cost curve typically assumes a long-run cost curve representing payments to all residual claimants. If some inputs are treated as fixed, then short-run returns to size or plant may be determined. Since the dichotomy between fixed and variable inputs is often artificial and since the costs of adjustment of moving from one plant size to others are frequently nontrivial, the choice of appropriate definitions of residual claimants is of more than academic interest (T. A. Miller 1984). Stefanou (1989b), and Fernandez-Cornejo et al. (1992) develop an explicit intertemporal optimization problem following Epstein and Denny (1983) allowing for costs of adjustment and then discuss alternative measures of returns to size. In a static setting, the elasticity of scale along the expansion path is equal to the increase in the cost function elasticity. Using a dynamic adjustment cost model, Stefanou (1989b) obtains the following representation of the scale elasticity

$$
\varepsilon_{LR} = \frac{[wx^* + \dot{K} \, J_K]}{[\lambda^* y(t)]}
$$
$$
= \frac{[wx^* + (I^* - \delta K) \, J_K]}{[\lambda^* y(t)]} \tag{52}
$$
$$
= \frac{LRAVC(t)}{SRMC(t)}
$$

where J is the optimal value function, I is investment, K is capital stock, x is a variable input, w is its price, $y(t)$ is target output level in each period, λ is the multiplier associated with the output constraint in the cost minimization problem , and $*$'s denote optimal levels.

In the static case the reciprocal of the cost elasticity is equal to the scale elasticity (equation 38). This equivalence no longer holds when the production process is dynamic and has adjustment costs. The reciprocal of the long-run cost elasticity is given by

$$\Psi_{LR} = \frac{J}{J_y y} = \frac{LRATC}{LRMC}$$

$$= \frac{wx^* + cK + \dot{K}^* J_K}{y(t)\lambda^* + y(t)\dot{K}^* J_{Ky}} \tag{53}$$

$$= \frac{wx^* + cK + \dfrac{dJ}{dt}}{y(t)\lambda^* + \dfrac{d(J_y)}{dt}}$$

This differs from the scale elasticity by the presence of fixed costs in the numerator and by the instantaneous change in marginal costs in the denominator. If there are no adjustment costs, the cases are the same; but if there are no adjustment costs, then the firm moves instantaneously to the cost minimizing solution and the dynamic nature of this problem no longer exists. The use of adjustment costs as a means of differentiating between a series of static fully flexible adjustment models and dynamic ones that adjust more slowly is but one way to capture the long run dynamics of the firm. Other models and their associated measures of scale and scope economies is a topic for future research.

Firms Facing Uncertainty

If the firm makes its decisions in an uncertain environment and is risk averse, then economies of size and scale may not be obtainable in the usual fashion. Chambers (1983), in analyzing a single output firm facing price uncertainty, shows that scale elasticity is given by

$$\varepsilon = \frac{\Sigma w_i x_i}{(p - \theta y)} \tag{54}$$

where $\theta = E[U'(\pi)e]/EU'(\pi)$ and e is the disturbance term in the price equation. This would seem to imply that information on preferences is needed. However, as pointed out by Flacco and Larson (1992), the information needed can be obtained using duality between the cost and production functions. The result of Flacco and Larson (1992) is a simple extension of the point that in many problems under uncertainty the standard notion of a cost function still exists. In other words, the problem can be separated into a two step problem where the firm minimizes cost in one step and then chooses the optimal output level in another step (Pope 1980). When the cost function is well defined, all the measures based on it are also well defined. In cases where this separation does not hold, such as with production uncertainty, the notion of returns to size and scale are more problematic since a clear notion of the optimal costs for a fixed output level exists since output is not certain. With uncertainty and expected profit maximization all firms will have zero profits in equilibrium, but positive profits are possible with risk averse decision makers. Increasing returns to scale may also be compatible with equilibrium since producers may find an optimum with marginal costs still below average costs due to their risk averse preferences.

Market Implications of Economies of Size and Scale and Contestable Markets

A few comments on the market implications of economies of size and scale seem in order before turning to the empirical measurement of these effects. If economies of size greater than one are clearly evident, then the industry will eventually tend toward monopoly and the need for government regulation. With returns to size less than one but with different firms possessing different costs structures there will be a tendency for higher cost firms to leave the industry. These effects, of course, will lead to changes in the size structure of the industry.

The work by Baumol, Panzar, and Willig (1982) has spawned numerous debates on the contestability of markets with few actors (Shepherd 1984; Teece 1980; Samuelson and Zeckhauser 1988; Panzar 1989). Baumol et al. (1982) contend that it is the potential entrants who can rapidly enter a market who control competition, others argue that the conditions for such ultra free entry are not mirrored by the real world. In particular, others argue that there are sunk costs associated with entry and exit that in effect do create barriers to entry and thus make markets not so contestable. While this debate is relevant in the vertically related agricultural sector, it is of doubtful significance in the primary agricultural sector due to rather insignificant barriers to entry for firms in other agricultural endeavors and even for new entrants. Furthermore, the seeming ability of small firms in the industry to alter output without significant changes in costs mitigates the effects of large and possibly more efficient firms.

Empirical Measurement of Economies of Size and Scale

Various empirical methods and data sources have been used to obtain estimates of the measures discussed above. While different questions require different methods and data, the lack of ideal data sets has often guided empirical research. There are four major types of studies that have appeared in the literature. The first two involve the application of positive methods to cross-section and aggregate time series data, respectively. The third approach analyzes the survivorship of firms over time to identify factors that are correlated with success. Economies of size are but one of the factors considered by such studies. The final method involves the normative or synthetic determination of economies of size by developing models of firms with similar but size varying technologies. The primary methods used in each area are presented with a minimal discussion of results. A review of empirical findings is contained in a companion chapter.

Positive Studies Using Cross-Section Data

Classification Analysis. There are a number of methods that can be used for analyzing cross-section data. The simplest is to simply plot average costs of production of firms producing different levels of output. The shape of the curve gives some information of economies of size. Given typical variation due to other factors it may be more reasonable to divide the sample into groups (percentiles) and determine average costs for groups in various size classes and then perform the analysis. The major difficulty with this approach is that it does not hold input prices fixed. If these are the same across firms, then no bias results; but otherwise, these should be controlled for by econometric estimation. The chapter by Ahearn, Whitaker, and Glaze in this volume is an example of this approach.

Cost Function Estimation. The most common approach in cross-section studies is to estimate a cost function. Early work using ad hoc specifications is represented by the book of Johnston (1960). Nerlove (1963) was the first to posit a dual cost function as an empirical way to measure economies of scale. More recent papers include a study of railroads by Caves, Christensen, and Swanson (1981), a study of Canadian farms by Fleming and Uhm (1982), and a study of crop farms by Cooke and Sundquist (1987). The major advantage here is that the size measures are easily computed from parameters of the cost function and any objective function consistent with cost minimizing behavior (including some cases of uncertainty) could have generated the data. The cost function has some drawbacks, however. Weaver (1984) argued that since most firms are assumed to be profit maximizers, changes in input prices will affect optimal output and returns to size. However, for a given set of input prices such comparative static issues are not relevant and the cost function provides the correct measure. At the industry level, input prices are not exogenous and so in time series studies instrumental variables should be used to correct for simultaneity problems (Berndt and Wood 1975). At the firm level, output may not be independent of the cost function disturbance (Walters 1963), since it may depend, like costs, on general economic conditions, weather, or entrepreneurial ability. Joint estimation of cost and output equations can eliminate these effects (Garcia and Sonka 1982). In cross-section studies, there may not be enough variation in input prices to obtain accurate parameter estimates. Furthermore, the use of firm data may lead to a lack of breadth in the sample so that tails of the distribution are not well represented. Stigler (1958) has criticized the use of cross-section data due to what he calls the "regression fallacy." Individual firms with similar resources and variable costs may exhibit different average total costs due to random influences on output such as weather or labor difficulties. Measures of capacity utilization as explanatory variables may mitigate the second effect. Alternatively, Walters (1960) has argued that the cost function is inherently stochastic and its expected value is the relevant measurement.

Production Function Estimation. An alternative approach is to estimate the firm's production function and then solve for optimal input demands. From these the shape of the average cost curve can be obtained by solving first-order conditions. The estimation of the production function has certain advantages and disadvantages relative to alternative methods. Cross-section data often has little variation in input prices and may lead to rather inaccurate estimates using cost function approaches. This problem is avoided by using output and input data. Furthermore, by making output the dependent variable the regression fallacy problem associated with random output is mitigated to some extent. The approach has several problems, however. These include the problems of simultaneous equation bias arising from optimizing behavior (Hoch 1958; Mundlak 1963; Zellner, Kmenta, and Dreze 1966), multicollinearity among inputs and difficulties in measuring the quantity of capital services. While the simultaneity problem may be overcome by assuming expected profit maximization, quantities of capital are more difficult to represent than capital prices. There are additional problems associated with the aggregation of inputs into groups. Finally, estimation of the production function means that the economies of size measure must be obtained by solving first-order conditions. This may involve the solution on nonlinear equations if realistic forms are used for the production function. An excellent example of this approach is Vlastuin, Lawrence, and Quiggin (1982).

Distance Function Estimation. A number of recent papers have approached the measurement of scale economies by directly estimating the distance function. The emphasis has generally been on efficiency rather than scale economies, but the methods hold potential for the measurement of size and scale. The papers by Grosskopf, Margaritis, and Valdmanis (1990) and Grosskopf and Hayes (forthcoming) are useful in this regard.

Profit Function Estimation. Just as the firm's technology can be obtained from the cost function, it can also be recovered using the profit function (McFadden 1978). While non-homothetic technologies can be modelled using the profit function and returns to size and scale determined (Weaver 1982), there is little point in using estimated parameters to measure returns to size since first-order conditions for profit maximization imply that price equals marginal cost and returns to size is just the ratio of cost to revenue. For a well-defined profit maximization problem to exist the production function must be concave and this implies decreasing returns to scale at the optimum. Thus, increasing returns to size or scale cannot be identified from a well-behaved profit function since the measure is costs divided by revenue and if costs are greater than revenue then the firm will not operate. The advantage of the cost function is that it allows measurement of returns to size when the decision choice may be other than profit maximization.

Frontier Functions. In estimating production, cost and profit functions, researchers typically assume that the error terms are symmetrically distributed with a zero mean. This implies (in the case of production) that some firms' production will lie above the estimated surface while others will lie below it. An alternative formulation (Farrell 1957) postulates that firms have different levels of economic efficiency and that the actual production function should be a frontier function such that no observations lie above it. Such production frontiers can either be deterministic (Aigner and Chu 1968) or stochastic (Aigner, Lovell, and Schmidt 1977). Deterministic frontiers can either be parametric or non-parametric. Parametric deterministic frontiers are typically derived using linear or quadratic programming to minimize the difference between each firm's production and the frontier subject to the constraint of a one-sided error. Stochastic frontiers allow deviations from the frontier to be either from inefficiency (a one-sided error deviation) or random factors (two-way deviations). They are usually estimated using maximum likelihood methods where the error term is composed of two parts. Frontier cost functions can also be estimated (Schmidt and Lovell 1979; Kopp and Diewert 1982) and used to estimate technical and allocative efficiency and also returns to size. If the purpose of a study is to understand structure tendencies in an industry and the way that they will change, the frontier function may provide a better estimate of the shape of the true long-run cost curve. The chief disadvantage of the frontier method is that outliers may significantly influence the shape of the resulting frontier. The literature on frontier functions is rapidly expanding, though most does not deal with the size issues. Two recent surveys are Schmidt and Bauer (1985-1986).

Uncertainty in Positive Studies. If producers must make choices in an uncertain environment, then measurement of size economies should take this uncertainty into account. There are several types of models where uncertainty may significantly affect size. In models where learning by doing (Arrow 1962) occurs, larger firms may be able to obtain better information than smaller firms and apply this information to greater output thus increasing its value. At the same time, both expansions and contractions of output provide information and these effects may tend to be ambiguous (Stefanou 1989a). Another way to understand returns to scale is then to measure the effects of learning by doing.

If firms are risk averse, then they may choose to increase in size to expand their portfolio and spread risk. Wilson and Eidman (1985), for example, have investigated size effects in swine production using stochastic dominance. They found that risk averse producers preferred smaller units while risk loving individuals preferred larger units. Nonlinear programming models could also be developed to analyze the effects of risk on firm size or enterprise choice.

Multiproduct Extensions. Most of the methods discussed above can be easily generalized to handle multiple outputs. Multiproduct cost functions are straightforward to estimate but provide information only on scale economies since no multiproduct analogue for returns to size exists. Recent papers estimating such functions include Cowing and Holtman (1983) on hospitals, Kim (1988) on water utilities, Deller, Chicoine, and Walzer (1988) on rural roads, and Akridge and Hertel (1986) on retail fertilizer plants. The survey paper by Panzar (1989) discusses a number of other applications. As mentioned by Kim (1987a), in order to compute the incremental measures using the translog (or other forms which implicitly imply all outputs are produced) cost function, costs where one output is zero must be approximated. Otherwise, the measures are all straightforward and easy to compute.

Multiproduct production functions are also possible although the lack of data on allocated inputs may hamper the estimation effort (Just, Zilberman, and Hochman 1983, Chambers and Just 1989). The estimation of multiproduct stochastic frontiers is not a well-developed procedure but recent work by Grosskopf (Grosskopf, Margaritis, and Valdmanis 1990; Grosskopf and Hayes forthcoming) is promising.

Causal Factors. Once measures of economies of size are estimated in cross-section studies, additional analysis may be undertaken to explain the results. For example, after estimating technical efficiency ratios for a cross-section of firms, regression analysis could be used to explain these ratios based on geographic location, demographic variables, farm size, or other such factors. Alternatively, factor analysis could be used to identify commonality among firms with similar efficiency or returns to size.

Aggregate Positive Studies

Returns to scale may be estimated from production, cost, or profit functions derived from aggregate time series data. Technological change is typically represented in such models using a time trend variable or some similar technique. This may not represent the discrete and firm varying nature of technical change but is a practical approximation. The returns to scale measure obtained is of course a short-run expression since many factors are typically held fixed in the analysis. Since technologies are firm specific, such studies only express general characteristics of the technology in the industry and say little about tendencies for firm growth or contraction. An example study is that by Ball and Chambers (1982) of the meat processing industry.

Returns to size measured from time series data is not meaningful in most situations since it measures points on various short-run cost curves and may not measure the long-run cost curve of the industry (Stefanou and Madden 1988). This is not a data problem or one that could be solved by better technique. Static models using time series data simply do not measure long-run relationships in most cases. For example, the short-run returns to size reported by Weaver (1984) using a multiproduct profit function may not be from points on a long-run cost curve. Furthermore, the fact that these returns are less than one is not surprising since they must be less than one for the conditions for profit maximization to hold.

Firm Survivorship or Growth Studies

An alternative way to analyze industry structure and tendencies is to track the rates of growth, contraction, and demise of firms. If particular sizes or types of firms stand the test of time and grow and/or prosper, this provides information on the future structure of the industry and the viability of other types of firms. Stigler (1958) originally used the method to examine the size concentration of firms over time, but it can also be used to explain changes in the size distribution over time. The technique is simple, is consistent with "Darwinian" views, and discovers characteristics of success. The major difficulty is that efficiency is given credit for all changes in firm distribution.

Many models attempt to explain the growth of firms over time. Much of this literature argues that the growth rate of firms is determined by random factors independent of firm size (i.e., Gibrat's law). This formulation typically leads to a skewed distribution of firms that is either log-normal or pareto. A number of studies have found this law of proportional growth to hold for large firms (Evans 1987a) while smaller firms have higher and more variable growth rates (Hall 1987). These models are almost devoid of economic content yet seem to fit the data well. Lucas (1978) uses Gibrat's law to determine an equilibrium size distribution of firms in his model of the economy where managerial talent differs across firms. Jovanovic (1982) has developed models of firm behavior that predict faster growth and more failures for smaller firms and slower growth for older firms irrespective of firm size. These models assume that firms learn about their efficiency by operation and the more efficient ones survive over time.

The basic equation used to test Gibrat's law is

$$\ln S_{it} = \alpha + \beta \ln S_{it-1} + e_{it} \qquad (55)$$

where S_{it} is the size of the ith firm in the tth period. Gibrat's law predicts that $\beta = 1$. While Gibrat's law has seemed to hold in many studies, these studies are not without flaws. Hall (1987) discusses problems with estimating size growth relationships. There may be selection attrition since firms may exit between sample periods. Serially uncorrelated measurement error in the size variable used can induce negative spurious correlation in growth rates and heteroskedasticity in growth rates may lead in inconsistent estimates of standard errors of regression coefficients. Methods are available for correcting each of these difficulties (Hall 1987). While only the first problem is probably serious, the others should at least be acknowledged. A paper by Shapiro, Bollman, and Ehrensaft (1987) tests this relationship for Canadian firms and rejects that it holds.

Some writers have postulated that firm size structures can be explained by risk preferences. More risk averse individuals may operate smaller firms while less risk individuals operate larger firms (Kihlstrom and Laffont 1979). While an equilibrium is efficient only if all entrepreneurs are risk neutral, market failure in the allocation of risks may prevent such an occurrence. In this case, a solution where less risk averse individuals operate larger firms may be a second best optimum.

In the efficient market literature, the difference in returns to large and small firms has been postulated to be due to different risk premiums for these firms. Early tests of this theory using the capital asset pricing model found that smaller firms had larger returns even after adjusting for risk premiums (Banz 1981). More recent work using multifactor pricing models found that risk adjustment explained most the return differences between large and small firms (Chan, Chen, and Hsieh 1985).

Another common technique used to analyze firm growth is the finite Markov chain. A finite Markov chain is one in which a population at time t has a distribution S^t over the discrete states, $S_1, S_2, ...S_n$, and in which the probability P_{ij} of moving from state S_i to state S_j depends only on the state S_i and not on prior states. In firm growth analysis, the states are various size categories of firms. The probabilities are all positive and sum to one across states. These probabilities may not be stationary over time but may change due to exogenous factors. Tests for examining stationarity are available (Amemiya 1985). The state of the process at time t+m can be determined by multiplying the state vector at time t by the transition matrix raised to the mth power. Using historical data, the transition matrices can be estimated and then used to project future size distributions. If the transition probabilities are not constant, then alternative methods for estimation must be employed.

Several studies have employed Markov chains to project farm size changes. Edwards, Smith, and Patterson (1985) analyzed changes in farm size using data from the 1974 and 1978 Census of Agriculture. They found that the percentage of mid-size farms decreased over the period. Garcia, Offutt, and Sonka (1987) applied similar techniques to samples of Illinois cash grain farms. They used both output (gross value) and acreage measures of size and found some differences in results. Transition matrices estimated using the output variable showed little change in size while those estimated using acreage showed some growth towards larger and smaller firms. The Office of Technology Assessment (U.S. Congress) used this technique to project U.S. farm sizes over the remainder of the century. They also suggest a decline in mid-sized farms. All of these studies must be viewed with some caution, however, since constant transition probabilities were used. The study by Garcia et al. (1987) tested for stability and found that transitions were stationary over a ten-year period.

As was suggested in the section on positive studies, follow-up analysis explaining the results of these models is also useful. Garcia et al. (1987) used discriminant analysis to explain whether firms grew or shrunk over their sample period. This allows economic reasoning to be applied to a basically noneconomic method of analysis. Logit and probit models could also be used for the same purpose.

Normative Models of Firm Size

Probably the oldest and most commonly used method for examining economies of size is the normative model of the firm where synthetic budgets are constructed using actual firm data or economic engineering estimates. These activities are then combined in an efficient manner to determine least cost production patterns for alternative output levels. The average cost curve is then easily obtained. Linear programming is usually the method used to obtain least cost solutions although engineering estimates are sometimes used. This approach has several advantages over other approaches. Accurate descriptions of firm technology are possible and can be easily combined with data on resources, product prices, and alternative resource uses. Frontier technology can also be represented using this technique. The difficulty is that it requires a myriad of assumptions about the firm's structure and behavior to be operational. Since the firms modelled are only representative of actual firms, it is difficult to verify the results of such studies. The costs to impute to certain fixed resources must also be determined when using this method. Given poorly defined labor markets it is often problematic to assign values to operator labor in such engineering approaches. Land charges are also difficult to assess in some situations. T. A. Miller (1984) provides an excellent discussion of these issues.

The issues related to multiproduct firms are of especial concern in normative studies since many farms produce multiple products. The standard practice is to represent output with gross revenue. Returns to scale uses the marginal costs of production to weight output levels. Since there is no returns to size in the multiproduct case, the use of revenue as an aggregator of output means that the researcher is measuring cost to revenue ratios and this measure is not independent of output prices and is therefore not a measure of technology. Furthermore the use of gross revenue as an aggregator is only consistent if the underlying technology exhibits constant returns to scale. Thus all normative studies will find constant costs unless the researcher has a priori information that can be used to restrict the model solutions (Chambers and Hallam 1992).

In the classic work by Madden (1967) on economies of size in farming, most studies used variants of this technique. The method has also proved popular since that time and was employed in a major study of U.S. field crop farming by Miller, Rodewald, and McElroy (1981).

Summary and Conclusions

This chapter has reviewed definitions and techniques useful in measuring economies of size, scale, and scope. The emphasis was on appropriate definitions and some of the common problems encountered. This chapter does not provide a cookbook for carrying out empirical studies but refers the reader to appropriate literature. The chapter also discusses other ways to analyze the size and structure of firms in an industry that may be independent of size or scale economies. These other factors which influence changes in structure may be as important or more important than economies of size.

Three areas seem to beg for further empirical analysis. There have been few studies of multiproduct agricultural firms using cross-section data and positive methods. The methods are straightforward and a compilation of results similar to those summarized by Madden (1967) for normative studies would be useful.

A need exists for better models and empirical analysis that explains the growth and contraction of individual firms over time. Economic explanations of Gibrat's law as it pertains to agriculture would be useful. There is also a need for models that can explain why firms may decrease in size as economic conditions change.

The estimation of frontier functions and subsequent analysis of the results would provide much better data on the effects of efficiency on the structure of an industry. Frontier functions can also provide a better measure of potential economies of size and scale as opposed to those associated with sample averages.

A few final comments on the measurement of size economies seem in order. Economies of size can be used to explain many phenomena or justify different economic policies. It is extremely important to know the purpose for which results will be used before undertaking studies. For example, if the purpose is to explain the viability of a particular size structure, economies of size are only one of many important factors. Efficiency, access to credit, producer preferences, and pecuniary economies may be more important than technical economies of scale.

There is a need for specific goals in applied structure research. If the goal is to demonstrate the viability of the family farms, then the methods are different than if the goal is to analyze the potential for monopolization of industry. Simple measurement without purpose may be an interesting theoretical exercise but analysis to answer specific questions seems much more useful.

Selected Bibliography on Size, Scale, and Efficiency

Acs, Zoltan J., and David B. Audretsch. 1987. "Innovation, Market Structure, and Firm Size." *Review of Economics and Statistics* 69: 567-74.

_____ . 1989. "Births and Firm Size." *Southern Economic Journal* 56: 467-75.

Acs, Zoltan J., and Steven C. Isberg. 1991. "Innovation, Firm Size and Corporate Finance: An Initial Inquiry." *Economics Letters* 35: 323-26.

Aigner, D. J., and S. F. Chu. 1968. "On Estimating the Industry Production Function." *American Economic Review* 58: 826-839.

Aigner, D., C. A. K. Lovell, and P. Schmidt. 1977. "Formulation and Estimation of Stochastic Frontier Production Function Models." *Journal of Econometrics* 6: 21-37.

Akridge, J. T., and T. Hertel. 1986. "Multiproduct Cost Relationship for Retail Fertilizer Markets." *American Journal of Agricultural Economics* 68: 928-938.

Allen, Bruce T. 1983. "Concentration, Scale Economies, and the Size Distribution of Plants." *Quarterly Review of Economics and Business* 23: 6-27.

Allen, Ralph C., and Jack H. Stone. 1988. "Managerial Productivity and Firm Size: A Comment on Oi's Managerial Time Allocation Model." *Economic Inquiry* 26: 171-73.

Aly, H. Y., K. Belbase, R. Grabowski, and S. Kraft. 1987. "The Technical Efficiency of Illinois Grain Farms: An Application of a Ray-Homothetic Production Function." *Southern Journal of Agricultural Economics* 19: 69-78.

Amato, Louis, and Ronald P. Wilder. 1985. "The Effects of Firm Size on Profit Rates in U.S. Manufacturing." *Southern Economic Journal* 52: 181-90.

Amel, Dean, and Luke Froeb. 1991. "Do Firms Differ Much?" *Journal of Industrial Economics* 39: 323-31.

Amemiya, T. 1985. *Advanced Econometrics*. Cambridge: Cambridge University Press.

Anderson, F. J. 1991. "Trade, Firm Size, and Product Variety under Monopolistic Competition." *Canadian Journal of Economics* 24: 12-20.

Anderson, J. R., and R. A. Powell. 1973. "Economies of Size in Australian Farming." *Australian Journal of Agricultural Economics* 17(1): 1-16.

Anderson, J. R., J. L. Dillon, and B. Hardaker. 1976. *Agricultural Decision Analysis*. Ames, IA: Iowa State University Press.

Antle, J. 1984. "The Structure of U.S. Agricultural Technology 1910-78." *American Journal of Agricultural Economics* 66: 414-421.

Aron, Debra J. 1988. "Ability, Moral Hazard, Firm Size, and Diversification." *Rand Journal of Economics* 19: 72-87.

Arrow, K. J. 1962. "The Economic Implications of Learning By Doing." *Review of Economic Studies* 29: 155-173.

Atkinson, S. F., and R. Halvorsen. 1984. "Parametric Efficiency Tests, Economies of Scale, and Input Demand in U.S. Electric Power Generation." *International Economic Review* 25: 647-662.

Audretsch, David B., and Zoltan J. Acs. 1991. "Innovation and Size at the Firm Level." *Southern Economic Journal* 57: 739-44.

Bagi, F. S. 1982. "Relationship between Farm Size and Technical Efficiency in West Tennessee Agriculture." *Southern Journal of Agricultural Economics* 14: 139-144.

Bagi, F. S., and C. J. Huang. 1983. "Estimating Production Technical Efficiency for Individual Farms in Tennessee." *Canadian Journal of Agricultural Economics* 31: 249-256.

Bailey, Elizabeth E. 1986. "Price and Productivity Change Following Deregulation: The U.S. Experience." *Economic Journal* 96: 1-17.

Bailey, E. E., and A. F. Friedlaender. 1982. "Market Structure and Multiproduct Industries." *Journal of Economic Literature* 20: 1024-1048.

Ball, V. Eldon. 1985. "Output, Input and Productivity Measurement in U.S. Agriculture, 1948-79." *American Journal of Agricultural Economics* 67: 475-86.

Ball, V. E., and R. G. Chambers. 1982. "An Economic Analysis of Technology in the Meat Products Industry." *American Journal of Agricultural Economics* 64: 699-709.

Banz, R. 1981. "The Relationship Between Return and Market Value of Common Stocks." *Journal of Financial Economics* 9: 3-18.

Bardhan, Pranab K. 1973. "Size, Productivity, and Returns to Scale: An Analysis of Farm-Level Data in Indian Agriculture." *Journal of Political Economy* 81: 1370-86.

Barkley, A. P. 1990. "The Determinants of the Migration of Labor Out of Agriculture in the United States, 1940-85." *American Journal of Agricultural Economics* 72: 567-73.

Barth, James R., Joseph J. Cordes, and Sheldon E. Haber. 1987. "Employee Characteristics and Firm Size: Are There Systematic Empirical Relationships?" *Applied Economics* 19: 555-67.

Bateson, Jeff, and Peter L. Swan. 1989. "Economies of Scale and Utilization: An Analysis of the Multi-plant Generation Costs of the Electricity Commission of New South Wales." *Economic Record* 65: 329-44.

Batra, R. N., and A. Ullah. 1974. "Competitive Firm and the Theory of Input Demand Under Uncertainty." *Journal of Political Economy* 82: 537-548.

Batte, M. T. 1980. "An Analysis of Scale Economies on Commercial Illinois Cash Grain Farms." Unpublished Ph.D. thesis, University of Illinois.

Battese, G. E., and T. J. Coelli. 1988. "Prediction of Firm-Level Technical Efficiencies with a Generalized Frontier Production Function and Panel Data." *Journal of Econometrics* 38: 387-398.

Bauer, P. W. 1990. "Recent Developments in the Econometric Estimation of Frontiers." *Journal of Econometrics* 46: 39-56.

Baumol, W. J. 1976. "Scale Economies, Average Cost, and the Profitability of Marginal Cost Pricing," in R. Grieson, ed., *Essays in Urban Economics and Public Finance.* Lexington.

_____ . 1977. "On the Proper Test for Natural Monopoly in a Multi-Product Industry." *American Economic Review* 68: 809-822.

Baumol, W. J., and Dietrich Fisher. 1978. "Cost Minimizing Number of Firms and Determination of Industry." *Quarterly Journal of Economics* 92: 439-468.

Baumol, W. J., J. C. Panzer, and R. D. Willig. 1982. *Contestable Markets and the Theory of Industry Structure.* New York: Harcourt Brace Jovanovich Inc.

Berger, Allen N., Gerald A. Hanweck, and David B. Humphrey. 1987. "Competitive Viability in Banking: Scale, Scope, and Product Mix Economies." *Journal of Monetary Economics* 20: 501-20.

Berndt, E. R., and M. A. Fuss. 1986. "Productivity Measurement with Adjustments for Variations in Capacity Utilization and Other Forms of Temporary Equilibrium." *Journal of Econometrics* 33: 7-29.

Berndt, E. R., and D. O. Wood. 1975. "Technology, Prices, and the Derived Demand for Energy." *Review of Economic Studies*: 259-268.

Berndt, E. R., and M. S. Khaled. 1979. "Parametric Productivity Measurement and Choice Among Flexible Functional Farms." *Journal of Political Economy* 87: 1220-1245.

Binswanger, H. P. 1974. "The Measurement of Technical Change Biases with Many Factors of Production." *American Economic Review* 64: 964-76.

Black, G. 1955. "Synthetic Method of Cost Analysis in Agricultural Marketing Firms." *Journal of Farm Economics* 37: 270-279.

Bohm, Voker. 1988. "Returns to Size vs. Returns to Scale: The Core with Production Revisited." *Journal of Economic Theory* 46: 215-19.

Boskin, M. J. 1988. "Tax Policy and Economic Growth: Lessons from the 1980's." *Journal of Economic Perspectives* 2: 71-97.

Bravo-Ureta, B. E. 1986. "Technical Efficiency Measures for Dairy Farms Based on a Probabilistic Frontier Function Model." *Canadian Journal of Agricultural Economics* 34: 399-415.

Bravo-Ureta, B. E., and L. Rieger. 1991. "Dairy Farm Efficiency Measurement Using Stochastic Frontiers and Neoclassical Duality." *American Journal of Agricultural Economics* 73: 421-28.

Brock, William A., and J. P. G. Magill. 1979. "Dynamics Under Uncertainty." *Econometrica* 47: 843-868.

Brock, William A., and David S. Evans. 1986. *The Economics of Small Businesses: Their Role and Regulation in the U.S. Economy.* New York: Holmes and Meier.

Broeck, J. van den, F. R. Forsund, L. Hjalmarsson, and W. Meeusen. 1980. "On the Estimation of Deterministic and Stochastic Frontier Production Functions." *Journal of Econometrics* 13: 117-138.

Brown, Charles, and James Medoff. 1989. "The Employer Size-Wage Effect." *Journal of Political Economy* 97: 1027-59.

Brown, R. S., D. W. Caves, and L. R. Christensen. 1979. "Modelling the Structure of Cost and Production for Multiproduct Farms." *Southern Economic Journal* 46: 256-273.

Brown, R., and L. Christensen. 1981. "Estimating Elasticities of Substitution in a Model of Partial Static Equilibrium: An Application to U.S. Agriculture, 1947 to 1974," in E. Berndt and B. Field, eds., *Modeling and Measuring Natural Resource Substitution.* Cambridge, MA: MIT Press.

Brueckner, J. K., and N. Raymon. 1983. "Optimal Production with Learning By Doing." *Journal of Economic Dynamics and Control* 6: 127-135.

Bruning, Edward R., and Michael Y. Hu. 1988. "Profitability, Firm Size, Efficiency and Flexibility in the U.S. Domestic Airline Industry." *International Journal of Transport Economics* 15: 313-27.

Buller, Orlan. 1971. "Analysis of Growth of Agricultural Firms." *Canadian Journal of Agricultural Economics* 19: 25-35.

Burdett, Kenneth, and Randall Wright. 1989. "Optimal Firm Size, Taxes, and Unemployment." *Journal of Public Economics* 39: 275-87.

Bureau of the Census. *Census of Agriculture*, Volume II, General Reports, Washington, DC, Department of Commerce, various issues.

Burgess, D. 1975. "Duality Theory and Pitfalls in the Specification of Technologies." *Journal of Econometrics* 3: 105-121.

Burke, Thomas P., and John D. Morton. 1990. "How Firm Size and Industry Affect Employee Benefits." *Monthly Labor Review* 113: 35-43.

Burtless, Gary, ed. 1990. "A Future of Lousy Jobs? The Changing Structure of U.S. Wages." Washington, DC: Brookings Institute, xiv, 242.

Byrnes, P., R. Färe, S. Grosskopf, and S. Kraft. 1987. "Technical Efficiency and Size: The Case of Illinois Grain Farms." *European Review of Agricultural Economics* 14-4: 367-382.

Callan, Scott J. 1988. "Productivity, Scale Economies and Technical Change: Reconsidered." *Southern Economic Journal* 54: 715-24.

Camacho, A. 1991. "Adaptation Costs, Coordination Costs and Optimal Firm Size." *Journal of Economic Behavior and Organization* 15: 137-49.

Capalbo, S. M. 1988. "Measuring the Components of Aggregate Productivity Growth in U.S. Agriculture." *Western Journal of Agricultural Economics* 13: 53-62.

Capalbo, Susan M., and Michael G. S. Denny. 1986. "Testing Long-Run Productivity Models for the Canadian and U.S. Agricultural Sectors." *American Journal of Agricultural Economics* 68: 615-25.

Carlson, S. 1939. *A Study in the Pure Theory of Production.* London: P. S. King and Sons.

Carman, H. 1972. "Changing Federal Income Tax Rates and Optimum Farm Size." *American Journal of Agricultural Economics* 54: 490-491.

Carter, H. O., and G. W. Dean. 1961. "Cost-Size Relationships for Cash-Crop Farms in a Highly Commercialized Agriculture." *Journal of Farm Economics* 43: 264-277.

Carter, H. O., W. E. Johnston, and C. F. Nuckton. 1980. *Farm-Size Relationships, With an Emphasis on California.* Giannini Foundation Project Report. Davis, CA: California Agricultural Experiment Station, University of California.

Castle, E. N. 1989. "Is Farming a Constant Cost Industry?" *American Journal of Agricultural Economics* 71: 574-582.

Caves, D. W., L. R. Christensen, and J. A. Swanson. 1981. "Productivity Growth, Scale Economies, and Capacity Utilization in U.S. Railroads, 1955-75." *American Economic Review* 71: 994-1002.

Caves, D. W., L. R. Christensen, and M. W. Tretheway. 1982. "Flexible Cost Functions for Multiproduct Firms." *Review of Economics and Statistics* 62: 477-481.

Cebenoyan, A. Sinan. 1988. "Multiproduct Cost Functions and Scale Economies in Banking." *Financial Review* 23: 499-512.

Chakravarty, Satya R. 1989. "The Optimum Size Distribution of Firms." *Mathematical Social Sciences* 18: 99-105.

Chambers, R. G. 1982. "Duality, the Output Effect, and Applied Comparative Statistics." *American Journal of Agricultural Economics* 64: 152-156.

_____ . 1983. "Scale and Productivity Measurement Under Risk." *American Economic Review* 73: 802-805.

_____ . 1984. "Some Conceptual Issues in Measuring Economies of Size." *Economies of Size Studies.* Ames, IA: Center for Agricultural and Rural Development.

_____ . 1988. *Applied Production Analysis: A Dual Approach.* Cambridge: Cambridge University Press.

Chambers, Robert G., and Utpal Vasavada. 1983. "Testing Asset Fixity for U.S. Agriculture." *American Journal of Agricultural Economics* 65: 761-69.

Chambers, R. G., and R. E. Just. 1989. "Estimating Multioutput Technologies." *American Journal of Agricultural Economics* 71: 980-95.

Chambers, R. G., and A. Hallam. 1992. "Empirical Studies of Size in Agriculture: Is Farming Really a Constant Cost Industry." Iowa State University Working Paper, Ames, IA.

Chan, K. C., N. Chen, and D. A. Hsieh. 1985. "An Exploratory Investigation of the Firm Size Effect." *Journal of Financial Economics* 14: 451-471.

Chavas, J. P., and T. L. Cox. 1988. "A Non-parametric Analysis of Agricultural Technology." *American Journal of Agricultural Economics* 70: 303-310.

Chelius, James R., and Robert S. Smith. 1987. "Firm Size and Regulatory Compliance Costs: The Case of Workers' Compensation Insurance." *Journal of Policy Analysis and Management* 6: 193-206.

Chen, Nai-Fu. 1988. "Equilibrium Asset Pricing Models and the Firm Size Effect," in Dimson, Elroy, eds., *Stock Market Anomalies*. Pp. 179-96. Cambridge, New York and Melbourne: Cambridge University Press.

Chester, Andrew. 1979. "Testing the Law of Proportionate Effect." *Journal of Industrial Economics* 27: 403-411.

Chien, Ying I., and Garnett L. Bradford. 1976. "A Sequential Model of the Farm Firm Growth Process." *American Journal of Agricultural Economics* 58: 456-65.

Cho, Dongsae. 1988. "Some Evidence of Scale Economies in Workers' Compensation Insurance." *Journal of Risk and Insurance* 55: 324-30.

Christensen, L. R., D. W. Jorgenson, and L. J. Lau. 1973. "Transcendental Logarithmic Production Functions." *Review of Economics and Statistics* 55: 28-45.

Christensen, L. R., and W. H. Green. 1976. "Economies of Scale in U.S. Electric Power Generation." *Journal of Political Economy* 84: 665-676.

Clark, Gregory. 1991. "Labor Productivity and Farm Size in English Agriculture Before Mechanization: A Note." *Explorations in Economic History* 28: 248-57.

Clark, Jeffery A. 1984. "Estimation of Economies of Scale in Banking Using a Generalized Functional Form." *Journal of Money, Credit and Banking* 16: 53-68.

Coate, Malcolm B. 1991. "The Effect of Dynamic Competition on Price-Cost Margins." *Applied Economics* 23: 1065-76.

Cochrane, W. W. 1979. *Development of American Agriculture: An Historical Analysis*. Minneapolis: University of Minnesota Press.

Cohen, Wesley M., Richard C. Levin, and David C. Mowery. 1987. "Firm Size and R&D Intensity: A Re-examination." *Journal of Industrial Economics* 35: 543-65.

Cohn, Elchanan, Sherrie L. W. Rhine, and Maria C. Santos. 1989. "Institutions of Higher Education as Multi-product Firms: Economies of Scale and Scope." *Review of Economics and Statistics* 71: 284-90.

Cooke, S. C., and W. B. Sundquist. 1987. "Cost Structures, Productivities, and the Distribution of Technology Benefits Among Producers for Major U.S. Field Crops." *Evaluating Agricultural Research and Productivity*. Proceedings of a Symposium, January 29-30, Atlanta, Georgia; Miscellaneous Publication 52-1987, Minnesota Agricultural Experiment Station, University of Minnesota.

_____ . 1989. "Cost Efficiency in U.S. Corn Production." *American Journal of Agricultural Economics* 71: 1003-1010.

Cowing, Thomas D., and Alphonse G. Holtman. 1983. "Multiproduct Short-Run Hospital Cost Functions: Empirical Evidence and Policy Implications From Cross-Section Data." *Southern Economic Journal* 49: 637-653.

Cowing, T. G., D. Reifschneider, and R. Stevenson. "A Comparison of Alternative Frontier-Cost Function Specifications" in Ali Dogramaci, ed., *Developments in Econometric Analyses of Productivity: Measurement and Modeling Issues*. Pp. 63-92. Boston: Kluwer-Nijhoff Publishing.

Cross, J. G. 1973. "A Stochastic Learning Model of Economic Behavior." *Quarterly Journal of Economics* 87: 239-266.

Daly, Rex F., J. A. Dempsey, and C. W. Cobb. 1972. "Farm Numbers and Size in the Future." in A. Gordon Ball and Earl O. Heady, eds., *Size, Structure, and Future of Farms*. Pp. 314-32. Ames, IA: Iowa State University Press.

Dawson, P. J., J. Lingard, and C. Woodford. 1991. "A Generalized Measure of Farm Specific Technical Efficiency." *American Journal of Agricultural Economics* 73: 1098-1104.

Day, L. M. 1981. "Research and the Family Farm: Implications for Agricultural Economics Research." *American Journal of Agricultural Economics* 63: 997-1004.

De Alessi, L. 1967. "The Short Run Revisited." *American Economic Review* 57: 450-461.

Dean, G. W., and H. O. Carter. 1962. "Some Effects of Income Taxes in Large Scale Agriculture." *Journal of Agricultural Economics* 44: 754-768.

de Janvry, A. 1972. "The Class of Generalized Power Production Functions." *American Journal of Agricultural Economics* 54: 234-237.

Deller, S. C., D. L. Chicoine, and N. Walzer. 1988. "Economies of Size and Scope in Rural Low-Volume Roads." *The Review of Economics and Statistics* 70: 459-65.

Deller, S. C., and C. Nelson. 1991. "Measuring the Economic Efficiency of Producing Rural Road Services." *American Journal of Agricultural Economics* 73: 194-201.

Deolalikar, Anil B. 1981. "The Inverse Relationship Between Productivity and Farm Size: A Test Using Regional Data From India." *American Journal of Agricultural Economics* 63: 275-79.

Dhrymes, P. J., and M. Kurz. 1964. "Technology and Scale in Electricity Generation." *Econometrica* 32: 287-315.

Diamond, P., D. McFadden, and M. Rodriquez. "Measurement of the Elasticity of Factor Substitution and Bias of Technical Change," in Melvyn Fuss and Daniel McFadden, eds., *Production Economics: A Dual Approach to Theory and Applications*, Vol. 2. Pp. 125-141. North-Holland Publishing Company.

Dixon, Bruce L., Marvin T. Batte, and Steven T. Sonka. 1984. "Random Coefficients Estimation of Average Total Product Costs for Multiproduct Firms." *Journal of Business and Economics Statistics* 2: 360-66.

Doering, O. C. 1978. "Appropriate Technology of U.S. Agriculture: Are Small Farms the Coming Thing? -- Introductory Comments." *American Journal of Agricultural Economics* 60: 293-294.

Duchatelet, Martine. 1982. "A Note on Increasing Returns to Scale and Learning by Doing." *Journal of Economic Theory* 27: 210-218.

Eakin, B. Kelly, Daniel P. McMillen, and Mark J. Buono. 1990. "Constructing Confidence Intervals Using the Bootstrap: An Application to a Multi-Product Cost Function." *Review of Economics and Statistics* 72: 339-44.

Economies of Size Studies. 1984. Ames, IA: Center for Agricultural and Rural Development.

Edwards, Clark. 1985. "Productivity and Structure in U.S. Agriculture." *Agricultural Economics Research* 37: 1-11.

Edwards, C., M. G. Smith, and R. N. Patterson. 1985. "The Changing Distribution of Farms by Size: A Markov Analysis." *Agricultural Economics Research* 37: 1-16.

Eginton, Charles W. 1980. "Impacts of Federal Tax Policies on Potential Growth in Size of Typical Farms." *American Journal of Agricultural Economics* 62: 929-39.

Ehrensaft, P., P. LaRamee, R. D. Bollman, and F. H. Buttell. 1984. "The Microdynamics of Farm Structural Change in North America: The Canadian Experience and Canada-U.S.A. Comparisons." *American Journal of Agricultural Economics* 66: 823-828.

Epstein, L. G., and M. S. G. Denny. 1983. "The Multivariate Flexible Accelerator Model: Its Empirical Restrictions and an Application to U.S. Manufacturing." *Econometrica* 51: 647-674.

Epstein, L. G., and A. J. Yatchew. 1985. "The Empirical Determination of Technology and Expectations: A Simplified Procedure." *Journal of Econometrics* 27: 235-258.

Ericson, R., and A. Pakes. 1989. "An Alternative Theory of Industry and Firm Dynamics." Columbia University, Department of Economics Working Paper #445.

Ethier, W. J. 1982. "National and International Returns to Scale in the Modern Theory of International Trade." *American Economic Review* 72: 389-455.

Ethridge, Don E., Sujit K. Roy, and David W. Myers. 1985. "A Markov Chain Analysis of Structural Changes in the Texas High Plains Cotton Ginning Industry." *Southern Journal of Agricultural Economics* 17: 11-20.

Evans, David S. 1985. *Entrepreneurial Choice and Success.* Washington, DC: U.S. Small Business Administration.

_____. 1987a. "The Relationship Between Firm Growth, Size, and Age: Estimates for 100 Manufacturing Industries." *Journal of Industrial Economics* 35: 567-81.

_____. 1987b. "Tests of Alternative Theories of Firm Growth." *Journal of Political Economy* 95: 657-674.

_____. 1987c. "Empirical Analysis of the Size Distribution of Farms: Discussion." *American Journal of Agricultural Economics* 69: 484-485.

Evans, David S., and Linda S. Leighton. 1989. "Why Do Smaller Firms Pay Less?" *Journal of Human Resources* 24: 299-318.

Evans, D. S., and J. J. Heckman. 1984. "A Test for Subadditivity of the Cost Function With an Application to the Bell System." *American Economic Review* 74: 615-623.

Faas, Ronald C., David Holland, and Douglas Young. 1981. "Variations in Farm Size, Irrigation Technology and After-Tax Income: Implications for Local Economic Development." *Land Economics* 57: 213-20.

Färe, Rolf. 1975. "A Note on Ray-Homogeneous and Ray-Homothetic Production Functions." *Swedish Journal of Economics* 77: 366-72.

_____. 1988a. "Returns to Scale and Size in Agricultural Economics: Comment." *Western Journal of Agricultural Economics* 13: 149-50.

_____. 1988b. *Fundamentals of Production Theory.* New York: Springer-Verlag.

Färe, R., L. Jansson, and C. A. Knox Lovell. 1985. "Modelling Scale Economies with Ray Homothetic Production Functions." *Review of Economics and Statistics* 67: 624-629.

Färe, R., S. Grosskopf, and C. A. Knox Lovell. 1986. "Scale Economies and Duality." *Zeitschrift fur Nationalolonomie* 46: 175-182.

Farrell, M. J. 1957. "The Measurement of Productive Efficiency." *Journal of Royal Statistics Society* 120: 253-281.

Farris, J. Edwin, and David L. Armstrong. 1965. "Economies in the Acquisition of Inputs – A Pilot Study." *Canadian Journal of Agricultural Economics* 13: 70-81.

Feder, Gershon. 1980. "Farm Size, Risk Aversion and the Adoption of New Technology Under Uncertainty." *Oxford Economic Papers, N. S.* 32: 263-83.

_____. 1982. "Adoption of Interrelated Agricultural Innovations: Complementarity and the Impacts of Risk Scale, and Credit." *American Journal of Agricultural Economics* 64: 94-101.

_____. 1985. "The Relationship Between Farm Size and Farm Productivity: The Role of Family Labor, Supervision and Credit Constraints." *Journal of Development Economics* 18: 297-313.

Feder, Gershon, and Gerald T. O'Mara. 1981. "Farm Size and the Diffusion of Green Revolution Technology." *Economic Development and Cultural Change* 30: 59-76.

Ferguson, C. E. 1971. *The Neoclassical Theory of Production and Distribution.* Cambridge: Cambridge University Press.

Fernandez-Cornejo, J., C. M. Gempesaw II, J. G. Elterich, and S. E. Stefanou. 1992. "Dynamic Measures of Scope and Scale Economies: An Application to German Agriculture." *American Journal of Agricultural Economics* 74: 329-42.

Flacco P., and D. Larson. 1992. "Nonparametric Measures of Scale and Technical Change for Competitive Firms Under Uncertainty." *American Journal of Agricultural Economics* 74: 173-76.

Fleming, Marion S., and I.H. Uhm. 1982. "Economies of Size in Grain Farming in Saskatchewan and the Potential Impact of Rail Rationalization Proposals." *Canadian Journal of Agricultural Economics* 1: 1-20.

Flinn, William L., and Frederick H. Buttel. 1980. "Sociological Aspects of Farm Size: Ideological and Social Consequences of Scale in Agriculture." *American Journal of Agricultural Economics* 62: 946-53.

Forsund, F. R., and L. Hjalmarsson. 1979. "Frontier Production Functions and Technical Progress: A Study of General Milk Processing in Swedish Dairy Plants." *Econometrica* 47: 883-899.

Forsund, F. R., C. A. K. Lovell, and P. Schmidt. 1980. "A Survey of Frontier Productions and of Their Relationship to Efficiency Measurement." *Journal of Econometrics* 13: 5-25.

Frank, Murray Z. 1988. "An Intertemporal Model of Industrial Exit." *Quarterly Journal of Economics* 103: 333-44.

Frisch, R. 1965. *Theory of Production*. Dordrecht: D. Reidel Publishing.

Fuller, J. David, and Yigal Gerchak. 1989. "Risk Aversion and Plant Size: Theory and Application to Tar-Sands Oil Plants." *Canadian Journal of Economics* 22: 164-73.

Fuss, Melvyn A. 1977. "The Structure of Technology Over Time: A Model for Testing the "Putty-Clay" Hypothesis." *Econometrica* 45: 1797-1821.

Fuss, M., and D. McFadden, eds. 1978. *Production Economics: A Dual Approach to Theory and Applications*. Amsterdam: North-Holland Publishing Company.

Garcia, P., S. T. Sonka, and M. Yoo. 1982. "Farm Size, Tenure, and Economic Efficiency on a Sample of Illinois Grain Farms." *American Journal of Agricultural Economics* 64: 119-123.

Garcia, P., and S. T. Sonka. 1984. "Methodological Issues in Assessing Economies of Size: Selected Positive Analytical Approaches," in *Economies of Size Studies*. Ames, IA: Center for Agriculture and Rural Development.

Garcia, P., S. E. Offutt, and S. T. Sonka. 1987. "Size Distribution and Growth in a Sample of Illinois Cash Grain Farms." *American Journal of Agricultural Economics* 69: 471-476.

Gardner, B. D., and R. D. Pope. 1978. "How Is Scale and Structure Determined in Agriculture?" *American Journal of Agricultural Economics* 60: 295-302.

Ghose, Ajit Kumar. 1979. "Farm Size and Land Productivity in Indian Agriculture: A Reappraisal." *Journal of Development Studies* 16: 27-49.

Gilligan, Thomas, Michael Smirlock, and William Marshall. 1984. "Scale and Scope Economies in the Multi-Product Banking Firm." *Journal of Monetary Economics* 13: 393-405.

Gold, B. 1982. "Changing Perspectives on Size, Scale, and Returns: An Interpretive Survey." *Journal of Economic Literature* 19: 5-33.

Goldberg, L. G., G. A. Hanweck, M. Keenan, and A. Young. 1991. "Economies of Scale and Scope in the Securities Industry." *Journal of Banking and Finance* 15: 91-107.

Gomulka, Stanislaw. 1990. *The Theory of Technological Change and Economic Growth*. P. 262. London and New York: Routledge.

Gorman, I. E. 1985. "Conditions for Economies of Scope in the Presence of Fixed Costs." *Rand Journal of Economics* 16: 431-436.

Gorman, W. M. 1953. "Community Preference Fields." *Econometrica* 21: 63-80.

Grabowski, R., and O. Sanchez. 1986. "Return to Scale in Agriculture: An Empirical Investigation of Japanese Experience." *European Review of Agricultural Economics* 13: 189-198.

Grabowski, R, and K. Belbase. 1986. "An Analysis of Optimal Scale and Factor Intensity in Nepalese Agriculture: An Application of a Ray-Homothetic Production Function." *Applied Economics* 18: 1051-63.

Grabowski, R., H. Y. Aly, S. Kraft, and C. Pasurka. 1990. "A Ray-Homothetic Production Frontier and Efficiency: Grain Farms in Southern Illinois." *European Review of Agricultural Economics* 17(4): 435-48.

Greene, W. H. 1980. "On Estimation of a Flexible Frontier Production Model." *Journal of Econometrics* 13: 101-15.

_____. 1982. "Maximum Likelihood Estimation of Stochastic Frontier Production Models." *Journal of Econometrics* 18: 285-89.

Greer, Douglas F., and Stephen A. Rhoades. 1976. "Concentration and Productivity Changes in the Long and Short Run." *Southern Economic Journal* 43: 1031-44.

Griffiths, W. E., and J. R. Anderson. 1982. "Using Time-Series and Cross-Section Data to Estimate a Production Function with Positive and Negative Marginal Risks." *Journal of the American Statistical Association* 77: 529-536.

Griliches, Z. 1957. "Specification Bias in Estimates of Production Functions." *Journal of Farm Economics* 39: 8-20.

_____. 1963. "The Sources of Measured Productivity Growth: United States Agriculture, 1940-60." *Journal of Political Economy* 71: 331-46.

Grosskopf, S. 1986. "The Role of the Reference Technology in Measuring Productive Efficiency." *Economic Journal* 96: 499-513.

Grosskopf, S., and V. Valdmanis. 1987. "Measuring Hospital Performance: A Nonparametric Approach." *Journal of Health Economics* 6: 89-107.

Grosskopf, S., D. Margaritis, and V. Valdmanis. 1990. "Nurse Productivity and Wages." *New Zealand Economic Papers* 24: 73-86.

Grosskopf, S., and K. Hayes. Forthcoming. "Local Public Sector Bureaucrats and their Input Choices." *Journal of Urban Economics*.

Guesnerie, R., and C. Oddou. 1988. "Increasing Returns to Size and Their Limits." *Scandinavian Journal of Economics* 90(3): 259-73.

Gunter, L., and K. T. McNamara. 1990. "The Impact of Local Labor Market Conditions on the Off-Farm Earnings of Farm Operators." *Southern Journal of Agricultural Economics* 22: 155-65.

Gupta, V. K. 1983. "Labor Productivity, Establishment Size, and Scale Economies, The Extent of Economies of Scale: The Effects of Firm Size on Labor Productivity and Wage Rates." *Southern Economic Journal* 49: 853-859.

Hall, B. F., and E. P. Leveen. 1978. "Farm Size and Economic Efficiency: The Case of California." *American Journal of Agricultural Economics* 60: 589-600.

Hall, B. H. 1987. "The Relationship Between Firm Size and Firm Growth in the U.S. Manufacturing Sector." *Journal of Industrial Economics* 35: 583-606.

Hall, R. E. 1973. "The Specification of Technology with Several Kinds of Output." *Journal of Political Economy* 81: 878-92.

Hallam, A. 1988. "Economies of Size: Theory, Measurement and Related Issues," in L. Robison, ed., *Determinants of Farm Size and Structure*.

_____. 1989. *Determinants of Farm Size and Structure*. Proceedings of the program sponsored by the NC-181 Committee on Determinants of Farm Size and Structure in North Central Areas of the United States, held January 7, 8, 9, and 10, 1989, in Tucson, Arizona. Ames, IA: Department of Economics, Iowa State University.

_____. 1990. *Determinants of Farm Size and Structure*. Proceedings of the program sponsored by the NC-181 Committee on Determinants of Farm Size and Structure in North Central Areas of the United States, held January 6, 8, and 9, 1990, in Albuquerque, New Mexico. Ames, IA: Department of Economics, Iowa State University.

_____ . 1991. "Economies of Size and Scale in Agriculture: An Interpretive Review of Empirical Measurement." *Review of Agricultural Economics* 13: 155-172.

Hallberg, M. C. 1969. "Projecting the Size Distribution of Agriculture Firms - An Application of a Markov Process with Non-Stationary Transition Probabilities." *American Journal of Agricultural Economics* 51: 289-302.

Halter, A. N., H. O. Carter, and J. G. Hocking. 1957. "A Note of the Transcendental Production Function." *Journal of Farm Economics* 39: 974-996.

Hammond, Christopher J. 1986. "Estimating the Statistical Cost Curve: An Application of the Stochastic Frontier Technique." *Applied Economics* 18: 971-84.

Hanoch, G. 1975. "The Elasticity of Scale and the Shape of Average Costs." *American Economic Review* 65: 429-97.

Hardwick, Philip. 1990. "Multi-product Cost Attributes: A Study of U.K. Building Societies." *Oxford Economic Papers, N. S.* 42: 446-61.

Hartley, Peter R., and Albert S. Kyle. 1989. "Equilibrium Investment in an Industry with Moderate Investment Economies of Scale." *Economic Journal* 99: 392-407.

Heady, E. O. 1946. "Production Functions From a Random Sample of Farms." *Journal of Farm Economics* 28: 989-1004.

_____ . 1952. *Economics of Agricultural Production and Resource Use.* Englewood Cliffs, NJ: Prentice-Hall.

Heady, E. O., G. L. Johnson, and L. S. Hardin. 1956. *Resource Productivity, Returns to Scale, and Farm Size.* Ames, IA: Iowa State College Press.

Heady, E. O., and R. D. Krenz. 1962. "Farm Size and Cost Relationships in Relation to Recent Machine Technology." Iowa Agr. and Home Econ. Expt. Sta. Res. Bul. 504.

Heady, E. O., and S. T. Sonka. 1974. "Farm Size, Rural Community Income, and Consumer Welfare." *American Journal of Agricultural Economics* 56: 534-542.

Hildebrand, J. 1960. "Some Difficulties With Empirical Results From Whole-Farm Cobb-Douglas-Type Production Functions." *Journal of Farm Economics* 42: 897-904.

Hildreth, C., and J. P. Houck. 1968. "Some Estimators for a Linear Model with Random Coefficients." *Journal of the American Statistical Association* 63: 583-595.

Hoch, I. 1958. "Simultaneous Equation Bias in the Context of Cobb-Douglas Production Function." *Econometrica* 26: 577-587.

_____ . 1962. "Estimation of Production Function Parameters Combining Time-Series and Cross-Section Data." *Econometrica* 30: 34-53.

_____ . 1976. "Returns to Scale in Farming: Further Evidence." *American Journal of Agricultural Economics* 58: 745-749.

Holland, D. 1978. "The Consequences of Acreage Limitations and Other Provisions of the Federal Reclamation Law: Discussion." *American Journal of Agricultural Economics* 60: 941-942.

Holmes, James M., Patricia A. Hutton, and Edward A. Weber. 1991. "Functional-Form-Free Test of the Research and Development/Firm Size Relationship." *Journal of Business and Economic Statistics* 9: 85-90.

Hornbaker, R. H., B. L. Dixon, and S. T. Sonka. 1989. "Estimating Production Activity Costs for Multi-Output Firms With a Random Coefficient Regression Model." *American Journal of Agricultural Economics* 71: 167-77.

Hottel, J. Bruce, and Robert D. Reinsel. 1976. *Returns to Equity Capital by Economic Class of Farm*. AER-347. Washington, DC: Economic Research Service, United States Department of Agriculture.

Huang, Cliff J. 1984. "Estimation of Stochastic Frontier Production Functions and Technical Inefficiency via the EM Algorithm." *Southern Economic Journal* 50: 847-56.

Hulten, Charles R. 1986. "Productivity Change, Capacity Utilization, and the Sources of Efficiency Growth." *Journal of Econometrics* 33: 31-50.

Idson, Todd L. 1990. "Establishment Size, Job Satisfaction and the Structure of Work." *Applied Economics* 22: 1007-18.

Ihnen, L., and E. O. Heady. 1964. "Cost Functions in Relation to Farm Size and Machinery Technology in Southern Iowa." Iowa Agr. and Home Econ. Expt. Sta. Res. Bul. 527.

Ijiri, Yuji, and Herbert A. Simon. 1977. *Skew Distributions and the Size of Business Firms*. Amsterdam: North-Holland Publishing Company.

Jensen, H. R. 1977. "Farm Management and Production Economics, 1946-70," in R. Martin, ed., *A Survey of Agricultural Economics Literature*. Vol. 1. Minneapolis: University of Minnesota Press.

_____. 1982. "Another Look at Economies of Size Studies in Farming." ER82-7. Department of Agricultural and Applied Economics, University of Minnesota.

Johnson, Paul R. 1964. "Some Aspects of Estimating Statistical Cost Functions." *Journal of Farm Economics* 46: 179-187.

Johnson, W. E. 1913. "The Pure Theory of Utility Curves." *Economic Journal* 23: 483-513.

Johnston, J. 1960. *Statistical Cost Functions*. New York: McGraw-Hill.

Jondrow, J., C. A. Knox-Lovell, J. S. Materov, and P. Schmidt. 1982. "On the Estimation of Technical Inefficiency in the Stochastic Frontier Production Function Model." *Journal of Econometrics* 19: 233-38.

Jorgenson, D. W., F. M. Gollop, and B. M. Fraumeni. 1987. *Productivity and U.S. Economic Growth*. Cambridge: Harvard University Press.

Jovanovic, B. 1982. "Selection and Evolution of Industry." *Econometrica* 50: 649-670.

Jovanovic, B., and R. Rob. 1987. "Demand-Driven Innovation and Spatial Competition Over Time." *Review of Economic Studies* 54: 63-72.

Just, R. E., and R. D. Pope. 1978. "Stochastic Specification of Production Functions and Economic Implications." *Journal of Econometrics* 7: 67-86.

_____. 1979. "Production Function Estimation and Related Risk Considerations." *American Journal of Agricultural Economics* 61: 67-86.

Just, Richard E., and David Zilberman. 1983. "Stochastic Structure, Farm Size and Technology Adoption in Developing Agriculture." *Oxford Economic Papers, N. S.*, 35: 307-28.

192

Just, R. E., D. Zilberman, and E. Hochman. 1983. "Estimation of Multicrop Production Functions." *American Journal of Agricultural Economics* 65: 770-780.

Kalter, R. J., and L. W. Tauer. 1987. "Potential Economic Impacts of Agricultural Biotechnology." *American Journal of Agricultural Economics* 69: 420-425.

Kass, David I. 1987. "Economies of Scale and Scope in the Provision of Home Health Services." *Journal of Health Economics* 6: 129-46.

Kennedy, Peter W. 1990. "The Market Provision of Club Goods in the Presence of Scale Economies." *Economica* 57: 515-24.

Khan, M. H. 1977. "Land Productivity, Farm Size and Returns to Scale in Pakistan Agriculture." *World Development* 5: 317-23.

Khan, M. H., and D. R. Maki. 1979. "Effects of Farm Size on Economic Efficiency: The Case of Pakistan." *American Journal of Agricultural Economics* 61: 64-69.

Kihlstrom, Richard, and Jean-Jacques Laffont. 1979. "A General Equilibrium Theory of Firm Formation Based on Risk-Aversion." *Journal of Political Economics* 89: 719-748.

Kilpatrick, Andrew, and Barry Naisbitt. 1988. "Energy Intensity, Industrial Structure and the 1970s' Productivity Slowdown." *Oxford Bulletin of Economics and Statistics* 50: 229-41.

Kim, H. Youn. 1986. "Economies of Scale and Economies of Scope in Multiproduct Financial Institutions: Further Evidence from Credit Unions: A Note." *Journal of Money, Credit, and Banking* 18: 220-26.

_____. 1987a. "Economies of Scale in Multi-product Firms: An Empirical Analysis." *Economica* 54: 185-206.

_____. 1987b. "Economies of Scale and Scope in Multiproduct Firms: Evidence from U.S. Railroads." *Applied Economics* 19: 733-41.

Kim, H. Youn, and Robert M. Clark. 1988. "Economies of Scale and Scope in Water Supply." *Regional Science and Urban Economics* 18: 479-502.

Kinnucan, H., U. Hatch, J. Molnar, and M. Venkateswaran. 1990. "Scale Neutrality of Bovine Somatotropin: Ex Ante Evidence from the Southeast." *Southern Journal of Agricultural Economics* 22: 1-12.

Kislev, Y., and W. Peterson. 1982. "Prices, Technology, and Farm Size." *Journal of Political Economics* 90: 578-595.

Klepper, Steven, and Elizabeth Graddy. 1990. "The Evolution of New Industries and the Determinants of Market Structure." *Rand Journal of Economics* 21: 27-44.

Kolari, James, and Asghar Zardkoohi. 1990. "Economies of Scale and Scope in Thrift Institutions: The Case of Finnish Cooperative and Savings Banks." *Scandinavian Journal of Economics* 92(3): 437-51.

Kopp, R. J. 1981. "The Measurement of Productive Efficiency: A Reconsideration." *Quarterly Journal of Economics* 96: 477-503.

Kopp, Raymond J., and Erwin W. Diewert. 1982. "The Decomposition of Frontier Cost Function Deviations into Measures of Technical and Allocative Efficiency." *Journal of Econometrics* 19: 319-31.

Krause, K. R., and L. R. Kyle. 1970. "Economic Factors Underlying the Incidence of Large Farming Units: The Current Situation." *American Journal of Agricultural Economics* 52: 748-761.

Kumar, M. S. 1985. "Growth, Acquisition Activity, and Firm Size: Evidence From the United Kingdom." *The Journal of Industrial Economics* 33: 327-338.

Kumbhakar, Subal C. 1987. "The Specification of Technical and Allocative Inefficiency in Stochastic Production and Profit Frontiers." *Journal of Econometrics* 34: 335-48.

_____. 1988a. "On the Estimation of Technical and Allocative Inefficiency Using Stochastic Frontier Functions: The Case of U.S. Class 1 Railroads." *International Economic Review* 29: 727-43.

_____. 1988b. "Estimation of Input-Specific Technical and Allocative Inefficiency in Stochastic Frontier Models." *Oxford Economic Papers, N. S.*, 40: 535-49.

Kumbhakar, Subal C., Basudeb Biswas, and DeeVon Bailey. 1989. "A Study of Economic Efficiency of Utah Dairy Farmers: A System Approach." *Review of Economics and Statistics* 71: 595-604.

Kuroda, Yoshimi. 1989. "Impacts of Economies of Scale and Technological Change on Agricultural Productivity in Japan." *Journal of the Japanese and International Economy* 3: 145-73.

LaDue, Eddy L. 1977. "Toward a More Meaningful Measure of Firm Growth." *American Journal of Agricultural Economics* 59: 210-15.

Lakonishok, Josef, and Alan C. Shapiro. 1986. "Systematic Risk, Total Risk and Size as Determinants of Stock Market Returns." *Journal of Banking and Finance* 10: 115-32.

Lambson, V. E. 1992. "Competitive Profits in the Long Run." *Review of Economic Studies* 59: 125-142.

_____. 1991. "Industry Evolution with Sunk Costs and Uncertain Market Conditions." *International Journal of Industrial Organization* 9: 171-196.

Lau, L. J. 1972. "Profit Functions of Technologies With Multiple Inputs and Outputs." *Review of Economic Statistics* 54: 281-289.

Lau, L. J., and P. A. Yotopoulos. 1971. "A Test for Relative Efficiency and an Application to Indian Agriculture." *American Economic Review* 61: 94-109.

_____. 1972. "Profit, Supply, and Factor Demand Functions." *American Journal of Agricultural Economics* 54: 11-18.

Lawrence, Colin. 1989. "Banking Costs, Generalized Functional Forms, and Estimation of Economies of Scale and Scope." *Journal of Money, Credit, and Banking* 21: 368-79.

Lee, Lung-Fei. 1978. "The Stochastic Frontier Production Function and Average Efficiency: An Empirical Analysis." *Journal of Econometrics* 7: 385-89.

_____. 1983a. "On Maximum Likelihood Estimation of Stochastic Frontier Production Models." *Journal of Econometrics* 23: 269-74.

_____. 1983b. "A Test for Distributional Assumptions for the Stochastic Frontier Functions." *Journal of Econometrics* 22: 245-67.

Leiby, James D. 1985. "An Analysis of Farm Size and Growth: The Case of Southern Dairy Farms." Ph.D. thesis, North Carolina State University.

LeVeen, P. E., and G. E. Goldman. 1978. "Reclamation Policy and Water Subsidy: An Analysis of the Distributional Consequences of Emerging Policy Choices." *American Journal of Agricultural Economics* 60: 929-934.

194

Levy, Brian. 1990. "Transactions Costs, the Size of Firms and Industrial Policy: Lessons from a Comparative Case Study of the Footwear Industry in Korea and Taiwan." *Journal of Development Economics* 34: 151-78.

Levy, David T. 1984. "Variation in the Concentration-Profit Relationship Across Industries." *Southern Economic Journal* 51: 267-73.

Lieberman, Marvin B. 1987. "Market Growth, Economies of Scale, and Plant Size in the Chemical Processing Industries." *Journal of Industrial Economics* 36: 175-91.

_____. 1990. "Exit from Declining Industries: "Shakeout" or "Stakeout"? *Rand Journal of Economics* 21: 538-54.

Lin, W., G. W. Dean, and C. V. Moore. 1974. "An Empirical Test of Utility vs. Profit Maximization in Agricultural Production." *American Journal of Agricultural Economics* 56: 497-508.

Lin, William, George Coffman, and J. B. Penn. 1980. "U.S. Farm Numbers, Size, and Related Structural Dimensions: Projections to Year 2000." ESCS, Technical Bulletin No. 1625. Washington: United States Department of Agriculture.

Lindsey, Charles W. 1981. "Firm Size and Profit Rate in Philippine Manufacturing." *Journal of Developing Areas* 15: 445-56.

Lins, David A. 1969. "An Empirical Comparison of Simulation and Recursive Linear Programming Firm Growth Models." *Agricultural Economics Research* 21: 7-12.

Lopez, R. E. 1980. "The Structure of Production and the Derived Demand for Inputs in Canadian Agriculture." *American Journal of Agricultural Economics* 82: 38-45.

_____. 1982. "Applications of Duality Theory to Agriculture." *Western Journal of Agricultural Economics* 7: 353-365.

Lopez, R. E., and F. L. Tung. 1982. "Energy and Non-energy Input Substitution Possibilities and Output Scale Effects in Canadian Agriculture." *Canadian Journal of Agricultural Economics* 30: 115-132.

Lowenberg-DeBoer, John, and Michael Boehlje. 1986. "The Impact of Farmland Price Changes on Farm Size and Financial Structure." *American Journal of Agricultural Economics* 68: 838-48.

Lucas, Robert E. 1978. "On the Size Distribution of Business Firms." *Bell Journal of Economics* 9: 508-523.

Luke Chan, M. W., and D. C. Mountain. 1983. "Economies of Scale and the Tornqvist Discrete Measure of Productivity Growth." *Review of Economics and Statistics* 65: 663-667.

Lund, P. J. 1983. "The Use of Alternative Measures of Farm Size in Analyzing the Size and Efficiency Relationships." *Journal of Agricultural Economics* 34: 187-189.

MacPhee, Craig R., and Rodney D. Peterson. 1990. "The Economies of Scale Revisited: Comparing Census Costs, Engineering Estimates, and the Survivor Technique." *Quarterly Journal of Business and Economics* 29: 43-67.

Maddala, G. 1971. "The Use of Variance Components Models in Pooling Cross-Section and Time-Series Data." *Econometrica* 39: 939-953.

Madden, J. P. 1967. *Economies of Size in Farming*. AER-107. Washington, DC: Economic Research Service, United States Department of Agriculture.

Mahmood, Moazam, and Nadeem-ul-Haque. 1981. "Farm Size and Productivity Revisited." *Pakistan Development Review* 2: 151-90.

Maindiratta, Ajay. 1990. "Largest Size-Efficient Scale and Size Efficiencies of Decision-Making Units in Data Envelopment Analysis." *Journal of Econometrics* 46: 57-72.

Marion, B. W. 1986. *The Organization and Performance of the U.S. Food System.* Lexington, MA: Lexington Books.

Marion, Bruce W., W. F. Mueller, R. W. Cotterill, F. E. Geithman, and J. R. Schmelzer. 1979. "The Price and Profit Performance of Leading Food Chains." *American Journal of Agricultural Economics* 61: 420-33.

Martin, W. E. 1978. "Economies of Size and the 160-Acre Limitation: Fact and Fantasy." *American Journal of Agricultural Economics* 60: 921-928.

Mayo, John W. 1984. "The Technological Determinants of the U.S. Energy Industry Structure." *Review of Economics and Statistics* 66: 51-58.

McClelland, J. W., M. E. Wetzstein, and W. N. Musser. 1986. "Returns to Scale and Size in Agricultural Economics." *Western Journal of Agricultural Economics* 11: 129-133.

_____. 1988. "Returns to Scale and Size in Agricultural Economics: Reply." *Western Journal of Agricultural Economics* 13: 151-52.

McElroy, F. W. 1969. "Returns to Scale, Euler's Theorem, and the Form of Production Functions." *Econometrica* 37: 275-279.

_____. 1981. "Scale and Cost Elasticity in the Multiproduct Case." *Atlantic Economic Journal* 9: 60-61.

McFadden, D. 1978. "Cost, Revenue, and Profit Functions," in M. Fuss and D. McFadden, eds., *Production Economics: A Dual Approach to Theory and Applications.* Amsterdam: North-Holland Publishing Company.

Meisner, J. C., and V. J. Rhodes. 1975. "Cattle Feeding Economies of Size Revisited." *Canadian Journal of Agricultural Economics* 23: 61-65.

Mester, Loretta J. 1987. "A Multiproduct Cost Study of Savings and Loans." *Journal of Finance* 42: 423-45.

Miller, Edward M. 1978. "The Extent of Economies of Scale: The Effects of Firm Size on Labor Productivity and Wage Rates." *Southern Economic Journal* 44: 470-87.

_____. 1984. "Extent of Economies of Scale: An Update." *Southern Economic Journal* 51: 582-87.

_____. 1986. "Determinants of the Size of the Small Business Sector: They Are Labor Productivity, Wage Rates and Capital Intensity." *American Journal of Economics and Sociology* 45: 390-402.

Miller, T. A. 1979. "Economies of Size and Other Growth Incentives." *Structure Issues of American Agriculture.* AER-438. Washington, DC: Economic Statistics Cooperative Service, United States Department of Agriculture, pp. 108-115.

_____. 1984. "Conceptual Issues in Economies of Size Studies: How Assumptions Drive Away Research." *Economies of Size Studies.* Ames, IA: Center for Agriculture and Rural Development.

Miller, T. A., G. E. Rodewald, and R. G. McElroy. 1981. *Economies of Size in U.S. Field Crop Farming.* AER-472. Washington, DC: Economic Statistics Service, United States Department of Agriculture.

196

Mintz, J. M. 1981. "A Note on Multiproduct Economies of Scale and Economies of Scope." *Economic Letters* 8: 29-33.

Mittelhammer, R. C., D. L. Young, D. Tasansanta, and J. T. Donnelly. 1980. "Mitigating the Effects of Multicollinearity Using Exact and Stochastic Restrictions: The Case of an Aggregate Agricultural Production Function in Thailand." *American Journal of Agricultural Economics* 62: 199-210.

Moghadam, Fatemet Etemad. 1982. "Farm Size, Management and Productivity: A Study of Four Iranian Villages." *Oxford Bulletin of Economics and Statistics* 44: 357-79.

Moles, Jerry A. 1978. "Appropriate Technology for U.S. Agriculture: Are Small Farms the Coming Thing? Discussion." *American Journal of Agricultural Economics* 60: 316-21.

Moll, Peter G. 1989. "The Rationality of Farm Size Growth: An Example from 'White' South Africa." *European Review of Agricultural Economics* 16: 345-57.

Moore, C. V., D. L. Wilson, and T. C. Hatch. 1982. *Structure and Performance of Western Irrigated Agriculture with Special Reference to the Acreage Limitation Policy of the U.S. Department of Interior.* Div. Agr. Sci. Bull. No. 1905, University of California.

Moore, C. V. 1982. "Impact of Federal Limitation Policy on Western Irrigated Agriculture." *Western Journal of Agricultural Economics* 7: 301-316.

Moschini, G. 1988. "The Cost Structure of Ontario Dairy Farms: A Microeconomic Analysis." *Canadian Journal of Agricultural Economics* 36: 187-206.

_____. 1990. "Nonparametric and Parametric Estimation: An Analysis of Multiproduct Returns to Scale." *American Journal of Agricultural Economics* 72: 589-98.

Muellbauer, J. 1975. "Aggregation, Income Distribution and Consumer Demand." *Review of Economic Studies* 62: 525-43.

Mulligan, James G. 1986. "Technical Change and Scale Economies Given Stochastic Demand and Production." *International Journal of Industrial Organization* 4: 189-201.

Munklak, Y. 1963. "Specification and Estimation of Agricultural Production Functions." *Journal of Farm Economics* 45: 433-443.

Murray, John D., and Robert W. 1983. "Economies of Scale and Economies of Scope in Multiproduct Financial Institutions: A Study of British Columbia Credit Unions." *Journal of Finance* 38: 887-902.

Naseem, Muhammad. 1975. "Credit Availability and the Growth of Small Farms in the Pakistan Punjab." *Food Research Institute Studies* 14: 65-80.

Nelson, Randy A. 1990. "Productivity Growth, Scale Economies and the Schumpeterian Hypothesis." *Southern Economic Journal* 57: 521-27.

Nelson, Robert H. 1972. "Economies of Scale and Market Size." *Land Economics* 48: 297-300.

Nerlove, M. 1963. "Returns to Scale in Electricity Supply." in C.F. Christ et al. eds., *Measurement in Economics: Studies in Mathematical Economics and Econometrics in Memory of Yehuda Grunfeld.* Stanford: Stanford University Press.

Nickell, Stephen. 1978. *The Investment Decisions of Firms*. Cambridge: Cambridge University Press.

O'Neill, Patrick B. 1991. "Concentration Trends and Profitability in U.S. Manufacturing: A Comment." *Applied Economics* 23: 717-19.

Ockwell, A. P., and R. L. Batterham. 1983. "The Influence of Credit on Farm Growth." *Review of Marketing and Agricultural Economics* 3: 247-64.

Olsen, J. A., P. Schmidt, and D. M. Waldman. 1980. "A Monte Carlo Study of Estimators of Stochastic Production Functions." *Journal of Econometrics* 13: 67-82.

Otha, Makota. 1975. "A Note on the Duality Between Production and Cost Functions: Rate of Returns to Scale and the Rate of Technical Progress." *Economic Studies Quarterly* 25: 63-65.

Padberg, D. I. 1962. "The Uses of Markov Processes in Measuring Change in Market Structure." *Journal of Farm Economics* 44: 189-199.

Page, J. M. 1984. "Firm Size and Technical Efficiency: Applications of Production Frontiers to Indian Survey Data." *Journal of Development Economics* 16: 129-152.

Pakes, A., and R. Ericson. 1987. "Empirical Implications of Alternative Models of Firm Dynamics," Columbia University, Department of Economics Working Paper.

Panagariya, Arvind. 1988. "A Theoretical Explanation of Some Stylized Facts of Economic Growth." *Quarterly Journal of Economics* 103: 509-26.

Panzar, J. C. 1989. "Technological Determinants of Firm and Industry Structure." in R. Schmalensee and R. D. Willig, eds., *Handbook of Industrial Organization*. Vol 1. Amsterdam: North Holland.

Panzar, J. C., and R. Willig. 1977. "Economies of Scale in Multi-Output Production." *Quarterly Journal of Economics* 91: 481-493.

_____. 1981. "Economies of Scope." *American Economic Review* 71: 268-272.

Patrick, G., and L. Eisgruber. 1968. "The Impact of Managerial Ability and Capital Structure on Growth of the Farm Firm." *American Journal of Agricultural Economics* 50: 491-506.

Peck, Richard M. 1989. "Taxation, Risk, and Returns to Scale." *Journal of Public Economics* 40: 319-30.

Perrakis, Stylianos, and George Warskett. 1986. "Uncertainty, Economies of Scale, and Barrier to Entry." *Oxford Economic Papers, N. S.,* Suppl. 38: 58-74.

Phillips, Joseph M., and Robert P. Marble. 1986. "Farmer Education and Efficiency: A Frontier Production Function Approach." *Economics of Education Review* 5: 257-64.

Pindyck, R. S., and J. J. Rotemberg. 1983. "Dynamic Factor Demands Under Rational Expectations." *Scandinavian Journal of Economics* 85: 223-238.

Pitt, M. M., and L. Lee. 1981. "The Measurement and Sources of Technical Inefficiency in the Indonesian Weaving Industry." *Journal of Development Economics* 9: 43-64.

Pope, R. D. 1980. "The Effects of Production Uncertainty on Input Demands," in D. Yaron and C. S. Tapiero, eds., *Operations Research in Agriculture and Water Resources*. Amsterdam: North Holland.

_____. 1982. "To Dual or Not to Dual." *Western Journal of Agricultural Economics* 7: 337-352.

Pope, R. D., and R. A. Kramer. 1979. "Production Under Uncertainty and Factor Demands for the Competitive Firm." *Southern Economic Journal* 46: 489-501.

Quiggin, John, and Bui-Lan Anh. 1984. "The Use of Cross-Sectional Estimates of Profit Functions for Tests of Relative Efficiency: A Critical Review." *Australian Journal of Agricultural Economics* 28: 44-55.

Raiffa, H. 1968. *Decision Analysis.* Reading, MA: Addison-Wesley.

Rao, V., and T. Chotigeat. 1981. "The Inverse Relationship Between Size and Land Holdings and Agricultural Productivity." *American Journal of Agricultural Economics* 63: 571-74.

Rao, T. V. S. Ramamohan, and Sarthi Acharya. 1976. "Total Factor Productivity of Small Scale Firms." *Economic Affairs* 21: 293-96.

Ratcliffe, Kerri, Bruce Riddle, and John Yinger. 1990. "The Fiscal Condition of School Districts in Nebraska: Is Small Beautiful?" *Economics of Education Review* 9: 81-99.

Rauch, James E. 1989. "Increasing Returns to Scale and the Pattern of Trade." *Journal of International Economics* 26: 359-69.

Raup, P. M. 1968. "Economies and Diseconomies of Large-Scale Agriculture." *American Journal of Agricultural Economics* 51: 1274-1283.

_____. 1978. "Some Questions of Value and Scale in American Agriculture." *American Journal of Agricultural Economics* 60: 303-308.

Ray, S. 1982. "Translog Cost Analysis of U.S. Agriculture." *American Journal of Agricultural Economics* 62: 490-498.

Reinsel, R. D., and E. I. Reinsel. 1979. "The Economics of Asset Values and Current Income in Farming." *American Journal of Agricultural Economics* 61: 1093-1097.

Revier, C. F. 1987. "The Elasticity of Scale, the Shape of Average Costs, and the Envelope Theorem." *American Economic Review* 77: 486-488.

Richardson, J., and G. D. Condra. 1981. "Farm Size Evaluation in the El Paso Valley: A Survival/Success Approach." *American Journal of Agricultural Economics* 63: 430-437.

Roberts, James. 1989. "A Micro-economic Analysis of Tea Production Using a Separable Restricted Profit Function." *Journal of Agricultural Economics* 40: 185-97.

Robison, L. 1988. *Determinants of Farm Size and Structure.* Proceedings of the program sponsored by NC-181 Committee on Determinants of Farm Size and Structure in North Central Areas of the United States, held January 16, 18, and 19, 1988, San Antonio, Texas. Michigan Agricultural Experiment Station, Journal Article No. 12899.

Robison, L., and P. J. Barry. 1986. *The Competitive Firm's Response to Risk.* New York: Macmillan.

Rodewald, Gordon E., and Raymond J. Folwell. 1977. "Farm Size and Tractor Technology." *Agricultural Economics Research* 29: 82-89.

Rosen, S. 1972. "Learning by Experience as Joint Production." *Quarterly Journal of Economics* 86: 366-382.

Rothschild, Michael, and Gregory J. Werden. 1979. "Returns to Scale from Random Factor Services: Existence and Scope." *Bell Journal of Economics* 10: 329-35.

Russell, R. Robert. 1985. "Measures of Technical Efficiency." *Journal of Economic Theory* 35: 109-26.

Ruttan, V. W. 1988. *Scale, Size, Technology and Structure: A Personal Perspective.* Staff Paper P88-1. St. Paul: University of Minnesota.

Ryan, J. G. 1976. "Growth and Size Economies Over Space and Time: Wheat-Sheep Farms in New South Wales." *Australian Journal of Agricultural Economics* 20: 160-178.

Saghafi, Massoud M., and Mohsen Attaran. 1991. "Concentration Trends and Profitability in U.S. Manufacturing: A Reply and Some New Evidence." *Applied Economics* 23: 721-22.

Sah, Raaj Kumar, and Joseph E. Stiglitz. 1987. "The Invariance of Market Innovation to the Number of Firms." *Rand Journal of Economics* 18: 98-108.

Samuelson, W., and R. Zeckhauser. 1988. "Status Quo Bias in Decision Making." *Journal of Risk and Uncertainty* 1: 5-59.

Sandler, T., and A. Swimmer. 1978. "The Properties and Generation of Homothetic Production Functions: A Synthesis." *Journal of Economic Theory* 18: 349-61.

Sandmo, A. 1971. "On the Theory of the Competitive Firm." *American Economic Review* 61: 65-73.

Schmidt, P. 1985-1986. "Frontier Production Functions." *Econometric Reviews* 4(2): 289-328.

Schmidt, P., and C. A. K. Lovell. 1979. "Estimating Technical and Allocative Inefficiency Relative to Stochastic Production and Cost Frontiers." *Journal of Econometrics* 9: 363-366.

Seckler, David, and Robert A. Young. 1978. "Economic and Policy Implications of the 160-Acre Limitation in Federal Reclamation Law." *American Journal of Agricultural Economics* 60: 575-588.

Seitz, W. D. 1970. "The Measurement of Efficiency Relative to a Frontier Production Function." *American Journal of Agricultural Economics* 51: 505-511.

Shaked, Avner, and John Sutton. 1990. "Multiproduct Firms and Market Structure." *Rand Journal of Economics* 21: 45-62.

Shalit, S. S., and U. Sankar. 1977. "The Measurement of Firm Size." *The Review of Economics and Statistics* 59: 290-298.

Shapiro, D., R. D. Bollman, and P. Ehrensaft. 1987. "Farm Size and Growth in Canada." *American Journal of Agricultural Economics* 69: 477-483.

Sharkey, William W., and Lester G. Telser. 1978. "Supportable Cost Functions for the Multiproduct Firm." *Journal of Economic Theory* 18: 23-37.

Shaw, Anthony B., and Richard C. daCosta. 1985. "Differential Levels of Technology Adoption and Returns to Scale in the Guyanese Rice Industry." *Canadian Journal of Agricultural Economics* 33: 99-110.

Shen, T. Y. 1970. "Economies of Scale, Penrose Effect, Growth of Plants and Their Size Distribution." *Journal of Political Economy Part I* 78: 702-16.

Shephard, R. W. 1953. *Cost and Production Functions.* Princeton: Princeton University Press.

———. 1970. *Theory of Cost and Production Functions.* Princeton: Princeton University Press.

———. 1973. *Indirect Production Functions.* Meisenheim am Glan: Verlag Anton Hain.

Shepherd, W. G. 1984. "Contestability vs. Competition." *American Economic Review* 74: 572-587.

Shertz, L. 1979. "Another Revolution in U.S. Farming." United States Department of Agriculture: AER Report #441.

Shoesmith, Gary L. 1988. "Economies of Scale and Scope in Petroleum Refining." *Applied Economics* 20: 1643-52.

Shorrocks, A. F. 1975. "On Stochastic Models of Size Distributions." *Review of Economic Studies* 42: 631-41.

Sidhu, Surjit S. 1974. "Relative Efficiency in Wheat Production in the Indian Punjab." *American Economic Review* 64: 742-51.

Sidhu, S., and C. Baanante. 1982. "Estimating Farming-Level Input Demand and Wheat Supply in the Indian Punjab Using a Translog Profit Function." *American Journal of Agricultural Economics* 63: 237-246.

Singh, Ajit, and Geoffrey Whittington. 1975. "The Size and Growth of Firms." *Review of Economic Studies* 42: 15-26.

Singh, Joginder, and D. S. Sidhu. 1975. "Farm Technology and Change in Factor Shares." *Economic Affairs* 20: 51-55.

Smith, E. G., J. W. Richardson, and R. D. Knutson. 1984. *Cost and Pecuniary Economics in Cotton Production and Marketing: A Study of Texas Southern High Plains Cotton Producers.* College Station, Texas: Texas Agricultural Experiment Station.

Smith, E. G., R. D. Knutson, and J. W. Richardson. 1986. "Input and Marketing Economies: Impact on Structural Change in Cotton Farming on the Texas High Plains." *American Journal of Agricultural Economics* 68: 716-720.

Smith, Vincent H. 1990. "The Effects of Changes in the Tax Structure on Agricultural Asset Replacement." *Southern Journal of Agricultural Economics* 22: 113-21.

Solow, R. M. 1956. "A Contribution to the Theory of Economic Growth." *Quarterly Journal of Economics* 70: 65-94.

Spence, A. M. 1981. "The Learning Curve and Competition." *Bell Journal of Economics* 12: 49-70.

Squires, Dale. 1987. "Long-run Profit Functions for Multiproduct Firms." *American Journal of Agricultural Economics* 69: 558-69.

_____. 1988. "Production Technology, Costs, and Multiproduct Industry Structure: An Application of the Long-run Profit Function to the New England Fishing Industry." *Canadian Journal of Economics* 21: 359-78.

Stanton, B. F. 1978. "Perspective on Farm Size." *American Journal of Agricultural Economics* 60: 727-37.

_____. 1984. "Changes in Farm Structure." Agricultural Economics Staff Paper No. 84-23. Ithaca, NY: Cornell University.

Stanton, B. F., and L. Kettunen. 1967. "Potential Entrants and Projections in Markov Process Analysis." *Journal of Farm Economics* 49: 633-42.

Stefanou, S. E. 1988. "Stochastic Learning-By-Experience and Returns to Scale." *Journal of Economic Dynamics and Control* 12: 137-43.

_____. 1989a. "Learning, Experience, and Firm Size." *Journal of Economics and Business* 41: 283-96.

_____. 1989b. "Returns to Scale in the Long Run: The Dynamic Theory of Cost." *Southern Economic Journal* 55: 570-79.

Stefanou, S., and J. P. Madden. 1988. "Economies of Size Revisited." *Journal of Agricultural Economics* 60: 727-37.

Stefanou, S. E., and S. Saxena. 1988. "Education, Experience, and Allocative Efficiency: A Dual Approach." *American Journal of Agricultural Economics* 70: 338-45.

Steindl, Joseph. 1965. *Random Processes and the Growth of Firms.* London: Griffin.

Stevenson, Rodney E. 1980. "Likelihood Functions for Generalized Stochastic Frontier Estimation." *Journal of Econometrics* 13: 57-66.

Stigler, G. J. 1958. "Economics of Scale." *Journal of Law and Economics* 1: 54-71.

Stiglitz, J. E. 1974. "Incentives and Risk Sharing in Sharecropping." *Review of Economic Studies* 41: 219-255.

_____ . 1975. "Incentives, Risk, and Information: Notes Toward a Theory of Hierarchy." *Bell Journal of Economics* 6: 552-579.

Stollsteimer, J. F., R. G. Bressler, and J. N. Boles. 1961. "Cost Functions From Cross-Section Data--Fact or Fantasy?" *Agricultural Economics Research* 13: 78-88.

Strickland, Roger P., Jr. 1973. "Alternative Analyses of Farm Growth." *Agricultural Economics Research* 25: 99-104.

Summer, D. A., and J. D. Leiby. 1987. "An Econometric Analysis of the Effects of Human Capital on Size and Growth Among Dairy Farms." *American Journal of Agricultural Economics* 69: 465-470.

Sutherland, A. 1983. "Farm Size: A Reply." *Journal of Agricultural Economics* 34: 191-195.

Swamy, P. A. V. B., and P. Tinsley. 1980. "Linear Prediction and Estimation Methods for Regression Models with Stationary Stochastic Coefficients." *Journal of Econometrics* 12(1980): 103-42.

Talukdar, Bikash Ranjan, and Biswa Nath Banerjee. 1984. "Farm Size and Productivity-A Case Study in Village Dakshin Duttapara of Haringhata Block, District Nadia (W. B.)." *Economic Affairs* 29: 106-12.

Taylor, Timothy G., H. Evan Drummond, and Aloisio T. Gomes. 1986. "Agricultural Credit Programs and Production Efficiency: An Analysis of Traditional Farming in Southeastern Minas Gerais, Brazil." *American Journal of Agricultural Economics* 68: 110-19.

Taylor, Timothy G., and J. Scott Shonkwiler. 1986. "Alternative Stochastic Specifications of the Frontier Production Function in the Analysis of Agricultural Credit Programs and Technical Efficiency." *Journal of Development Economics* 21: 149-60.

Teece, D. J. 1980. "Economies of Scope and the Scope of the Enterprise." *Journal of Economic Behavior and Organization* 2: 233-273.

Teitel, Simon. 1975. "Economies of Scale and Size of Plant: The Evidence and the Implications for the Developing Countries." *Journal of Common Market Studies* 13: 92-115.

Telser, Lester G. 1991. "Industry Total Cost Functions and the Status of the Core." *Journal of Industrial Economics* 39: 225-40.

Tew, B. W., S. Spurlock, W. Musser, and B. R. Miller. 1980. "Some Evidence on Pecuniary Economies of Size for Farm Firms." *Southern Journal of Agricultural Economics* 12: 151-154.

Thomas, Lacy Glenn. 1990. "Regulation and Firm Size: FDA Impacts on Innovation." *Rand Journal of Economics* 21: 497-517.

Timmer, C. Peter. 1971. "Using a Probabilistic Frontier Production Function to Measure Technical Efficiency." *Journal of Political Economy* 79: 776-794.

Tolley, George S. 1970. "Management Entry Into U.S. Agriculture." *American Journal of Agricultural Economics* 52: 485-493.

Tribe, M. A., and R. L. W. Alpine. 1987. "Sources of Scale Economies: Sugar Production in Less Developed Countries." *Oxford Bulletin of Economics and Statistics* 49: 209-26.

Tripathy, Niranjan, and Richard L. Peterson. 1991. "The Relationship Between OTC Bid-Ask Spreads and Dealer Size: The Impact of Order-Processing and Diversification Costs." *Journal of Financial Research* 14: 117-27.

Trosper, R. L. 1978. "American Indian Relative Ranching Efficiency." *American Economic Review* 68: 503-516.

Tweeten, Luther. 1983. "The Economics of Small Farms." *Science* 219: 1037-1041.

Tyler, William G., and Lung-Fei Lee. 1979. "On Estimating Stochastic Frontier Production Functions and Average Efficiency: An Empirical Analysis with Colombian Micro Data." *Review of Economics and Statistics* 61: 436-38.

U.S. Congress, Office of Technology Assessment. 1985. *Technology, Public Policy, and the Changing Structure of American Agriculture: A Special Report for the 1985 Farm Bill.* OTA-F-272, Washington, DC, March.

Varian, H. R. 1978. *Microeconomic Analysis.* New York: W. W. Norton.

_____ . 1984. "The Non-Parametric Approach to Production Analysis." *Econometrica* 52: 579-597.

Viner, J. 1952. "Cost Curves and Supply Curves." in K. E. Boulding and G. J. Stigler, eds., *AEA Readings on Price Theory.* Chicago: R. D. Irwin, Inc.

Viscusi, W. Kip. 1983. "Frameworks for Analyzing the Effects of Risk and Environmental Regulations on Productivity." *American Economic Review* 73: 793-801.

Vlastuin, C., D. Lawrence, and J. Quiggin. 1982. "Size Economies in Australian Agriculture." *Review of Marketing and Agricultural Economics* 51: 27-50.

Waldman, Donald M. 1982. "A Stationary Point for the Stochastic Frontier Likelihood." *Journal of Econometrics* 18: 275-79.

_____ . 1984. "Properties of Technical Efficiency Estimators in the Stochastic Frontier Model." *Journal of Econometrics* 25: 353-64.

Walker, Odell L., Mike. L Hardin, Harry P. Mapp Jr., and Clint E. Roush. 1979. "Farm Growth and Estate Transfer in an Uncertain Environment." *Southern Journal of Agricultural Economics* 11: 33-44.

Walters, A. A. 1960. "Expectations and the Regression Fallacy in Estimating Cost Functions." *Review of Economics and Statistics* 42: 210-215.

_____ . 1963. "Production and Cost Functions: An Econometric Survey." *Econometrica* 31: 1-66.

Weaver, R. D. 1982. "Specification and Estimation of Consistent Sets of Choice Functions," in Gordon C. Rausser, ed., *New Directions in Econometric Modelling and Forecasting in U.S. Agriculture.* Chapter 6. New York: North-Holland.

_____ . 1983. "Multiple Input, Multiple Output Production Choices and Technology in the U.S. Wheat Region." *American Journal of Agricultural Economics* 65: 45-56.

_____. 1984. "On the Usefulness and Proper Measurement of Returns to Size and Scale," in *Economics of Size Studies*. Ames, IA: CARD.

Weersink, A., and L. W. Tauer. 1991. "Causality between Dairy Farm Size and Productivity." *American Journal of Agricultural Economics* 73: 1138-1145.

Weimar, M. R., A. Hallam, and L. D. Trede. 1988. "Economies of Size, Tax Reform and Profitability of Alternative Midwestern Feedlot Systems." *Journal of the American Society of Farm Managers and Rural Appraisers* 52: 11-18.

Weiss, Andrew, and Henry J. Landau. 1984. "Wages, Hiring Standards and Firm Size." *Journal of Labor Economics* 2: 477-99.

West, Jerry G. 1979. "Agricultural Economics Research and Extension Needs of Small-Scale, Limited-Resource Farmers." *Southern Journal of Agricultural Economics* 11: 49-56.

Willig, Robert D. 1979. "Multiproduct Technology and Market Structure." *American Economic Review* 69: 346-51.

Wilson, Paul N., and Vernon R. Eidman. 1985. "Dominant Enterprise Size in the Swine Production Industry." *American Journal of Agricultural Economics* 67: 279-88.

Wohlgemuth, Melody J. 1988. "The Relation Between Firm Size and the Informational Content of Earnings." *Quarterly Journal of Business and Economics* 27: 135-48.

Wu, Craig, C. 1979. "Price-Output Uncertainty and Allocative Efficiency: An Empirical Study of Small-Scale Farms." *Review of Economics and Statistics* 61: 228-33.

Yaron, D., and U. Horowitz. 1972. "A Sequential Programming Model of Growth and Capital Accumulation of a Farm Uncertainty." *American Journal of Agricultural Economics* 54: 441-51.

Yotopoulos, Pan A., and Lawrence J. Lau. 1973. "A Test for Relative Economic Efficiency: Some Further Results." *American Economic Review* 63: 214-23.

Yotopoulos, P. A., and L. J. Lau, eds. 1979. "Resource Use in Agriculture: Applications of the Profit Function." *Food Research Institute Studies* 17(1).

Zellner, A., J. Kmenta, and J. Dreze. 1966. "Specification and Estimation of Cobb-Douglas Production Function Models." *Econometrica* 34: 784-795.

Zulauf, C. R. 1986. "Changes in Selected Characteristics of U.S. Farms During the 1970s and Early 1980s: An Investigation Based on Current and Constant Dollar Sales Categories." *Southern Journal of Agricultural Economics* 18: 113-122.

8

Empirical Studies of Size, Structure, and Efficiency in Agriculture

Arne Hallam

The structure of firms within an industry influences the structure of the industry and firms are influenced by that industry structure. The viability and profitability of firms are partially determined by elements associated with firm size, firm structure and the character of the industry in which they compete. The joint determination of firm structure and industry structure is important in analyzing the effects of factors that influence the evolution of firms and industry performance over time. The relevance of firm size and structure cannot be determined independently of industry structure because the causal factors are not unidirectional.

In many cases, however, the size or structure of firms at a point in time can be described or measured somewhat independently of industry dynamics. By measuring economies of size or scale for a particular industry or class of firms, a useful snapshot of the current situation can be obtained. The measurements obtained can then be used in two distinct ways. The effects that a particular measured structure may infer for the future of the firm and the industry can be analyzed. For example, if increasing returns to scale for a particular technology are discovered, then inference about firm growth and changes toward a more concentrated industry structure may be hypothesized. Alternatively, measurements of industry structure can be used as dependent variables in the analysis of factors that may influence the structure of firms and the industry. For example, measurements of firm efficiency may be obtained by cross-section analysis. These factors can then be related to other variables associated with the firm such as age, ownership pattern, or degree of diversification. Of course simultaneity problems must be addressed in such analyses.

This chapter takes the snapshot and reviews previous studies that have attempted to measure economies of scale, size and scope in agriculture. The chapter also considers studies of firm efficiency and firm growth or survivorship. The purpose is to investigate previous empirical work in order to discover central tendencies in the areas of size and structure in agriculture. The chapter considers normative studies first and then examines positive investigations. The positive studies are broken down by functional structure used to conduct the analysis. Research investigating firm efficiency is then examined, followed by the review of studies analyzing firm viability and growth. Pecuniary economies are mentioned briefly.

Normative Studies of Size and Scale

Normative studies usually measure economies of size by either synthetic construction of optimal firm plans using budget and technology data or through the use of programming models. Sometimes size economies are inferred by simply comparing average costs of production for firms producing different levels of output. The classic monograph by Madden (1967) summarizes a variety of studies based on linear programming. Rather than discuss specific studies, since they are analyzed in Madden (1967), some general conclusions will be drawn. For crop farms, the general trend is for the cost curve to decline over some range but flatten out at moderate output levels and decline little thereafter. Studies by Ihnen and Heady (1964) and Heady and Krenz (1962) found minimum average costs could be obtained at sizes of less than 500 acres. A variety of studies conducted in California, of which the ones by Carter and Dean (1962) are representative found larger farm sizes associated with minimum cost production but still found few economies for farm sizes above 700 acres.

Economies of size for beef feedlots discussed in Madden (1967) were generally found to be more significant however. In almost all studies average costs declined as size increased. The decline in costs was usually associated with the utilization of larger and more efficient equipment such as mills or feeding equipment.

The results of these early studies also found some economies of size for dairy farms but the cost curve levelled off for sizes considered very small by today's standards (<100 cows).

A more recent comprehensive study was conducted by Miller et al. (1981) using linear programming and data from the 1978 Firm Enterprise Data System (FEDS). For each of seven regions they developed representative farm models and minimized the cost of meeting certain revenue targets associated with farm size as defined by census data. In each region they considered three to four enterprises with a fixed machine component and fixed operator labor. Producers were allowed to rent land in addition to owned land. While economies of size were found in all regions the extent of these economies was limited. For example in all regions farms that were 33 percent of the size of the most efficient farm attained at least 90 percent of the resource return rate of the most efficient farms.

Moore (1982), in response to concern about possible changes in the Federal 160 acre limitation, investigated economies of size for farms in 18 Federal irrigation districts in the Western United States. Short run average costs curves were developed using linear programming and then the long run average cost curve was obtained by tracing the envelope of these curves. All of these curves declined rapidly up the point where gross sales exceeded $100,000 and were relatively flat thereafter. The acreage needed to achieve 98 percent of the efficiency of the most cost effective farm sizes was relatively low ranging from 200 acres in Oklahoma to 1350 acres on a heavy soil in the Imperial Valley of California. The same economies were attained on a farm of 375 acres on a light soil in the same area. The conclusion was that few advantages were obtained by increases in farm size for farms receiving Federal water.

In an investigation of optimal feedlot configuration Weimar et al. (1988) using economic engineering techniques found substantial economies of size for beef feedlots in the corn belt. The cost savings were generally associated with feed handling and waste handling equipment. Recent changes in the tax laws favored larger feedlots (5,000-10,000 head) due to lower tax brackets, though changes in depreciation laws mitigated these effects.

Most normative studies of crop farms, both early and recent, while identifying economies of size, have given little evidence that the cost curve deviates significantly from a sagging "L" shape. This result may be influenced in part by the size measure used since these studies typically use gross revenue as an output measure and crop specific economies may be blurred in optimal enterprise choice. Studies of livestock farms have generally identified economies of size or scale with larger firms having advantages both in production and other facets of the business (Meisner and Rhodes 1975). These conclusions are consistent with the overall growth in the size of livestock operations over time while the growth of crop farms has been less significant. In this sense the normative models of three decades past have been accurate predictors of industry dynamics. While normative studies have been immensely popular in investigating economies of size, the number of such studies reported in professional journals has declined dramatically in recent years.

Cross-section Studies of Size and Scale

The most popular alternative to normative analysis in investigating size economies is the direct estimation of size measures using cross section or time series data. Since cross-section results are the most straightforward to interpret and apply they will be discussed first.

Cost Function Studies

Given data on firm level output, inputs, and input prices, an industry cost function can be obtained by assuming a similar technology across firms. Differences in technology are controlled for through the use of other explanatory variables such as soil type in the case of crop farms or farmer education. The book by Johnston (1960) documents early empirical efforts in this area. The seminal work of Nerlove (1963) is another example in this area.

An example of this technique is a recent paper by Fleming and Uhm (1982) that estimates a cost curve for Saskatchewan grain farmers. Average production cost per kilogram of grain is regressed on kilograms of production, input prices, amount of livestock on the farm and other control variables. Average costs were found to fall as output initially increases and then decline much more slowly over a wide range of outputs. These results are consistent with normative studies.

Developments in duality theory led to a proliferation on articles measuring the cost structure of firms and implied size economies in the general economics literature. Many of these studies considered aggregate data and did not analyze returns to size (Berndt and Wood 1975). Given the historical interest for size economies in agriculture a number of studies estimating cost functions using cross section data might have been expected. Such is not the case however. One example is a recent paper by Cooke and Sundquist (1987, 1989). Using enterprise specific data they estimate cost functions for corn, soybeans, wheat and cotton. The data is taken from surveys conducted by USDA. This study assumes constant returns to size within three size categories but compares rates of cost efficiency across sizes to investigate economies of size. The study finds evidence of size economies between medium and large and between large and very large firms in corn, soybeans and cotton but mixed evidence for wheat. These size economies are not large in percentage terms, however (1-12 percent) and do not imply significantly different industry structure than earlier results.

The lack of papers applying single product cost functions is a natural outcome of the application of theory to empirical data. Most agricultural firms produce several products. While the production processes for these firms may often be non-joint and allow the estimation of enterprise specific functions, the data typically available does not allocate inputs by enterprise and so multiproduct cost estimation is inevitable. Early research on multiproduct cost structures did not consider returns to size, however (R. Hall 1973). While returns to size is an unambiguous measure in single output firms, it is not well defined for multiproduct firms (Chapter 7). Thus the measurement of returns to size, scale and scope for these firms lagged until the development of a theory of cost structures for these firms (Baumol, Panzar, and Willig 1982). The three measures applicable to multiproduct firms are economies of scope, product specific economies of scale, and multiproduct economies of scale which measures the effect of increased production on ray average cost.

Three recent papers apply these multiproduct techniques to the agricultural industry. A recent paper by Akridge and Hertel (1986) considers the retail fertilizer industry. They find that multiproduct scale economies exist for the average firm in a cross-section sample of Indiana and Illinois fertilizer plants. They consider six output categories: dry fertilizer, fluid fertilizer, anhydrous ammonia, chemicals, custom services and other sales. Variable inputs were aggregated into labor and energy while fixed inputs were represented by management, plant and equipment and other inputs. Due to data limitations economies of scope were computed between anhydrous ammonia and all other outputs. Economies of scope were identified, with the cost of producing anhydrous ammonia and the other five outputs being 84.8 percent lower in joint production. Product specific scale economies were found in anhydrous production. At average output, marginal cost was over $12.00 less than average incremental cost. Product specific economies of scale were not found for the other five outputs so that overall economies of scale were primarily due to economies of scope. Analysis of long run equilibrium conditions imply that the average plant in the sample overinvested in plant and equipment. The major limitation of the paper is the inability to compute more specific scale measures since many plants did not produce all outputs.

A paper by Moschini (1988) analyzes the structure on Ontario dairy farms using a hybrid-translog multiproduct cost function. Three output categories are considered in the analysis; namely, milk, livestock products and crops, and other products. The four inputs considered are labor, feed, intermediate inputs and capital. A service price is computed for capital. The farms are further classified by various demographic and production technique variables. The results imply increasing returns to scale for a wide range of output levels. The returns to scale become closer to one only for the largest firms in the sample. The hypothesis that milk production is non-joint with the other outputs was not rejected.

Therefore average incremental cost for milk, holding other outputs constant, is a reasonable measure of economies of size. Average incremental cost for milk production is "L" shaped but implies increasing returns over a sizeable range of production with most of the firms still not capturing the significant economies of scale. The results of this study are consistent with previous studies of the dairy industry but imply increasing returns for larger firms than the studies summarized in Madden. In a more recent paper Moschini (1990) uses nonparametric kernal density estimation to estimate overall returns to scale for the same data set and finds a considerable degree of scale economies thus adding support to the idea of scale economies in dairy production.

Work by Deller, Chicoine, and Walzer (1988) considers economies of size and scope in rural low-volume roads. The paper analyzes the production of paved, bituminous and gravel/earth roads in townships in the states of Illinois, Ohio, Minnesota and Wisconsin. The inputs are labor, earth graders, trucks, and the types of materials needed for each road type. Exogenous variables considered included utilization and government road aid. Multiproduct economies of scale, which the authors call economies of size, were identified with the average measure being .466 (a positive number indicates economies of scale). These results indicate that the total cost of maintaining roads in two average size townships could be reduced by 50 percent with a merger of townships. Economies of scope were also evident in the sample. This implies that specialization in specific road types is probably not of value.

One additional line of work should be mentioned. Dixon, Batte, and Sonka (1984) and Hornbaker, Dixon, and Sonka (1989) have developed a way to estimate average production costs for multiproduct firms using cost and output data. By generalizing the Hildreth and Houck (1968) random coefficients model along a line suggested by Swamy and Tinsley (1980) they are able to estimate individual enterprise activity costs. While not explicitly recovering size and scale data the method could be used to estimate production costs for samples of differing size firms.

The cost function studies, while few in number, seem to imply increasing returns to size or scale for livestock producers at current production levels for many firms, but much less significant returns to size for crop farms. While these studies have found that firm characteristics are good independent variables in explaining differences across firms, no theory of how these characteristics affect optimal decision rules is presented.

Production Function Studies

While economies of size are typically measured using the cost function, economies of scale have traditionally been measured using the production function. A number of early studies are summarized and critiqued by Hoch (1976). Hoch (1976) also presents the results of a study involving California dairy herds. Milk production was postulated as a function of feed, capital, cow service flow, labor and operating costs. Multicollinearity among variables led to the combining of all non-feed inputs. Analysis of covariance was used with effects for firm, year, month, breed and membership in the Dairy Herd Improvement Association (DHIA). The Cobb-Douglas functional form was used. Ignoring firm effects, returns to scale was close to one. When firm effects were included the elasticity sum fell to less than one for all market-milk samples and was greater than one for manufacturing-milk samples. The study not only implied that long run returns to scale may be decreasing due to inherently fixed factors but pointed out some of the difficulties in interpreting empirical results. The results on returns to scale for these dairy herds are very different from those found when measuring returns to size as should be expected. This difference should be kept in mind when analyzing returns to scale (size is not defined) for multiproduct firms.

Much of the impetus for more flexible functional forms was the inability of simple forms to measure variable returns to scale and variable elasticities of substitution. The paper by Zellner and Revankar (1969) addresses many of these issues. Flexible functional forms associated with the advent of duality theory also led to increased interest in returns to scale (Christensen and Green, 1976; Christensen, Jorgensen, and Lau, 1973). A somewhat unique approach is that proposed by Fare, Jansson, and Lovell (1985), who advocate the use of ray-homothetic production functions. These functions, while less flexible than true second order approximations, are parsimonious in the use of parameters and generate more plausible results than general homothetic functions. When reestimating a production function for the transportation equipment industry using Zellner and Revankar's (1969) data they find that although there is little statistical support for ray-homotheticity over homotheticity or ray homogeneity, the implied returns to scale over the output and input range are much more plausible using the ray homothetic form. Applications of this approach in agriculture have been rather few, exceptions being a recent paper by Grabowski and Sanchez (1986) on Japanese agriculture and another by Grabowski and Belbase (1986) on Nepalese agriculture. Both papers, however, used aggregate rather than cross section data and thus apply more to representative firms than actual ones. The former paper found increasing returns to scale in Japanese agriculture over the period 1874-1940.

The cost function, as discussed above, is the most common method for measuring returns to size. An alternative approach is to estimate the firm's production function and then solve for optimal input demands.

From these the shape of the average cost curve can be obtained by solving first-order conditions. The direct estimation of the production function for determining size economies has certain advantages and disadvantages relative to alternative methods and has been primarily applied by Australian researchers (Anderson and Powell 1973). The approach has several problems, however. These are discussed elsewhere (Chapter 7). One of the most troubling computationally is that the economies of size measure must be obtained by solving first-order conditions. An excellent recent example of this approach is Vlastuin, Lawrence, and Quiggin (1982). They estimate a translog production function for the New South Wales Wheat/Sheep zone using cross section data. Output was estimated as a function of labor, livestock, materials, capital and land. When operator and hired labor were excluded from the right hand side the sample exhibited constant returns to scale but exhibited the standard "L" shape when they were included. To investigate economies of size the sample was divided into five size groups based on total output. Scale elasticities were slightly greater than one for each group in 1976-77 but less than one in all but one group in 1966-67. In the latter sample returns to scale increased with size classification. Cost curves were not obtained by solving first order conditions but by plotting cost as a function of output. The derived curves were rather flat as expected given nearly constant returns to scale for the case of excluded family labor. The difference in the shape of the curves would imply that size economies are primarily derived from making full use of fixed operator and family labor. This finding should be tested against U.S. data as it has important implications for public policy related to off-farm employment.

Profit Function Studies

Just as the firm's technology can be obtained from the cost function, it can also be recovered using the profit function (McFadden 1978). For a well-defined profit maximization problem to exist, the production function must be concave and this implies decreasing returns to scale at the optimum. Thus, increasing returns to size or scale cannot be identified from a well-behaved profit function since the measure is costs divided by revenue and if costs are greater than revenue then the firm will not operate.

A recent innovative paper by Squires (1987) estimates economies of scope, overall and product specific economies of scale, and measures of capacity utilization for the New England otter trawl industry. The paper extends a model of partial equilibrium due to Brown and Christensen (1981) to multiproduct firms in a profit function framework. The model estimates variable input demands as functions of quasi-fixed factors but also endogenously determines the levels of those quasi-fixed factors. For the multiproduct firm increasing ray returns to scale may be compatible with profit maximization since some other division of outputs between smaller firms may be more cost effective. Thus the profit function can be used in this case to measure ray returns to scale. Economies of scope can be determined by testing for the subadditivity of costs. While economies of scope due to allocable fixed factors will vanish in the long run they can be observed in the short run. Product specific economies of scale can be estimated by examining incremental marginal costs.

The paper by Squires (1987) uses panel data from 1980 and 1981 to estimate a profit function for otter trawl vessels. Economies of scope between different types of fish output are not found. Decreasing product specific returns to scale are found in roundfish and flatfish with increasing returns in the "all others" category. Overall ray returns to size is found to be decreasing. While the results of the paper are not of specific interest to agricultural researchers, the technique is interesting and should be considered. In particular the technique is important in the case where some factors are fixed in the short run and where multiproduct returns to size and scale are important.

Aggregate Studies of Size and Scale

While the most natural way to measure economies of size and scale is to use cross section data on individual firms, the use of aggregate time series data to represent either an aggregate technology or a representative firm has also been used extensively in the literature. Aggregate time series studies usually assume some type of technological progress in the underlying technology. The scale or size measure calculated in such studies will represent short run returns to scale or size since the observations are from points on a short run production, cost or profit function. Measures of returns to scale will summarize general tendencies of the industry while measured returns to size are probably not meaningful since they come from points on different short run cost curves. Technical change will also influence measures of scale economies since changes in productivity may be difficult to separate from movements along a non-homothetic production function (Diamond et al. 1978). Thus in measuring and comparing factor productivity the effects of returns to scale and non-neutral technical change must be considered. These drawbacks from using aggregate data have not dissuaded many researchers (e.g. Berndt and Khaled 1979), however. A number of recent papers investigating returns to size or scale using aggregate data are considered here.

A recent paper by Chan and Mountain (1983) investigates returns to scale in Canadian agriculture using annual data from 1952-1977. They found that the hypothesis of a Hicks neutral technology with no technological change could not be rejected by the data. They did, however, find increasing returns to scale over the sample period. Their "naive" estimates of factor productivity were consequently revised downwards due to the identified increasing returns to scale.

Ray (1982) estimated a two output cost function for U.S. agriculture using annual data from 1939-77. He used a translog form that allowed for a non-homothetic technology with Hicks-neutral technical change. He found overall scale economies indicating diminishing returns but that returns to scale have increased over time. Family labor was not considered as an input in the study and its exclusion may have biased the results.

Ball and Chambers (1982) applied a non-homothetic, non-neutral translog cost function to the U.S. meat products industry using annual data from 1954-76. A single output aggregate was used along with data on four input categories i.e., capital equipment, capital structures, labor, energy, and intermediate materials. The results show that over the latter years of the sample the industry was characterized by increasing returns to scale. The results also indicate negative technical progress over the same period. While these results might be confounding, the existence of increasing returns has important policy implications as the industry continues to become more concentrated.

Weaver (1983) estimated returns to size for North and South Dakota producers using data from 1950-70. He estimates a multiproduct translog profit function allowing for non-homotheticity and biased technical change. While he finds decreasing returns to size over the sample period, this is not surprising since as discussed earlier returns to size is just costs divided by revenue and this will be less than one for the rational firm. The estimated cost revenue ratios range from .59 to .74 over the sample period.

There have been a variety of profit function studies that have investigated the non-homotheticity of aggregate technology. Antle (1984) and Capalbo (1988) estimated aggregate profit functions for U.S. agriculture in order to investigate technical change and productivity growth. Antle (1984) made a strong case for non-homotheticity of the technology, but his results may be biased since his quantity and price series are not from the same source and thus non-homotheticity may be reflecting aggregation or measurement errors. Both Antle (1984) and Capalbo (1988) found decreasing returns to scale, but that is all that can be found using the profit function and so the results are not surprising.

In general the results of aggregate studies are mixed. Most have identified non-homothetic technologies but both increasing and decreasing returns to size and scale have been recognized. Aggregate studies on specific crop production have not been undertaken so no direct comparison to normative and cross-section results is possible. Most aggregate studies have been undertaken for a purpose other than the estimation of size economies and so less emphasis has been placed on the accurate measurement of such economies. This should be a caution to the prospective user of such research.

In summarizing studies relating to economies of size and scale, two general conclusions can be drawn. There do not seem to be significant economies of scale in the production of individual crops, at least at average firm size. The cost curve does fall, but remains rather flat over acreages compatible with the "average" family farm. Cost curves for livestock producers fall more sharply and over a larger range of output sizes. The lack of data on input allocations by multiproduct firms has hampered the estimation of size economies, especially in the case where production may be non-joint (Just et al. 1983, Chambers and Just 1989). The overriding impression from analyzing these studies is either that not enough work has been done on documenting the presence of size economies using actual data or such analysis is not relevant enough for general interest and analysis. Indeed, many studies (e.g. Hall and LeVeen 1978) were undertaken with specific policy implications in mind.

Studies of Firm Efficiency

Studies of firm efficiency can be divided into two major types: those that estimate average functions and test for optimizing behavior and economic efficiency, and those that estimate frontier functions and compare firms to the frontier. The first type is represented by the work of Lau and Yotopoulos (1971) and their colleagues and has primarily been applied in developing countries while the second approach follows early work of Farrell (1957) and later extensions to the stochastic case by Aigner, Schmidt and others.

Relative Efficiency Models

The relative efficiency approach proposes estimation of a profit function for a cross section of firms with a dummy variable to represent differences in firm efficiency due to technical or pricing inefficiency in maximizing profits. Most studies have used the Cobb-Douglas normalized profit function. The original study by Lau and Yotopolous

(1971) found that small farms in India were more efficient than large farms. The approach has also been used to investigate efficiency differences due to other factors with mixed results (Yotopoulos and Lau, 1979). A more recent paper by Kahn and Maki (1979) considered the effects of farm size on efficiency for farms in two regions of Pakistan. They found that larger farms were more economically efficient than smaller farms and that the sample farms exhibited increasing returns to scale. Both large and small firms were found to maximize profits given the output and input prices. In analyzing the findings, the authors attempt to explain reasons for the differences. Possible candidates are information availability, tenancy, and subsidized input purchases. Such an attempt to identify factors correlated with efficiency differences is important for policy analysis and will be considered in more detail later.

Trosper (1978) considered relative ranching efficiency between Indian and non-Indian ranchers in the Northern Great Plains using the Lau and Yotopolous (1971) method. He found that both groups profit maximize to the same degree but that per acre output for the Indians was lower due to less capital per acre.

Of more direct relevance, Garcia et al. (1982) applied the Lau and Yotopoulos (1971) method to a sample of Illinois cash grain farms. Variable inputs were categorized as hired labor and cash expenditure items (fertilizer, seed etc.). Fixed inputs used were family labor, land, and non-land capital. Other farm specific variables were also used to control for differences across firms. The hypothesis of equal economic efficiency across firms with more and less than 700 acres was not rejected at the 5 percent level. The firms did not display equal relative pricing efficiency for hired labor. Constant returns to scale were not rejected for the entire sample. One interesting explanation for the findings of the study is that all farms in the sample participated in the Illinois Farm Business Farm Management Association and thus management differences between the two samples may have been minimized.

No general implications can be drawn from these efficiency studies. The testing of such efficiency hypotheses across various regions and firm types would seem a useful endeavor.

Frontier Functions

In estimating production and cost functions, researchers typically assume that the error terms are symmetrically distributed with a zero mean. This implies that some firms' production will lie above the estimated surface while others will lie below it. An alternative formulation (Farrell) postulates that firms have different levels of economic efficiency and that the actual production function should be a frontier function such that no observations lie above it. Such production frontiers can either be deterministic (Aigner and Chu 1968) or stochastic (Aigner, Lovell, and Schmidt 1977). Deterministic frontiers can either be parametric or non-parametric. Parametric deterministic frontiers are typically derived using linear or quadratic programming to minimize the difference between each firm's production and the frontier subject to the constraint of a one-sided error. Stochastic frontiers allow deviations from the frontier to be either from inefficiency (a one-sided error deviation) or random factors (two-way deviations). They are usually estimated using maximum likelihood methods where the error term is composed of two parts. Frontier cost functions can also be estimated (Schmidt and Lovell 1979; Kopp and Diewert 1982) and used to estimate technical and allocative efficiency and also returns to size. Recent survey papers on this topic include Seiford and Thrall (1990), Bauer (1990), and Hallam (1993).

A comparison of deterministic and stochastic specifications using data from Swedish dairy plants is made in the paper by van den Broek et al. (1980). Parameters of the deterministic model are obtained by solving a linear programming problem while parameters of the stochastic model are obtained from maximum likelihood estimates both with an exponential efficiency distribution and with a composed error term. The maximum likelihood estimates are similar to the linear programming estimates but differ from those in the composed error model. The composed error model, in general, behaved more like an average rather than a· frontier estimator. The elasticity of scale function for the composed error model implied higher optimal scale than the linear programming or one sided maximum likelihood models. The general conclusion is that the results obtained are somewhat method dependent and so caution in interpreting the results of different studies should be made.

Before discussing specific applications to agriculture, two additional papers should be mentioned. Page estimates deterministic translog production frontiers for Indian manufacturing firms. The paper is interesting in the use of a flexible functional form and also in attempting to explain the effects on non-measured inputs on technical efficiency. After computing measures of technical efficiency for four industries (shoes, printing, soap and tools), the author develops a regression model to explain differences between firms based on factors such as literacy, labor force experience, capacity utilization, firm age and firm size. In all but the tool industry, firm size is not highly correlated with efficiency. The approach adopted in this study could be of general use in identifying factors correlated with agricultural efficiency.

Pitt and Lee (1981) apply the stochastic method of Aigner et al. (1977) to time series cross-section data on the Indonesian weaving industry. With this approach individual firm efficiency is not estimable, but mean efficiency can be recovered. Alternative specifications of the error term over time are considered. In order to investigate individual firm effects, separate firm intercepts obtained from the analysis of covariance are regressed on three firm characteristics. The results indicate that larger firms are more efficient than smaller firms and younger firms are more efficient than older firms. Where pooled data is available the methods proposed allow estimation of efficiency factors using the stochastic frontier method.

Different techniques as applied agriculture are illustrated by considering some recent papers investigating technical efficiency. The paper by Hall and LeVeen (1978) on farm size in California uses the non-parametric approach as developed by Farrell (1957) and extended by Seitz (1970) to investigate efficiency for small fruit and vegetable farms. While farms of all sizes were close to the efficiency locus, small farms were relatively farther from it. A statistical test rejected the hypothesis that large and small farms had equal efficiency. The study also found that pricing efficiency was more important than technical efficiency for the firms in the sample.

An early application of the stochastic technique to U.S. farms is the paper by Bagi (1982) that investigates the relationship between size and efficiency for a sample of a Tennessee farms. The study uses a Cobb-Douglas production function with a composed error term. The symmetric disturbance is normal and the one-sided disturbance is a truncated normal. The sample of 193 farms was divided into crop and mixed crop and livestock farms and further divided into small and large classes. Size divisions were made both on acreage and sales. Average technical efficiency was larger for crop farms than mixed farms. Small and large crop farms had similar levels of technical efficiency while large mixed farms had superior technical efficiency. Results between the different definitions of size were not significant. Overall levels of efficiency showed possible gains in output of from 15-25 percent given current input levels.

In a related study, Bagi and Huang (1983), using a decomposition technique due to Jondrow et al. (1982), investigated individual firm technical efficiency using the same data set and a translog production function. The translog function fit the data better than a Cobb-Douglas. Results indicate that the one-sided error dominated the symmetric error for mixed farms but that the errors were of about equal size for crop farms. Technical inefficiency was between 22 percent and 24 percent for both types of farms. There was wide variation in the level of technical efficiency for farms within the sample.

Bravo-Ureta (1986) investigates technical efficiency for a group of New England dairy farms using the deterministic approach of Aigner and

Chu (1968) as modified by Timmer (1971). The data is from a cross-section of 222 New England dairy herds in 1980. The Cobb-Douglas production function is used for estimation. Using the approach suggested by Timmer (1971), sample observations were deleted from the programming model until the estimates stabilized. This resulted in deletion of 4 percent of the observations. Average efficiency in sample was 82 percent. Returns to size for the sample was slightly larger than one. While data allowing estimation of factors correlated with efficiency was not available, a simple test of independence showed no relationship between firm size and efficiency. In a more recent paper, Bravo-Ureta and Rieger (1991) estimate a Cobb-Douglas stochastic production frontier to obtain a stochastic cost function and then use the decomposition procedures of Kopp and Diewert (1982) to allocate it to technical, economic and allocative factors. They also analyze the effects of a variety of socioeconomic factors on firm efficiency. The efficiency levels are not markedly affected by farm size, farmer education or experience.

Data from the Illinois Farm Business Farm Management Association records is the basis for two recent studies applying different methods of analysis. The first study by Aly et al. (1987) uses corrected least squares (Greene, 1980b) to estimate a deterministic statistical frontier. Rather than positing a two-part composed error, a single sided error is used. The method estimates the production function and shifts the constant term up until no residual is positive and a least one is zero. While all error is attributed to technical efficiency, the method has some advantages over the stochastic frontier approach. The technology can usually be specified more flexibly and multiple outputs can be handled more easily. The study uses a ray-homothetic production function to represent technology. This allows returns to scale to vary with output and input levels. A cross-section of data on grain farms from 1982 was used for the analysis. Gross revenue was used as an output measure. The average technical efficiency was 58 percent. Of the total efficiency loss about 60 percent was due to technical inefficiency and 40 percent due to scale inefficiency. Differences in efficiency were compared for different farm sizes. Larger farms tended to be more technically efficient than smaller firms when size was measured both in terms of acreage and sales classes. The most striking result of the paper is not the differences in efficiency between classes as the overall level of inefficiency of all farms.

A study by Byrnes et al. (1987) extends the non-parametric approach of Farrell (1957) to the case of non-constant returns to scale without strong disposability of inputs. In this way technical inefficiency can be divided into scale, congestion and pure technical components. The sample consisted of 107 cash grain farms in Illinois. Multiple outputs were explicitly accounted for by specifying a multiproduct production possibility set. Efficiency for the average farm in the sample was only 4 percent below maximum efficiency. The largest loss in efficiency was due to scale inefficiency. Pure technical efficiency was very high with some inefficiency due to congestion of inputs. Given the results on scale inefficiency, the sample was grouped based on size. Using 700 acres as dividing line between large and small farms, smaller farms were more efficient overall. For smaller farms, congestion was the major source of efficiency while scale inefficiencies were the major problem for larger farms. Most large farms were producing in a region of decreasing returns to scale. Statistical tests of the differences between technical efficiencies showed that the differences were not significantly different.

Using a similar data set, Grabowski et al. (1990) estimated a ray-homothetic production frontier and analyzed the effects of size on efficiency. Larger firms had a higher rate of returns to scale than smaller firms. The farms as a group produced at 82 percent of their potential output. But based on size, smaller firms were more technically efficient given their returns to scale than larger firms. The authors postulate that this results may be related to differing risk attitudes.

Kumbhakar et al. (1989) use data on Utah dairy herds to simultaneously investigate the effects of firm size on technical, allocative and economic efficiency. They conclude that all firms are inefficient but that larger firms are much less inefficient than smaller firms. They also find that education increases efficiency while off farm income seems to reduce it.

A final paper should be mentioned though it deals with Australian dairy farms. Battese and Coelli (1988) extend the results of Jondrow et al. (1982) to panel data sets in order to estimate individual firm efficiency levels using the stochastic frontier approach. While the approach requires specific distributional assumptions it allows for the prediction of individual firm effects. Schmidt (1985) has pointed out, however, that estimates of the one-sided errors developed in this way are not consistent in the usual manner.

The results of these frontier function studies of efficiency provide little clear guidance about firm size and firm efficiency except perhaps some evidence for improved efficiency among larger firms in dairy production. Smaller firms were found to be less, equally and more efficient than larger firms. This type of work applied to cross section data on many firms and across many regions would provide more definitive conclusions and policy recommendations.

Models of Firm Growth and Survival

An alternative way to analyze industry structure and tendencies for change is to track the rates of growth, contraction, and demise of firms.

If particular sizes or types of firms stand the test of time and grow and/or prosper, this provides information on the future structure of the industry and the viability of other types of firms.

A popular technique used to analyze firm growth is the finite Markov chain. A finite Markov chain is one in which a population at time t has a distribution S^t over the discrete states, $S_1, S_2, ...S_n$, and in which the probability P_{ij} of moving from state S_i to state S_j depends only on the state S_i and not on prior states. In firm growth analysis, the states are various size categories of firms. Using historical data, transition matrices can be estimated and then used to project future size distributions. If the transition probabilities are not constant, then alternative methods for estimation must be employed.

Studies by Padberg (1962), Hallberg (1969), and Stanton and Kettunen (1967) are early examples of application of the technique to agriculture. Several recent studies have employed Markov chains to project farm size changes. Ethridge, Roy, and Myers (1985) investigated structural changes in the Texas cotton ginning industry using both stationary and non-stationary transition probabilities. They consider five size and four activity groups. Activity groups are new entrant, dead, inactive and active. In the non-stationary analysis wages, electricity costs, lagged capacity utilization, lagged local production, and technological change variables were used as explanatory factors. All but capacity utilization and lagged production were significant. Projections to the year 1999 indicated a movement away from smaller firms and toward larger firms with a decline in the total number of firms. Changes in technology accelerated this trend.

Garcia, Offutt, and Sonka (1987) applied the Markov chain technique to a sample of 161 Illinois cash grain farms for which data was available from 1976-85. They used both output (gross value) and acreage measures of size and found some differences in results. Transition matrices estimated using the output variable showed little change in size while those estimated using acreage showed some growth towards larger and smaller firms. They tested for stability and found that transitions were stationary over a ten-year period. They then used the estimated probabilities to project the size distribution for 1990 and 2000. While the estimates based on gross product showed little change in farm structure, that based on acreage implied fewer middle sized farms and a slight tendency for a bimodal distribution. The authors then used discriminant analysis to explain whether firms grew or shrunk over their sample period. The general conclusion was that farm growth is fairly unpredictable.

Two earlier studies using aggregate U.S. data were carried out by Daly et al. (1972) and Lin et al. (1980) using upper triangular transition matrices. Recently Edwards, Smith, and Patterson (1985) analyzed changes in farm size and numbers using data from the 1974 and 1978 Census of Agriculture. They use acreage as a measure of size. They found that the percentage of mid-size farms decreased over the period but that long run projections show no major shift to a bimodal distribution. They also conclude that the future will be more like the present than the present is like the past. The Office of Technology Assessment (U.S. Congress 1985) used Markov chains to project U.S. farm sizes over the remainder of the century. They used the deflated gross value of farm income as a measure of size. They suggest a rather dramatic decline in mid-sized farms. Both of these studies must be viewed with some caution, however, since constant transition probabilities were used.

These Markov chain studies are important in that they provide a way to characterize past growth and project future change. The results are not strong enough to justify the assumption of a bi-modal distribution although many reasonable scenarios are compatible with such a distribution. The development of better such projection models with non-stationary probabilities explained by economic factors could provide useful characterizations of the future.

An alternative method of analyzing firm growth and survival is through simulation. By simulating the growth and survival of hypothetical firms over time, information on success probabilities and factors related to success can be obtained. Two studies in this regard will be mentioned. Patrick and Eisgruber (1968) develop a farm simulator for an Indiana crop and livestock farm based on 1964 data. The model includes consumption and multiple goals for the farm family. They project the model out for 20 years under a variety of alternative managerial and economic environment scenarios. Managerial ability was found to be the major determinant of firm growth. Interest rates were important in explaining firm growth and survivability. The availability of long-term credit was also important in determining firm growth.

A more recent paper by Richardson and Condra (1981) uses a dynamic Monte-Carlo simulation-programming model to analyze the projected survival and success of four alternative farm sizes in the El Paso Valley. While the model by Patrick and Eisgruber (1968) simulated alternative environments it did not consider the environments to be random. The Richardson and Condra (1981) paper postulated probability distributions for a variety of economic factors and then used Monte-Carlo techniques to compute a cumulative probability of success/failure for the farm classes. Alternative managerial strategies were not explicitly considered as in Patrick and Eisgruber (1968). The results indicate that initial equity is an important factor in explaining success. Furthermore, straight cash lease farms had little chance of success. Another interesting result is that for a 50-50 chance of success the farms needed at least 100 percent equity in 640 acres of farmland. The rather negative conclusions of this study for smaller farms should be validated with other data sets and techniques.

There has been recent renewed interest by the economics profession in various forms of Gibrat's law (Evans 1987a, 1987b; B. Hall 1987). Gibrat's law postulates that the growth rate of firms is determined by random factors independent of firm size. The basic equation used to test Gibrat's law is

$$\ln S_{it} = \alpha + \beta \ln S_{it-1} + e_{it} \tag{1}$$

where S_{it} is the size of the ith firm in the tth period. Gibrat's law predicts that $\beta = 1$.

Evans (1987a) found this law of proportional growth to hold for large firms while Hall found that smaller firms had higher and more variable growth rates. A recent paper by Shapiro, Bollman, and Ehrensaft (1987) tests this relationship for Canadian farms between 1966 and 1981. They use acres, gross sales and capital value as alternative measures of size. They found that small farms grow faster than large farms and thus reject Gibrat's law using any of the three measures. They also found that the variance of the growth rate was inversely related to farm size so that larger firms had more stable growth. Factors contributing to the faster growth of smaller firms could be investigated.

Some authors have developed analytical and econometric models to explain changes in size and structure of firms over time. Sumner and Leiby (1987) in a series of papers analyze the effects of human capital (managerial ability), size and certain control variables on the growth of firms in a sample of Southern dairies. They use number of cows as a measure of firm size. They find that size increases with operator age for low ages but decreases with over twenty years of experience. Herd size also increases with experience for lower and middle experience levels. Schooling had a positive effect on size. There was a strong positive relationship between a variable representing improved management practices and firm growth. The results of this study, while supporting the hypothesis of management effects on size, also point out the need for analysis of the lifecycle of the firm, especially the sole proprietor firm.

Kislev and Peterson (1982) argue that changes in farm size can be explained by changes in the machine-labor ratio. They postulate a model in which input prices, nonfarm income, and technology explain farm size. They use acres as a measure of farm size. They develop an equilibrium model of the agricultural sector based on a two-level production function (mechanical and biological), a demand for food equation, a land rental price equation, and a fixed supply of labor per farm. By combining empirical analysis with previous estimates they attempt to validate their model. They find that relative factor price changes explain 99 percent of the change in both the machine labor ratio and farm size over the 1930-70 period. They further predict that with increases in the price of capital that the trend to larger farms may decline.

These later studies point out an important area of empirical research. Rather than attempting just to measure size or scale economies, they attempt to explain the factors that lead to changes in firm structure and develop structural models that can be tested and validated using actual observations on firms. While the underlying models in the papers are not fully developed or estimated they point a useful direction for future research. Measurement in conjunction with theory and underlying economic structure may be of more value than measurement alone.

Other Approaches

There are a variety of papers that have addressed firm size issues using other techniques. A couple will be mentioned. A paper by Weersink and Tauer (1991) uses Granger causality techniques to test whether dairy firm size causes productivity differences or whether productivity differences cause differences in firm size. They conclude that size has tended to cause differences in productivity rather than vice versa.

A much cited recent study by Kinnucan et al. (1990) analyzed in an ex ante fashion the effects of a variety of variables on the adoption of bovine somatotropin (bst) in dairy herds. They found in their survey of producers in the Southeast a significant positive relationship between firm size and knowledge of bst and also between firm size and how soon the producer would adopt it if became available. They found a positive but insignificant relationship between size and the intensity of adoption. They then argued that adoption will not be scale neutral. This effect may be short run, however.

Pecuniary Economies and Firm Size

In addition to economies related to size and scale, firms may attain economies in the purchase of inputs or sale of outputs. Such economies may result from technological economies due to large transactions in the marketing process, price discrimination, transactions costs, timing of input purchases, vertical integration and spatially separated markets. Two recent studies have investigated such pecuniary economies. Tew et al. (1980) consider sales by a large Georgia agricultural supply firm. They regress the price of the product sold on the quantity sold, and a series of control (dummy) variables. They find that for seven of the inputs that lower prices are implied by larger quantities. For the other eight inputs the relationship was not statistically significant. When the implied price discounts were substituted into production budgets the effect on variable costs was negligible however.

Smith et al. (1986, 1984) investigated pecuniary economies for a sample of cotton farms in the Texas Southern High Plains. They use the composite firm approach (budgets from actual farm records) in comparing costs for firms ranging from 50 to 5,570 acres. Their sample includes 98 farms stratified by size for the 1980 crop year. Agribusiness firms surveyed indicated a preference for uniform pricing of products in order to maintain good customer relations and few price discounts were discovered. The study also found few economies due to optimal timing of input purchases. The authors did find, however, economies due to vertical integration either through cooperatives or direct ownership. Producers cited lack of volume discounts as a reason for participating in vertical integration. The study did find significant pecuniary economies associated with cotton marketing. The cotton price received by the largest firms was 7 percent higher than that received by the smallest producers. These economies could not be traced to specific marketing strategies.

Both of these studies provide evidence of some pecuniary economies. The existence of such phenomenon for agricultural firms in general should be verified or refuted. Given that few economies of size seem to exist for crop farms, explanations for the gradual growth in size may rely on such external economies.

Summary and Conclusions

This chapter has reviewed empirical studies which investigate economies of size, scale and scope in agriculture. The chapter also discusses the related issues of firm efficiency, firm growth and firm survival. The general conclusion is that while some economies of size or scale may exist for livestock farms, that significant economies, at least as conventionally defined, do not exist for most crop production activities. While differences in efficiency and growth paths differ among firms, few of these seem to be directly related to economies of size and scale. Nevertheless, the structure of agriculture is changing rapidly. The factors that are bringing about these changes are myriad but a clear theory of how they influence firm size and structure is not yet present.

The lack of definitive statements about actual technical economies affects the researchers ability to use empirical findings as a springboard for new theoretical forays. The three areas for further research mentioned by this author in Chapter 7 seem relevant after this journey through the literature. There have been few studies of multiproduct agricultural firms using cross-section data and positive methods. The methods are straightforward and a compilation of results similar to those summarized by Madden for normative studies would be useful. A need exists for better models and empirical analysis that explains the growth and contraction of individual firms over time. The estimation of frontier functions and subsequent analysis of the results would provide much better data on the effects of firm efficiency for the structure of an industry.

The research agenda for the future should include not only descriptive or predictive analysis, but improved models that explain behavior and ability to follow decision rules. The differences in firm efficiency may be more a reflection of commitment to perceived goals than the result of any underlying differences in technology or economic environment.

References

Aigner, D., C. A. K. Lovell, and P. Schmidt. 1977. "Formulation and Estimation of Stochastic Frontier Production Function Models." *Journal of Econometrics* 6: 21-37.

Aigner, D. J., and S. F. Chu. 1968. "On Estimating the Industry Production Function." *American Economic Review* 58: 826-839.

Akridge, J. T., and T. Hertel. 1986. "Multiproduct Cost Relationship for Retail Fertilizer Markets." *American Journal of Agricultural Economics* 68: 928-938.

Aly, H. Y., K. Belbase, R. Grabowski, and S. Kraft. 1987. "The Technical Efficiency of Illinois Grain Farms: An Application of a Ray-Homothetic Production Function." *Southern Journal of Agricultural Economics* 19(1): 69-78.

Anderson, J. R., and R. A. Powell. 1973. "Economies of Size in Australian Farming." *Australian Journal of Agricultural Economics* 17(1): 1-16.

Antle, J. 1984. "The Structure of U.S. Agricultural Technology 1910-78." *American Journal of Agricultural Economics* 66: 414-421.

Bagi, F. S., and C. J. Huang. 1983. "Estimating Production Technical Efficiency for Individual Farms in Tennessee." *Canadian Journal of Agricultural Economics* 31: 249-256.

Bagi, F. S. 1982. "Relationship between Farm Size and Technical Efficiency in West Tennessee Agriculture." *Southern Journal of Agricultural Economics* 14: 139-144.

Ball, V. E., and R. G. Chambers. 1982. "An Economic Analysis of Technology in the Meat Products Industry." *American Journal of Agricultural Economics* 64: 699-709.

Battese, G. E., and T. J. Coelli. 1988. "Prediction of Firm-Level Technical Efficiencies with a Generalized Frontier Production Function and Panel Data." *Journal of Econometrics* 38: 387-398.

Bauer, P. W. 1990. "Recent Developments in the Econometric Estimation of Frontiers." *Journal of Econometrics* 46: 39-56.

Baumol, W. J., J. C. Panzer, and R. D. Willig. 1982. *Contestable Markets and the Theory of Industry Structure.* New York: Harcourt Brace Jovanovich Inc.

Berndt, E. R., and M. S. Khaled. 1979. "Parametric Productivity Measurement and Choice Among Flexible Functional Farms." *Journal of Political Economy* 87: 1220-1245.

Berndt, E. R., and D. O. Wood. 1975. "Technology, Prices, and the Derived Demand for Energy." *Review of Economics and Statistics* 57: 259-268.

Bravo-Ureta, Boris E. 1986. "Technical Efficiency Measures for Dairy Farms Based on a Probabilistic Frontier Function Model." *Canadian Journal of Agricultural Economics* 34(3): 399-415.

Bravo-Ureta, B. E., and L. Rieger. 1991. "Dairy Farm Efficiency Measurement Using Stochastic Frontiers and Neoclassical Duality." *American Journal of Agricultural Economics* 73: 421-28.

Broeck van den, J., F. R. Forsund, L. Hjalmarsson, and W. Meusen. 1980. "On the Estimation of Deterministic and Stochastic Frontier Production Functions." *Journal of Econometrics* 13: 117-138.

Brown, R., and L. Christensen. 1981. "Estimating Elasticities of Substitution in a Model of Partial Static Equilibrium: An Application to U.S. Agriculture, 1947 to 1974," in E. Berndt and B. Field, eds., *Modeling and Measuring Natural Resource Substitution.* Cambridge, MA: MIT Press.

Byrnes, P., R. Fare, S. Grosskopf, and S. Kraft. 1987. "Technical Efficiency and Size: The Case of Illinois Grain Farms." *European Review of Agricultural Economics* 14(4): 367-382.

Capalbo, S. M. 1988. "Measuring the Components of Aggregate Productivity Growth in U.S. Agriculture." *Western Journal of Agricultural Economics* 13: 53-62.

Carter, H. O., and G. W. Dean. 1961. "Cost-Size Relationships for Cash-Crop Farms in a Highly Commercialized Agriculture." *Journal of Farm Economics* 43: 264-277.

Carter, H. O., and G. W. Dean. 1962. "Cost-Size Relationships for Cash Crop Farms in Imperial Valley, California." Calif. Agr. Expt. Sta., Giannini Found. Res. Rpt. 253.

Chambers, R. G. 1988. *Applied Production Analysis: A Dual Approach.* Cambridge: Cambridge University Press.

———. 1984. "Some Conceptual Issues in Measuring Economies of Size." *Economies of Size Studies.* Ames, IA: Center for Agricultural and Rural Development.

Chambers, R. G., and R. E. Just. 1989. "Estimating Multioutput Technologies." *American Journal of Agricultural Economics* 71: 980-95.

Chan, M. W. Luke, and D. C. Mountain. 1983. "Economies of Scale and the Tornqvist Discrete Measure of Productivity Growth." *Review of Economics and Statistics* 65: 663-667.

Christensen, L. R., and W. H. Green. 1976. "Economies of Scale in U.S. Electric Power Generation." *Journal of Political Economy* 84: 665-676.

Christensen, L. R., D. W. Jorgensen, and L. J. Lau. 1973. "Transcendental Logarithmic Production Functions." *Review of Economics and Statistics* 55: 28-45.

Cooke, S. C., and W. B. Sundquist. 1987. "Cost Structures, Productivities, and the Distribution of Technology Benefits Among Producers for Major U.S. Field Crops." *Evaluating Agricultural Research and Productivity.* Proceedings of a Symposium, January 29-30, 1987, Atlanta, GA; Miscellaneous Publication 52-1987, Minnesota Agricultural Experiment Station, University of Minnesota.

_____. 1989. "Cost Efficiency in U.S. Corn Production." *American Journal of Agricultural Economics* 71: 1003-1010.

Cowing, T. G., D. Reifschneider, and R. Stevenson. "A Comparison of Alternative Frontier-Cost Function Specifications," in Ali Dogramaci, ed., *Developments in Econometric Analyses of Productivity: Measurement and Modeling Issues.* Pp. 63-92. Boston: Kluwer-Nijhoff Publishing.

Daly, Rex F., J. A. Dempsey, and C. W. Cobb. 1972. "Farm Numbers and Size in the Future," in A. Gordon Ball and Earl O. Heady, eds., *Size, Structure, and Future of Farms.* Pp. 314-32. Ames: Iowa State University Press.

Deller, S. C., D. L. Chicoine, and N. Walzer. 1988. "Economies of Size and Scope in Rural Low-Volume Roads." *The Review of Economics and Statistics* pp.459-465.

Diamond, P., D. McFadden, and M. Rodriquez. 1978. "Measurement of the Elasticity of Factor Substitution and Bias of Technical Change," in Melvyn Fuss and Daniel McFadden, eds., *Production Economics: A Dual Approach to Theory and Applications.* Vol. 2, pp. 125-141. North-Holland Publishing Company.

Dixon, Bruce L., Batte, Marvin T., and Steven T. Sonka. 1984. "Random Coefficients Estimation of Average Total Product Costs for Multiproduct Firms." *Journal of Business and Economic Statistics* 2(4): 360-66.

Edwards, C., M. G. Smith, and R. N. Patterson. 1985. "The Changing Distribution of Farms by Size: A Markov Analysis." *Agricultural Economics Research* 37: 1-16.

Ethridge, Don E., Sujit K. Roy, and David W. Myers. 1985. "A Markov Chain Analysis of Structural Changes in the Texas High Plains Cotton Ginning Industry." *Southern Journal of Agricultural Economics*: 11-20.

Evans, D. S. 1987a. "The Relationship Between Firm Size, Growth, and Age: U.S. Manufacturing, 1976-1982." *Journal of Industrial Economics* 35: 567-581.

_____. 1987b. "Tests of Alternative Theories of Firm Growth." *Journal of Political Economy* 95: 657-674.

Fare, R., L. Jansson, and C. A. Knox Lovell. 1985. "Modelling Scale Economies with Ray Homothetic Production Functions." *Review of Economics and Statistics* 67: 624-629.

Farrell, M. J. 1957. "The Measurement of Productive Efficiency." *Journal of Royal Statistics Society* 120: 253-281.

Farris, J. Edwin, and David L. Armstrong. 1965. "Economies in the Acquisition of Inputs – A Pilot Study." *Canadian Journal of Agricultural Economics* 13: 70-81.

Fleming, Marion S., and Ihn H. Uhm. 1982. "Economies of Size in Grain Farming in Saskatchewan and the Potential Impact of Rail Rationalization Proposals." *Canadian Journal of Agricultural Economics* 1: 1-20.

Forsund, F. R., C. A. K. Lovell, and P. Schmidt. 1980. "A Survey of Frontier Productions and of Their Relationship to Efficiency Measurement." *Journal of Econometrics* 13: 5-25.

Garcia, P., S. E. Offutt, and S. T. Sonka. 1987. "Size Distribution and Growth in a Sample of Illinois Cash Grain Farms." *American Journal of Agricultural Economics* 69: 471-476.

Garcia, P., S. T. Sonka, and M. Yoo. 1982. "Farm Size, Tenure, and Economic Efficiency on a Sample of Illinois Grain Farms." *American Journal of Agricultural Economics* 64: 119-123.

Grabowski, R., and K. Belbase. 1986. "An Analysis of Optimal Scale and Factor Intensity in Nepalese Agriculture: An Application of a Ray-Homothetic Production Function." *Applied Economics* 18: 1051-63.

Grabowski, R., and O. Sanchez. 1986. "Return to Scale in Agriculture: An Empirical Investigation of Japanese Experience." *European Review of Agricultural Economics* 13: 189-198.

Grabowski, R., H. Y. Aly, S. Kraft, and C. Pasurka. 1990. "A Ray-Homothetic Production Frontier and Efficiency: Grain Farms in Southern Illinois." *European Review of Agricultural Economics* 17(4): 435-48.

Greene, W. H. 1980b. "On Estimation of a Flexible Frontier Production Model." *Journal of Econometrics* 13: 101-15.

Hall, Bronwyn H. 1987. "The Relationship Between Firm Size and Firm Growth in the U.S. Manufacturing Sector." Paper No. 1965, National Bureau of Economic Research Work. *Journal of Industrial Economics* 35: 583-606.

Hall, B. F., and E. P. LeVeen. 1978. "Farm Size and Economic Efficiency: The Case of California." *American Journal of Agricultural Economics* 60: 589-600.

Hall, R. E. 1973. "The Specification of Technology with Several Kinds of Output." *Journal of Political Economy* 81: 878-92.

Hallam, A. 1993. "A Brief Overview of Nonparametric Methods in Economics." *Review of Agricultural and Resource Economics.* Forthcoming.

Hallberg, M. C. 1969. "Projecting the Size Distribution of Agriculture Firms - An Application of a Markov Process with Non-Stationary Transition Probabilities." *American Journal of Agricultural Economics* 51: 289-302.

Heady, E. O., and R. D. Krenz. 1962. "Farm Size and Cost Relationships in Relation to Recent Machine Technology." Iowa Agr. and Home Econ. Expt. Sta. Res. Bul. 504.

Hildreth, C., and J. P. Houck. 1968. "Some Estimators for a Linear Model with Random Coefficients." *Journal of the American Statistical Association* 63: 583-595.

Hoch, I. 1976. "Returns to Scale in Farming: Further Evidence." *American Journal of Agricultural Economics* 58: 745-749.

Hornbaker, R. H., B. L. Dixon, and S. T. Sonka. 1989. "Estimating Production Activity Costs for Multi-Output Firms With a Random Coefficient Regression Model." *American Journal of Agricultural Economics* 71: 167-77.

Ihnen, L., and E. O. Heady. 1964. "Cost Functions in Relation to Farm Size and Machinery Technology in Southern Iowa." Iowa Agr. and Home Econ. Expt. Sta. Res. Bul. 527.

Johnston, J. 1960. *Statistical Cost Functions.* New York: McGraw-Hill.

Jondrow, J., C. A. K. Lovell, I. S. Materov, and P. Schmidt. 1982. "On the Estimation of Technical Inefficiency in the Stochastic Frontier Production Function Model." *Journal of Econometrics* 19(2/3): 233-38.

Just, R. E., D. Zilberman, and E. Hochman. 1983. "Estimation of Multicrop Production Functions." *American Journal of Agricultural Economics:* 770-780.

Kahn, M. H., and D. R. Maki. 1979. "Effects of Farm Size on Economic Efficiency: The Case of Pakistan." *American Journal of Agricultural Economics* 61: 64-69.

Kinnucan, H., U. Hatch, J. Molnar, and M. Venkateswaran. 1990. "Scale Neutrality of Bovine Somatotropin: Ex Ante Evidence from the Southeast." *Southern Journal of Agricultural Economics* 22: 1-12.

Kislev, Yoav, and Willis Peterson. 1982. "Prices, Technology, and Farm Size." *Journal of Political Economy* 90: 578-595.

Kopp, R. J., and W. E. Diewert. 1982. "The Decomposition of Frontier Cost Function Deviations Into Measures of Technical and Allocative Efficiency." *Journal of Econometrics* 19: 319-331.

Krause, K. R., and L. R. Kyle. 1970. "Economic Factors Underlying the Incidence of Large Farming Units: The Current Situation." *American Journal of Agricultural Economics* 52: 748-761.

Kumbhakar, Subal C., Basudeb Biswas, and DeeVon Bailey. 1989. "A Study of Economic Efficiency of Utah Dairy Farmers: A System Approach." *Review of Economics and Statistics* 71: 595-604.

Lau, L. J., and P. A. Yotopoulos. 1971. "A Test for Relative Efficiency and an Application to Indian Agriculture." *American Economic Review* 61: 94-109.

Lin, William, George Coffman, and J. B. Penn. 1980. "U.S. Farm Numbers, Size, and Related Structural Dimensions: Projections to Year 2000." ESCS, Technical Bulletin No. 1625. Washington: U.S. Dept. of Agriculture.

Madden, J. P. 1967. *Economies of Size in Farming.* AER-107. Washington, DC: Economic Research Service, U.S. Department of Agriculture.

McFadden, D. 1978. "Cost, Revenue, and Profit Functions" in M. Fuss and D. McFadden, eds., *Production Economics: A Dual Approach to Theory and Applications.* Amsterdam: North-Holland Publishing Company.

Meisner, Joseph C., and V. James Rhodes. 1975. "Cattle Feeding Economies of Size Revisited." *Canadian Journal of Agricultural Economics* 23: 61-65.

Miller, Thomas A., Gordon E. Rodewald, and Robert G. McElroy. 1981. *Economies of Size in U.S. Field Crop Farming.* AER-472. Washington, DC: Economic Statistics Service, U.S. Department of Agriculture.

Moschini, Giancarlo. 1988. "The Cost Structure of Ontario Dairy Farms: A Microeconomic Analysis." *Canadian Journal of Agricultural Economics* 36: 187-206.

_____. 1990. "Nonparametric and Parametric Estimation: An Analysis of Multiproduct Returns to Scale." *American Journal of Agricultural Economics* 72: 589-98.

Moore, Charles V. 1982. "Impact of Federal Limitation Policy on Western Irrigated Agriculture." *Western Journal of Agricultural Economics* 7: 301-316.

Nerlove, M. 1963. "Returns to Scale in Electricity Supply," in C.F. Christ et al., eds., *Measurement in Economics: Studies in Mathematical Economics and Econometrics in Memory of Yehuda Grunfeld.* Stanford: Stanford University Press.

Padberg, D. I. 1962. "The Uses of Markov Processes in Measuring Change in Market Structure." *Journal of Farm Economics* 44: 189-199.

Page, J. M. 1984. "Firm Size and Technical Efficiency: Applications of Production Frontiers to Indian Survey Data." *Journal of Development Economics* 16: 129-152.

Patrick, G., and L. Eisgruber. 1968. "The Impact of Managerial Ability and Capital Structure on Growth of the Farm Firm." *American Journal of Agricultural Economics* 50: 491-506.

Pitt, M. M., and L. Lee. 1981. "The Measurement and Sources of Technical Inefficiency in the Indonesian Weaving Industry." *Journal of Development Economics* 9: 43-64.

Ray, S. 1982. "Translog Cost Analysis of U.S. Agriculture." *American Journal of Agricultural Economics* 62: 490-498.

Richardson, J., and G. D. Condra. 1981. "Farm Size Evaluation in the El Paso Valley: A Survival/Success Approach." *American Journal of Agricultural Economics* 63: 430-437.

Schmidt, P. 1985-86. "Frontier Production Functions." *Econometric Reviews* 4(2): 289-328.

Schmidt, P., and C. A. K. Lovell. 1979. "Estimating Technical and Allocative Inefficiency Relative to Stochastic Production and Cost Frontiers." *Journal of Econometrics* 9: 363-366.

Seitz, W. D. 1970. "The Measurement of Efficiency Relative to a Frontier Production Function." *American Journal of Agricultural Economics* 51: 505-511.

Shapiro, D., R. D. Bollman, and P. Ehrensaft. 1987. "Farm Size and Growth in Canada." *American Journal of Agricultural Economics* 69: 477-483.

Seiford, L. M., and R. M. Thrall. 1990. "Recent Developments in Data Envelopment Analysis: The Mathematical Programming Approach to Frontier Analysis." *Journal of Econometrics* 46: 7-38.

Smith, E. G., R. D. Knutson, and J. W. Richardson. 1986. "Input and Marketing Economies: Impact on Structural Change in Cotton Farming on the Texas High Plains." *American Journal of Agricultural Economics* 68: 716-720.

Smith, E. G., J. W. Richardson, and R. D. Knutson. 1984. *Cost and Pecuniary Economics in Cotton Production and Marketing: A Study of Texas Southern High Plains Cotton Producers.* College Station, TX: Texas Agricultural Experiment Station.

Squires, D. 1987. "Long-Run Profit Functions for Multiproduct Firms." *American Journal of Agricultural Economics* 69: 558-569.

Stanton, B. F., and L. Kettunen. 1967. "Potential Entrants and Projections in Markov Process Analysis." *Journal of Farm Economics* 49: 633-42.

Sumner, D. A., and J. D. Leiby. 1987. "An Econometric Analysis of the Effects of Human Capital on Size and Growth Among Dairy Farms." *American Journal of Agricultural Economics* 69: 465-470.

Sumner, Daniel A., and James D. Leiby. 1987. "An Econometric Analysis of Size and Growth: Size/Growth Relationships, Selection, Entry/Exit, and Human Capital." Department of Economics and Business, North Carolina State University.

Swamy, P. A. V. B., and P. Tinsley. 1980. "Linear Prediction and Estimation Methods for Regression Models with Stationary Stochastic Coefficients." *Journal of Econometrics* 12: 103-42.

Tew, Bernard W., et al. 1980. "Some Evidence on Pecuniary Economies of Size for Farm Firms." *Southern Journal of Agricultural Economics* 12: 151-154.

Timmer, C. Peter. 1971. "Using a Probabilistic Frontier Production Function to Measure Technical Efficiency." *Journal of Political Economy* 79: 776-794.

Trosper, R. L. 1978. "American Indian Relative Ranching Efficiency." *American Economic Review* 68: 503-516.

U.S. Congress, Office of Technology Assessment. 1985. *Technology, Public Policy, and the Changing Structure of American Agriculture: A Special Report for the 1985 Farm Bill.* OTA-F-272, Washington, DC.

Vlastuin, C., D. Lawrence, and G. Quiggin. 1982. "Size Economies in Australian Agriculture." *Review of Marketing and Agricultural Economics* 51: 27-50.

Weaver, R. D. 1983. "Multiple Input, Multiple Output Production Choices and Technology in the U.S. Wheat Region." *American Journal of Agricultural Economics* 65: 45-56.

Weersink, A., and L. W. Tauer. 1991. "Causality between Dairy Farm Size and Productivity." *American Journal of Agricultural Economics* 73: 1138-1145.

Weimar, M. R., A. Hallam, and L. D. Trede. 1988. "Economies of Size, Tax Reform and Profitability of Alternative Midwestern Feedlot Systems." *Journal of the American Society of Farm Managers and Rural Appraisers* 52: 11-18.

Yotopoulos, P. A., and L. J. Lau. 1979. Eds. "Resource Use in Agriculture: Applications of the Profit Function." *Food Research Institute Studies* 17(1).

Zellner, A., and N. Revankar. 1969. "Generalized Production Functions." *Review of Economic Studies* 36: 241-50.

Zulauf, C. R. 1986. "Changes in Selected Characteristics of U.S. Farms During the 1970s and Early 1980s: An Investigation Based on Current and Constant Dollar Sales Categories." *Southern Journal of Agricultural Economics* 18: 113-122.

9

Why Are Some Farms More Successful than Others? A Review

Glenn Fox, Philip A. Bergen, and Ed Dickson

"What experience and history teach is this - that people and governments never have learned anything from history, or acted on principles deduced from it."

Georg Wilhelm Friedrich Hegel, *Philosophy of History*

"History is only a confused heap of facts."

Lord Chesterfield, *Letters*, Feb. 5, 1750

Background

In Chapter 2 of this volume, Stanton describes the multiple dimensions of the concept of farm structure. Generally the study of farm structure is conducted at an aggregate level. The farm sector or industries within the farm sector which produce particular commodities form the basic unit for analysis. Within that unit of analysis the distribution of sizes, incomes, the structure of resource ownership, or other aspects of structure are considered. In each the analysis represents a snapshot which characterizes a particular dimension of farm structure at a moment in time. Each of these snapshots is itself a product of a history of managerial decisions of farm operators. Some of these farm operators have been more successful, others have been less successful, but the periodic characterization of structure, measured at an industry or sectoral level, needs to be seen as a sequence of stop action photographs of the dynamic evolution of populations of farm firms.

A long tradition of empirical farm management research in the United States and Canada has emphasized the study of the performance and evolution of farm businesses at the firm level. The purpose of this chapter is to assess the consistency and relevance of this research program for the study of farm structure, particularly for the study of the size distribution of farm firms. The chapter concludes with some suggestions to enrich the interface between empirical farm management studies and farm structure studies in the future.

According to legend,[1] the bulk of research activity of agricultural economists, and the lion's share of class time in agricultural economics undergraduate programs in North America for the first half of this century was devoted to addressing the question posed in the title of this chapter. This question is important for a number of reasons. To the farm manager, knowledge of actions which have enhanced the viability of other farms could be of value in strategic planning. The development of public policy would benefit from a separation of systemic, that is sector- or industry-wide, factors which act on farm incomes, from firm specific factors. The traditional motivation for income transfer farm policies has derived much force from the perception that incomes and rates of return in farming are chronically and universally depressed. The fact that some farms have done well even when circumstances have been adverse, has not been emphasized. To the economist, study of the factors which make some farms more successful than others offers the opportunity to assess the performance of alternative theories of the behavior of firms.

In spite of all this, however, research on this topic came, more or less, to an abrupt halt in about 1950. The published record gives little indication as to the reason.[2] The oral history points to a divisive conflict between the "theorists" and the "empiricists," which occurred at Black Duck, Minnesota in 1949.[3] Shortly thereafter, the thrust of farm management and production economics research shifted away from the investigation of success factors and toward the development of farm-level linear programming models and the estimation of biological production functions. The study of what makes some farms more successful than others suffered from an extended period of neglect. The 1980s, however, witnessed a resurgence of interest in this topic. Unfortunately, researchers in the current era have shown limited sensitivity to the issues that delimited the battle lines in 1949. In addition, the development of both the theoretical and the empirical apparatus of economic research since 1950 has revealed conceptual and practical matters that would have added to the stock of armaments on both sides at Black Duck.

A brief history of the exchange of views is instructive. Schultz (1939) is often identified as an early critic of the empiricist tradition in farm management research. He argued that the accumulated results of past work identifying profitability factors were of little value to a farm operator faced with constantly changing technical and pecuniary conditions. One interpretation of this criticism is that farm management research had not adopted the language and concepts of the emerging neoclassical theory of the firm. His references to Black, Knight, Kaldour and Hicks would support this view. A second plausible interpretation, however, is that farm management research had been conducted without a model of entrepreneurship. The neoclassical theory of the firm, particularly in its static mode, has never offered such a model. (See Baumol (1968) or Kirzner (1973).) Schultz (1939, 583) also questioned the usefulness of efforts to estimate agricultural production functions and to establish values of input-output coefficients for agricultural production practices. Fifty years later, Schultz (1939) seems to offer little comfort to either camp.

Heady (1948), a protagonist for the "theorists," is less equivocal. He offers a two-fold critique of the then current practice in farm-management research. First, results of many of the empirical profitability studies suggest increasing returns, often without any acknowledged limit, to certain actions. Heady saw failure to recognize the law of eventually diminishing returns as a fatal flaw. Furthermore, he pointed out that many of the variables used to explain differences in profitability were variations on single-factor productivity ratios. He correctly pointed out that using these ratios as measures of "efficiency" made little economic sense. Shaw and Wright (1955) reiterated this point, emphasizing that many studies confounded overlapping and interdependent single factor productivity ratios. In addition, they point out that longitudinal studies often paid inadequate attention to changes in prices and technology over time. Wilcox (1947), a supporter of empirical farm management research, acknowledged that many studies relied on non-random samples and often failed to clearly identify a cause and effect relationship between actions and success.

Engene (1950) conceded that farm management research had made little use of economic theory, but suggested that the static neoclassical theory of the firm was of little use to the question of what makes some farms more successful than others. His argument was based on the idea that the theory is untested and that its assumptions are unrealistic. Continued evolution of the theoretical framework was necessary. It may be the product of excessively kind hindsight, but Engene seems to flirt with the idea that the static neoclassical theory of the firm cannot explain the existence of a distribution of profits in an industry, which is the very phenomenon that empirically oriented farm management researchers sought to rationalize. Glenn Johnson (1950) also called for a more completely developed theory of the firm. It is possible to interpret

Johnson's early remarks as a call for a theory of entrepreneurship, which he has continued to articulate (Johnson 1988).

An Indicative Review of the Literature

The empirical literature on this topic is enormous.[4] We have selected a small sample of studies in three time periods to illustrate the evolution of procedures and to assess the consistency of results obtained in this extended research program. Studies have been grouped into time periods delimited by the years 1900 to 1948, 1948 to 1980 and 1980 to the present. Research in this genre has defined farm success generally in terms of profitability, that is the ability to generate returns to operator labor and equity, or viability, meaning the ability to meet current and future financial obligations or growth, characterized as increases in the real value of total assets, production or net worth over time.

1900 to 1948

Farm management research during this period focused primarily on profitability as the measure of farm success. Table 9.1 outlines the key characteristics of several typical studies. One of the earliest studies is Warren and Burritt (1909). They cross classified[5] survey data of 178 New York State farms to investigate their profitability. Their major finding was that the most profitable farms were larger in size, producing more revenue but controlling their cash expenses at about the same level as the least profitable farms. In a more extensive study, Boss et al. (1917) cross-classified farm record data from Minnesota farms between 1912 and 1913 to identify factors associated with higher operator labor income. In general, the authors identified increased farm size and productivity[6] of land, livestock and labor as associated with increased operator labor income. Further, income also increased as more of the above factors were above their respective sample average. Warren (1932) believed that "by studying large numbers of farms, the real reasons for success and their relative importance usually stand out clearly." Using cross-classification analysis, he evaluated a total of 1789 New York farm records for the years 1908, 1918 and 1928, for factors contributing to higher labor income. He concluded that farms with higher labor income had the following characteristics: presence of vegetable and/or dairy enterprises, higher than average crop and livestock productivity, larger than average farm size with more efficient[7] labor, capital and horse use and farm operators with a higher than average education. Warren emphasized that success

TABLE 9.1 Empirical Studies Evaluating Success Factors: 1900-1948

Author(s)	Location Commodity Class(es) and Time Period	Empirical Method	Success Criterion	Success Factors Evaluated	Results (Success Increasing Factors)
Warren and Burritt (1909)	New York Farms 1906-1907	Cross classification of survey data	Percent Return on Investment	Enterprise Type Farm Size Capital Items Expense Items	Larger farm size and control of cash expenses.
Boss et al. (1917)	Minnesota Farms 1912-1913	Cross classification of survey data	Operator Labor, Income	Farm Size Input Productivity - land - labor - horses Input Effidiency - livestock	An increase in each factor evaluated.
Warren (1932)	New York Farms 1908, 1918, 1928	Cross classification of farm business records	Operator Labor, Income	Enterprise Type Farm Size Input Productivity - land - livestock - labor Tenure Operator Education	Vegetable and/or dairy farms with above average size, input productivity and operator education.
Engene (1943)	Minnesota Dairy Farms 1928-1938	Multiple correlation and regression analyses using farm management records	Operator Labor, Income	Farm Size Enterprise Type Input Productivity - land - labor Input Intensity - feed	All factors evaluated should be assessed for improvement with respect to operator income.

was associated with better than average performance on each variable. Engene (1943) used correlation and regression analyses to assess the relative importance of eight management factors on operator labor earnings of Minnesota dairy farms for 1928 to 1938. His results indicated that higher operator labor income occurred when all eight factors were above average.

1948 to 1980

Interest in success factor studies declined sharply particularly at the beginning of this period. Some research on farm viability was conducted toward the end of the period. Table 9.2 outlines the key characteristics of typical studies on profitability and viability.

Profitability

Mosher and West (1952) used linear regression analysis to evaluate factors associated with the net farm income of Illinois mixed grain and livestock farms between 1936 and 1945. The four most important factors, in order of importance, were crop and livestock productivity, labor intensity[8] and market price. Their results also emphasized the importance of as many factors as possible be above the survey average for greater net farm income. Later, Pearse (1966) used factor analysis on Missouri farm records for the years 1961 to 1964. Regression analysis indicated that return to operator labor and return to capital increased with cropping intensity, herd size, labor intensity and productivity and machinery efficiency. Luckham (1976) focused on identifying financial ratios associated with profitability of Virginia dairy farms. Discriminant analysis identified the operating expense to income ratio and the debt to asset ratio as key financial ratios. The first ratio, which measures cost control, was negatively related to profit while the second ratio, which measures financial leverage, was positively related to profit. This study emphasized that cost control remains one of the farm manager's most important means of increasing profitability.

Viability

Krause and Williams (1971) evaluated behavioral-personality variables of South Dakota farm operators in relation to growth in net worth. The variables included motivational, ability and biographic characteristics. Their regression results indicate that most of the thirteen variables considered were statistically significant in predicting the success or failure of a farm. The author's conclusions were that personality characteristics,

TABLE 9.2 Empirical Studies Evaluating Success Factors: 1948-1980

Author(s)	Location Commodity Class(es) and Time Period	Empirical Method	Success Criterion	Success Factors Evaluated	Results (Success Increasing Factors)
1) Profitability					
Mosher and West (1952)	Illinois Mixed Grain and Livestock Farms 1936-1945	Linear regression analysis using farm management records	Net Farm Earnings	Input Productivity - land - livestock Marketing Factors - price efficiency Input Intensity - labor - machinery - buildings - fertilizer	Above average crop and livestock productivity and below average labor intensity.
Pearse (1966)	Missouri Dairy, Grain, Livestock and Mixed Farms 1961-1964	Factor and regression analyses using farm management records	Operator's Labor, Income, Return to Capital	Farm Size Enterprise Type(s) Input Intensity Input Productivity Financial Structure (67 variables in total)	Increased cropping intensity, herd size, labor intensity, and productivity and machinery efficiency.
Luckham (1976)	Virginia Grade A Dairy Farms	Discriminant analysis using farm financial records	Return to Equity, Return to Total Assets	Financial Structure - liquidity - leverage - asset structure Cost Control Farm Size Financial Performance - profitability	Cost control and increased financial leverage.

including those of farm operators' wives, were significant in determining the success of a farm. Osburn (1978) evaluated personal and business characteristics of U.S. farmers who had defaulted on loans in 1971. Using a generalized analysis of variance he found that default loan losses increased with increased total debt, debt to collateral ratio, education, experience and outside business interests. The positive relationship between greater financial failure and education and experience was attributed to those farmers being greater risk takers, a feature often assumed to be associated with success.

1980 to Present

The latest time period has seen a resurgence of interest in success factors research, including a wide variety of profitability studies, as well as studies on viability[9] and farm growth. Table 9.3 outlines selected studies conducted since 1980.

Profitability

Cunningham (1982) evaluated the effects of five production and twelve business management factors of New York egg farms on labor and management income per operator. Regression analysis identified farm size, income per hen and investment per hen as significant as determinants of income per operator. Correlation analysis also revealed that hen productivity and feed efficiency were related to income per hen. Laband and Lentz (1983) tested for the importance of intergenerational transfers of human capital on the earnings of U.S. farms. Using a logistic regression model with 1965 U.S. household cross-section data they found that second generation farmers consistently earned more income than first generation farmers. Further, the greater earnings were due to experience and intergenerational transfers of farming knowledge and skills and not the transfer of physical assets or formal education. Korth (1984) used a variety of statistical techniques to identify factors related to the financial success of Nebraska beef-hog, grain and dairy farms between 1978 and 1982. He considered seventy-one variables and found that two-thirds of the statistically significant factors were in the categories of volume of production, efficiency and expense structure. Return on investment was used as the main measure of success. Cunningham-Dunlop (1986) assessed the effect of education on farm profits in Canada. She concluded that the net returns to schooling in Canadian agriculture have been positive. This implies formal education is an important factor in farm profitability, contrary to the findings of Laband and Lentz (1983). Wood et al. (1987) focused on identifying farm management measures

TABLE 9.3 Empirical Studies Evaluating Success Factors: Post 1980

Author(s)	Location Commodity Class(es) and Time Period	Empirical Method	Success Criterion	Success Factors Evaluated	Results (Success Increasing Factors)
1) Profitability					
Cunningham (1982)	New York Poultry Farms 1975-1979	Correlation and regression analyses using farm management records	Labor and Management Income per Operator	Marketing Factors Farm Size Input Intensity - labor - feed - total Input Productivity	Increased farm size, income per hen and decreased investment per hen.
Laband and Lentz (1983)	U.S. Farms 1965	Logistic regression model using household survey data	Human Capital Earnings of Second Generation Farmer	Personal Characteristics Organizational Structure Farm Factors	Increased experience and intergenerational transfer of farming knowledge and skills.
Korth (1984)	Nebraska Beef-Hog, Grain and Dairy Farms 1978-1982	Correlation, regression and covariance analyses using farm management records	Return on Investment, Return on Equity, Return on Management, Net Farm Income	Farm Size Financial Structure Financial Performance Input Intensity Expense Structure Personal Characteristics Macroeconomic Factors	Factors related to volume of production, efficiency and expense structure.
Cunningham-Dunlop (1986)	Canadian Farms 1971-1981	Variable Profit Function Model using aggregate Canadian data	Marginal Profits	Personal Characteristics - education level	Increased formal education.

Study	Data	Method	Dependent Variable	Factors	Results
Wood et al. (1987)	North Dakota Crop Farms 1922-1984	Correlation and regression analysis using farm records and interview data	Returns to Labor and Management	Input Productivity - land - labor Farm Size Input Intensity Marketing Factors Agricultural Policies Crop & Machinery Management Factors	Increased total assets, government payments, crop yield and labor efficiency and decreased machinery costs.
Sonka et al. (1988)	Illinois Grain Farms 1976-1983	Logit model using farm management records	Management Return per Acre, Rate of Return on Non-land Investment	Farm Size Enterprise Type Input Productivity - land Expense Structure - interest rates - total cash exp. Marketing Factors - price	Higher product price and crop yield. Also found considerable year to year variability in the rank of farms according to profit.
Burton and Abderrezak (1988)	Kansas Farms 1985	Linear regression using projected farm management data	Expected Net Farm Income	Farm Size Financial Structure Organizational Structure - degree of ownership Enterprise Type Off-Farm Factors Personal Characteristics	Increased farm size, financial efficiency and decreased ownership of real estate and machinery.

(continued)

TABLE 9.3 (continued)

Author(s)	Location Commodity Class(es) and Time Period	Empirical Method	Success Criterion	Success Factors Evaluated	Results (Success Increasing Factors)
2) Viability					
Shepard and Collins	U.S. Farms 1910-1978	Regression analysis using aggregate U.S. data	Farm Bankruptcy Rate	Farm Size Financial Structure Farm Income Agr. Policies Macroeconomic Factors	Financial leverage, farm size and farm income variability. The effect of each factor varied in sign and weight depending on the time period.
Ellinger and Barry (1987)	Illinois Farms 1972, 1976, 1980 and 1985	Cross classification and regression analyses using farm management records	Rate on Return on Assets/Equity, Debt-to-Asset Ratio	Tenure Farm Size	Increased proportion of leased land.
Garcia et al. (1982)	Illinois Grain Farms (1977)	Unit Profit Function Model with seemingly unrelated regression analysis using farm management records	Economic Efficiency	Farm Size (growth) Organizational Structure - tenancy	Farm size growth is *not* due to increased economic efficiency of increased farm size.
3) Growth					
Shapiro et al. (1987)	Canadian Farms 1966-1981	Regression, variance and logit analyses using aggregate data	Farm Size, Growth	Random Environmental Factors	Stochastic environmental factors do *not* determine farm firm growth.

Garcia et al. (1987)	Illinois Grain Farms 1976-1985	Discriminant analysis using farm management records	Farm Size, Growth	Financial Performance - management returns Input Intensity - management Production Risk Input Productivity - land Agricultural Policies	Increased management intensity and earnings and use of government programs.
Sumner and Leiby (1987)	Southern U.S. Dairy Farms 1983	Cohort and regression analyses using sample survey data	Farm Size, Growth	Farm Size Personal Characteristics - age - experience - schooling	Farm growth rate decreased with increased operator experience and farm size.

that explain variations in the operator labor and management income of North Dakota crop farms. Their analysis identified total assets, machinery cost, government payments, crop yield and labor efficiency as significantly correlated to operator labor and management earnings. Total assets and machinery cost were negatively correlated and together with crop yield explained most of the variation in operator earnings. Sonka et al. (1988) evaluated eight management factors on Illinois grain farms with respect to management return per acre and rate of return on non-land investment. Using a logit model, they determined that market price received and crop yield were more associated with financial success than farm size or types of crop enterprise. However, superior performance in all factors tended to result in greater financial success. Further, this study determined that there was considerable year to year variability in the profitability and rank of farms according to profit. Burton and Abderrezak (1988) used the expected 1985 production and financial plans of Kansas farmers to identify statistically significant farm characteristics related to expected profit. Using linear regression, they found farm size, proportion of non-ownership of real estate and machinery, production and financial efficiency[10] to be the statistically significant factors positively related to expected profit.

Viability

Shepard and Collins (1982) tested several factors thought to be determinants of U.S. farm bankruptcies for the 1910-1978 time period. They found that prior to World War II, increased financial leverage coincided with a higher frequency of failure while increased farm size reduced it. However, for the post-war period, increased farm size and varying farm incomes were the major factors associated with failure. Overall, there was no evidence that agricultural support payments had any effect on the farm bankruptcy rate while in contrast, bankruptcy was strongly correlated with the non-agricultural bankruptcy rate. This implies greater importance of macroeconomic factors as opposed to agricultural policies in farm viability. Ellinger and Barry (1987) used cross classification[11] and regression analyses on Illinois farm records to assess the relationship between land tenure and solvency and profitability. Their results indicate that as the proportion of leased land increased, both the debt-to-asset ratio and the rates of return to assets or equity increased. However, most of the variation in the debt-to-asset ratio and return to assets or equity was not explained by either farm size or land tenure.

Growth

Garcia et al. (1982) attempted to determine if differences in economic efficiency[12] on Illinois grain farms, could explain farm growth. They used a unit profit function estimated with seemingly unrelated regression analysis. Their results reject the hypothesis that farm growth is due to increased economic efficiency with increased farm size. Also, using a sample of Illinois grain farm records Garcia et al. (1987), used discriminant analysis to identify factors associated with changes in farm size. Although there was considerable randomness in farm growth behavior, the key factors associated with farm size growth were management intensity, management earnings and use of government programs. Shapiro et al. (1987) tested Gibrat's Law[13] as a model of the farm growth in Canada. Since the size distribution of farms is usually approximately log normal, the assumption was that stochastic environmental factors determine farm firm growth. However, they, and Fox and Dickson (1989), found that farm growth patterns were inconsistent with Gibrat's Law. Sumner and Leiby (1987) studied the relationship between human capital[14] and the growth and size of southern U.S. dairy farms. Using cohort and regression analyses, the clearest result was that increased operator experience reduces the farm growth rate. Also, growth rates decreased with farm size which was related to operator age.

Discussion

Depressed commodity prices, high interest rates, burgeoning farm program costs and high rates of farm business failure in the early 1980s rekindled long dormant interest in what makes some farms more successful than others. While the issues motivating this work remain important, the achievements of this inductive-empiricist research program continue to be disappointing. Nine decades of success factors research has failed to produce a robust or consistent explanation of differences in farm performance. It is particularly distressing that current research in this area has failed to address the contentious issues raised at the Black Duck meeting.

In our judgment, the resolution of four major problems is a necessary condition for progress on the success factors research agenda.

Success Is Multi-Dimensional

Success can be characterized as annual income or returns, growth rates of sales, assets or equity, ability to withstand adverse market conditions or low levels of income risk. Different farm operators place different weights on the different dimensions, and a single farmer's emphasis on different measures of success can change over his career. Limited empirical evidence suggests that correlations among different measures of success are low, and that some factors do a good job of explaining variations in one measure of success but not of others.

The Theoretical Agenda

The neoclassical theory of the firm has not yet produced a theory of industry structure for an atomistic industry. It cannot offer an explanation of distribution of profits, growth rates or probabilities of success. Until such an explanation is available, success factors research will be subject to criticism as "measurement without theory". Work by Panzar and Willig (1978); Lucas (1978); and Oi (1983) as well as Oxley (1988) and Bergen (1989) shows promise, but more needs to be done. Much of the failure of the current configuration of the neoclassical theory of the firm to produce an explanation of the observed distributions in firm sizes in particular industries can be traced to the trivial role which that theory ascribes to entrepreneurship. Kirzner (1973) has been one of the most articulate critics of this omission.

The Simultaneity Problem

Success factors research has shown no sensitivity whatsoever to the statistical problems of using endogenous variables to explain variations of endogenous variables in the context of a single equation model. For example, milk production per cow might be used to explain differences in net farm income across farms. Milk production per cow and profits are themselves functions of input use decisions which are driven by, among other things, input and output prices. The emergence of the dual approach to empirical production research has been motivated in large measure by simultaneity problems such as these. Heady questioned the use of single factor productivity ratios as "efficiency" measures in farm management research forty years ago. Use of these variables may be a more serious error than even he fully appreciated.

Design of the Estimating Model

Current work in success factors research often still does not reflect the existence of (eventually) diminishing returns and examples of redundant or at least overlapping explanatory variables abound. Part of this problem arises from the lack of a coherent consistent theoretical framework to guide the development of empirical models. In the interim, more careful efforts to identify less interdependent sets of explanatory variables are needed.

Conclusion

The purpose of this chapter was to assess the potential contribution of empirical farm management success factors research for the study of farm structure. The emphasis that the success factors research program has placed on tracing the history and dynamics of the evolution of particular farm businesses and the emphasis that it has also placed on the multiple dimensions of farm business success has the potential to enrich our understanding of the adjustment paths that link periodic snapshots of industry structure. Success factors research has also documented the extent to which the intra-industry position of particular firms changes over time, even when the parameters of aggregate distributions remain relatively stable. Much of the potential of this research, however, has not been realized. Even this brief review of selected literature has illustrated that few robust results have been generated. There is much which remains unknown about the reasons for some farms being more successful than others. Nevertheless, important opportunities remain for fruitful collaboration between analysts of farm structure and the practitioners of success factor studies. Perhaps the Black Duck conference should be reconvened.

Notes

Research support from the Ontario Ministry of Agriculture and Food is gratefully acknowledged.

1. This legend was recounted to Glenn Fox by Glenn Johnson in a Mexican restaurant in San Antonio in January, 1988. None of the authors of this chapter assume any responsibility for the summary of the legend as it was originally told, or for the faithfulness of the account presented here.

2. Jensen (1977, 9-10) provides a rare but important account of this change in emphasis.

3. Johnson (1955) describes the views held by the protagonists at this meeting.

4. This chapter is not intended to be an exhaustive survey. Jensen (1977) and Korth (1984) offer more complete reviews.

5. Cross-classification refers to sorting farms into groups by, for example, total sales. Averages of physical and financial variables such as acreage and net farm income, respectively, are calculated for each group. The averages are then compared across groups. This type of analysis is also referred to as sorting, tabulation, comparative and factor analyses in the literature.

6. That is single-factor productivity.

7. Efficiency was defined as the increase of a single-factor productivity ratio.

8. Intensity is defined as the quantity of input per unit of another input. An example is labor cost per hectare.

9. Although there has been considerable research on farm viability since 1980, most have been simulation type studies making projections about future viability. These have not been included in this review. We refer the interested reader to publications by Richardson listed at the end of the chapter.

10. Calculated as gross farm income divided by the sum of operating expenses and depreciation.

11. Ellinger and Barry (1987, 110) used the Jerkey-Kramer method to test for statistically significant differences between pairs of cell means.

12. Economic efficiency refers to the degree of technical and pricing efficiency as well as the extent to which the farms are profit maximizers.

13. Gibrat's Law states that if the number of firms in an industry is fixed, and all firms are exposed to random shocks in proportion to their size, the long run size distribution of firms will resemble that of a log normal distribution.

14. Measured as age, experience, schooling and the use of specific dairy management techniques.

References

Baumol, W. J. 1968. "Entrepreneurship in Economic Theory." *American Economic Review* 58: 64-71.

Bergen, Philip A. 1989. "Technology, Relative Input Prices and Growth in the Size of Ontario Dairy Farms Between 1961 and 1986." M.Sc. thesis, University of Guelph, Department of Agricultural Economics and Business.

Boss, A., A. J. Benton, and W. L. Cavert. 1917. "A Farm Management Study in Southeastern Minnesota." Bulletin 172. St. Paul: University of Minnesota, Agricultural Experiment Station.

Burton, R. O., and A. Abderrezak. 1988. "Expected Profit and Farm Characteristics." Contribution No. 89-32-D. Manhattan: Kansas State University, Agricultural Experiment Station.

Cunningham, D. L. 1982. "A Five-Year Analysis of New York Egg Farm Management Factors." A.E. Res. 82-18. Ithaca: Cornell University, Department of Poultry and Avian Sciences.

Cunningham-Dunlop, C. 1986. "The Effect of Education on Farm Profit and Productivity in Canada." M.Sc. thesis, University of British Columbia, Department of Agricultural Economics.

Ellinger, P. N., and P. J. Barry. 1987. "The Effects of Tenure Position on Farm Profitability and Solvency: An Application to Illinois Farms." *Agricultural Finance Review* 47: 106-118.

Engene, S. A. 1943. "New Light on Factor Analysis." *Journal of Farm Economics* 25: 477-486.

_____. 1950. "Limitations of Economic Theory - Discussion." *Journal of Farm Economics* 32: 1120-1121.

Fox, G., and E. J. Dickson. 1989. "Farm Growth in the Ontario Dairy Industry." Paper presented to the NC-181 Regional Committee Meeting, Tuscon, AZ.

Garcia, P., S. E. Offutt, and S. T. Sonka. 1987. "Size Distribution and Growth in a Sample of Illinois Cash Grain Farms." *American Journal of Agricultural Economics* 69: 471-476.

Garcia, P., S. T. Sonka, and M. S. Yoo. 1982. "Farm Size, Tenure and Economic Efficiency in a Sample of Illinois Grain Farms." *American Journal of Agricultural Economics* 64: 119-123.

Heady, E. O. 1948. "Elementary Models in Farm Production Economics Research." *Journal of Farm Economics* 30: 201-225.

Jensen, H. R. 1977. "Farm Management and Production Economics, 1946-1970," in L. R. Martin, ed., *A Survey of Agricultural Economics Literature*. Vol. 1. University of Minnesota Press.

Johnson, G. L. 1950. "Needed Developments in Economic Theory as Applied to Farm Management." *Journal of Farm Economics* 32: 1140-1156.

_____ . 1955. "Results from Production Economics Analysis." *Journal of Farm Economics* 40: 206-222.

_____ . 1988. "Farm Managerial Inquiry: Past and Present Status and Implications for the Future." Paper presented at NC-181 "Determinants of Farm Size and Structure," held at San Antonio, TX.

Kirzner, I. M. 1973. *Competition and Entrepreneurship*. Chicago: The University of Chicago Press.

Korth, B. D. 1984. "Factors for Determining Financial Success for Farm Managers." M.Sc. thesis. Lincoln: University of NE.

Krause, K. R., and P. L. Williams. 1971. "Personality Characteristics and Successful Use of Credit by Farm Families." *American Journal of Agricultural Economics* 53: 619-624.

Laband, D. N., and B. F. Lentz. 1983. "Occupational Inheritance in Agriculture." *American Journal of Agricultural Economics* 65: 311-314.

Lucas, R. E. 1978. "On the Size Distribution of Business Firms." *Bell Journal of Economics* 9: 508-523.

Luckham, W. R. 1976. "Financial Ratios for Grade A Dairy Farms in Virginia." *Farm Credit Administration Research Journal.*

Mosher, M. L., and V. I. West. 1952. "Why Some Farms Earn So Much More Than Others." Bulletin 558. Urbana: University of Illinois, Agricultural Experiment Station.

Oi, W. Y. 1983. "Heterogeneous Firms and the Organization of Production." *Economic Inquiry* 21: 147-171.

Ontario Milk Marketing Board. "Ontario Dairy Farm Accounting Project - Extension Report." Annual.

Osburn, D. D. 1978. "An Analysis of Factors Influencing Loan Losses Among Production Credit Association." *Farm Credit Administration Research Journal* 3: 18-24.

Oxley, J. C. 1988. "An Ex Ante Analysis of Technological Change: Bovine Somatotropin and the Ontario Dairy Industry." M.Sc. thesis, University of Guelph, Department of Agricultural Economics and Business.

Panzar, J. C., and R. D. Willig. 1978. "On the Comparative Statics of a Competitive Industry with Inframarginal Firms." *American Economic Review* 68: 474-478.

Pearse, R. A. 1966. "A Study of Relationships Used in Farm Record Analysis." Research Bulletin 911. Columbia: University of Missouri, Agricultural Experiment Station.

Perry, G. M., M. E. Rister, J. W. Richardson, and J. Sij. 1986. "The Impact of Crop-Share Arrangements and Crop Rotations on Upper Gulf Coast Rice Farms: A Survival Approach." Texas Agricultural Experiment Station, Bulletin, B-1530.

Nixon, C. J., and J. W. Richardson. 1985. "Implications of Tax Policy Changes on Agriculture, Rural Communities, and Businesses." Increasing Understanding of Public Problems and Policies--1985. Oak Brook, Illinois: Farm Foundation.

Richardson, J. W., and C. J. Nixon. 1986. "Description of FLIPSIM V: A General Firm Level Policy Simulation Model." Texas Agricultural Experiment Station, Bulletin, B-1528.

Richardson, J. W., and E. G. Smith. 1985. "Impacts of Farm Policies and Technology on the Economic Viability of Texas Southern High Plains Wheat Farms." Texas Agricultural Experiment Station, Bulletin, B-1506.

Schultz, T. W. 1939. "Theory of the Firm and Farm Management Research." *Journal of Farm Economics* 21: 570-586.

Shapiro, D., R. D. Bollman, and P. Ehrensoft. 1987. "Farm Size and Growth in Canada." *American Journal of Agricultural Economics* 69: 477-483.

Shaw, H. R., and P. A. Wright. 1955. "Alternative Methods of Farm Management Analysis." *Canadian Journal of Agricultural Economics* 3: 67-80.

Shepard, L. E., and R. A. Collins. 1982. "Why Do Farmers Fail? Farm Bankruptcies 1910-1978." *American Journal of Agricultural Economics* 64: 609-615.

Sonka, S. T., R. H. Hornbaker, and M. A. Hudson. 1988. "Managerial Performance and Income Variability for a Sample of Illinois Cash Grain Producers." *North Central Journal of Agricultural Economics*.

Sumner, D. A., and J. D. Leiby. 1987. "An Econometric Analysis of the Effects of Human Capital on Size and Growth Among Dairy Farms." *American Journal of Agricultural Economics* 69: 465-470.

Warren, G. F., and M. C. Burritt. 1909. "The Incomes of 178 New York Farms." Bulletin 271 Ithaca: Cornell University, Agricultural Experiment Station.

Warren, S. W. 1932. "Factors for Success on Dairy and General Farms in Northern Livingston County, New York." Bulletin 242. Ithaca: Cornell University, New York State College of Agriculture.

Wilcox, W. W. 1947. "Research in Economics of Farm Production." *Journal of Farm Economics* 29: 632-643.

Wood, M. A., R. G. Johnson, and M. B. Ali. 1987. "Performance Factors and Management Practices Related to Earnings of East Central North Dakota Crop Farms." Ag. Econ. Report No. 224. Fargo: North Dakota State University, Department of Agricultural Economics.

10

Use of Firm-Level Agricultural Data Collected and Managed at the State Level for Studying Farm Size Issues

George L. Casler

Data from individual farm financial records is available in at least two dozen states. These farm record programs are sponsored by three types of groups: (1) farm management associations, (2) departments of agricultural economics and Cooperative Extension and (3) vocational-technical school programs. In several states there is cooperation between the farm management associations and agricultural economists in the collection and analysis of data from individual farm records. A list of the organizations responsible for collecting and publishing the data in each state appears in an appendix at the end of the chapter. Much of this effort is primarily related to extension farm management programs, but in some cases the data is the basis for research studies. This chapter is primarily concerned with (1) the use of this firm level data as a basis for studying issues such as farm size and structure, (2) whether the data could be made consistent to facilitate comparisons of net returns across states and (3) presentation of data from three states relating farm size to measures of profitability as an example of the use of farm record data in studying farm size issues.

Farm Record Data

The history of farm record data collection as part of an extension-type effort varies greatly among states. Some states appear never to have been involved in such activity while others have been continuously involved for several decades. A few states (universities) have started new data collection efforts in recent years but, perhaps more significantly, several (Purdue, Ohio State, Wisconsin) largely discontinued such efforts after 1983. However, Purdue restarted their efforts in 1987. Some of the farm record efforts have been in close cooperation with independent and largely farmer-financed farm management associations. The largest of these efforts is in Illinois. A combination of farm management fieldmen and college staff summarized and analyzed 7,192 records for 1990. It is probably fair to state that the farm records and analysis programs in most states are a blend of education and service to the farmers involved and a source of information to be used in extension programs with other farmers and in teaching programs at various universities and colleges. While the data have been used for research, probably in no state was that the original purpose for collecting the data.

Use of farm record data for research purposes lies on a somewhat shaky foundation: in no state are the records collected on a random sample basis. Rather, data is collected from farmers who voluntarily agree to participate in these educational-service programs. Nevertheless, researchers have used the data for a variety of studies, many of which relate to the relationship between various management factors or variables such as farm size and measures of net returns from operating the business. A purist could argue that the non-random sample negates or at least seriously impairs the validity of the results. However, many researchers argue or apparently believe that, even though the records, on the average, come from farms that are above average in size and are operated by above-average managers, the results are useful and the conclusions probably wouldn't be much different if the record data came from a random sample of farms of the same farm type.

With the exception of a few states such as Illinois and Kansas, the number of farm records available in any one year may be small enough that valid analysis is limited, particularly if the researcher wants to study farms of a particular type on similar soil resources, or with a particular type of equipment, or livestock housing system. In addition, because farmers do not necessarily participate on a continuous basis, numbers become even more limited if the desire is to study the same farms over a period of years. The numbers situation leads to the question of combining farms from several states to study issues such as costs or net returns by farm size. An immediate problem of such a data combination is that each state (really the data collectors therein) has its own idea of how the data should be collected and analyzed. For example, the measures of net returns and the way they are calculated are extremely variable among states. Whether such differences could be resolved, so that every state uses the same procedures in the future, is questionable.

The inconsistencies among states appear in several items, such as methods of: (1) asset valuation, (2) handling appreciation of assets, (3) handling inventory changes, (4) calculating depreciation, (5) handling changes in accounts receivable and payable, (6) calculating "value of farm production," (7) calculating interest on assets and production expenses, and (8) calculating the value of operator's labor and management and non-operator family labor. In addition, some states publish data for the total farm business, including the landlord's share, while others publish only the data for the operator's share. Most of these inconsistencies are the apparent result of the notions of economists in the various states about these issues. It is clear that we have agreed upon neither what to measure nor how to measure it. Methods of charging depreciation, interest, and operator and family labor and methods of asset valuation for several states are shown in Table 10.1. Some of this information can be ascertained from the publications, but some of it was obtained by personal communication with the authors.

The matter of publishing the data for the total business, including the operator and landlord shares vs. publishing only the operator share, appears to be a particular problem and is related partly to the prevalence of tenant operators in some states. Illinois has chosen to publish in the annual Summary of Illinois Farm Business Records the combined operator-landlord shares, although this is not clearly pointed out in the bulletin. The operator's share is published for only one item which is net farm income. Operator and landlord shares are published in a separate publication (Scott 1986), and operator-only data are presented in a third publication. These publications are less widely distributed. Minnesota (Olson 1990) and Indiana publish only the operator's share. Missouri has chosen to publish, in a two column format, the numbers for the operator and for the total business, with the difference being the landlord's share (Hein 1989). This writer suggests that when a "management return" or "labor and management return" is being computed, the computation should be for the person who is managing the business and that, in most cases, it is the operator. However, in some share rental situations it is possible that the landlord or the landlord's representative exerts substantial (or even total) managerial control over the business. The Missouri procedure appears to solve the reporting problem by publishing both the operator share and total business -- the choice of which is the important data is left to the reader.

One problem in studying the data published by the various states is that the publications frequently do not fully describe the procedures used to compute the various measures of net returns. For example, it is not

TABLE 10.1a Methods Used for Depreciation, Asset Valuation, Interest Charges and Unpaid Labor Charges, Corn Belt States and New York

	Illinois 1990	Iowa 1988	Michigan 1990	Minnesota and Indiana 1989	Missouri 1988	New York 1990
No. of farms	7192	3000	404	Minn. = 261	313	395
Years of data	66	62	c/			37
Depreciation						
Real estate	tax	10% of C.V.	tax	Indirect	tax	tax
Machinery	tax	10% of C.V.	tax	Indirect	tax	tax
Interest						
Interest paid	For operator NFI	Yes	For NFI	Yes	Yes	Yes
Interest on equity	No	6%	6%	6%	No?	5% real
Interest on total						
Land	4.5%a/	No	Yes	No	8%	No
Other	10%b/	No	Yes	No	8%	No
Asset valuation						
Land	Market	Market	Market (agr.)	Marketd/	Market	Marketf/
Buildings	Cost-tax depr.	Cost-econ. depr.	Cost-tax depr.	Cost-tax depr.	Cost-	Marketf/
Equipment					tax depr.	Marketf/

Dairy & Breeding Livestock	Market	Market	Market trend	Market	Market	Market f/
Labor						
Operator	1350/mo.	1200/mo.	6.00/hr.	15,000/yr.	e/	g/ 1250/mo.
Family	1350/mo.	700/mo.	6.00/hr.			

a/ Land charge-net rent, revised annually based on average landlord net rents.
b/ Revised annually.
c/ Continuous set of records since 1928. Computer readable tapes of data since 1983.
d/ The Southwestern Association uses a conservative market value. The Southeastern Association uses original cost.
e/ Operator labor is valued by each operator. Therefore, management return is included in return to capital.
f/ Market values are used in calculating interest on equity. Year-to-year changes in market values of real estate, equipment and livestock are labelled appreciation and excluded from the calculation of labor and management income.
g/ For calculating return on investment, each farmer estimates the value of his labor and management.

TABLE 10.1b Methods Used for Depreciation, Asset Valuation, Interest Charges and Unpaid Labor Charges, Selected States

	Kansas 1990	Kentucky 1988	N. and S. Dakota c/ 1988	Nebraska 1989	Pennsylvania 1988	Oklahoma 1989
No. of farms	2043	580	N = 343 S = 183	95	888	141
Years of data	61	26	16	14		
Depreciation						
Real estate	tax	tax	tax	tax	tax	Indirect
Machinery	tax	tax	tax	tax	tax	Indirect
Interest						
Interest paid	Yes	No	Yesd/	No	Yes	Yes
Int. on equity	10%	b/	d/	No	6%	No
Interest on total						
Land	No	5%	7%	6%	No	f/
Other	No	11%	7%	12%	No	f/

Asset valuation						
Land	Market	Conserv. market	Market	Market	Market	Market
Buildings	Cost-tax depr.	Cost-tax depr.	Cost-tax depr.	Cost-tax depr.	Cost-tax depr.	Market
Equipment	Cost-tax depr.	Cost-tax depr.	Cost-tax depr.	Cost-tax depr.	Cost-tax depr.	Market
Dairy & Breeding Livestock	Market	Conserv. market	Market	Constant	Market	Market
Labor						
Operator	15000/yr.a/ 1250/mo.	1250/mo.	No 20/day	1100/mo.	1000/mo.e/ 3.55/hr.	4.00/hr.
Family	1250/mo.	1250/mo.		750/mo.		4.00/hr.

a/ In calculating return to capital, a management charge of 10 percent of gross farm income is added to the labor charge.

b/ Interest paid is deducted from interest on land at 5 percent plus interest on other capital at 11 percent to obtain interest on equity.

c/ Beginning in 1989, N. Dakota adopted the Minnesota system using Finpack.

d/ In calculating returns to the operator, interest paid is subtracted from interest on total capital at 7 percent to obtain interest on equity.

e/ A management charge of 5 percent of cash receipts is included in computing cost of producing milk.

f/ For calculating return to unpaid labor and management, the rate on one-year treasury notes is charged on total capital.

always clear whether assets are valued based on market value, book value (cost less depreciation) or something else. Some of the implications of asset valuation relative to computing net returns are discussed in the next section.

Asset Valuation

The market values of farm assets frequently are quite different from the book values. For example, the market value of land is likely to be substantially greater than the book value (cost) if the land was purchased 20 or more years ago, but less than book value if it was purchased in the late 1970s. Market values of machinery are likely to be higher than book values if rapid depreciation has been used for income tax purposes, and inflation tends to make the divergence greater. Farmers who report on the cash basis for tax purposes have no basis or book value in raised livestock. Most farm record systems value raised animals at market or perhaps at some kind of modified market value in the case of breeding stock. This is done even in systems that use book value for assets such as land, buildings and machinery.

Asset valuation procedures affect the charge for equity capital and for total capital in systems that do not include interest paid in the expenses. Asset valuation also affects the calculation of return on equity and return on total assets.

Those who argue for using market values as the basis for the calculation of interest charges and return on investment believe that the opportunity cost of equity capital should be based on the amount of money that is invested in the farm business that could earn a return if it were invested elsewhere.[1]

Appreciation of Assets

In recent years, many analysts have argued that appreciation of assets should not be included in calculating net returns from the year's operation of a farm business. For example, if the value of the land increases $20,000 during the year, this $20,000 should be considered ownership income rather than operating income. Similarly, if the value of a herd of breeding stock increases $5,000 during the year due to a change in the general level of cattle prices, this $5,000 should not be included in annual operating income. The same concept can be applied to depreciable assets such as machinery and buildings, but the mechanics are more difficult. For example, the depreciation on a tractor that is charged to the income statement should reflect using up a year in the life of that tractor. Take a simple example in which a machine has an initial cost of $12,000 and is expected to provide services for 12 years. With straight line and no salvage value, each year's depreciation would be $1,000. After six years, the adjusted basis or book value would be $6,000. However, during a period where machinery prices were rising at 5 percent per year, a new machine at the end of year six would cost $16,081 and the value of the used machine would likely be greater than if there were no inflation. Rather than reducing the depreciation to reflect the effects of inflation, "real" depreciation should be charged to the income statement and appreciation should be credited to the ownership account. The difficult part is to know how to calculate "real" depreciation. In practice, those who calculate appreciation on machinery use income tax depreciation as a proxy for real depreciation. A comparison of income tax adjusted basis with market value at both the beginning and the end of the year allows appreciation to be calculated. With rapid depreciation for income tax purposes, it is likely that both depreciation and appreciation are overstated.

Using market values for all assets and including the change in inventory values in the calculation of measures of net return has the potential of distorting such measures because of fluctuations in asset values. The Coordinated Financial Statements procedure of Frey and Klinefelter (1980) seeks to separate the income from operating the farm from the gains (or losses) from owning the assets by using a two-column valuation procedure on the balance sheet. One column is market value and the other is a cost- (or modified cost-) based valuation. This procedure as currently used does not actually use the cost-based values for all assets. For example, raised breeding stock and a number of other assets are valued at market rather than at cost. In addition, use of adjusted basis from income tax records for valuation of depreciable assets and the accompanying depreciation as a charge on the income statement may overstate the depreciation charge in the early years of asset life if rapid depreciation is being used for tax purposes.

Of the farm record systems reviewed, only one (New York) explicitly calculates and publishes appreciation. It is likely that many of the other systems keep appreciation on land out of the net return calculations by not including the change in land values in changes in inventories. If market values are used for some of the net return calculations, the changes in market values are done "between years."

In the systems where machinery depreciation is calculated from the changes in market values of the machinery, any inflation in used machinery prices, which some people consider to be appreciation, results in the depreciation charge being lower than it otherwise would be.

Depreciation

The method used to calculate depreciation can affect the net income and other measures of profitability. The two common methods of depreciation used in farm record systems are (1) income tax and (2) net figure derived from (beginning inventory + purchases) - (ending inventory + sales) with inventories being at market value. A variation on the second method is to use a standard percentage, such as 10 percent, of beginning plus new. One might think that distortion of income caused by the use of income tax rapid depreciation would be only temporary and minor -- depreciation can be taken only once. For example, five-year rapid depreciation under the accelerated cost recovery system (ACRS) would lead to a high depreciation charge in the early 1980s, but this would be offset by no depreciation on these items once the five-year period is over. However, in an inflationary period, it is likely that use of income tax depreciation, whether rapid or straight line, will result in a higher depreciation charge than using a market value approach.

An example which illustrates the depreciation charges calculated by different methods is shown below, using the 1990 Cornell dairy farm business summary data:

A. Average machinery depreciation from income tax = $16,624
   Appreciation on machinery = $1,793

B. Decline in market value
   Example:

Beginning value	$123,423	Ending value	$137,111
+ Purchases	29,859	+ Sales	1,340
	$153,282		$138,451

   Depreciation = $153,282 - 138,451 = $14,831
   Note that appreciation equals the difference between depreciation calculated by methods A and B.

C. Standard percentage of market value, beginning plus purchases
   Example:
   153,282 x 10% = 15,328

Accounts Receivable and Payable

Most farmers report on the cash rather than accrual basis for income tax purposes. A true financial picture of a business requires accrual accounting. All of the farm record systems reviewed included changes in inventories in calculations of net returns. Some of the systems specifically list the changes in accounts receivable and payable and changes in prepaid expenses. It is not clear whether the remaining systems make these adjustments. To the extent that changes in these items are significant, net returns are distorted if such changes are not accounted for.

Value of Farm Production

The purpose of calculating value of farm production is unclear to this author. This measure is not calculated in the farm record systems of several of the states. For the systems where it is calculated, in general, value of farm production is total receipts minus purchased livestock and purchased feed.

Value of farm production apparently is intended to be some sort of "value added" concept. Its origin may go back to a time when purchased inputs such as fertilizer, pesticides and fuel were minimal and purchased livestock and feed were the major inputs acquired from off the farm. As currently calculated, value of farm production has little relevance as a value-added concept.

Interest

A few systems do not include interest paid as an expense, but charge interest at standard rates for all farms. One argument for using this procedure is that it allows comparisons among farms independent of debt levels. While debt level is subject to a measure of managerial control, debt level is at least partly a function of items such as a farm operator's stage in the life cycle of the business and how much was inherited from others.

Those who argue that interest paid should be a farm expense believe that a true measure of net income from operating the business can be obtained only by including interest paid in farm expenses. That belief is hard to argue against.

This writer would like to see both calculations - that is, a net income calculated by including interest paid, and another measure calculated by using a standard interest charge on all the capital used by each farm business. The latter calculation would facilitate comparisons of managerial results that are not based on debt level, something that is partly a function of things over which the operator has no control.

Some states use interest actually paid and interest on equity at a standard rate for some of the profitability calculations while others use a standard charge on all capital, regardless of whether it is equity or debt.

The example below illustrates the varying interest charges that result, depending (A) on the level of debt and equity and (B) on using a standard charge on all capital. In (A), equity capital is charged at a real rate of 5 percent.

A. Debt and equity
   Example: $500,000 assets
   "Net" before interest = $60,000

	100% equity		100% debt
	$60,000		$60,000
Interest on $500,000: @ 5% real =	25,000	@ 10% paid =	50,000
Net farm income	$35,000		$10,000

B. Standard interest charge on all capital rather than interest paid plus interest on equity.
   Example: $500,000 @ 8% = $40,000

In (A) for a farmer with 100 percent equity, the interest charge is $25,000 but $50,000 if the farmer has all debt. In (B), with a standard charge of 8 percent, the interest is $40,000.

Interest on Equity

There appears to be agreement that an opportunity cost charge should be made for the use of equity capital in the business. The disagreement is over the level of the charge. In the business summaries reviewed, charges ranged from around 5 percent to 12 percent. A variety of arguments, stated or implied, are used to support the level of interest rate used. Some are intended to be "real" rates while others clearly are intended to be nominal rates. For example, the Cornell system uses a 5 percent real rate on equity capital. This rate is intended to represent the long-term average rate of return, after removing the effect of inflation, that could be earned in non-farm investments of comparable risk. It is argued that in addition to this real rate, the farm operator benefits from appreciation of assets in a way similar to benefits from investing in the stock market. To charge a nominal rate based on current market interest rates would, in a sense, be double counting.

In reality, interest on equity could be charged at either real or nominal rates and the charge could be based on either market value or book value of assets. The varying combinations that could be used would lead to large variations in the charge for equity capital. There does not seem to be a compelling theoretical argument saying that any one procedure is the correct one. However, this author believes that market values of assets should be used as the basis for calculating equity and charging interest on equity, assuming that one believes in opportunity costs. He also believes that equity capital should be charged at a real rate rather than at a nominal rate.

Value of Operator Labor and Management and Family Labor

A variety of methods are used by the various systems to value operator labor and management. Several states use a standard hourly rate on all farms, sometimes explicitly based on something like the going rate for hired labor. The hours to which the rate is applied must be an estimate because few farmers keep records of hours actually worked. Some states use a standard charge per month, such as $1,000 or $1,200^2 as the labor and management charge. New York does not use any of these standard charge procedures for valuing operator labor and management. Instead, each operator is asked to estimate the combined value of his/her labor and management. If there is more than one operator, a value is obtained for each.

The value of operator labor and management is used to help calculate measures of net return such as return on investment or return on equity. A higher charge for labor and management results in a lower total return to assets or equity and therefore a lower rate of return. One advantage of using a standard charge procedure is that every farm is treated the same way albeit an arbitrary way. In the Cornell procedure, each operator could influence the rate of return by the value assigned to labor and management.

The Importance of Imputed Costs

It is important to point out that the methods used to calculate imputed costs (depreciation, interest on equity or total assets, and value of operator labor and management) have a large impact on measures of profitability because these items make up a large proportion of total costs. For example, in the case of 1990 Illinois northern and central grain farms, in computing management returns ($16,918 on average) the imputed charges for interest on non-land capital ($18,717), land charge-net rent ($67,856) and operator and unpaid family labor ($17,010) total $103,583 or 82 percent as much as all other costs including depreciation. If depreciation, which is also an imputed or at least allocated cost, is included with imputed costs, the total imputed costs are 1.12 times all other costs, not including depreciation. Thus, in the computation of management returns in this example the imputed costs are nearly as important as, or, if depreciation is included, more important than, the costs that can be accurately measured. If interest on land (land charge-net rent) was charged at 4 percent rather than 4.5 percent, the average management return would be $26,257 rather than $18,717. If the interest charge was 5.5 percent rather than 4.5 percent, the average management return would be $3,638 as opposed to $18,717.

The intent here is not to say that Illinois is doing something wrong -- it is only to illustrate the importance of the imputed costs in some of the profitability calculations. Similar examples could be drawn from the calculations made in other states. (What is the appropriate interest charge on land? Clearly the interest rate on mortgage loans in most cases is above 5 percent.)

Perhaps there is one consolation if such data are being used to study farm size issues: if the procedures are used consistently on all farms being studied, the level of imputed charges may not affect the relationships between farm size and profitability.

Contrast of the Methods of Several States

Data from the 1990 New York dairy farm business summary (DFBS) are used in Tables 10.2 through 10.5 to illustrate the differing procedures and results obtained by using the procedures of Illinois, Kansas, Michigan and New York. These states were chosen to illustrate a range of procedures in calculating measures of net income. One difficulty in making the calculations was to know whether appreciation was included or excluded. Therefore, it was assumed that appreciation is excluded from the calculations. For a more detailed analysis covering the systems of 12 states, see the publication by Casler (1990) listed in the references.

While there are several differences in the systems, only a few will be discussed here. In calculating Net Farm Income, Michigan, New York and Kansas include interest paid as an expense, but Illinois does not. In calculating Labor and Management Income, New York uses interest paid and 5 percent real interest on equity, while Illinois uses 4.5 percent on land and 10 percent on all other capital. In 1990, Michigan used 6.0 percent on equity capital. Kansas uses interest paid and 10 percent on equity capital. New York separates appreciation on land, machinery and livestock in making the profitability calculations but net farm income and return on capital are calculated with and without appreciation. Appreciation is not included in the Illinois calculations. No other state system calculates any measure of profitability both with and without appreciation.

Availability of Data

The data for the state-supervised farm record systems are collected on a confidential basis. Therefore, data must be handled in a way to

TABLE 10.2 Calculation of Measures of Net Returns, Average for 1990 New York Dairy Farm Business Summary

	Without Appreciation	With Appreciation
Total Accrual Receipts	328,849	338,401
Total Operating Expense	252,163	
Expansion livestock	4,056	
Machinery depreciation	16,624	
Building depreciation	8,986	
Total Accrual Expenses	281,829	281,829
Net Farm Income	47,020	56,572
Less: Unpaid family labor @ $1250/month	3,538	3,538
Return to operator labor, management and equity	43,482	53,034
Less: Real interest @ 5% on 471,322 equity	23,566	
Labor and management income	19,916	
Labor and Management income per operator (1.39 operators)	14,328	

(continued)

TABLE 10.2 *(continued)*

	Without Appreciation	With Appreciation
Return to operator labor, management and equity	43,482	53,034
- Value of operator labor and management (1.39 operators)	<u>30,613</u>	<u>30,613</u>
Return on equity capital	12,869	22,421
+ Interest paid	<u>19,914</u>	<u>19,914</u>
Return on total capital	32,783	42,335
Rate of return on equity capital (471,322)	2.7%	4.8%
Rate of return on total capital (703,477)	4.7%	6.0%

TABLE 10.3 Calculation of Measures of Net Income, Illinois System, Using 1990
New York Data

Value of farm production (net of purchased feed and livestock)	239,579
- Total operating expense, except feed, livestock and interest	142,979
- Depreciation	<u>25,610</u>
Net farm income	70,990
- Unpaid family labor, 2.83 mos. @ $1,350/month	3,820
- Interest on all capital (land @ 4.5%, all other @ 10%)	<u>61,563</u>*
Labor and Management Income	5,607
- Value of operator labor (16.68 mos. @ $1,350)	<u>22,518</u>
Management Return	-16,911
Net farm income	70,990
- Operator and family labor @ $1,350/month	<u>26,338</u>
Capital and management earnings	4,652
÷ Total investment (703,477)	
Rate earned on investment	6.3%

*An assumption was made that one-half the real estate on the average NY
dairy farm is land.

Note: In the Illinois system the calculations include the landlord's as well as the
operator's share. The New York data do not include any share-rented
farms. The data used are the New York "without appreciation" numbers.
The format used on this page is the economic analysis of the farm
business under which management return is calculated. The farm
operator also receives an income statement measuring net farm income
from an accounting standpoint.

TABLE 10.4   Calculation of Measures of Net Income, Michigan System, Using 1990 New York Data

Value of production (Receipts less purchased feed and livestock)		239,579
Expenses except feed, livestock and interest paid	168,589	
+ Interest on all capital @ 7.0%[*]	49,244	
+ Value of operator and family labor	29,268[†]	
Total costs		247,101
Management income		-7,523
+ Value of operator labor		29,268
Labor income[‡]		21,745
Management income		-7,523
+ Interest @ 7.0%[*]		49,244
Return on owned (total) capital		41,721
÷ Average owned (total) capital		703,477
Rate earned on owned capital[§]		5.9%

[*]Varies from year to year depending on interest rates actually paid by farmers.
[†]Operator labor 3,000 hrs. x 1.39 = 4,170 hrs. @ $6.00 = $25,020.
  Family labor 708 hrs. @ $6.00 = 4,248.
[‡]Conceptually equal to New York's labor and management income.
[§]Return on capital includes management.
Note 1: The Michigan system does not calculate appreciation. It is not clear whether price changes on livestock are included in inventory changes. It is assumed here that Michigan calculates depreciation the same way Cornell does and that appreciation of livestock and real estate is excluded from the income calculations.
Note 2: The above calculations are the standard procedure used for all types of farms in the Michigan system. For dairy farms only, the Michigan system also calculates Net Farm Income about the same way that Cornell does except that appreciation is not specifically separated.

TABLE 10.5   Calculation Measures of Net Income, Kansas System, Using 1990
New York Data

Gross farm income, including inventory change	328,849
- Cash operating expense (including interest paid)	256,219
- Depreciation	25,610
= Net farm income	47,020
- Interest on 471,322 equity @ 10%	47,132
- Unpaid family labor	3,538
= Return to labor and management	-3,650
= Return to labor and management (per operator)	-2,626
Net farm income	47,020
+ Interest paid	19,114
- Charge for operator labor ($15,000 per operator)	20,850
- Value of unpaid labor	3,538
- Management charge (10% of gross income)	32,885
= Return to capital	8,861
÷ Total capital managed, including the value of rented land*	
= Rate earned on 703,477 total capital	1.3%
Return to capital	8,861
- Interest paid	19,114
= Return on net worth	-10,253
÷ Net worth	471,322
= Percent return on net worth	-2.2%

*This calculation is made based on total capital owned because the value of
rented land is not known in the New York System. Appreciation is not included.

maintain confidentiality. In many states, the data are available for use by researchers at the university but usually under rather strict procedural guidelines. Researchers from other states would be able to gain access to the data for research purposes only by making individual arrangements with the person in charge of the data gathering project. See the list of organizations responsible for collecting and publishing state farm record data at the end of this chapter. In some cases, access to the data is limited by the nature of the arrangements between the university and the farm business management associations.

Tentative Conclusions

Anyone who would like to combine data from two or more states to study issues such as farm size and structure is faced with a rather formidable task. In addition to obtaining permission to use the data, a researcher would be faced with the task of reformulating data to make it consistent in terms of charges for items such as depreciation, interest, operator labor and family labor. Some of this may be difficult or impossible because the necessary data may not exist in the record files.

Considering the non-random character of the data along with the inconsistencies among systems, perhaps researchers should seek another source of data. However, many of the record systems likely have a higher level of detail than the USDA data and would therefore be quite useful in studying details and cause and effect relationships related to farm size and structure. The data may also be more accurate because the record data are less dependent on recall than are survey data such as those collected by the USDA. In cases where data are needed that are not part of the record system, participants in the record system could be surveyed to obtain the missing data, and the large amount of data already available on the record systems would not have to be re-collected. The non-random nature of the collection of the record system does not make the data unusable. Moreover, much information can be obtained from the data, even if the data are not randomly selected. Results from research based on farm record data may be considered hypotheses to be tested using data that are collected by means of random sampling. Use of data that are already available is less expensive than collecting new data, and results based on the farm record data may be used to guide the collection of data from random samples, therefore lowering the costs of collecting those data.

A number of people believe that a standard procedure for farm business summaries should be used by all groups who sponsor farm record systems. A standard procedure would facilitate making comparisons among states and systems as well as allowing research using data from more than one state. Conversations with persons involved with the data in several states suggest that it will not be easy to get the various states to conform to a standard procedure. One reason for not changing is to maintain continuity with past data. Another is difficulty of getting agreement on a "correct" procedure to handle items such as imputed costs and asset valuation procedures. One person suggested that it might be easier to get the various systems to agree to apply a standard set of procedures to the data stored in the computer than to change the published data. Published data for each state would continue to follow past procedures, but there would also be a data set consistent across states that could be used for research purposes. If this could be done by just changing items such as the interest rate charged on equity capital, conformance could be easily achieved. However, some changes likely would require changes in the basic data collection. For example, if the standard procedure was to use market values of assets, a system that used book values would also need to collect market values. Nevertheless, the merit of this approach should be studied.

Farm Financial Standards Task Force

A Farm Financial Standards Task Force (FFSTF) sponsored by the American Bankers Association with a membership of nearly 50 persons from the academic community, financial institutions and other interested groups spent two years working toward a set of standard procedures for farm financial reports. A report containing the recommendations of the task force was issued in May 1991. The recommendations are too numerous and long to include here, but a brief review is presented below. This author believes that no state farm record system produces reports which are consistent with the task force recommendation. Groups who sponsor farm record programs should seriously consider adoption of the standard procedures resulting from the task force.

The FFSTF recommendations relate to some but not all of these issues raised in this chapter. The task force made recommendations on asset valuation, depreciation, value of operator and family labor (or family living withdrawals), calculation of net farm income, but not on appropriate imputed charges for equity or total capital. One must recognize that this task force is primarily interested in financial reporting to lenders and therefore may have little interest in some measures of net return such as labor and management income that are of interest to workers in research, extension and teaching.

The FFSTF tried to move toward GAAP accounting but stopped short in several areas. With respect to asset valuation, it recommended, in general, that market values be used but that cost (less depreciation) also be used for land, depreciable assets and marketable securities (similar to the Coordinated Financial Statements Approach). The underlying objective is to be able to separate changes in net worth into (1) the portion earned from operating the business (retained earnings) and (2) the portion due to valuation changes.

The FFSTF did not specify a single method of depreciation, such as tax depreciation. It recognized that tax depreciation methods available in the early 1980s allowed depreciation at a much more rapid rate than economic depreciation and that currently available methods are less likely to do this. It recommended that any method that aims to distribute the cost over the estimated life in a systematic and rational manner be used. It rejected methods that use a percentage, such as 10, of the current market value because of the potential for the total depreciation to exceed the original cost of the asset.

Net farm income was defined as return to "Operator and Unpaid Family Labor, Management and Equity Capital." Interest paid is included in the calculation.

The FFSTF did not recommend that standard charges be used for operator and family labor (or labor and management). It recognized that a charge must be made for this resource in order to calculate return on assets or return on equity and that, from an economic viewpoint, the charge should be the opportunity cost. For a financial analysis, it recommended that withdrawals for family living be used as the value of operator and family labor and management.

Task Force on Commodity Costs and Returns

In 1991, the American Agricultural Economics Association established a Task Force on Recommendations for Commodity Costs and Returns Measurement. The charge to this task force is: "To recommend standardized practices for generating costs and returns estimates for agricultural commodities after a careful examination of the relevant economic theory and the merits of alternative methods." While the focus of this task force is on commodity costs and returns rather than whole-farm measures of net return, it will wrestle with many of the same issues as did the Financial Standards Task Force. In addition, the task force necessarily will deal with issues of charges for owner equity and unpaid operator and family labor and management.

The Relationship of Farm Size to Profitability

This section presents data from several state farm record systems which show the relationship between farm size and several measures of net returns from the farm operation. The data are taken from the farm record systems in Illinois, Michigan, and New York.

Illinois

The Illinois data presented here are for northern and central Illinois grain farms with soil ratings of 86 to 100 (Table 10.6). Therefore, these data represent a relatively homogeneous group of farms in terms of type and soil quality. The total number of farms represented ranges from 780 in 1985 to 936 in 1990.

Six measures of net returns are shown in Table 10.6 because no one measure of profitability is necessarily superior to other measures. Each measure is described below. All measures are for the total farm business, including the landlord's share unless otherwise noted.

Net farm income includes the return to the farm and family for unpaid labor, the interest on all invested capital and the returns to management. The 1985-90 data clearly show that, on the average, larger farms in this set have larger net farm incomes than smaller farms. However, one must recognize that larger farms have larger interest charges and that interest has not been deducted in computing net farm income. The 1985-90 data also clearly indicate that the operator's share of net farm income is greater, on the average, on the larger farms. One should recognize that this is a per farm number and that larger farms are more likely to have more than one operator than are smaller farms.

Labor and management income per operator is total net farm income, less the value of family labor and the interest -- including net rent -- charged on all capital invested. This figure, as the residual return to all unpaid operator's labor and management efforts, is then divided by the months of unpaid operator labor and multiplied by 12 to reflect income for one operator on multiple-operator farms. The data tend to indicate that the labor and management income per operator is greater on the larger farms in this group. However, in a "poor" year such as 1988 the labor and management income per operator bears little relationship to farm size.

Management return is the residual surplus after a charge for unpaid labor and the interest or land charge on capital are deducted from net farm income. The unpaid labor charge includes operator as well as other family labor. Interest on land is charged at a rate that represents the long-run rate of return on land (4.5 percent in 1990, but the rate varies from year to year). Interest on non-land capital is charged at a rate (10 percent in 1990) that represents the cost of operating capital.

The 1985-90 data do not show a clear relationship between farm size and management returns. In 1985, 1987, 1989, and 1990 there was a

TABLE 10.6 Measures of Farm Profitability by Farm Size, Northern and Central Illinois Grain Farms with Soil Rating of 86-100, 1985-90

Acres	1985	1986	1987	1988	1989	1990
Net Farm Income (before interest)						
180-339	31,587	27,439	38,090	26,652	41,991	37,249
340-799	76,243	64,500	81,618	55,403	89,885	89,768
800-1199	144,578	125,138	143,643	92,452	153,699	154,941
1200 & over	245,645	179,149	235,475	167,818	258,584	259,442
Net Farm Income, Operator's Share						
180-339	16,370	8,004	18,645	9,951	21,769	17,981
340-799	29,447	18,297	33,729	19,221	38,586	36,311
800-1199	49,386	35,369	53,156	22,628	59,133	59,331
1200 & over	83,940	38,755	79,911	47,245	102,622	99,819
Labor and Management Income per Operator						
180-339	7,348	-6,384	8,145	-5,860	8,802	4,472
340-799	22,384	783	21,115	-6,842	26,001	24,591
800-1199	44,750	13,836	37,284	-15,090	43,382	40,424
1200 & over	75,780	2,471	51,228	-7,889	58,320	60,386
Management Returns						
180-339	-9,365	-19,168	-5,303	-19,341	-4,917	-10,291
340-799	5,389	-12,652	6,984	-21,588	11,638	9,026
800-1199	29,966	-429	23,924	-30,983	29,469	25,972
1200 & over	62,036	-16,372	51,236	-25,739	58,723	56,690

Rate Earned on Investment

Acres						
180-339	4.25	2.40	4.59	2.22	4.40	3.47
340-799	5.66	4.46	6.21	3.60	6.09	5.76
800-1199	6.40	5.55	6.87	3.94	6.55	6.24
1200 & over	7.33	5.07	7.32	4.71	6.87	6.67

Management Returns per Acre

180-339	-37.35	-71.38	-20.23	-72.54	-18.21	-40.13
340-799	10.41	-24.37	13.14	-40.79	21.56	16.66
800-1199	32.94	-.47	25.92	-33.82	31.77	27.93
1200 & over	39.60	-11.36	35.15	-17.65	39.40	37.76

rather dramatic positive relationship. In 1986 and 1988 management return was not related to farm size.

Rate earned on investment is net farm income minus unpaid labor divided by total farm investment. For this group of farms in the 1985-90 period, larger farms earned a greater average rate of return on investment than did smaller farms.

Management return per acre is the total management return divided by tillable acres. During the time period studied, the average management return per acre was clearly greater on the larger farms than on the smaller farms.

The six measures presented here, in general, indicate that larger farms have greater net returns than smaller farms.

New York

Data from approximately 400 New York dairy farms are used to illustrate the relationship between farm size, as measured by number of cows and several measures of profitability (Table 10.7). These farms are scattered throughout the state. Farms on which the operator owns no real estate are excluded, as are farms with crop sales greater than 10 percent of milk sales. Measures of net income are described below.

Net farm income is the return per farm to operator and unpaid family labor, management, and equity capital. Interest paid has been included in expenses. Note that this definition of net farm income is different from the one used in Illinois. The 1985-90 data for this group of New York dairy farms clearly indicate that there is a strong positive relationship between farm size (number of cows) and net farm income. If the net farm income was adjusted for operators per farm, the relationship would be somewhat less dramatic.

Labor and management income per operator is the return to operator labor and management after deducting a charge for non-operator unpaid family labor and a 5 percent real interest charge on equity capital from net farm income. The relationship between farm size and labor and management income per operator is not as consistent as the farm size/net income relationship. However, there is considerable evidence that, on average, the larger farms have higher labor and management incomes than the smaller farms.

Percent return on equity capital without appreciation is the return to equity capital (net farm income minus a charge for operator and unpaid family labor) divided by the average equity capital for the year.

Percent return on equity capital, including appreciation is calculated as above except that changes in the values of assets due to price level changes are included in the return to equity capital. There is a strong

TABLE 10.7 Relationship of Farm Size to Measures of Net Return, New York Dairy Farms, 1985-90

	1985	1986	1987	1988	1989	1990
Number of Cows						
			Net Farm Income, Without Appreciation			
Under 40	5,569	6,845	11,140	12,875	13,766	10,520
40 to 54	9,759	7,644	15,546	15,005	20,201	16,710
55 to 69	12,975	16,164	17,099	19,823	29,428	24,803
70 to 84	16,637	15,600	26,024	30,326	31,871	31,670
85 to 99	23,932	19,361	34,773	38,682	43,983	43,717
100 to 149	28,491	39,080	41,411	47,404	59,493	56,628
150 to 199	33,028	33,630	52,589	52,624	70,376	81,181
200 to 299	51,786	42,881	81,414	69,533	109,814	104,096
300 and over	131,638	123,246	208,798	233,809	291,433	227,064
			Labor and Management Income per Operator			
Under 40	-3,689	-2,533	1,228	2,119	1,828	-811
40 to 54	-508	-2,186	4,429	2,782	5,646	255
55 to 69	-541	1,361	1,362	2,415	8,055	4,224
70 to 84	-320	-1,372	6,573	8,313	8,459	7,652
85 to 99	2,911	378	12,999	13,710	12,705	10,895
100 to 149	3,464	8,981	10,501	13,886	21,038	17,901
150 to 199	4,355	3,696	12,244	10,480	18,259	24,271
200 to 299	10,367	4,803	27,968	17,676	43,897	30,241
300 and over	48,423	42,319	99,693	110,437	149,485	83,880

(continued)

TABLE 10.7 (continued)

Number of Cows	1985	1986	1987	1988	1989	1990
	Percent Return on Equity Capital, Without Appreciation					
Under 40	-6.6	-8.8	-4.6	-4.3	-4.6	-7.4
40 to 54	-6.4	-8.6	-3.2	-4.0	-1.6	-4.6
55 to 69	-4.0	-3.7	-2.8	-2.6	0.3	-1.5
70 to 84	-2.3	-2.8	0.4	0.5	0.7	-0.1
85 to 99	-0.6	-2.1	2.5	2.9	2.9	1.6
100 to 149	0.0	1.5	1.7	2.8	4.4	3.7
150 to 199	0.6	0.7	2.8	2.5	4.2	5.0
200 to 299	2.8	1.8	6.2	3.9	7.9	5.4
300 and over	8.7	7.1	12.7	13.4	15.1	9.9
	Percent Return on Equity Capital, Including Appreciation					
Under 40	-7.1	-3.2	1.8	0.0	1.4	-5.6
40 to 54	-7.0	-2.1	5.3	2.8	5.7	-1.6
55 to 69	-5.3	0.1	2.4	2.2	6.4	1.3
70 to 84	-1.4	2.5	6.3	5.2	6.4	1.6
85 to 99	-1.8	4.2	8.8	9.1	8.3	3.2
100 to 149	0.1	7.3	6.5	7.6	10.3	4.5
150 to 199	-1.1	5.3	11.4	9.4	9.0	5.5
200 to 299	1.8	5.1	11.3	7.2	12.2	8.0
300 and over	7.7	10.6	18.2	16.8	20.6	13.9

relationship in this set of New York dairy farm data between farm size and rate of return on equity capital, with or without appreciation.

Michigan

Data on net returns by farm size for Michigan dairy farms are shown in Table 10.8. Descriptions of the measures are presented below.

Labor income is the return to the operator for the year's labor and management after deducting an interest charge for all the capital owned. Labor and management income would be a more descriptive label. The evidence on the relationship between dairy farm size (number of cows) and labor income is mixed. In 1987 through 1990 there was a strong positive relationship. However, in 1985 and 1986 the average labor income for all groups was substantially negative and there was little difference among the results for the three larger herd size groups.

Rate earned on owned capital is the return to owned capital, after deducting a charge for the operator's labor (but not management) divided by the total capital owned. There was a strong positive relationship between farm size and rate earned on owned capital in all years.

In general, the 1985-90 data from Illinois, New York and Michigan presented here indicate that large farms are more profitable than small farms. However, the 1985-86 Michigan labor income data and the 1988 (severe drought year) Illinois data on management returns do not support this statement.

Conclusions

Data from over 14,000 farms are available from records collected by a combination of agricultural economics departments and farm management associations. While these data are not from random samples, they do represent a large number of farms over a very long period of time. Researchers would face some difficulty in combining data from various states because of inconsistencies in data collection procedures. However, the data could be useful in studying farm size issues. Data from farms in Illinois, Michigan and New York, in general, show a strong tendency for larger farms to be substantially more profitable than smaller farms.

TABLE 10.8   Relationship of Farm Size to Profitability, Michigan Dairy Farms, 1985-90

Number of Cows*	1985	1986	1987	1988	1989	1990
			Labor Income			
Less than 50	-16,047	-7,433	-82	963	8,453	2,513
50 to 74.9	-28,645	-19,957	2,882	649	28,640	26,805
75 to 99.9	-29,342	-12,087	10,321	6,678	37,496	47,018
100 or more	-29,053	-12,429	22,114	10,845	74,551	63,131
			Rate Earned on Owned Capital			
Less than 50	-3.07	-3.19	-.41	-.16	4.31	1.89
50 to 74.9	-.85	-1.66	3.95	3.35	9.88	9.52
75 to 99.9	.50	1.28	6.00	4.86	10.12	11.27
100 or more	3.96	3.69	8.18	6.55	12.17	11.25

*For 1989 and 1990, the four size groups are: less than 65; 65 to 99.9; 100 to 149.9; and 150 or more.

Appendix

Organizations Responsible for Collecting and Publishing State Farm Record Data

Illinois:	Illinois Farm Business Management Associations, Illinois Cooperative Extension Service and Department of Agricultural Economics, University of Illinois, Urbana, Illinois 61801. Attention: Dale Lattz.
Indiana:	Department of Agricultural Economics, Purdue University, West Lafayette, Indiana 47907. Attention: Donald Pershing.
Iowa:	Iowa Farm Business Associations (data collection) and Iowa State University Cooperative Extension Service (compilation and publication), Ames, Iowa 50011. Attention: Michael Duffy.
Kansas:	Kansas Farm Management Associations and K-Mar-105 Association, Department of Agricultural Economics and Cooperative Extension Service, Kansas State University, Manhattan, Kansas 66506-4026. Attention: Larry Langemeier.
Kentucky:	Department of Agricultural Economics, Cooperative Extension Service and Kentucky Farm Management Groups, Inc., University of Kentucky, Lexington, Kentucky 40546-0276. Attention: Fred Justus.
Michigan:	Department of Agricultural Economics, Michigan State University, East Lansing, Michigan 48824. Attention: Sherrill Nott.
Minnesota:	Southeastern and Southwestern Farm Business Management Association, University of Minnesota Institute of Agriculture and Cooperative Extension, Department of Agricultural Economics, University of Minnesota, St. Paul, Minnesota 55108. Attention: Kent Olson.
Missouri:	Department of Agricultural Economics and Missouri Extension Division, University of Missouri, Columbia, Missouri 65211. Attention: Norlin Hein.
Nebraska:	Nebraska Farm Business Association in cooperation with Cooperative Extension, University of Nebraska, Lincoln, Nebraska 68583-0719. Attention: Gary Bredensteiner.

282

| New York: | Department of Agricultural Economics, Cornell University, Ithaca, New York 14853. Attention: Linda Putnam. |

North Dakota: Department of Agricultural Economics, North Dakota State University, Fargo, North Dakota 58105. Attention: Andy Swenson.

Oklahoma: Oklahoma State University, Department of Vocational-Technical Education, Agricultural Education Division, 1500 West 7th Avenue, Stillwater, Oklahoma 74074-4364. Attention: Verlin Hart.

Pennsylvania: Data collected by Pennsylvania Farmers Association, publication prepared by Pennsylvania State Cooperative Extension, Pennsylvania State University, University Park, Pennsylvania 16802. Attention: Larry Jenkins.

South Dakota: State Office of Adult Vocational-Technical Education and State Department of Agriculture, Department of Education and Cultural Affairs, Richard F. Knapp Building, Pierre, South Dakota 57501-2293. Attention: Ed Mueller.

Notes

1. The amount that would be available for alternative investments should be adjusted for the tax that would be paid on the sale of farm assets, but seldom is.

2. In several systems all farms have one operator, according to the published data. Some of these farms must have more than one operator. Apparently, any operators in excess of one are counted as hired labor and such labor valued with a procedure not explained in the publication.

References

American Bankers Association. 1991. "Recommendations of the Farm Financial Standards Task Force." The American Bankers Association, Washington, DC.

Bredensteiner, Gary. "1989 Farm Analysis, Nebraska Farm Business Association." University of Nebraska, Lincoln, NE.

Brodek, Virginia. 1987. "Michigan Farm Business Analysis Summary - All Types of Farms, 1985." Agricultural Economics Report Number 486. Department of Agricultural Economics, Michigan State University, East Lansing.

Casler, George L. 1990. "Firm Level Agricultural Data Collected and Managed at the State Level." Staff Paper No. 90-5, Department of Agricultural Economics, Cornell University, Ithaca, NY.

Duffy, Michael, and Kyle Stephens. 1988. "1987 Iowa Farm Costs and Returns." Iowa State University Cooperative Extension Service, FM-1789, Ames, IA.

"Farm Incomes and Production Cost Summary from Illinois Farm Business Records, 1989." 1990. AE-4566, Department of Agricultural Economics, College of Agriculture, University of Illinois at Urbana-Champaign.

Frey, Thomas L., and Danny A. Klinefelter. 1980. *Coordinated Financial Statements for Agriculture*. 2d. ed., Agrifinance, Skokie, IL.

Hastings, Janice E. 1990. "1989 Agrifax Dairy Farm Summary, Farm Credit Banks of Springfield." Springfield, MA.

Hein, Norlin A. 1989. "Missouri Farm Business Summary: 1988." University of Missouri-Columbia, Extension Division, FM 8990.

Helt, Lawrence F. "North Dakota Farm Management Annual Report, 1988." Farm Analysis Center, Bismarck State Community College, 1500 Edwards Avenue, Bismarck, ND.

Justus, Fred. 1990. "The Kentucky Farm Business Management Program 1990 Annual Summary." Agri. Economics-Extension Series No. 94, University of Kentucky Cooperative Extension Service and Agricultural Experiment Station and Kentucky Farm Management Groups, Inc.

Langemeier, Larry N., and Frederick D. DeLano. 1991. "Kansas Farm Management Associations. The Annual Report, 1990." Department of Agricultural Economics, Cooperative Extension Service, Kansas State University, Manhattan, KS.

McSweeney, William T., and Larry C. Jenkens. 1989. "1988 Pennsylvania Dairy Farm Business Analysis." Extension Circular 374, College of Agriculture, Cooperative Extension, Pennsylvania State University, University Park, PA.

Nott, Sherrill B. 1991. "Business Analysis Summary for Specialized Michigan Dairy Farms, 1990 Telfarm Data." Agricultural Economics Report No. 554, Department of Agricultural Economics, Michigan State University, East Lansing, MI.

Oklahoma Farm Business Management. "Annual Report 1986." Oklahoma State Department of Vocational and Technical Education, Stillwater, OK.

Olson, Kent D. 1990. "Southwestern Minnesota Farm Business Management Association: 1989 Annual Report." Economic Report ER90-4. Department of Agricultural and Applied Economics, Institute of Agriculture, St. Paul, MN.

Scott, John T. "Operator and Landlord Shares, 1986." Department of Agricultural Economics, University of Illinois at Urbana-Champaign.

Schluckebier, Lynn. "1988 Statewide Annual Report, Farm and Ranch Business Management, April 1989." State Office of Vocational Technical Education, State Department of Agriculture, and Agricultural Education Department, South Dakota State University.

Smith, Stuart F., Wayne A. Knoblauch, and Linda D. Putnam. 1991. "Dairy Farm Management Business Summary: New York, 1990." A.E. Res. 91-5. Department of Agricultural Economics, Cornell University Agricultural Experiment Station. New York State College of Agriculture and Life Sciences, Cornell University, Ithaca, NY.

University of Illinois. 1991. "1990 66th Annual Summary of Illinois Farm Business Records." Circular 1316. College of Agriculture, Cooperative Extension Service, Urbana-Champaign.

11

Micro-Level Agricultural Data Collected and Managed by the Federal Government

James D. Johnson

The concepts and methods used to specify a research problem govern the types and level of specificity of data needed for use in the analysis. Readily available data often do not coincide with needed data. Differences between needed and available data exist for a variety of reasons, including the changing nature of policy issues and research problems, and the underlying purpose for collecting primary data. Most micro-level agricultural data collected by the Federal Government are designed to support programmatic or legislative functions. Data are rarely, if ever, collected within the Federal establishment solely to support a research activity. Exceptions may include data collected by cooperators in research agreements, and the Census of Agriculture. The Census is a special case since it exists to provide a comprehensive cross-sectional view of U.S. agriculture and to fill informational voids, particularly those that exist for smaller geographic areas such as counties.

The incongruity between data needs and availability may require that necessary data be collected, research objectives be altered, or less than ideal data be used in analysis. Information about the scope and coverage of firm and other micro-level data existing within the agricultural research community may lessen the need for primary data collection or project adjustments, particularly if data sharing is a feasible alternative. The focus of this chapter is on micro-level agricultural data (including proprietary farm business or establishment data) collected and managed by the Federal Government. Specific attention is given to data that may be useful in analyses related to the structure of agriculture and to the measurement of the costs, returns, and financial performance of farm businesses.

Data Used in Analyses of Farm Structure, Costs and Returns

To identify data useful in analyses related to the structure of farm businesses, changes in structural attributes over time, and the measurement of costs and returns, a variety of articles and reports were reviewed to gain a perspective about the types of data that have been used. The term structure, as applied to agriculture, has been defined in several ways. For example, former Secretary of Agriculture Bergland (1979) defined structure as, "how farming is organized, who controls it, and where it is heading." Sundquist (1978) in a similar statement took structure to mean "essentially the number, size, and decision making control that prevails for firms engaged in the production and marketing phases of our natural food and fiber industry." A number of other definitions incorporating a greater or lesser degree of specificity exist in the literature. For example, Stanton (1984, 84) has noted that most discussion of farm structure issues revolves around

(1) the number and size distributions of farms, (2) the ownership and control of farm land and production decisions, (3) tenure patterns and business arrangements for farming, (4) the control of markets in which farmers buy production inputs and sell their products, and (5) the financial systems and credit arrangements available to producers and farm operators.

Babb (1979) and Brinkman and Warley (1983) discussed farm sector structure from the vantage point of causal factors and sector performance measures. Babb (1979) identified dimensions of farm structure to include (1) the number and size distributions of farms by commodity, type of farm and geographic region, (2) degree of specialization in production and organization of the firm, (3) ownership and control of productive resources, (4) barriers to entry and exit, and (5) socioeconomic characteristics of farm operators and resource owners. Babb's (1979) listing of factors affecting structure and his judgment of the relative importance of how a factor may affect a specific dimension of structure is shown in Table 11.1. Brinkman and Warley (1983)[1] extended Babb's (1979) listing of structural attributes to include discussion of financial structure, market information and control, ideological conformity and

TABLE 11.1 Importance of Factors Affecting Farm Structure[a]

Factors Affecting	Dimensions of Structure				
	No. and Size of Farm	Specialization	Owner-control	Entry Barriers	Socio-economic
A. Variation in input prices	2	2	3	4	4
B. Technology	1	1	4	2	3
C. Economics of size	1	2	3	1	3
D. Variation in commodity prices	2	2	3	4	4
E. Risk and expectations	2	2	3	4	4
F. Price-cost margin	2	3	3	3	4
G. Exchange arrangements	3	3	1	2	1
H. Capital requirements	2	4	2	1	2
I. Taxes	3	4	2	2	2
J. Goals of the farmer	2	4	3	4	2
K. Managerial ability	3	4	3	2	2
L. Alternative opportunities	3	4	3	2	1
M. Macroeconomic policies[b]					

[a]The number of each combination of factors and dimensions of structure indicates the relative importance of the factor in influencing structure; 1 means great importance and 4 means little importance.
[b]These policies and their impacts are so diverse that no attempt was made to rank their importance.
Source: Data drawn from Babb (1979).

relation to rural community. Their discussion of structural attributes and causative factors is similar to Babb's (1979), except they provided a discussion of, and noted trade-offs that exist among performance variables. For example, they noted how increasing farm size and enterprise specialization may be consistent with increased efficiency of resource use, but be incompatible with resilience to market instability, the vitality of rural communities, or resource conservation. In Chapter 2 of this book Stanton proposes that structure relates to farm businesses, farm households, and agricultural resources, and that the distribution, ownership, and decisionmaking structure of these are the key elements of the structure of agriculture.

The focus of this chapter is on identifying existing sources of micro or firm level data useful in analyses of farm structure issues. Whether the causal relationships among the factors and dimensions of structure portrayed in Table 11.1, or in other definitions, are accepted is not of primary importance. The main issue is what the hypothesized relationships suggest about the types of data needed to analyze structural attributes of farm businesses and farm households as they exist and change over time.

Data Needed for Structural Analysis

As discussed in the introduction to this book, many factors influence agricultural structure and the factors which influence changes in the structure of agriculture are dynamic with regard to how they affect a dimension of structure, and with regard to how the factor itself may adjust. This indicates that two types of data may need to be used in analyses of farm structure issues. The first category includes data that reflect the outcomes of adjustments that have taken place. This category includes data that, for example, show how farm resources are organized and managed at a point in time; the ownership and use of resources; the distribution of income and wealth among households which provide resources to farm firms; and the division of household resources among farm and non-farm activities. These data may be the result of either cross-sectional or time series observation; they also may include both processed (aggregated) and micro (firm) level measurements.

The second main type of data includes data designed to monitor ongoing adjustments that are taking place in farm businesses and their associated households. These data reflect temporal linkages--past situation to current situation; and current situation to some planned future course with regard to investment, production, consumption, saving, and other factors. Since these data address temporal issues, they would tend to be obtained through the use of either longitudinal and/or experimental data collection efforts (experimental microdata are defined as data that reveal operator characteristics and attitudinal perspectives about firm and household activities that may affect dimensions of structure).

Data Used by Members of Professional Associations

In 1989 the Economic Statistics Committee of the American Agricultural Economics Association (AAEA) surveyed approximately 6,200 members of seven professional associations, such as the AAEA, the Community Development Society, and the Rural Sociological Association, which have a strong interest in issues related to the structure of agriculture (Hushak, Chern, and Tweeten 1989). Survey respondents were asked to report whether they had used any of approximately 225 data sets (plus given the opportunity to identify other data sets), made available by a wide variety of national and international sources.

For data sets that had been used, respondents were asked to report whether each data set was either not important, somewhat important, or very important for future use. To pare the list to a subset of data and information sources that would be most useful for a specific research interest, survey respondents were also asked to identify up to 10 data sets they considered most important from among those that they had used. Key data sets, identified by respondents, who selected structure of agriculture as their primary area of research interest are given in Table 11.2. The data sets shown in Table 11.2 are produced by the Departments of Commerce (Bureau of the Census, Bureau of Labor Statistics) and Agriculture (National Agricultural Statistics Service, Economic Research Service).

Data sets identified by survey respondents contain data elements which either directly measure or are close proxies for the dimensions of . structure indicated by Babb (1979) and Stanton (1984). For example data related to farm number and size, specialization and ownership, as well as data on the socioeconomic characteristics of farm operators (and perhaps their households) may be obtained from the Census of Agriculture as well as from a variety of USDA surveys, statistical bulletins, and other reports. Analysts may also obtain data on factors such as technology, economies of size, price-cost margins, and other factors that may affect various dimensions of farm structure from data sources highlighted in Table 11.2. For example, data on capital requirements or price-cost margins can be obtained either from the farm finance survey, the farm cost and returns survey, or the Economic Indicators reports. Another feature of data sets identified in Table 11.2 is that they tend to reflect the outcomes of adjustments that have taken

TABLE 11.2   Data Sets Used by AAEA Members and Rural Social Scientists by Primary Areas of Focus, Structure of Agriculture

Data Set[a]	Rural Social Scientists		AAEA
	Ten Most Important Data Sets	Used and Very Important Data Sets	Used and Very Important Data Sets
	Percent of Respondents		
Census of Agriculture	46.4	49.7	80
Census of Population and Housing	24.8	45.8	58
Current Population Survey	16.0	45.1	62
The County and City Data Book	13.1	26.2	23
Farm Cost and Returns Survey: Farm Financial Statistics	12.9	32.1	64
Economic Indicators: National Financial Summary	11.9	46.4	64
Production & Efficiency Statistics	10.6	44.4	67
Farm Population	10.6	44.4	51
County Business Patterns	10.4	21.0	35
Number of Farms	9.8	35.3	68
Farm Finance Survey	9.6	21.4	43
Census of Manufacturers	9.0	23.9	35
Major Uses of Land in the U.S.	8.6	19.0	49
Economic Indicators: Cost of Production	7.0	43.8	64
Agriculture Land Value Statistics	5.9	46.4	45
Consumer Prices and Price Indexes	4.9	41.8	56

[a]Data sets are listed in order based on the percent of respondents ranking the data set among the 10 most important.

Source: The Economic Statistics Committee, AAEA.

place. As discussed later, it may be possible to link observations from one survey period to the next for some data sets (e.g. Census of Agriculture), but none of the data sets containing farm business data have been explicitly designed to monitor on-going adjustments taking place in the structure and organization of farm businesses and their associated households. This is expected since most Federal data sets are collected either to be used in implementing or monitoring programmatic activities, or to provide a cross-sectional perspective about farm and other rural business, and household characteristics. The remaining sections of this chapter focus on describing the firm or micro-level data collected (prepared) and managed by Federal agencies. Data sets identified in Table 11.2 are fairly inclusive of Federal sources of farm business data. Thus, data set content and accessibility is emphasized.

Micro- and Firm-Level Data Collected and Managed by Federal Agencies

Data that federal agencies collect and maintain on farm production activities, organizational characteristics of farms and socio-economic characteristics of farm operators and their households can be obtained primarily from the Bureau of the Census and USDA. Data from these agencies may be accessed either in published or electronic form for aggregated data, or through approved access to individual farm records. Some micro-data may also be available from administrative records (e.g., data from the Internal Revenue Service (IRS) or from the Agricultural Stabilization and Conservation Service (ASCS)). These data tend to be available as processed data contained in published reports issued by these agencies. From a conceptual perspective, administrative data from either IRS or ASCS may not match well with definitions of businesses being used in structural analyses (e.g., IRS data may be published as data for tax filers or households and ASCS data for persons or farms defined for commodity program purposes). This difference in reporting unit is a major difference between farm economic data reported by Census and USDA, and data reported by other sources such as the IRS. Since most structural analyses focus on the farm as a business unit, administrative records data may have a limited usefulness for these analysis.

Federal data sources also include data collected (or assembled) by cooperators in research agreements. Some of these data may be pertinent to the analysis of structural issues (e.g. the 1980 survey of farm families conducted jointly by Mississippi State University and the U.S. Department of Agriculture). Since these data may not be collected or managed directly by Federal agencies involved in the contractual arrangement, they are not discussed except to alert analysts interested in structural issues that pertinent data other than that available from ongoing surveys and statistical series may be available.

Census

The Census of Agriculture contains an extensive amount of data on sources of farm cash earnings and expenses along with a wide variety of data useful in classifying farm businesses by primary type of commodity produced, sales volume, tenure, ownership, land use, and characteristics of the operator (Table 11.3).

Summary cross-tabulations of data are published for farms by concentration of market value of agricultural products sold, tenure of operator, type of organization, age and principal occupation of the operator, size of farm in acreage, size measured by value of product sold, and by standard industrial classification of farm. For each major classification, data are reported for farms and land in farms, market value of product sold, farm production expenses, land use, machinery and chemical uses, tenure, and operator characteristics. Comparable data are published for the U.S. and each State. The Census also publishes county-level data on the characteristics of farms and farm operators. Many dimensions of structure such as the number and size of farm, ownership, specialization, and socio-economic characteristics can be assessed through the use of Census data. Thus, it is not surprising that the Census of Agriculture received the largest response as a data set both used and important from structure analysis respondents to the AAEA data needs survey.

The Census does not, however, provide enough data to develop a complete income or balance sheet statement. Nor does the Census attempt to sort out ownership of production or the sharing of revenues or expenses among participants in the farm operation. All items are allocated to the operation as a farm business unit. One operator household is assumed to be associated with each farm operation. Thus, the data user must exhibit some caution in interpreting published data cross-tabulations for conclusions related to farm structure issues. For example, the 1987 Census indicates that farms with sales in excess of $1 million had an average value of land and buildings in excess of $3.9 million. It would be easy to move from a statement which said that *farms* with over $1 million in sales, on average, had in excess of $3.9 million in assets to a statement which said *farmers* with over $1 million in sales, on average, had in excess of $3.9 million in assets. Yet, the explanation sections of published Census volumes clearly indicate that the value of land and buildings is the value controlled as opposed to the value owned

TABLE 11.3 Farm Business Collected and Maintained at the Federal Level

Source of Data	Population Covered	Coverage and Scope of Data	Data Access
Census of Agriculture	Farm firms with sales (or expected sales) of $1,000	Aimed at universal coverage of operators of farm firms	Hard copy and electronic media
		No separation of ownership of production; sales, or expenses; data obtained on tenure; land use; crop mix; crop and livestock sales; CCC loans and payments; cash farm related income, major cash production expenses, and estimated market value of land and machinery (expenses, assets and farm-related income estimated from a 1 in 6 sample of the in-scope Census)	Customized cross-tabs may be purchased Is possible to obtain access to micro, film level data by working through Census under a secure working agreement
		Incomplete income and balance sheet statements	Collection on a five year cycle
Census Follow-on or Special Surveys (Agricultural Economics and Land Ownership Survey (AELOS)	Same as Census	List sample of in-scope Census operations (45,000) and landlords reported by operators (40,000); collection of data on land operated, land use, operating expenses and capital expenditures, value of products sold, debts and assets of firm land ownership and acquisition	Access same as for Census Collection usually follows Census
		Aggregated firm balance sheet and net cash operating income statements feasible; no separation of ownership on operator's report; landlords report operating expenses, debts and assets; land ownership and acquisition;	

matching of landlord and operator data incomplete; limited contractual information; no enterprise information

Hard copy and electronic media

USDA

Economic Indicators of the Farm Sector (ECIFS)[2]

Same as Census

Aimed at providing comprehensive estimates of the income, expenses, assets, and liabilities of the farm sector. Estimates of cash receipts include separate estimates for approximately 120 crop and livestock commodities. Expenses are estimated for intermediate products such as feed and fuel, labor, interest, net rent, capital consumption, and property taxes. Liabilities are estimated by type of lender, and assets by type of asset (e.g., real estate, crop and livestock inventory).

Estimates are also developed for sources of farm earnings other than receipts from marketings (e.g., Government payments and non-farm sources of income for the farm operator household).

All estimates are developed for the U.S. and each State.

Estimates are distributed to size of farm classifications based on benchmarks from the Census of Agriculture.

(continued)

294

TABLE 11.3 (continued)

Source of Data	Population Covered	Coverage and Scope of Data	Data Access
National Agricultural Statistics Service (farm numbers, land in farms)	Same as Census	National and State annual estimates of total number of farms and land in farms; national estimates of the distribution of farms and land in farms by sales class; annual estimates of average size of farm (acres) by sales class.	Hard copy and electronic media
Farm Costs and Returns Survey	As of 1987, same definition as Census	List and area frame sample aimed at coverage of farm operations	Hard copy of data crosstabs
		Separation of ownership of production, sales, and expenses among operators, landlords, and contractors; detailed listing of expenses by component, sales by major commodity or group, and other farm earnings by sources; disaggregated balance sheet, including term of loan; tenure, farm organization, land use and crop mix; detailed production practices and input application for major enterprises; construction of cash and accrual based income statement; firm and household balance sheet; household labor allocation; farm and enterprise (budgets) cost and returns relationships.	Research, technical and statistical reports On site data access for approved research projects
Administrative Records (IRS)	Tax-filers	Individuals that reported farm income expenses	Hard copy data (Statistics of Income) but very limited data on farmers

Data Collected (assembled) by Cooperators in Research Agreements	Data dependent upon research objectives; ranges from participants in farm record/account systems to respondents to project specific surveys	Varies	Data items reported in the tax schedules	Public use data tape includes general tax information
			Data for farm sole proprietors, excludes partnerships and corporations	Special tax samples on a very limited basis
				Depends on how data are collected and local institution

by the farm business. Other estimates such as those for expenses, sales, and net return are also for the farm as a unit rather than for the operator. Thus, analyses that focus on ownership, control, or barrier to entry issues such as the concentration of income and wealth need to consider the multiple parties that participate in the modern farm business operation. Consideration must be given, in particular, to the distribution of asset ownership and to the sharing of income produced by the business.

Special or follow-on surveys to the current in-scope Census of Agriculture extend the types of data available to analyze characteristics of farm business operations as well as provide enhanced data on farm operator households (Table 11.4). The 1979 Finance Follow-On Survey provided data on operator and landlord expense sharing for production as well as data on assets and liabilities held by operators and landlords. This survey also provided considerable detail on the characteristics of operators and the off-farm income earned by operators and members of their families. The follow-on to the 1987 Census of Agriculture (the Agricultural Economic and Land Ownership Survey) will provide comparable data for the 1988 calendar year. The AELOS also asked for considerable detail on the sources of off-farm income earned by the operator and members of his/her household as well as for data on size of household, and assets owned by the farm operator households. AELOS collected the data needed to develop an aggregated balance sheet and a net cash operating income statement for the farm business. In addition, AELOS was designed to provide considerable detail on the characteristics of land ownership and the time frame in which land was either acquired or sold.

Both Census and Census follow-on survey data are available through print and electronic media. Raw survey data are processed (aggregated) to avoid disclosure and breech of rules that require confidentiality of respondent data. Standard data products released from the Census of Agriculture or special Census surveys provide data that describe selected structural attributes of farm operations at a point in time. Since public releases tend to be designed to provide historic continuity, emerging data needs may not be adequately addressed by published data. Some of the more specific needs, such as for data on super size or highly specialized farming operations, can be met through the purchase of customized cross tabulations of data. The 1987 Census does contain, however, a new table that addresses directly the issue of concentration of production in U.S. agriculture. Data are published for the fewest number of farms that account for 10, 25, 50, and 75 percent of sales.

To more fully address some questions concerning the distributions of farm operations and corresponding characteristics of farm operators, access to micro-data files or individual farm records may be necessary.

TABLE 11.4 Sources for Data to Supplement/Complement Firm Business Data

Source of Data	Population Covered	Coverage and Scope of Data	Data Access
Census of Agriculture	Farm firms with sales (or expected sales) of $1,000	Operators of farm firms characteristics of operators–residence, principal occupation, days of off-farm work, years of operation, age, sex, race	Same as for firm level data
Census Follow-on or Special Surveys (AELOS)	Same as Census	Data on households associated with a list sample in-scope farm operations and reported landlords	
		Characteristics and occupation of landlord, residence, race, age, occupation (e.g., farming, private business, government, etc.) portion of business income from farming	
		Operator household assets and liabilities; operator and spouse type of off-farm work, education, off-farm earnings by source, and size of household	
USDA (FCRS) data	Same as Census	Operators and operator households associated with farm operations	Same as for firm level data
		Operator household assets and liabilities (in conjunction with farm operation assets and liabilities); number of households, size of the operation household, percent of net farm income received by the operator household, education and off-farm work of operator and	

(continued)

TABLE 11.4 (continued)

Source of Data	Population Covered	Coverage and Scope of Data	Data Access
		spouse, major occupation, off-farm earnings by source, hours of farm work, household consumption expenditures, operator opinions about operation, expansion or contraction of the business (some of the detailed household data such as that for the spouse have been enumerated for one year; expectations are that it will be enumerated again for future years).	
Survey of and Program Participation (SIPP)	Households in the civilian noninstitutionalized population	Adults in households obtained from a stratified sample of the noninstitutional resident population of the U.S. (sample is large enough to facilitate metro, nonmetro comparisons) Data for labor force participation, employment, sources of income, sources of assets, annual income and taxes, net worth, housing conditions, health and disability; data designed to provide cross-section as well as longitudinal view of respondents	Quarterly income reports public-use microdata files (designed to maintain confidentiality)
Current Population Survey	Civilian noninstitutional population: primary sampling unit is the housing unit	Coverage includes data related to the demographic and employment characteristics of the rural population and persons living on farms Data include household, demographic, employment and income variables	Hard copy publications Public use microdata file

This access has been accomplished by Economic Research Service staff working with the Bureau of the Census. Peterson (1985) used the 1979 Follow-On Financial Survey to the 1978 Census of Agriculture to examine the financial position of specialized dairy farms by size of operation and age of the operator. Hatch (1982) accessed 1978 Census of Agriculture micro-data files to develop twenty typical farm descriptive data sets. In Hatch's (1982) work, a computer routine was designed to perform sequential data sorts using farmer responses to the Census of Agriculture. Desired farm characteristics were determined by successively eliminating farms from the population of interest by means of user specified criteria. The system was designed to prevent disclosure of information by aggregating data among categories. Moreover, the computer routines were run by Census personnel on their computer system. Hatch (1982) received processed output; albeit output that met his specific research needs.

The Agriculture Division, Bureau of the Census, has also created a longitudinal data set from the control files of the three most recent Census' of Agriculture. The control files contain a limited number of variables providing information on tenure, acres, sales, type of organization, standard industrial code, and the age, sex and principal occupation of the operator. Edwards, Smith, and Peterson (1985) applied Markov analysis to the longitudinal data set for 1974-78 to evaluate changes in size by acres per farm during 1974-78. Data sets such as the longitudinal file created by the Census hold promise for use in addressing structural issues that focus on the causes or process of adjustment rather than on the outcome of the adjustment process. For example, what are the factors that contribute to firm growth as opposed to what is the average size of farm operation. As formulated, the Census longitudinal data set contains a limited amount of structural attribute information. As the data set is enhanced, its ability to provide a perspective about factors that contribute to firm adjustments will be expanded.

Census data have also been used to assess differences in costs and returns for farms in different economic size classes. The data do not, however, reflect all costs of production incurred by the business and refer to an average farm expense within each size class. There is no way of knowing from the published data the distribution of expenses among farms within a group—e.g., whether a particular expense or receipt is large because all farms have the item or because some farm has an unusually large amount of the item. Moreover, data are for the entire farm operation. As near as one can come to a specific crop or livestock enterprise is to obtain special cross-tabulations that focus on the commodity of interest. Thus, while it is possible to compare the average cost-returns relationship among different sizes of farms using Census data, it is difficult to determine why costs vary. Obtaining customized cross-tabulations or access to the micro-data would help mitigate these difficulties.

USDA Economic Statistics

The USDA prepares a wide variety of farm production and economic statistics for use in monitoring the sector's performance. Many of these data series, such as those which measure the number and size farms, or the incomes, assets and liabilities of farmers and their households, provide data useful in gauging characteristics of agricultural businesses, farm households, and agricultural resources. Data are available to describe the size and income distribution of farm businesses, the ownership and control of agricultural resources, and to a lessor extent the attitudes of farmers, farm households, and resource owners. USDA economic statistics have not been prepared primarily for research purposes (structural analyses or otherwise), but have been collected and processed to serve legislative and agency needs. To better serve research needs and to provide a more comprehensive perspective about sector performance and the well-being of the sector's participants, estimates of aggregate sector indicators have been disaggregated to provide as much size, type, and geographic detail as possible. For example, USDA's mandate is to develop an estimate of aggregate sector income, treating all of agriculture as one farm. The Economic Research Service has moved beyond this mandated official estimate of sector-wide earnings to provide as much distributional information as possible given the underlying estimation procedures and data. USDA economic data may be obtained either as processed (aggregated) data series or as microdata which retains the characteristics of each firm or household in the sampled population. As will be discussed later, access to USDA microdata for research use is available only under very explicit conditions designed to conform with the confidentiality pledge granted survey respondents.

Aggregated Statistics. The primary source of farm business economic statistics made available by USDA is the series of bulletins comprising the Economic Indicators of the Farm Sector (ECIFS). This series includes five bulletins which contain time series data on income from farm and nonfarm sources, production efficiency, and crop and livestock cost and returns. Income and balance sheet statements are prepared for the U.S. and for each state.[2] For the U.S., income and balance sheet estimates are distributed by size and type of farm. The size classifications are based on value of sales, comparable to those used in the Census of Agriculture. Components of data published by sales classification include gross income, Government payments, production expenses, net income, and off-farm income. Asset, debt, and equity measurements are also published by size class. The size classification data published in the ECIFS series are not annually observed data for farm businesses of specified size, but are instead aggregate sector measurements that have been distributed to size classes based on distributors from the most recent Census of Agriculture. The National Agricultural Statistics Service provides an observed annual estimate of the number of farms and land in farm by size classification. The farm number data are used to compute the per farm averages published in the ECIFS series.

Microdata from Farm Surveys. The Department is required to produce estimates of the costs of producing major crop and livestock commodities, net income of the farm sector (for inclusion into the National Income and Gross National Product data series published by the U.S. Department of Commerce), and the balance sheet of the farm sector. For costs of production, the mandate is legislatively determined. For the financial statistics the mandate is derived from memoranda of understanding among agencies of the executive branch of government. To develop statistically and conceptually sound estimates of farm business production and financial statistics, requires a firm and enterprise level data collection system designed to accurately measure a population of interest and to provide the data elements needed to properly implement economic and accounting concepts. The Farm Costs and Returns Survey (FCRS) is currently the survey vehicle through which USDA collects economic data for use in its farm business, operator household, and enterprise analyses (Johnson and Baum 1986).

The FCRS is a probability-based, nationwide survey of approximately 26,000 farm and ranch operators. Since the survey is designed to generate an expanded annual estimate of total expenses incurred in the production of farm commodities, all facets of agricultural production activities are represented in the sample design.

The FCRS features a combined list and area frame sampling approach where the list is stratified by economic size and other attributes. For the area frame, all land area in the U.S. is divided into strata based on suspected land use. From these strata, sampling units of generally one to two square miles in size are selected for field enumeration. The area frame accounts for incompleteness in the list frame. The multiple frame sampling approach uses desirable attributes of both list (focused data collection from desired sub-populations of farm businesses/households) and area frames (coverage of changes that occur in the population of farms due to entry, exit, and other adjustments).

Both list and area frame samples are selected in replicates and designated for use with specific versions of the FCRS questionnaire

(Figure 11.1). The expenditure version (depicted by the left side of the panel shown in Figure 11.1) is designed to collect detailed data on farm operating characteristics, cost structure, business and household income structure, and the capital structure of the farm business. Expenditure version sample allocations are made to all list and area strata and are designed to maximize precision in the expansions for farm production expenditures (the estimates of total expenses by component from the FCRS underlie the Department's estimate of farm sector net income). The cost of production and other (e.g., farm operator household characteristics) versions of the FCRS questionnaires (depicted in the right hand side of the panel shown in Figure 11.1) are designed to produce precise cost of production estimates for a specific commodity, and representative coverage of other desired information.

The FCRS is designed to provide data needed to analyze farm financial performance by type and size of operation, and geographic location. In addition to this primary function, the survey also provides observed data useful in analyses of production technology and adjustment (competitive position of farms and enterprises, efficiency, and distributions of production by cost structure and economic characteristics), farm organization and finance (number, size, and operating and financial structure), and the well-being of farm-related households (income and wealth level and composition, farm production, and labor allocation). Thus, the FCRS is an annual source of firm level data for many of the structural attributes identified by Babb (1979), Brinkman and Warley (1983), and Stanton (1984) (e.g., number, size, type, ownership, tenure, specialization, concentration, financial structure, etc.).

The FCRS provides data as repeated cross-sections. Thus, it is not capable of addressing temporal, longitudinal issues related to the process of farm business-farm household adjustment over time. The FCRS does, however, address the issues of resource ownership and use by separating the way that farm resources are organized and managed from their ownership. The FCRS is also designed to obtain household data showing how labor is divided between farm and non-farm uses, investment in the farm business, and the distribution of farm business earnings among the various units that provide resources used in production (farm operator households, hired labor households, landlords, contractors, and other lessors of inputs). The FCRS explicitly accounts for the earnings that go to the farm operator household (some of these household data such as the division of labor between farm and nonfarm use have been collected only for one year; other data such as household income, assets, liabilities, and consumption have been collected more than once). In addition to these firm level structural data, the FCRS also obtains enterprise production and cultural practice data.[3] This allows an enterprise budget to be developed using observed input data. Budgets can be developed by size of farm, degree of specialization, and other attributes that may affect costs and returns.

In contrast to other federal sources of firm level data, the FCRS has a direct micro-data access policy. It is possible for researchers outside the Federal establishment to access and use the FCRS. Access is restricted, however, to on-site use for research projects that have been jointly approved by the Economic Research Service and the National Agricultural Statistics Service. Access is restricted to ERS, University and other public interest analysts for research purposes that are designed to serve the public. Analysts that want to use the FCRS are required to write a project proposal which specifies the items of information required, explain the need for the data, identify methods of analysis or techniques to be used and level of reliability required, and indicate the level of interpretation planned. If the planned use is approved, the user must sign forms which require him/her to maintain strict confidentiality of all microdata. Publications have to be cleared by USDA prior to release (screening for disclosure).

Summary

Federal agencies collect firm level agricultural data to support programmatic or legislative functions. Several sources of firm level data currently exist, including the Census of Agriculture, Census follow-on or special surveys, the USDA's Farm Costs and Returns Survey, administrative data from the Internal Revenue Service, and data collected by cooperators in research agreements. Each of these data sources obtains data potentially useful in analyses related to farm structure, cost, and returns. Depending upon a specific research project's design and scope, data from one or more Federal Source may be pertinent. A primary focus of the Census, for example, is to provide information on selected structural attributes of farm operations. Moreover, the Census is large enough to provide statistically reliable data at the county level. USDA's farm economic survey is designed to support statistically sound estimates of farm production expenses and cost of production estimates for major farm commodities (grain, oilseeds, cotton, livestock and dairy). A by-product of this work is the collection of the data needed to conduct whole farm analyses. Both Census and Farm Costs and Returns data are available in hard copy and through electronic media. Moreover, in what appears to be a little used option, micro-data may be accessed under certain approved (stringent) circumstances. Use of the micro-data option, either through the acquisition of customized tabulations or through the

304

FIGURE 11.1 The Farm Costs and Returns Survey

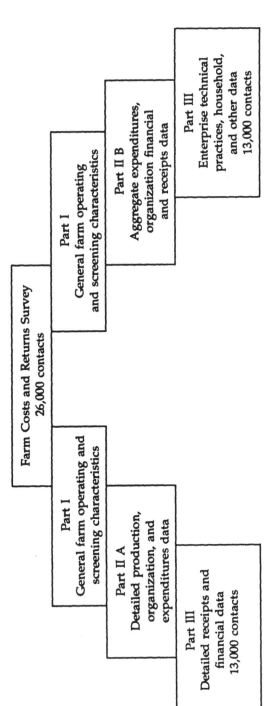

(continued)

FIGURE 11.1 (*continued*)

Part I: Screening and general farm operating characteristics:
- Land use
- Crop acreages, yields, and so forth
- Farm business and financial organization

Part II: Farm production expenditures, receipts and financial data, including items such as:
- Whole-farm expenses by type or category
- Livestock inventory, sales, and purchases
- Crop receipts, inventory, and so forth
- The farm business balance sheet
- Distribution of earnings and expenses among operators, landlords, contractors

Part III: Modular sections for specific detail
- A. Detailed information needs for special and key variables and data items relating to production activities and whole farm expenses
- B. Data on particular types or categories of farm organizational characteristics and technical practices used in crop and livestock production

submission of specialized computer routines, may extend the usefulness of Federal firm level agricultural data in structural, or cost and returns analyses. For example, see the analyses of production cost-size relationships undertaken by Ahearn, Whittaker, and El-Osta in Chapter 6 of this volume.

Notes

1. Brinkman and Warley's (1983) list of performance variables included allocative efficiency, contribution to national economic goals, income parity, income and wealth distribution, income adequacy, stability, resilience-vulnerability, adaptability and flexibility, ease of access, control, resource conservation and environmental quality, rural community health, rural-urban balance, balanced regional development, farm-rural values, government expenditure-revenue, and federal-provincial relations.

2. For a complete description of the procedures used to estimate farm income, see "Major Statistical Series of the U.S. Department of Agriculture, Farm Income," USDA, Agriculture Handbook No. 671, Vol. 3, Nov. 1988. Volume II of this same series (issued in May 1989) describes procedures used to estimate the balance sheet.

3. For a complete description of the procedures used to estimate commodity costs and returns estimates published by USDA, see Ahearn, Whittaker, and El-Osta, Chapter 6, this volume and "The National Commodity Costs and Returns Estimates." by Mitchell Morehart, James Johnson, and Hosein Shapouri, *Costs and Returns for Agricultural Commodities*, Ahearn and Vasavada, eds., Westview Press, 1992.

References

Babb, E. M. 1979. "Some Causes of Structural Change in U.S. Agriculture." *Structure Issues of American Agriculture.* U.S. Dept. of Agr., Econ., Stat., and Coop. Service, Agricultural Economic Report 438, pp. 51-60.

Bergland, Bob. 1979. "Foreword." *Structure Issues of American Agriculture.* U.S. Dept. of Agr., Econ., Stat., and Coop. Service, Agricultural Economic Report 438.

Brinkman, G. L., and T. K. Warley. 1983. *Structural Change in Canadian Agriculture.* Report to the Development Policy Directorate, Regional Development Branch, Agriculture Canada.

Edwards, Clark, Matthew G. Smith, and R. Neal Peterson. 1985. "The Changing Distribution of Farms by Size: A Markov Analysis." *Agricultural Economics Research* 37(4).

Hatch, Thomas, C. Cole Gustafson, Kenneth Baum, and David Harrington. 1982. "A Typical Farm Series: Development and Application to a Mississippi Delta Farm." *Southern Journal of Agricultural Economics*, pp. 31-36.

Hushak, L. J., Wen S. Chern, and Luther Tweeten. 1989. "Priorities for Data on Agriculture and Rural Areas: A Survey of Agricultural and Rural Social Scientists." The Economic Statistics Committee, AAEA.

Johnson, James, and K. Baum. 1986. "Whole Farm Survey Data for Economic Indicators and Performance Measures." *Agricultural Economics Research* 38 (3).

Peterson, R. N. 1985. *U.S. Dairy Farmers in 1979: Financial Characteristics by Operator Age and Dairy Size*. ERS Staff Report No. AGES840824, Economic Research Service, U.S. Dept. Agr.

Stanton, B. F. 1984. *Changes in Farm Structure: The United States and New York, 1930-1982*. Cornell Agricultural Economics Staff Paper, No. 84-23.

_____ . "Farm Structure: Concept and Definition" (this volume).

Sundquist, W. B. 1970. "Changing Structure of Agriculture and Resulting Statistical Needs." *American Journal of Agricultural Economics* 52(3).

12

Technology and Its Impact on American Agriculture

Marvin T. Batte and Roger Johnson

Technological change is generally viewed as a major determinant of structural change. The mechanization of farming as an outgrowth of the industrial revolution has been largely responsible for the dramatic increase in farm size over the last 70 years. A study by the Office of Technology Assessment concludes that "technology and associated economies of size, specialization, and capital requirements have had an important influence on structural change in agriculture" (U.S. Congress, OTA 1986, 117).

This chapter begins with a brief historical review of technological change and its impact on farm structure. The classical diffusion model and the induced innovation hypothesis are explored. Models of firm-level and aggregate (sector-level) impacts of technology adoption are presented. Following this is a description of emerging agricultural technologies, a discussion of differences in these from previous technologies, and suggested impacts of the emerging technologies on future farm structure.

Impacts of Technological Change on Structure

Tweeten (1986, 1) has characterized technological change in agriculture into three distinct revolutions. The first featured the wheel and simple hand tools, irrigation and domestication of plants and animals. This revolution "lasted for thousands of years and was still underway when the first white settlers came to America". The second revolution began with the industrial revolution in Britain during the late

18th century. It featured cheap steel and steel farm implements, railroads, and steam power. The third revolution began about 1920. Primarily mechanical, chemical, and biological in nature, it featured rural electrification, chemical fertilizers and pesticides, improved plant and animal genetic material, and the modern tractor and its complement of machinery. These technologies, especially those of the third revolution, embodied a tremendous potential to substitute capital for labor and/or land.

Many changes occurred in agriculture during the third revolution. Farms consolidated; The number of U.S. farms decreased by 55 percent since 1920, while average farm size increased by 2.4 times (U.S. Department of Commerce). Farmers were "drawn" from the farm by higher wages in the nonfarm sector. Remaining farmers adjusted to incorporate the new technology, typically by enlarging farm size and specializing production.

Technological change in agriculture also corresponds to changes in the control of key agricultural resources. Many mechanical innovations were embodied in durable capital assets. Capital requirements often exceeded farm equity. Increased use of debt and leasing was common through this period. Contract farming became common in selected commodities. Furthermore, profits from the adoption of technology were often bid into prices for other assets, most notably land, further increasing the capital requirements of farming.[1] Technological change also affected consumers of agricultural food and fiber. Americans today spend a lower percentage of their income for food than at any time in our history.[2]

Models of Technology and Structure

Technological change is generally understood to be any change in the method of production that results in a change in the output-to-input ratio. Hayami and Ruttan (1985, 86) regard technological change as "any change in production coefficients resulting from the purposeful resource-using activity directed to the development of new knowledge embodied in designs, materials or organizations". Technological change does not necessarily result in an inward movement of the isoquant. The new technology may "twist" the isoquant, moving it inward over some regions of input use and outward over other regions. Tweeten (1989, 102) observes that "Technology is not free, it is a form of capital. It requires scarce resources to produce and is durable, giving off services over several periods. Hence the decision to develop or use technology is economic."

Figure 12.1 graphically illustrates concepts of technological change and efficiency. Isoquant I_1 represents an existing technology that allows substitution of capital (K) and labor (L) in the production of a constant output (Q_0). Presuming individuals A and B are producing the same quantity of output, individual A operates with an inferior technology because the combination of inputs lies off the efficient isoquant. Individual A is *technically inefficient*. Individual B is producing with the correct technology, but, B has not chosen the correct set of inputs given prevailing factor prices (indicated by the isocost line XX'). Therefore, B displays *allocative* (price) inefficiency.

Technological change is said to be neutral with respect to a factor (Hicksian neutrality) when the ratio of inputs is constant for both the old and new technologies. If the relative use of factor i decreases with the technical change, the change is said to be biased and is i-saving. A technical change that results in an increase in the relative use of factor i is said to be biased and i-using.

Isoquants I_2, I_3, and I_4 (each with identical output, Q_0) represent substitution of labor and capital under three new technologies. Technology I_2 represents a Hicks neutral technical change. With constant relative factor prices, this technology combines the two inputs, labor and

FIGURE 12.1 Technical and allocative inefficiencies and technical change.

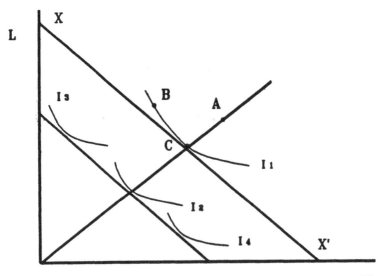

capital, in the same ratio as did the technology (I_1) that it replaced. Technology I_3 is capital-saving (labor-using). This technology, again with constant relative factor prices, calls for an input combination that employs relatively less capital than did the existing technology. Similarly, technology I_4 is labor-saving (capital-using). The relative use of labor in the input mix is reduced.

To assess the relative importance of labor-saving and land-saving technology on productivity in U.S. farming, Swanson and Sonka (1980) used data on productivity and efficiency (1940-77) for the U.S. farming sector. They separated farm output into its two technological components. Acreage per labor unit was used as an index of labor-saving technology, and output per acre as an index of land-saving technologies. Their results show a quadrupling of land area farmed per unit of labor since 1940, but the index of farm output per acre increased only about one-half as much. They concluded that during this period labor-saving technologies have been more influential than land-saving technologies when describing the changing productivity of agriculture.

Technological change also may be biased with regard to the scale characteristics of the production process. Scale-biased technical change alters the range of production over which returns to scale of a given degree can be realized (Stevenson 1980). This may change the minimum-average-cost output level for the firm. Scale bias may occur because certain technological changes are better tailored to firms of a particular size. This often is true for capital-embodied technologies. For example, many mechanical technologies in agriculture were embodied in a durable capital asset. Often the capacity of this asset exceeded that of the typical farm. Thus, larger farms realized lower costs of production with the new technology than did smaller farms. The large-scale bias of the mechanical technologies often were magnified by tax and credit policies that reduced the cost of capital inputs. Scale bias also could result simply because farms of one size tend to innovate at a faster rate than do farms of another size. In the latter case, a scale bias effect results even though the technology is not inherently biased.

Models of Innovation, Adoption, and Technology Diffusion

New technologies are developed and adopted by the population of producers over time. The nature of the innovation process should be treated as endogenous to the system. The structural reorganization that follows widespread adoption of a technology is a function both of the nature of the technology and the manner in which it spreads among producers. Various models of technology development and diffusion have been developed to explain these phenomenon.

The classical diffusion model depicts technology diffusion as comprised of several stages: invention, innovation, adoption, diffusion, and structural reorganization. Inventions are ideas perceived as new by an individual. Schumpeter (1939) identified innovation as the essential function of the entrepreneur: Innovation is the incorporation of invention into technology, thereby successfully establishing the superiority of the new technology in practice. Adoption occurs when the entrepreneur accepts a new idea full-scale and continues its use. Diffusion is the process by which innovation is transferred among producers. Diffusion progresses by stages – awareness, interest, evaluation, trial, and adoption. Finally, structural reorganization occurs because of changes within the system due to the adoption of technology.

Solo (1972) compares the transfer of technology to current flowing between two poles, the source and the recipient, with the one exercising a positive and the other a negative magnetic force. On the one side there must be the motivation, competence, and organization to press it onward (reveal, disseminate, promote). And, on the other, there must be the motivation, competence, and organization to seek superior technology, to evaluate its claims, and to draw it into diverse new applications.

Five characteristics of innovations influence the length of the diffusion period (Rogers 1983, 15-16): (1) relative advantage – the degree to which an innovation is superior in profitability, convenience and other ways to ideas it supersedes, (2) compatibility – the degree to which an innovation is consistent with existing values, past experiences and needs of adopters, (3) complexity – the degree to which an innovation is difficult to understand and use, (4) trialability – the degree to which an innovation may be experimented with on a limited basis, and (5) observability – the degree to which the results of an innovation are visible to others. Of these, relative advantage, complexity, and compatibility are the most influential in determining the rate of technology adoption.

Havelock (1971) groups the major theoretical and empirical studies of knowledge dissemination and adoption into three categories corresponding to the principal models, methods, and orientations employed by their authors. These three categories are: (1) the "Research, Development, and Diffusion" Perspective (supply of innovations), (2) the "Social Interaction" Perspective (supply-demand interaction), and (3) the "Problem Solver" Perspective (demand for innovations). Each viewpoint contributes to an understanding of the total dissemination and utilization process.

The Research, Development, and Diffusion Perspective (RD&D) posits a user population that can be reached effectively and influenced through a process of dissemination, provided that this dissemination is preceded by an extensive process of research and development. A major feature of this perspective is the assumption of a clearly defined target audience; a specified passive consumer, who will accept the innovation if it is delivered on the right channel, in the right way, and at the right time. Agriculture, with the Land Grant Universities and the Cooperative Extension Service, is a prototype for the RD&D model.

The Social Interaction Perspective (S-I) assumes the existence of a diffusible innovation as a precondition for any analysis of the diffusion process. Therefore, S-I proponents are somewhat indifferent to the value of the innovation or to the type of scientific and technical know-how. The flow of adoption for a social system through time is the primary concern. The user's position in the network is emphasized, and S-I researchers have gone on to identify a variety of positions in the social system that need to be considered in the diffusion process: The "innovator", who tends to be on the fringe of his home system because he has so many links with the other outside systems; the "laggard," who is isolated and is peripheral to the main streams of interpersonal relations; and the "early majority," who adopt quickly because of their positions near the center of the influence network and their proximity to the leadership.

The third major perspective, the Problem Solver Perspective (P-S), rests on the primary assumption that knowledge utilization is a part, and only a part, of a user's problem-solving process that begins with a need, and ends with the satisfaction of that need. Five major points of P-S are: (1) the user is the starting place, (2) diagnosis precedes solution identification, (3) the outside helping role is nondirective, (4) the importance of internal resources, and (5) user-initiated change is the strongest. It is a psychological and "user-oriented" approach to the problem of diffusion and adoption.

Economists view the rate of adoption as dependent on the difference in profitability between the new and old technology and perceived risk. The profitability of a given technology is greatly influenced by the relative scarcity of the resource being replaced. The United States, where labor has been scarce relative to land and capital, has led in mechanical innovations. On the other hand, Japan, with very limited land, has led in biological and chemical processes leading to increased yields.

The theory of induced innovation (Hayami and Ruttan 1985) interprets the process of technical change as endogenous to the economic system. There are multiple paths of technological development. Induced innovation views technical change as a dynamic response to changes in resource endowments and to growth in demand. "Technology can be developed so as to facilitate the substitution of relatively abundant (hence cheap) factors for relatively scarce (hence expensive) factors in the economy (Hayami and Ruttan 1985, 73)." Both the adoption of innovations by farmers and the development of technological innovations by research institutions are driven by relative factor prices that reflect underlying changes in resource endowments. The fact that real wages in manufacturing have stabilized over the last 15 years may account in part for recent farm size stabilization (Ruttan 1988).

Firm-Level Models of Technology Impacts

The relationship of farm size to adoption depends on fixed costs of adoption, risk preferences, credit constraints, human capital, and labor considerations (Feder, Just, and Zilberman 1985). The influences of these factors therefore vary by area, technology, and over time.

The induced innovation model provides that technological advances will occur to save the relatively more scarce resource. Technological advances in the U.S. this century have tended to be labor-saving. Many of these technologies, particularly the mechanical technologies, have been embodied in indivisible, durable assets. Often the capacity of the new technology (machine complement) substantially exceeds that of the existing farm. Hence, immediately upon adopting of the technology, the farmer is presented with an incentive to expand the farm business size. By more fully utilizing the capacity of the machinery set, average fixed costs can be reduced, thereby reducing average total costs.

Even technologies that are completely divisible often have structural consequences. Technologies such as pesticides and livestock growth stimulants are highly divisible and are often described as size neutral. However, higher adoption rates by larger farms can, in part, be explained by learning cost (Feder and Slade 1984). The fixed cost of learning to use a complex technology can be justified by the number of units over which it can be used. When learning costs are considered, even divisible inputs have a fixed component that renders the technology scale biased. Only when learning costs are negligible can a technology be truly size neutral.

Larger firms also may experience pecuniary economies of size. Pecuniary economies, associated with reduced input costs, are important for certain types of production (Smith, Richardson, and Knutson 1984). Most pecuniary economies are savings achieved by input firms because of volume sales, which can be passed on to farmers. If pecuniary economies are important, farmers may expand business size to capture pecuniary economies of large size. Here technology is not the motive for expansion, but rather "enables" growth.

Risk perceptions may influence farmers' willingness to expand business size. A new technology may either increase or decrease the variability in yield. Still, improvement in yield, even with a higher variance, may result in a lower probability of losses below the original mean or a target income level. However, farmers' technology choices are based on their subjective probabilities of gains and losses and hence on acquired information regarding new technologies. Because information acquisition costs are invariant with output, the economics of information acquisition favor larger farms. Furthermore, it is likely that both the ability to bear risks and farmers' willingness to assume risks are related to farm business size. Because larger farms are typically more leveraged than small farms, they face greater exposure to financial risk. On the other hand, innovators, who tend to be younger, better educated, and farming larger units, are likely less risk averse than their later adopting counterparts.

Credit constraints tend to limit the adoption of technology. Credit constraints are especially limiting for indivisible technologies but also have been found to limit use of divisible technologies (Feder, Just, and Zilberman 1985). Capital in the form of accumulated savings or access to capital markets is typically more available to larger farms. Thus, credit availability may create a size bias in the technology adoption pattern.

Labor is more abundant and has a lower opportunity cost on smaller, family farms. Therefore, larger farms have greater incentives to adopt labor saving technologies. Yield-increasing technologies tend to increase harvest-time labor shortages, which could hinder adoption by larger farms where labor is relatively more scarce (Feder, Just and Zilberman 1985). Although seasonal labor shortages have been found to have a negative effect on high-yield technology adoption in developing countries, labor bottlenecks probably are less important for highly mechanized United States agriculture.

The literature on diffusion of innovations suggests strong relationships between technology adoption patterns and individual and business characteristics. Rogers (1983) summarized the research literature on this topic and drew the following generalizations about variables related to adopter behavior: (1) Earlier adopters are not different from later adopters in age, (2) earlier adopters have more years of education than later adopters, (3) earlier adopters have larger sized business units than later adopters, and (4) earlier adopters have more specialized operations than later adopters.

Technological change usually is suggested to be of prime importance when explaining the changing farm structure. For instance, Tweeten (1984) found technological change to be the overwhelming source of farm size growth in each decade during the period 1940-1979. However, dissenters exist. Kislev and Peterson (1982, 578) examined the out-migration of farm labor and the growth of farm size in the U.S. during the period 1930 to 1970. They hypothesize that the size of the family farm is explained by changes in three principal factors: input prices, nonfarm income and technology. They conclude that input price is the most important of these, explaining "virtually all of the growth in the machine-labor ratio and in farm size over the 1930-70 period by changes in relative factor prices without reference to 'technological change' or 'economies of scale'." However, input prices are expected to be closely related to technological change. Relative scarcity of a factor (and hence high relative input price) provides an incentive for innovation to save that factor (induced innovation). And, once widespread adoption of a new technology occurs, relative input prices are expected to change due to altered factor productivity (and, thus, derived demand) for the factors of production.

Industry-Level Effects of Technological Change

Industry long-run cost curves are usually drawn as static. However, in a dynamic context, technology adoption affects long-run cost curves. To the extent that a new technology reduces average total costs, the industry long run average cost curve will shift downward. However, because larger firms may tend to adopt technologies more quickly than do their smaller firm counterparts, the industry long-run cost curve may "twist" downward, reflecting the lower costs of the early adopter, large farm.

The process of economic adjustment to farm technological advance is explained by the theory of the treadmill (Cochrane 1979, 387-390). Farmers have incentives to adopt new technologies that are profitable by virtue of increased productivity or reduced cost. The innovative and early adopters of technology reap pure profits which may be bid into farmland or other inputs. As others also adopt the technology, the increased supply of product, coupled with an inelastic aggregate demand, results in a lower commodity price, eliminating the profit. Farmers are caught in a treadmill, rushing to adopt a new technology and capture the "adoption rents." Conversely, the laggards are forced into a situation of economic loss, and eventually, out of business.

Innovators and early adoptors, to the extent that they correctly identify appropriate new technologies, can profit in the short run. However, there are risks faced by innovators. Adoption of inappropriate technologies may produce substantial capital losses, especially for capital-embodied technologies. New capital-embodied technologies may be high-cost due to low manufacturing volume and high production costs.

Often there is a learning cost associated with complex technologies. Innovators must pay this learning cost due to their pioneering behavior.

Ruttan (1977) found that the differential adoption rates of divisible technologies disappear once the adoption process is sufficiently advanced. The economic advantage for the early adopters still makes it possible for them to acquire more wealth and bid resources away from the laggards. The farm size structure is affected not only by technological change but also by the rate of introduction of these changes.

As Teigen (1986, 362) observes, the technological treadmill holds implications for several groups in society.

> For farm sector observers, the message is that while farmers are indeed upon a technological treadmill, structural change relieves the pressures on farmers when technological change exceeds the growth of demand. But, structural change wherein firms enter or leave the industry requires a functioning off-farm economy which can provide the entrants or absorb the leavers in an orderly fashion. The chronic farm sector characteristics of rising output, falling product prices, flat total industry net income, falling farm numbers, rising land values, and rising output per farm worker all can be explained by a framework of technological and structural change.

Public Policy and Agricultural Externalities

Outputs of the agricultural production process include not only goods (commodities), but also a number of bads -- waste products, pollution, and chemical residues on food. The negative impacts or costs of these bi-products often are realized beyond the farm, affecting people in proximity to the farm or society at large (Atwood and Hallam, this volume).

Many agricultural technologies developed in this century have the potential for important external costs. Intensive use of chemical fertilizers and pesticides can result in surface and groundwater contamination and pollution. There are health concerns associated with the use of animal growth stimulants and antibiotics. Irrigation contributes to groundwater depletion and salinization. Increased enterprise specialization of the past few decades may be associated with increased soil erosion due to continuous row crop rotations. Concentration of livestock feeding activities into large units compounds manure disposal problems and may increase external costs. Furthermore, structural changes (apparent as fewer farm and rural families) resulting from technological advance are viewed by some Americans as an undesirable societal cost. Undoubtedly, emerging technologies will produce unforeseen external costs as well as benefits. Much concern has been voiced concerning the safety of the new biotechnologies.

Concomitant with this increased awareness of society of the external costs of agriculture has been the diminished number of voters in farming. Farmers now represent a small portion of U.S. political constituency. The agendas of other groups increasingly will be incorporated into farm policy. Recent farm policies have incorporated important goals for soil conservation. Iowa legislators proposed taxing nitrate fertilizers to reduce fertilizer application rates and nitrate contamination of the groundwater. Farmer and consumer groups have emerged with legal challenges of biotechnology releases. Concerns about global warming, antibiotic use in animals, and animal rights also could emerge as constraints on the use of technology.

The imposition of taxes, regulations, or incentives on agricultural producers by government may become an important determinant of future technological change. The sustainable agriculture technologies have been suggested as farming techniques that are less detrimental to the environment and more healthful. The Agricultural Productivity Act, passed by Congress in 1985 as part of the Food Security Act (P.L. 99-198), provided authority and funding to conduct sustainable agriculture research and education programs.

To the extent that public policy mandates or encourages (through taxes or incentives) adoption of environmentally benign production technologies, agricultural structure may be modified in ways not foreseen. Changes in relative factor prices may induce innovations much different from those witnessed in the past. And, the structural consequences of those technologies also may be different.

Emerging Agricultural Technologies

Tweeten (1988, 184) suggests that four lessons have been learned from past technological revolutions in agriculture: (1) the pace of technological change is accelerating among technological revolutions, but technology within each revolution shows diminishing returns, (2) the revolutions are occurring more frequently, (3) modern revolutions originate from increasingly sophisticated science and are more successful with close cooperation between public basic research and private applied research and development, and (4) the origins and control of technological revolutions increasingly originate outside agriculture although they massively influence agriculture. These observations, if true, hold consequences for future agricultural technologies. The following sections will identify some emerging technologies of consequence for agriculture and likely structural consequences of their adoption.

Mechanical Technologies

Mechanical technology initially was applied to reduce labor requirements in crop production. Beginning about 1820, horse-drawn machines were developed to till the soil, plant, harvest, and transport field crops (Cochrane 1979, 200-201). After the turn of the century, gasoline and later diesel-powered tractors began replacing horses as the power source. The mechanization of animal agriculture occurred after World War II with the advent of large-scale confined poultry, dairy, and animal feeding facilities.

Further mechanization in field crop farming is ongoing but is being increasingly constrained by physical limits. Total size and weight of tractors and self-propelled harvesters are limited by weight and width restrictions on public roads. The optimum width of tillage, planting, and harvesting equipment is constrained by time losses in turning, adjusting, and loading seeds and chemicals. The complexity and costs involved in hitching and transporting limit the maximum economic size of tractor-drawn equipment. In some soils compaction problems also serve as a limit on machinery size.

In the livestock sector further concentration of animals may not be economical. Additional labor savings for confinement feeding may not offset the costs associated with increased transportation of feed and manure disposal. The risks from contagious diseases are still a limiting factor for confined poultry and hogs. Dairy herds in the lake states and northeast have considerable potential for further mechanization and enlargement in herd size (U.S. Congress, OTA 1986).

Further mechanization opportunities are greatest for fruit and vegetable operations, and robotic farming (U.S. Congress, OTA 1986). Mechanical harvesting of fruits and vegetables is most applicable if the produce is to be used for processing where effects of mechanical handling will not be evident. Potatoes and tomatoes are now mechanically harvested, as are many other processed vegetables. Research is still underway to automate the harvesting of cauliflower, lettuce, okra, and asparagus. The mass removal of fruit from tree crops by mechanical or oscillating-air tree shakers is proceeding although for some fruits, including oranges, hand picking is still the most economical method. A lowering of real interest rates and a rise in the average wage for farm workers would speed the development and adoption of this technology.

Robotic applications are likely to be first used in high-value labor-intensive horticultural crops like oranges. Robotics also may have applications in animal agriculture. Reductions in cost and increases in speed of operation may make robotic technology economical in the 1990s (U.S. Congress, OTA 1986).

Engine improvements can be expected to improve energy efficiency. Losses in efficiency to cooling can be overcome through the development of special ceramic engines that withstand high operating temperatures. Exhaust gasses can be utilized through a turbo-compound unit. This device applies exhaust gasses directly to the drive train of the vehicle. Electronic engine controls are used to improve fuel efficiency. This technology automatically optimizes fuel delivery under changing conditions based on information from sensors on the engine and implement (U.S. Congress, OTA 1986). Increases in fuel efficiency will likely be offset by the rising real cost of energy.

Tweeten (1988) suggests a "fifth" technological revolution may feature fusion energy. Such developments promise substantial change in relative factor prices, especially for energy intensive inputs such as fuel and chemical inputs. This could induce a series of innovations in agricultural production with much potential for agricultural change. Comparative advantage for particular commodities may shift among regions in response to reduced energy prices.

Impacts of Emerging Mechanical Technologies. Mechanization in the United States has been accompanied by a major structural change toward larger farming units and a decline in the number of farms. It is not clear whether machinery technology has led to farm enlargement, or if labor shortages induced the development of mechanization. In either case, mechanization has been the enabling factor for farm enlargement. For machinery technology to be a direct determinant of farm growth it must be accepted that (1) smaller farmers cannot mechanize effectively and are induced to quit farming for economic reasons or (2) many farmers cannot economically obtain the machinery capacity they require (either through ownership or rental of machinery services), and are compelled to expand farm size to utilize the capacity of machinery available at a low per unit cost (Donaldson and McInerney 1973). These arguments hinge on the premise of lumpy machinery inputs and economies of machine size, i.e., larger machines have lower cost per unit of capacity.

As noted, emerging machinery technology has the most potential in fruit and vegetable production, a type of agriculture already largely organized into large production units using hired labor. The effect of further mechanization, therefore, will be a further substitution of capital for labor with little effect on the size of operating units. Crop and livestock farms using family labor will continue to grow in size, but the rate of change will likely be slower than in the past.

Specialization will continue to be favored by adoption of mechanical technology due to the size of enterprise needed to spread fixed costs of ownership. The risks hindering specialization have been reduced by improved control of plant and animal diseases. Price stabilization from farm price and income supports and disaster relief plus steady growth in the economy further encourage specialization.

Continued development and adoption of mechanical technologies may be accompanied by more separation of ownership and control of resources. The effects of mechanical technologies on vertical coordination and control, market access and barriers to entry were summarized by the Office of Technology study as follows: "No significant change in vertical coordination or market access is expected from technologies in the mechanical group for either crop or livestock production." "The impact of mechanical technologies on barriers to entry in the long term is expected to be a slight increase for crop production but no significant increase for livestock production." (U.S. Congress, OTA 1986, 128-130).

Chemical and Biological Technologies

An overview of emerging chemical and biological technologies provides perspective of their potential future impacts on agriculture. Herbicides, insecticides, and fertilizers are discussed first. This is followed by discussion of the sustainable agriculture technologies and the emerging biotechnologies.

Emerging trends in herbicide development and use are reducing their impact on health and the environment (Nalewaja 1990). One trend is the development of chemicals that are highly active against targeted weeds but no more toxic to humans than previous materials. This provides weed control with extremely low dosages. Another development is the switch from pre-emergence to post-emergence herbicides. Unlike pre-emergence herbicides, post-emergence materials are only applied if weed density exceeds a threshold level. Also, existing herbicides are being made to work more effectively through the use of additives such as wetting agents, through better timing of application, and through banding.

Making crops resistant to specific herbicides is another approach receiving increasing attention. Compounds called protectants or safeners applied to the seed or soil are being developed to make a crop resistant to a particular herbicide (U.S. Congress, OTA 1986). Also, biotechnology is being used to develop crops resistant to herbicides known to be more benign to the environment.

Entomological research has de-emphasized the search for toxic substances to kill insects in favor of Integrated Pest Management (IPM) and Total Population Management (TPM) strategies. IPM is based on the use of natural enemies and host-plant resistance supplemented by other suppression methods. TPM is based on an attack on the total population of a pest in a large area rather than field by field suppression. However, farmers remain highly dependent upon chemical controls except in selected situations where either IPM or TPM have been successful (Perkins 1988, 171).

Although a major contributor to growth in crop productivity in the past, commercial fertilizer is likely to play a smaller role in the future. The fertilizer industry spends less than 10 percent of the research and development budget of the entire chemical industry (U.S. Congress, OTA 1986, 61). According to the OTA study, "No revolutionary or radically new products or processes appear to be near commercialization."

Research and development is being done to maximize effectiveness, minimize cost, and protect the environment. About half the nitrogen and 80 percent of the phosphate applied is not used by the crop. Inhibitors, seed coatings, and placement procedures are being developed to reduce nutrient losses through leaching or fixation in the soil. Improved processes are being developed to decrease the energy requirements to produce and transport fertilizers. Technology for the production of ammonia from coal is in the advanced stage of development (U.S. Congress, OTA 1986).

Adjusting fertilizer and pesticide rates within a field to make more effective use of the materials and minimize contamination from over-application is gaining in popularity. Methods of sampling and testing as the applicator covers the field are being developed.

The sustainable agriculture technologies have been suggested as farming techniques that are less detrimental to the environment and more healthful. The reduction of chemical inputs is typically expressed as a primary goal of sustainable agriculture. Sustainable technologies typically employ techniques that allow substitution of labor, machinery services, management, crop rotations, and integrated pest management practices for agricultural chemicals, particularly nitrate fertilizers and pesticides. Reduced chemical use is proposed by two fronts: Environmentalists who are concerned with the high levels of pollution and other external costs of existing technologies, and farmers who are seeking input combinations that have greater allocative efficiency.

New biotechnologies hold substantial promise to modify agricultural production relationships. Emerging biotechnology uses recombinant DNA, cell fusion, and other genetic engineering techniques. Natural recombination of genetic information is the basis for evolution. However, recombinant DNA technology has made it possible to direct the nature of genetic recombination, and to shorten the time required for genetic change. These biotechnology techniques are likely to give rise to new production methods and products, and to change the nature of farming.

Plant biotechnologies with commercial potential include development of plants with disease, herbicide, and pest resistance; tolerance to drought, salinity, or other harsh environmental conditions; enhanced nutrient or other market desired characteristics; perenniality for annuals; and nitrogen fixation for cereals. Animal biotechnologies with immediate promise include feed additives, natural hormones, and monoclonal antibodies produced using recombinant DNA techniques. Genetically engineered soil microorganisms have been developed to increase the efficiency of nitrogen fixation by legumeous plants; to combat soil-borne plant diseases; or to increase soil nutrients available to plants. Direct-fed microbials, animal feed additives that contain organisms believed to contribute to intestinal microbial balance, are being developed to aid in the digestion of feed and enhance animal health.

Impacts of Emerging Chemical and Biological Technologies. The voluntary or regulation-forced movement away from chemical controls of crop and animal pests to more complex control strategies may have an effect on farm structure. On the one hand, the fixed cost of learning and implementing complex technology favors large farms. On the other hand, medium sized units, where a full-time farmer supplies both management and labor, are very adept at decision making requiring close coordination between monitoring biological phenomena and daily or hourly decision making responses.

The chemical and biological technologies will tend to encourage more vertical coordination and control up the marketing chain. This would be more important for livestock than for crop production. For example, more contracted livestock production would be encouraged to obtain products with uniform specification (U.S. Congress, OTA 1986). In addition, this group of technologies may slightly reduce market access for livestock producers. Finally, no significant changes in entry barriers is expected from the chemical/biological technologies for either livestock or crop production (U.S. Congress, OTA 1986).

Studies of sustainable and conventional farms have found conflicting results on the issue of profitability (Cacek and Langner 1986; Batte 1991; Goldstein and Young 1987; Helmers, Langemeier, and Atwood 1986). The mix of inputs employed in production change with the adoption of sustainable agriculture technologies. Although costs of chemical fertilizers, herbicides, and insecticides will decrease, costs associated with labor, machinery, and management may increase. Additionally, yield reductions may accompany the reduced use of chemical fertilizers and pesticides. More importantly, the mix of farm outputs will change due to the importance of crop rotations. Such rotations include both nonmarketable crops (e.g., green manure crops) and low-valued crops. Should widespread adoption of sustainable agriculture technologies occur, aggregate output for many agricultural commodities will decrease and commodity prices will rise. For instance, if the price elasticity of demand for corn is -0.3 as estimated by Gardiner and Dixit (1986), a one percent reduction in national output raises corn price by 3.3 percent. This would serve as an additional source of revenue to producers, but would increase the cost of food and fiber to consumers. When these output effects are incorporated into the profitability analysis, it is unclear whether sustainable agriculture farms are more profitable than their conventional agriculture counterparts.

The emerging biotechnologies hold implications for the size distribution of farm firms. Kalter and Tauer (1987) suggest that biotechnology developments will continue the long-term decline in real agricultural prices. This, coupled with the fact the continuous technological change often results in chronic excess resources, may result in low returns for agriculture. Kalter and Tauer (1987) point out that this low return is only an average for all producers. In fact, the distribution of returns will be much more dispersed with higher rates of technological change. The early adopting farmers likely will receive returns above that average, with laggards earning below average returns. Thus, even though biotechnologies are likely to be embodied in divisible inputs and should be inherently size neutral, farmers will still be placed on the treadmill, with early adopting farmers expanding by cannibalizing the laggards.

Recent evidence suggests that the adoption patterns for biotechnologies will not be largely different than those of previous technologies. Zepeda (1990) developed an *ex ante* adoption model for bovine somatotropin based on producer intentions. Data were from a random sample of California Grade A milk producers. She concluded that potential adoption of bST for this group was about 44 percent. Factors found to increase significantly the probability of bST adoption were increases in herd size, education, industry involvement, and computer use. Operator age was negatively associated with the probability of bST adoption.

Marion and Wills (1990), Larson and Kuchler (1990) also expect a more gradual and less extensive adoption of Bovine Somatotropin than suggested by the OTA and other studies. Marion and Wills (1990) suggest that the economic feasibility of bST will be very sensitive to (1) the production response to bST, (2) milk prices, (3) price of bST, and (4) farmers' required return to adopt the new technology. Larson and Kuchler (1990) recognize that bST adoption also implies use of a changed mix of inputs and suggest that previous studies have ignored some components of total costs associated with bST adoption. Their conclusion is that bST adopters may not have the incentive to increase milk yields as much as has been obtained in experimental studies. This would imply smaller changes in aggregate output associated with a given adoption percentage, smaller impacts on milk prices and profitability, and thus a reduced incentive to adopt.

Other biotechnologies are expected to alter the regional comparative advantage and the spacial location of production. For instance, biotechnological advances in the form of pest and disease resistance will have the greatest relative impact in the south where pest pressures are greatest. Other technologies may have a differential performance increase based on the quality of management. Weersink and Tauer (1989) have suggested that bovine somatotropin (bST) adoption will result in some concentration of dairy production in the traditional milk producing regions at the expense of reduced production in other regions of the U.S. The OTA study, on the other hand, suggested that although there would be pressure for herds size to increase in all regions, the greatest concentration of dairy production would occur in the Southwest, Southeast, and Northwest regions.

Many of the emerging biotechnologies are cost-reducing, output-increasing and relatively scale-neutral. Yet, these technologies may have substantial impact on the mix of commodities produced and on aggregate output prices. As an example, biotechnology research is underway to increase the efficiency of nitrogen fixation in legumes by modifying either the plant, the nitrogen fixing bacteria, or both. Such modifications would make legumes in rotation more competitive with commercial nitrogen fertilizers. The technology likely would result in more acreage devoted to legumes, say soybeans, and fewer acres devoted to corn or other crops. This would result in downward pressure on soybean prices and upward pressure on corn prices.

Other biotechnologies promise new products or modified characteristics for existing commodities. Although porcine somatotropin (pST) impacts physical feeding efficiency (feed conversion may increase up to 24 percent and growth rates by 16-18 percent), it also impacts pork quality. Hogs administered pST have 20-30 percent less backfat and the area of the loin-eye muscle increases 1-5 square centimeters (Sterling). Such product changes, to the extent that they are consistent with consumer tastes and preferences, will be important in the battle among commodities for consumer expenditures.

Environmental concerns and restrictions may slow introduction of biotechnology into U.S. agriculture. For instance, it is possible that before environmental release of recombinant organisms into the environment, approval would need be received from the USDA's Animal and Plant

Health Inspection Service (APHIS) and Food Safety and Inspection Service (FSIS), the Environmental Protection Agency (EPA), and the Food and Drug Administration (FDA) (Payne). Furthermore, the patent process for new biotechnologies requires an average of about 4.5 years from application to approval (Morgan). There also promises to be expensive, time-consuming court battles instigated by consumer and other special interest groups.

Information Technologies

Information technologies, broadly defined, include all those developments designed to measure, store, retrieve, process and communicate data or information. Its output, information, is an intermediate product in the production relationship.

Several information technologies are relevant to farm firms. We have categorized these technologies into four groups based on their application. Although this categorization implies separability, these technologies may be applied as a system.

Communications Technologies. The new communications technologies allow for more rapid and reliable transfer of information. Included here are both (1) hardware developments such as fiber optic cables, satellite communication and computer hardware, and (2) information dissemination networks. The latter includes computerized information delivery systems, satellite relay of information (broadcasts or signals targeted to individual subscribers), computer software, and the network of consultant services (both private and public) that have developed to facilitate the interpretation and application of information.

Data Processing/Storage/Retrieval Technologies. The modern information collection/storage/retrieval/processing technologies are primarily computer based. Large scale integration of circuits is largely a bi-product of the large public research expenditures through the National Aeronautic and Space Administration (NASA). This basic research was integrated quickly into the commercial sector and was important in the development of microcomputer technologies. Within a decade of their introduction, microcomputer systems have become extremely capable, possessing processing speeds and primary and secondary storage capacities previously found only on expensive mainframe computers. Because explicit ownership costs have been reduced greatly, microcomputers allow distribution of processing power to the small firm level.

Although hardware developments are necessary for firm level computer processing systems, a more important component of this technology is embodied in computer software. The earliest computers were programmed with hard-wired circuits programmable only by engineers. This was followed by early generations of programming languages that could be used only by individuals well educated in the use of languages.

Computer languages are now in their fourth and fifth generations. Modern computer languages are more highly structured than were the earlier generation languages. They typically are more "macro" in nature -- meaning that the computer does more of the tasks of memory, input, and output management given a few simple instructions by the user -- and frequently are presented via a graphical interface. This has resulted in a computer environment operable by individuals with little training. Furthermore, much of this development has occurred for microcomputer technology.

Artificial Intelligence/Expert Systems. The fifth generation of languages, as usually defined, includes artificial intelligence (AI) and expert systems. AI remains in development stages. The idea of artificial intelligence is a simple one. Computers are programmed to process data and draw conclusions in the same manner as human decision makers. However, the process of developing such software systems has proven difficult. The primary difficulty has been modelling the human thought process. Humans use elaborate intuition and qualitative judgments when making decisions. On the other hand, computer languages developed to date have required that all alternatives be judged using logical comparisons (IF-THEN-ELSE structures). As a result, complex problems must be modelled individually. Each problem requires a specific computer program. Development costs frequently outweigh the benefits of computerizing the solution unless the decision and its parameters are identical for many decision makers or the decision is repeated frequently. AI programming, if successful, will allow complex decisions to be modelled with a general purpose software. Development costs for individual decisions would be lower. Thus, farm firms could automate a larger number of decisions of greater complexity.

Automated Systems. Automated systems technologies include robotics, process control, and automated data collection. Much like the machine revolution, these technologies embody new ways of substituting capital for labor. Robotics and process control involve the merging of computer technologies with electro-mechanical control devices. Process control requires either real-time computer processing or stand-alone microprocessors. The development of low-cost microcomputers opens the door for economically feasible process control for small business. Modern dairy technologies provide an example of process control in agriculture. Technology exists to allow automatic identification and feeding of individual cows using transponders for identification and automated feed dispensers. The quantity of milk produced can be sensed automatically by a device in the milking pipeline, and recorded on computer media with cow identification number. These data can be processed, allowing input from the herd manager, to adjust the ration fed to the cow in subsequent feeding periods. Datta et al. (1984) reported research results of an on-line milk monitoring system for milk conductivity, milk temperature and milk yield. Using a statistic calculated from milk conductivity, a method of mastitis detection was established by which three consecutive observations of milk conductivity above a critical value would be indicative of mastitis. Application of such a system would allow the manager useful information for herd health and feeding decisions without substantial time commitments for data collection and processing.

Impacts of Emerging Information Technologies

During the past decade, information options available to farmers changed substantially. However, adoption of this technology has lagged far behind its development. Various projections made nearly a decade ago forecasted that most commercial farmers would be using computers by 1990. These projections have proven to be grossly inaccurate even though computer hardware and software capabilities have improved remarkably. Recent estimates of computer adoption rates range from a mere 3 percent of farmers (Willimack 1989) to just over 25 percent (Putler and Zilberman 1988). Adoption rates are higher for large than for small farms.

The information technologies will impact the agricultural sector differently than previous technologies. Information technologies primarily facilitate the processing of data into information, allowing improved efficiency in decision-making. As such, information technologies do not modify the physical production process (isoquant shape or location). Instead, information technologies allow better choice from among the decision variables, and thus more efficient allocation of inputs.

The separation of information and production technologies is useful when considering changes in agricultural structure that will occur with adoption of information technologies. Using this dichotomy, information technologies have few characteristics in common with other technologies introduced in the agricultural sector. They usually do not change the production process, but instead allow more control of existing production technologies.[3]

There are several reasons to argue that adoption of information technologies will progress differently than previous agricultural innovations. With the large, indivisible durable capital investments of machines during the mechanical revolution, economies of size created an incentive to expand farm size. However, many information technologies require small fixed explicit outlays. Ownership costs are minor. And, most information service networks are priced on a variable cost basis. This argues for a small incentive for business expansion.

On the other hand, there may be substantial implicit costs that make total costs of adoption large and thus favor adoption by larger firms. Information technologies often are complex. Unlike the tractor, which was simple to operate and could quickly be mastered, the computer and its complement of software may require lengthy periods of training for the typical manager. Furthermore, larger firms can better afford specialized labor in the form of a computer operator or information specialist.

Many information technologies are management intensive. Information is not employed directly as an input in the production process as are fertilizer, seed and fuel, but rather as an intermediate input that is useful if its interpretation allows other inputs to be regulated in an economically more efficient manner. The quality of management, therefore, influences how the firm will respond to a given piece of information. Thus, if management quality and innovativeness are correlated with any attribute of farm structure, we may see differential adoption and structural change.

Information technologies will be varied in their effect on the firm. Some will be labor saving: For instance, the use of computers for data collection and processing and through robotics. Other technologies may be output increasing. That is, more accurate and timely information can mean better regulation of the production process and greater output. Other information technologies may be both labor using and output increasing. Labor-saving technologies such as the tractor and its complements have influenced farm structure in the U.S. more than have output-increasing technologies such as hybrid seed. Thus, the most important impact of computers will be through robotics and process control which save labor.

The literature on diffusion of innovations indicates that early adopters often are large farmers who eagerly seek new knowledge, and as such, highly value information (Rogers 1983). Several recent studies have estimated farmers' computer adoption rates. Willimack (1989), using data from the 1987 Farm Costs and Returns Survey, found that less than three percent of U.S. farmers use a computer to maintain farm records. Her analyses focused on the farm record-keeping function and employed the census ($1,000 sales) definition of a farm, and thus probably understates the incidence of computer use by commercial farmers.

Other studies that have focused on commercial-sized farm businesses have found higher adoption rates. *Farm Futures* magazine conducted a telephone survey of a random sample of its subscribers and found that 31 percent owned computers. Lazarus and Smith (1988) found that 15 percent of New York dairy farmers enrolled in the Farm Business Summary and Analysis program in 1986 owned computers. A follow-up study (Lazarus, Streeter, and Jofre-Giraudo 1989) followed a panel of record-keeping farmers over a four-year period and found an increasing cumulative adoption pattern: 3.6, 5.6, 8.7, and 11.7 percent of the panel owned computers in the years 1984-87, respectively. Putler and Zilberman (1988) surveyed farmers in Tulare County, California and found that just over 25 percent owned computers in 1986. Batte, Jones, and Schnitkey (1990) found 24 percent of Ohio commercial farmers had used computers in some aspect of farm management. Computer use was most commonly reported by mixed-livestock farmers (30 percent) and least frequently used by specialized cash grain farmers (18 percent). Putler and Zilberman (1988) also found higher adoption rates among livestock producers. However, Willimack (1989) found higher adoption rates for crop farmers than for livestock producers.

Computer adoption rates by farmers have been shown to vary along operator and business characteristics. Survey results by Willimack (1989), Lazarus and Smith (1988), *Farm Futures Magazine*, and Batte, Jones, and Schnitkey (1990) found evidence of an inverse relationship between adoption rates and farmer age. These studies, with Putler and Zilberman (1988) also found positive relationships between adoption rates and education level, and between adoption rates and business size. Willimack (1989) also found important differences in adoption percentages by region of the nation.

Conclusions

Much structural change has occurred within U.S. agriculture during the past century, a period of accelerating technological change. Technical change in the U.S. has displayed a decided labor-saving bias. The result has been excess labor within agriculture. The adjustment process has been painful for those involved with agriculture. Many people were drawn off the farm by better employment opportunities and, simultaneously, remaining farmers expanding their businesses so as to more completely realize the advantages of the new technology. Technology has lowered real food costs and freed millions of workers to produce goods and services such as food processing, health care, entertainment, and other items associated with quality of life and an affluent society.

The rapid innovation in mechanical technologies earlier this century resulted in rapid structural change because they were capital embodied, indivisible technologies, and hence exhibited a strong size bias. The chemical and biological technologies that followed were less size biased, yet still created incentives for structural change.

Future technological developments promise continued structural change, especially in the number and distribution of farms by size. Although the chemical and biological technologies largely will be embodied in variable inputs, and thus will not be greatly size biased, costs of learning to use the new technology are lower per unit of output for larger farmers. The biological and information technologies also are proving to be management intensive. Successful implementation of these technologies will require careful regulation of a number of other inputs. To the extent that management quality is correlated with farm size, economic performance and adoption incentives will be greater for these farms. Finally, studies continue to demonstrate that operators of larger farms and more highly educated farmers more quickly adopt new technologies. Even for scale-neutral technologies, profits captured by early adopting farmers will be used to acquire additional resources at the expense of the laggard farmers, leading to further concentration in farming.

Capital-embodied, labor-saving technologies continue to hold the greatest potential for structural change. The combination of mechanical and information technologies (robotics or process control) allows for continued substitution of capital for labor. Furthermore, these technologies will also display a size bias due to the capital embodiment of the technology.

Increased public involvement through the regulation of agriculture is a factor that might slow, or even reverse, the consolidation of farms. For instance, federally sponsored incentives encouraging adoption of sustainable farming practices would dramatically alter the relative demands for labor and other resources. It might alter the maximum size of farm feasible or restrict the individual's ability to work part-time off the farm. However, there is little evidence to suggest that such restrictions would have a large impact on structural change.

The pace of technological change as apparent in productivity growth and farm labor and population adjustments has substantially slowed. The trend rate of increase in crop and livestock output per unit of all production inputs currently is only about half the trend rate of 1950. Farm numbers fell from 2.2 million in 1982 to 2.1 million in 1987, a modest reduction indeed given the intense financial stress of that period.

The relative distribution among large, medium, and small farms remained almost unchanged. A concern is that real food prices will rise in the 1990s and thereafter if the pace of technological change does not quicken -- especially given environmental concerns that may slow adoption of biotechnologies.

Notes

The authors wish to thank Luther Tweeten and Loren Tauer for helpful comments on an earlier draft of this chapter.

1. As Tweeten (1989) has observed, with adoption of labor-saving, scale-biased technologies usually comes a concomitant expansion of business size. Pure profits arising from the technology adoption are bid into farmland prices. However, output-increasing, scale-neutral technological change does not necessarily alter business size, and the corresponding change in input prices is expected to be smaller.

2. Although the efficiency of agriculture is one reason for the low real cost of food, another is the high per capita income of consumers.

3. Exceptions arise in the area of automated systems. These represent new processes by which capital and labor can be combined in the physical production process, and thus represent new production technologies.

References

Batte, Marvin T. 1991. "A Growing Debate: Organic Agriculture in Ohio." *Ohio's Challenge* 4(3), Department of Agricultural Economics and Rural Sociology, The Ohio State University.

Batte, Marvin T., Eugene Jones, and Gary D. Schnitkey. 1990. "Computer Use by Ohio Commercial Farmers." *American Journal of Agricultural Economics* 72: 935-945.

Cacek, Terry, and Linda L. Langner. 1986. "The Economic Implications of Organic Farming." *The Journal of Alternative Agriculture* 1: 25-29.

Cochrane, Willard W. 1979. *The Development of American Agriculture: A Historical Analysis.* Minneapolis: University of Minnesota Press.

Datta, A. K., H. B. Puckett, S. L. Spahr, and E. D. Rodda. 1984. "Real Time Acquisition and Analysis of Milk Conductivity Data." *Transactions of the ASAE* 27(4): 1201-1205.

Donaldson, G. F., and J. P. McInerney. 1973. "Changing Machinery Technology and Agricultural Adjustment." *American Journal of Agricultural Economics* 55: 829-39.

Farm Futures. 1988. Milwaukee, WI: AgriData Resources, Inc., 16(5).

Feder, Gershon, and Roger Slade. 1984. "The Acquisition of Information and the Adoption of New Technology." *American Journal of Agricultural Economics* 66: 312-20.

Feder, Gershon, Richard E. Just, and David Zilberman. 1985. "Adoption of Agricultural Innovations in Developing Countries: A Survey." *Economic Development and Cultural Change* 33: 255-98.

Gardiner, Walter, and Praveen Dixit. 1986. "Price Elasticity of Export Demand: Concepts and Estimates." U.S. Department of Agriculture, Economic Research Service, AGES 860408, Washington, DC.

Goldstein, Walter A., and Douglas L. Young. 1987. "An Agronomic and Economic Comparison of a Conventional and a Low-input Cropping System in the Palouse." *American Journal of Alternative Agriculture* 2: 51-56.

Grabanski, Raymond L. 1990. "The Role of Farm Size in Technology Adoption." Unpublished M.S. Thesis. Department of Agricultural Economics, North Dakota State University, Fargo, ND.

Havelock, Ronald G. 1971. *Planning for Innovation Through Dissemination and Utilization of Knowledge.* Center for Research on Utilization of Scientific Knowledge and the Institute for Social Research, University of Michigan, Ann Arbor, MI.

Hayami, Yujiro, and Vernon W. Ruttan. 1985. *Agricultural Development: An International Perspective.* Revised and expanded edition. Baltimore: Johns Hopkins University Press.

Helmers, Glenn A., Michael R. Langemeier, and Joseph Atwood. 1986. "An Economic Analysis of Alternative Cropping Systems for East-central Nebraska." *American Journal of Alternative Agriculture* 1: 153-158.

Hicks, J. R. 1932. *The Theory of Wages.* Macmillian.

Kalter, Robert J., and Loren W. Tauer. 1987. "Potential Economic Impacts of Agricultural Biotechnology." *American Journal of Agricultural Economics* 69: 420-425.

Kislev, Yoav, and Willis Peterson. 1982. "Prices, Technology, and Farm Size." *Journal of Political Economy* 90: 578-95.

Larson, Bruce A., and Fred Kuchler. 1990. "Technical Possibilities and Economic Realities of Bovine Growth Hormone." *Journal of Production Agriculture* 3: 174-179.

Lazarus, William F., and T. R. Smith. 1988. "Adoption of Computers and Consultant Services by New York Dairy Farmers." *Journal of Dairy Science* 70: 1667-75.

Lazarus, William F., Deborah Streeter, and Eduardo Jofre-Giraudo. 1989. "Impact of Management Information Systems on Dairy Farm Profitability." Paper presented at the annual meeting of the American Agricultural Economics Association, Louisiana State University, Baton Rouge, July 30 - August 2.

Marion, Bruce W., and Robert L. Wills. 1990. "A Prospective Assessment of the Impacts of Bovine Somatotropin: A Case Study of Wisconsin." *American Journal of Agricultural Economics* 72: 326-336.

Morgan, Robin. "The Recombinant DNA Controversy: Past and Present Concerns." University of Delaware, Department of Animal Science and Animal Biochemistry, unpublished working paper.

Nalewaja, John. 1990. Professor of Crop and Weed Sciences, North Dakota State University. Personal interview.

334

Payne, John H. "Biotechnology Regulation in the United States." United States Department of Agriculture, unpublished working paper.

Perkins, John, H. 1988. "The Future History of Biotechnology: A Commentary," in *Biotechnology, Biological Pesticides and Novel Plant-Pest Resistance for Insect Pest Management*. Proceedings of Conference July 18-20. Insect Pathology Resource Center, Boyce Thompson Institute for Plant Research. Cornell University, Ithaca, NY.

Putler, Daniel S., and David Zilberman. 1988. "Computer Use in Agriculture: Evidence from Tulare County, California." *American Journal of Agricultural Economics* 70: 790-802.

Rogers, Everett. 1983. *Diffusion of Innovations*. New York: Macmillan Publishing Company.

Ruttan, Vernon W. 1977. "The Green Revolution: Seven Generalizations." *International Development Review* 19: 16-23.

_____. 1988. "Scale, Size, Technology and Structure: A Personal Perspective." *Determinants of Farm Size and Structure*. Proceedings of NC-181, Michigan State University.

Schumpeter, J. 1939. *Business Cycles*. McGraw-Hill.

Smith, E. G., J. W. Richardson, and R. D. Knutson. 1984. *Cost and Pecuniary Economies in Cotton Production and Marketing: A Study of Texas Southern High Plains Cotton Producers*. Texas Agr. Bulletin B-1475, Texas Agricultural Experiment Station, Texas A&M University, College Station.

Solo, Robert A. 1972. "Sociologist, Anthropologist, Economist," in Robert A. Solo and Everett M. Rogers, ed., *Inducing Technological Change for Economic Growth and Development*. East Lansing: Michigan State University Press.

Sterling, Lesa G. "Biotechnology on the Farm." University of Delaware. Unpublished working paper.

Stevenson, Rodney. 1980. "Measuring Technological Bias." *The American Economic Review* 70: 162-173.

Swanson, Earl R., and Steven T. Sonka. 1980. "Technology and the Structure of U.S. Agriculture," in *Farm Structure: A Historical Perspective on Changes in the Number and Size of Farms*. U.S. Senate, Committee on Agriculture, Nutrition, and Forestry.

Teigen, Lloyd D. 1986. "Technological Diffusion: Effects on Productivity, Structure, Firms, and Markets," in Joseph J. Molnar, ed., *Agricultural Change: Consequences for Southern Farms and Rural Communities*. Boulder: Westview Press.

Tweeten, Luther. 1984. *Causes and Consequences of Structural Change in the Farming Industry*. NPA Food and Agriculture Committee Report No. 2.

_____. 1986. "Agricultural Technology – The Potential Socio-Economic Impact." Department of Agricultural Economics, Oklahoma State University, Paper AE8680.

_____. 1988. "Agricultural Technology – The Potential Socio-economic Impact." *Research in Domestic and International Agribusiness Management* 8: 183-212.

_____. 1989. *Farm Policy Analysis*. Boulder: Westview Press.

U.S. Congress, Office of Technology Assessment. 1986. *Technology, Public Policy, and the Changing Structure of American Agriculture*. OTA-F-285 (Washington, DC: U.S. Government Printing Office).

U.S. Department of Commerce, Bureau of the Census. *Census of Agriculture* 1920 and 1987. Washington DC: Government Printing Office, 1920 and 1987.

Weersink, Alfons, and Loren W. Tauer. 1989. "The Distributional Impacts of Technical Change on the U.S. Dairy Sector." Department of Agricultural Economics, Cornell University, A.E. Staff Paper 89-22.

Willimack, Diane K. 1989. "The Financial Record-Keeping Practices of U.S. Farm Operators and their Relationship to Selected Operator Characteristics." Paper presented at the annual meeting of the American Agricultural Economics Association, Louisiana State University, Baton Rouge, July 30 - August 2.

Zepeda, Lydia. 1990. "Predicting Bovine Somatotropin Use by California Dairy Farmers." *Western Journal of Agricultural Economics* 15: 55-62.

13

Government Commodity Program Impacts on Farm Numbers

Luther G. Tweeten

One objective of farm commodity programs has been to maintain the family farm system as evidenced by the following quote from the Food and Agriculture Act of 1977:

> Congress hereby specifically reaffirms the historical policy of the United States to foster and encourage the family farm system of agriculture in this country. Congress firmly believes that the maintenance of the family farm system of agriculture is essential to the social well-being of the Nation and the competitive production of adequate supplies of food and fiber.

The quote begs the basic question: Do farm commodity programs help to maintain the system of family farms? This chapter attempts to answer the question.

Although considerable attention has been given to government commodity programs (see Robison 1988), their impact on farm structure remains controversial and in need of additional analysis. For manageability, I arbitrarily narrow the impact on structure to farm numbers. However, because size and numbers are highly (inversely) correlated, an analysis of farm numbers is implicitly an analysis of size. I do not attempt to predict numbers of farms by size or type. Some of these issues have been ably addressed by Richardson et al. (1988) and Headley (1988).

Brief Literature Review

Three Conflicting Conclusions

Various social scientists have reached three conflicting conclusions: In the absence of government commodity programs, the U.S. today would have (a) fewer and larger farms, (b) more and smaller farms, and (c) the same number of farms.

Programs Reduce Farm Numbers. Many social scientists contend that commodity programs have reduced the number of family farms. Quance and Tweeten (1972, 35-36) in 1972 contended that the economic stability provided by commodity programs reduced numbers and increased size of farms. They stated that "Commodity programs reduce uncertainties and unleash the larger farmer to use his efficiency to out-compete the small farmer. Programs providing capital and security allow a given equity to be leveraged further." Quance and Tweeten (1972) also noted that programs encourage expansion of farms to utilize efficiently their machinery, labor, and other overhead as acres are diverted by government programs.

Willard Cochrane (1985, 1008) is the most outspoken current advocate of the position that commodity programs increase size and reduce the number of farms. In 1985 he noted that "Maintaining the present level of price and income support helps [the moderate-size family type farm] some, but it helps their large, aggressive neighbors a lot more." He contended that commodity programs have outlived their usefulness because, in net, they contribute to loss of family farms.

The conclusion that farm programs reduce farm numbers remains very much alive. The following is a 1989 quote from Swanson (1989, 15):

> The public still appears to associate farming with rural well-being, and to believe that the farm programs of the past fifty years have helped farm families. In fact, the evidence is that these programs have facilitated the decline in the number of family farms.

Programs Preserve Farms. Although they did not view the commodity program influence as decisive or large, both Gardner (1978, 842) and Stanton (1985, 327) concluded that government programs retarded structural change, i.e. presumably programs slowed movement to fewer, larger farms. Richardson et al. (1988, 154) also contended that commodity programs preserve family farms:

> Based on the farm survival approach, mid-size farms having low off-farm incomes, high debts, and a high proportion of rented land benefit

the most from farm programs. Without farm program benefits, it is this class of farms that is most likely to be forced out of business.

The authors reached this conclusion on the basis of farm firm simulations and on data showing direct payments by size of farm.

Programs Have No Impact. Spitze et al. (1980, 67) concluded that "on net, the mass of data, evidence, and professional judgments provides little basis for any conclusion other than that government price and income payment policy has generally been neutral in its effect on farms of varying size." After reviewing theory and empirical data, I (Tweeten, 1984, 33) and Sumner (1985, 284) concurred with that conclusion.

Methodologies for Measuring Structural Impacts of Policy

At least three methodologies can be utilized to evaluate the impact of commodity programs on farm structure: (1) representative farm firm analysis, (2) judgment estimates based on theory and scattered empirical observations, and (3) statistical inference from time series data. The first approach most often has concluded the programs save family farms, the second often has concluded the opposite, and a combination of (1) and (2) has led to the conclusion of no net impact. The third, a new approach, is used in this study.

Farm Firm Analysis. Farm data by economic class of farm have been widely used to judge the impact of commodity programs on farm structure. A common error of laypersons is to conclude that larger payments to large farms than to smaller farms give large farms the competitive edge and cause fewer, larger farms. This is comparable to concluding that wheat causes concentration of production on large farms because wheat receipts on larger farms exceed receipts on smaller farms. The appropriate measure of scale effect is payment per unit of farm output.

In 1988, direct payments were a major component of farm income on mid-size farms, less important on small farms, and relatively unimportant on large farms (Table 13.1). Payments are a large portion of farm income on mid-size farms because these farms produce enterprises covered by commodity programs and because they have high participation rates. Benefits are relatively less important on small farms because they have lower participation rates and receive more of their farm income in-kind. Payments are low relative to income (but high absolutely) on large farms because they emphasize enterprises such as livestock, fruits, and vegetables not covered by commodity programs and because payment limitations reduce participation and benefits. In short, Table 13.1 data clearly illustrate the importance of commodity programs to mid-size

TABLE 13.1 Direct Government Payments as a Share of Selected Farm Economic Indicators by Sales Class of Farms, 1988

Item	Sales Class of Farms ($1,000)								
	500+	250-500	100-250	40-100	20-40	10-20	5-10	Under 5	Total
Direct government payment per farm ($)	40,238	31,978	21,118	11,283	5,730	2,331	1,010	374	1,697
Payments as share of: (%)									
Cash receipts	2.2	9.0	13.7	17.5	20.0	16.1	13.7	18.1	9.6
Gross farm income	2.1	7.8	11.5	13.6	14.3	10.1	6.8	4.8	8.0
Net farm income	5.7	24.6	42.3	64.8	87.7	88.5	89.2	*	29.0
Total farm income, all sources	5.4	21.9	31.2	35.1	22.1	10.0	3.8	1.2	14.3

*Income from farming these small farms is negative.

Source: U.S. Department of Agriculture (September 1989).

farms but do not show the dynamic impacts of programs on farm structure.

Large farms can persevere in an unstable economic environment by pursuing sophisticated risk strategies, while small farms persevere by cushioning farm setbacks with off-farm income. Mid-size farms especially need government payments. However one farm-firm dynamic simulation made a case that commodity programs result in fewer, larger farms because of high savings and investment rates out of positive transitory income (Tweeten et al. 1984, 21-28). The shortcomings of simulation is that results depend on assumptions analysts build into the model.

Informed Judgments of Impacts. Because forces influencing farm structure are many, varied, and conflicting, and because data are not available with and without farm programs, numerous analysts have despaired of empirically determining the impact of farm commodity programs on farm structure. The alternative is to rely mainly on deductive judgments and scattered data.

A dilemma of this approach is that the various theories, scattered data, and fragmented judgments provide no weights to reach consensus from conflicting evidence. Not surprisingly, this approach mostly has brought the conclusion that commodity programs have no or small net impact on farm structure (see Spitze et al. 1980, 67; Tweeten 1984, 33; Sumner 1985, 284).

Statistical Inference. Because simulation of farm firm growth and survivability and reviews of theory and fragmented data provide conflicting results, the issue of whether farm commodity programs increase, decrease, or leave unchanged farm size and numbers must be resolved empirically from a more complete model. Modest, single equation models with government payments an explanatory variable have been estimated by Gale (1989) and by Shepard and Collins (1982). Shepard and Collins (1982, 614), based on least squares statistical analysis of farm bankruptcy rates, concluded that "there is no evidence that agricultural support payments since World War II [to 1978] have induced, deferred, or reduced farm failures." As part of a pooled state cross-sectional and time series analysis of effects of prices and income on the number of farms, Gale (1989) included real government payments per farm as an explanatory variable. Because payments per farm are greater for states with large farms than for states with small farms, a positive association between farm size and government payments would be expected. Yet, Gale (1989, 19) found only a "small role of government payments" in explaining farm numbers.

Time series covering many years and wideranging programs differing considerably in scope and form are now available for statistical inference.

Thus, more comprehensive statistical models incorporating more variation in data can be used to infer whether programs influence farm size and numbers. This chapter mines this third approach.

I do not attempt to examine the impact of programs by sales class because (a) the issue has been treated at least conceptually elsewhere (Richardson et al. 1988; Tweeten, 1984), (b) time series data by farm size classes are inadequate or require substantial massage to provide the empirical base for multivariate analysis, and (c) the definition of family farm is arbitrary. Regarding the latter, using a broad definition 95 percent of farms are of the family type and have been for years (see Tweeten 1984, 8). Hence trends in numbers of all farms provide insights into what is happening to family farms broadly defined. Although data and methodology of this chapter are intended to improve on prior studies, results must be interpreted with caution and need to be supplemented by additional analysis as more refined data become available.

Conceptual Model

The initial conceptual model is driven by the assumption that farm operators and their families are utility maximizers and that equilibrium occurs when farm returns equal nonfarm returns. In this milieu, a small, economically inefficient farm (measuring efficiency in the narrow context of returns equal to opportunity costs of resources) can coexist with a large, economically efficient farm if the operator of the small farm is willing to pay for the farm way of life out of off-farm income. At the margin, the small farmer and the big farmer are in equilibrium because they equate marginal *social* (not just economic) returns with marginal *social* costs. Adding the psychic, nonmonetary benefits of rural living to the private monetary returns from farming brings the *total* social return on small farms up to that on large farms or other alternatives. It is not possible to conclude that one operator is more socially desirable or rational than the other. Of course, public subsidies differing among farm types and sizes can distort social efficiency decisions. I abstract from that issue, herein.

In equilibrium, operator labor-management return in the farm sector L equals labor-management return in the nonfarm sector L´. Farm size measured in output (real sales) is assumed to be determined by the labor-output ratio. Farm food and fiber output and use are relatively unresponsive to price in the short run (Tweeten and Quance 1969; George and King 1971). Hence, larger output per farm implies fewer farms in the nation. We can extract considerably more information by decomposing the determinants of farm size as measured by annual output per farm, (S) into various components:

$$S = (S/X)(X/L)(L/L')L' \qquad (1)$$

where:

S/X is productivity measured by the ratio of aggregate farm output to aggregate input X.[1] If aggregate land area and output demand are highly inelastic, then additional improved resource-neutral, output-increasing technology as measured by S/X increases farm size and decreases the number of farms.

L/X is the factor share of farm labor. It is essentially the inverse of aggregate output (aggregate input) per unit of labor. Changes in the variable over time reflect scale-biased, labor-substituting technology apparent in economies of farm size. *Ceteris paribus*, a change in technology giving rise to economies of size and a falling share of labor causes farm size as measured by capital or land to increase and farm numbers to decline.

L/L´ is the ratio of operator and labor income per family on the farm to that in the nonfarm sector. For many years after the 1930s, farm income substantially lagged nonfarm workers income. As farm income expanded to approach nonfarmers' incomes, farm size expanded. The ratio approaches 1.0 in equilibrium and hence can be dropped from (1) in the long-run. Data for recent decades indicate that farmers' and nonfarmers' per family income are somewhat comparable. Greater off-farm employment opportunity reduces pressure to expand farm size to achieve the equilibrium farm and off-farm income equal to L´, other things equal. A higher proportion of farmers' income from off-farm sources FY (see Table 13.2) reduces need for income from the farm and for expansion of farm size as L´ grows.

L´, off-farm earnings, influences farm size because, other things equal, an increase in off-farm income raises the opportunity cost of farming and hence causes farm size to grow and numbers to fall to "keep up with the Jonses."

Because equation (1) is an identity, the coefficient of each variable is hypothesized to approach unity (absolute value) in a double-log equation.

TABLE 13.2 Variable Definitions and Sources

Variable[a]	Definition (annual data from 1950 to 1987)	Source
ATFY	Permanent income measured by past 5-year moving average of deflated (by GNP price) farm income from farm and off-farm sources, million 1988 dollars.	U.S. Department of Agriculture, September 1989, P. 40 and earlier issues; Council of Economic Advisors, p. 312.
DTFD	Dummy interaction with transitory income. Variable is dummy of 1 for negative transitory income times DTFY, zero elsewhere, million 1988 dollars.	See ATFY for data source.
DTFY	Transitory income measured as deviation of total farm income t-1 from ATFY, million 1988 dollars.	See ATFY for data source.
D	Debt-to-asset ratio, percent.	U.S. Department of Agriculture, September 1989, p. 58.
E	Excess farm capacity as percent of total capacity. Estimated by Tweeten for 1987 to be 4 percent. Sum of acreage diversions, subsidized exports, and net stock accumulation as percent of estimated normal farm output.	Dvoskin.
FAR	Farm numbers, 100.	U.S. Department of Agriculture, September 1989, p. 8.
FY	Ratio of net farm income to total farm income of farm operators from all sources, percent.	U.S. Department of Agriculture, September 1989.
G	Direct government payments, million 1982 dollars − deflated by GNP deflator.	U.S. Department of Agriculture, September 1989, p. 43; Council of Economic Advisors, p. 312.
i	Real interest rate measured by non-real estate farm interest rate less GNP deflator inflation rate, percent.	U.S. Department of Agriculture, 1989 and earlier issues; Council of Economic Advisors, p. 312.
NW	Farm real net worth, billion 1982 dollars.	U.S. Department of Agriculture, September 1989, p. 58; Council of Economic Advisors, p. 312.

(continued)

TABLE 13.2 *(continued)*

Variable[a]	Definition (annual data from 1950 to 1987)	Source
PF	Factor terms of trade measured by real price received per unit of output produced by farm production resources. Parity ratio times productivity rate, 1977=100.	U.S. Department of Agriculture, August 1989 and June 1989; Council of Economic Advisors, p. 421.
PMPL	Ratio of farm machinery price to farm labor price, 1988=1.0.	U.S. Department of Agriculture, June 1989; Council of Economic Advisors, p. 421.
POP	Farm Population, 1,000. Old farm definition used through 1981. New definition used after 1981 but dummy variable added to independent variable to allow change in intercept.	Council of Economic Advisors, 1989, p. 420.
SX	Productivity rate defined as output of crops and livestock per unit of all production inputs, 1977=100.	U.S. Department of Agriculture, August 1989, p. 50; 1989, p. 29.
TI	Tractor inventory, constant dollar real value. Tractor inventory deflated by index of prices paid by farmers for tractors and self-propelled vehicles, million 1988 dollars.	U.S. Department of Agriculture, September 1989, p. 65; U.S. Department of Agriculture, June 1989.
V	Coefficient of variation of net farm income (deflated by GNP price index) as ratio of standard deviation of past 5-year net farm income to past 5-year mean of net farm income.	U.S. Department of Agriculture, September 1989, p. 40; Council of Economic Advisors, p. 312.
XL	Farm operator, family, hired labor factor share, percent.	U.S. Department of Agriculture, August 1989, p. 38; Also unpublished worksheets from Economic Research Service, U.S. Department of Agriculture.

[a]Variables A for diverted acres and G' for total commodity program outlays omitted because they are not included in subsequent statistical tables.

The percentage change in farm size as measured by sales is equal to the sum of percentage changes in each of the right-hand-side variables in (1). Tweeten (1984) used such methodology to estimate historic sources of change in farm size and projected changes to the future.

Deterministic identity equation (1) is a useful baseline but needs considerable modification for multivariate statistical analysis. In empirical analysis the necessity to substitute proxy variables for those above and the inevitability of errors in variables and omitted variables (all cannot be included due to multicollinearity) makes the relationship in (1) inexact. Because data are more adequate to measure farm size in area rather than in real sales per farm over time, farm size is proxied by number of farms FAR. Size and numbers are closely (negatively) correlated because aggregate acres are quite fixed.

The variable S/X (abbreviated to SX) measuring output-increasing technology is expected to have only a modest impact on farm size, which preferably is measured in resources or acres. However, SX might be retained because an increase in aggregate multifactor productivity SX which expands output and reduces prices may drive marginal farms and excess labor out of agriculture, reducing farm numbers.

The factor share of labor (in abbreviated notation XL) represents labor-saving technology which has changed the face of rural America. It is such an important element that a related variable, tractor inventory (TI), is introduced to more fully account for scale-biased, labor-saving technology which has so radically altered the structure of agriculture and reduced farm numbers.[2] The price of machinery relative to labor is sometimes used in place of XL and TI, although the latter are better suited to a two-step analysis where XL and TI are regressed on commodity program variables.

Closing the gap between farm and nonfarm income substantially changed farm size prior to 1950 -- the first year of data used in this study. The disequilibrium variable L/L' has been less important in recent decades and is dropped from the analysis to reduce multicollinearity.

After variables measuring commodity programs as well as prices are introduced into (1), the statistical form is depicted as

$$FAR = f(p, XL, TI, i, g, \varepsilon) \qquad (2)$$

where p refers to a vector of price variables, XL is factor share of labor, TI is tractor inventory, i is real interest rate, g is commodity program variables, and ε is random error. An increase in real price of farm commodities may cause farmers to try to expand aggregate national output and acreage, or for economically stronger farms to squeeze out weaker farms. Price is measured by real factor terms of trade PF, the real commodity price received per unit of resources. PF is commodity terms of trade (parity ratio) times the productivity ratio.

Prices and government commodity programs also may influence FAR indirectly through behavioral relationships explaining X_i as a function of relative prices, financial conditions, commodity programs, and other variables. Hence a two-step model is proposed. The direct influence of p and g is found from statistical estimates of equation (2). The indirect influence of p and g on FAR is found by regression of the right-hand-side variables X_i in (1) on price (p), commodity program (g), other variables (d), and random error μ as below:

$$X_i = f(p, g, d, \mu) \tag{3}$$

After statistical estimation the various X_i are substituted into (2) to determine the full impact of commodity programs. Predicted values of (3) could be used in estimating (2) in a recursive formulation to avoid simultaneous equation bias. However, experience suggests that statistical efficiency loss more than offsets any gains from reduced bias in the recursive system of equations.

Commodity Programs

The analysis is restricted to 38 years of annual data (1950 through 1987) because earlier data are nonexistent, inaccurate, or from a different structure. As noted above, commodity programs are assumed to enter the equation explaining farm size either (a) directly as a shifter of the equation or of the independent variables listed earlier explaining farm numbers or (b) indirectly through equations explaining right-hand-side variables in (1).

The following variable sets (with simple correlation coefficients between them) alternatively measure farm programs:

$$\begin{array}{lll}
(a) & G \text{ and } A & r = .83 \\
(b) & G \text{ and } E & r = .33 \\
(c) & G' &
\end{array}$$

where G is government payments to farmers, A is acres withheld from production by government diversion programs, G' is overall government spending on farm price and income supports, and E is excess capacity measured by the proportion of farm output diverted from markets by acreage diversion, export subsidy, and stock accumulation programs. These variables are used in the alternative sets rather than all at once to test various hypotheses because G and G′ considerably overlap and A and E considerably overlap.

Prices, labor share, tractor inventories, and commodity program variables are presumed to influence farm numbers and not vice versa. Real interest rates can directly influence farm numbers through credit restraint (see Gustafson and Barry, this volume), or can indirectly influence numbers by changing land values.[3] Higher land values are hypothesized to retard entry into farming. Values for all variables are for the current year unless otherwise indicated in subsequent tables.

In summary, the direct influence of government programs is measured by regressions of FAR on selected core variables and the program variables listed above. Then core variables are regressed on the program variables. Results are substituted into the equation for farm numbers FAR to determine the full effects of commodity programs on farm numbers.

Specific Hypotheses

Focus is primarily on the following six hypotheses, not necessarily mutually exclusive. The first two hypotheses are related to the direct effect of government payments. Other hypotheses work through intervening variables.

Hypothesis 1: An increase in the rate of excess production capacity E (or diverted acres A) increases the size and reduces the number of farms. Farms of optimal size before diversion find they have redundant labor and machinery after diversion, hence must acquire more land to realize economies of size.

Hypothesis 2: An increase in G or G′ increases the size and reduces the number of farms. Farm payments and receipts enhanced by government programs reduce internal and external capital rationing, causing farm size to grow. Farms consolidate to grow in size.

Hypothesis 3: Quance and Tweeten in 1972 contended that government programs provide stability, allowing risk-averse investors to leverage a given equity further. This hypothesis is tested by introducing equity (net worth NW) or the equity ratio (D) into the equation for FAR and into equations explaining core explanatory variables -- along with program variables as before. The hypothesis is that income variance V (reduced by programs) and the leverage ratio D or NW covary positively. That is, a given net worth or equity ratio results in fewer and larger farms in the presence of greater stability from government involvement in farming. An insignificant coefficient of NWV or DV would suggest rejection of the hypothesis that more stable farm income under commodity programs reduces the number of farms as a given net worth is leveraged further.

Hypothesis 4: The Cochrane hypothesis is that farm commodity programs provide stability and capital to increase *productivity* of farming. Increased productivity increases output and reduces farm prices and receipts, driving marginal farms out and bringing farm consolidation in. This hypothesis is tested by including SX in the equation explaining FAR and also regressing SX on variables measuring government programs. The Cochrane (1985) hypothesis is similar to the Quance-Tweeten (1972) hypothesis 3 but the programs operate through productivity rather than investment.

Hypothesis 5: The permanent income hypothesis holds that farm commodity programs influence farm size and numbers primarily through the intermediary variable -- investment (see Tweeten et al. 1984, 21-28). Government programs reduce investment because they stabilize farm income. The propensity to invest out of permanent income is low, out of transitory positive income is high, and out of transitory negative income is near zero, the reasoning goes. For a given average farm income, greater income stability under government programs reduces positive transitory income and hence investment. Less investment means less substitution of capital for labor and hence means more and smaller farms, *ceteris paribus*.

Hypothesis 6: Government supports have opposite impacts in the short and long run. A strong case can be made that commodity programs maintain or increase farm numbers and hold down the size of farms *in the short run* because they increase survivability of marginal farms that would otherwise fail or voluntarily exit when facing unfavorable economic conditions. Large farms emphasize fruits, vegetables, livestock feeding, and other enterprises not covered by commodity programs. Small to medium-size farms are especially prominent in grains, soybeans, dairy, and tobacco enterprises covered by commodity programs. Other forces associated with government programs and expressed in the above hypotheses work to reduce farm numbers on the average *in the long run*. Thus hypothesis 6 holds that commodity programs may have opposite affects on size structure by length of run. Long- and short-run coefficients are estimated by Koyck-Nerlove distributed lag adjustment models.

Empirical Results

Each of the tables reporting empirical results has three main equations:

1. The first equation measures the contribution of prices alone to farm numbers. Of concern is whether prices alone can account for variation in farm numbers.
2. The second equation measures the contribution of government commodity programs alone. Of interest is whether government programs or prices account for more variation in farm numbers.
3. The last equation is considered to be the most adequately specified equation and is used to calculate elasticities. The equation is estimated both by ordinary least squares and autoregressive least squares because autocorrelation in residuals was found to be a problem in the former.

Farm Numbers Equations

1. Prices alone inadequately account for changes in farm numbers FAR over time despite significant coefficients on prices in the distributed lag equation 3.1 (Table 13.3). The adjusted coefficient of determination (R^2 = .519) is much lower than for the more completely specified equation 3.3.
2. Government program variables G for government payments and E for excess farm production capacity performed best but explain a small portion of variation in farm numbers (equation 3.2). The coefficient of the program variable G measuring government direct payments is statistically significant at the .01 level. Variables measuring diverted acres and total inflation-adjusted spending on farm programs performed less well and are not included.
3. The more completely specified equation 3.3 accounts for a large proportion of the variance in farm numbers. Coefficients of all variables are significant at the .02 level or better.
4. The sign of PF reverses from equation 3.1 to 3.3. The expected positive sign in the more adequately specified equation 3.3 indicates that an increase in farm real prices (factor terms of trade) increases the number of farms. All variables in 3.3 display signs consistent with economic logic. The sign of the coefficient on either PF or ln PF was the same in equation 3.3, but only the latter variable is shown because it more fully explained FAR as judged by the R^2.
5. Except for variable lnXL, elasticities are of modest size in equation 3.3. In equation 3.3, lnXL and lnTI replace the related machinery-labor price ratio. The variable i indicates that higher real interest rates increase farm numbers perhaps because they reduce land prices, easing entry barriers.

TABLE 13.3   Statistical Results of Least Squares Regression of Farm Numbers (FAR) on Selected Independent Variables with U.S. Annual Data from 1950 to 1987

Equation	Constant	PF	PMPL	lnXL	lnTI	i	E	G	FAR(t-1)	R^{2b}	DW
					Independent Variables[a]						
3.1											
Coef.	13518	-58.239	-4876.4						0.00163	0.519	
s.e.	1652	15.802	1218.5						0.00405		0.719
pr>|t|	<.01[c]	<.01	<.01						0.69		
SR Elast.		-2.46	-2.23								
3.2											
Coef.	4298						-21.954	-0.1508	-0.00807	0.239	
s.e.	321						53.068	0.0434	0.00408		0.153
	<.01						0.68	<.01	0.06		
							-0.05	-0.45			
3.3											
	-3527	448.36[d]		3194.7	-458.81	20.991	-25.625	-0.0299		0.992	
	2818	263.61		98.74	179.18	8.357	6.111	0.0077			1.285
	0.22	0.1		<.01	0.02	0.02	<.01	<.01			
		0.19		1.36	-0.20	0.08	-0.06	-0.09			
3.3A[e]											
Coef.	-3807	259.28		3162.3	-341.16	17.689	-15.755	-0.0286		0.993	
s.e.	2940	241.63		118.9	206.67	10.026	5.978	0.0090			
SR Elast.		0.11		1.35	-0.15	0.07	-0.04	-0.08			

[a]See Table 13.2 for variable definitions and sources.
[b]In this and other tables in this chapter, the R^2 is adjusted for degrees of freedom.
[c]"<.01" means that probability of getting a larger t-value is less than 1 percent when sampling from a population in which the true coefficient is zero.
[d]Variable lnPF.
[e]Estimated by autoregressive least squares; first-order autoregressive coefficient 0.505. The Durbin-Watson (DW) coefficient is inappropriate for the equation.

6. Equation 3.3A is estimated by autoregressive least squares because of evidence for autocorrelated disturbances as indicated by the Durbin-Watson (DW) coefficient. Results were not fundamentally different in equations 3.3 and 3.3A.

Results for an equation explaining farm population are included as Table 13.8 at the end of this chapter. Outcomes were similar to those in Table 13.3 as might be expected.[4]

Farm Labor Factor Share Equations

1. Measured by statistical significance and adjusted R^2, prices alone do not adequately explain variation in labor factor share (equation 4.1, Table 13.4).
2. Commodity program variables alone in equation 4.2 do not account for much variation in labor share. Commodity programs explain much less than do prices.
3. The coefficient of the machinery-labor price ratio PMPL in equation 4.3 is highly significant but the sign is inconsistent with logic. The variable is replaced in 4.4 by an alternative form, machinery inventory TI in year t-1.
4. All coefficients have expected signs and high statistical significance in equation 4.4. Elasticities are relatively large compared to those from equations for FAR and TI.
5. Each 1 percent increase in government payments is projected to directly reduce farm numbers -.66 percent according to equation 4.4.
6. Equation 4.4A is the only instance in the entire analysis where autoregressive least squares substantially changes results. Specifically, the coefficient of G changes from a significant negative to an insignificant positive. That is very important in subsequent evaluation of the overall impact of programs because G potentially has a major impact on farm numbers through labor share given the large coefficient of XL on FAR in Table 13.3.

   The pivotal importance of the equation for XL and the significant first-order autoregressive coefficient of 0.99 in equation 4.4A prompted a further modification -- estimation of equation 4.4 in first differences. (The time variable was dropped because the intercept in the first difference equation is equivalent to the coefficient of the time variable.) The estimated coefficient of 0.00007 for G and a standard error of 0.00005 in the first difference equation were very near the results from equation 4.4A. Based on all specifications of variables and functional

TABLE 13.4 Statistical Results of Least Squares Regression of Farm Labor Share (XL) on Selected Independent Variables with U.S. Annual Data from 1950 to 1987

Equation	Constant	Independent Variables[a]							R^{2b}	
		PF	PMPL	TI(t-1)	TIM	E	G	XL(t-1)		DW
4.1										
Coef.	96.89	-0.4548	-32.98					0.0024	0.526	
s.e.	12.11	0.1182	8.80					0.0044		0.786
pr>|t|	<.01	<.01	<.01					0.59		
SR Elast.		-3.00	-2.35							
4.2										
	28.24					-0.1645	-0.00095	-0.00079	0.230	
	2.26					0.3934	0.00032	0.00041		0.170
	<.01					0.68	<.01	0.06		
						-0.06	-0.44			
4.3										
	75.76	-0.3585[b]	0.6539[b]		-0.2138		-0.00102	0.00603	0.773	
	7.53	0.0926	0.2296		0.0532		0.00018	0.00232		1.219
	<.01	<.01	0.01		<.01		<.01	0.01		
		-2.37	0.05		-1.20		-0.47			
4.4										
	97.22	-0.4104		-0.00072	-0.0697		-0.00144		0.818	
	6.73	0.0692		0.0016	0.0070		0.00016			1.078
	<.01	<.01		<.01	<.01		<.01			
		-2.71		-1.43	-0.39		-0.66			
4.4A[c]										
Coef.	71.03	-0.0017		-0.00002	-0.6612		0.00006		0.993	
s.e.	7.29	0.0127		0.0004	0.0915		0.00006			
SR Elast.		-0.011		-0.04	-3.70		0.028			

[a]See Table 13.2 for variable definitions and sources.
[b]For t-1.
[c]Estimated by autoregressive least squares; first-order autoregressive coefficient 0.99.

forms in Table 13.4, my judgment was that equation 4.4A coefficients are the most reliable for use in the concluding section to measure the impacts of government payments on FAR through labor share XL.

7. Variables ATFY, DTFY, and DTFD included to test the permanent income hypothesis and variable V directly and in interaction with other variables performed so poorly they are not listed in Table 13.4. Although improper specification may be the reason for unsatisfactory performance of variables measuring impacts of reduced instability on structure, other factors such as uncertain continuity in programs inherent in the political process (see Sumner 1985) and the tendencies of programs to focus on income enhancement rather than economic stabilization could also be factors.

Tractor Inventory Equations

The influence of government programs on farm numbers through tractor investment is estimated from equations in Table 13.5. Tractor inventory is a proxy for and is closely correlated with all farm production asset inventory -- results can be interpreted accordingly.[5]

1. As in previous equations, prices alone do not perform well in explaining tractor inventory (equation 5.1).
2. Government programs account for more variation in TI (equation 5.2) than prices (equation 5.1) but only the coefficient of G is statistically significant in 5.2.
3. The ability of the permanent income hypothesis to explain tractor inventory is tested in equation 5.3 with disappointing results.
4. The hypothesis that stability as measured by V allows a given net worth to be more leveraged to increase tractor inventory is tested in equation 5.4. Using V alone or interacting with D or NW gave unacceptable results. NW performs better alone (equation 5.5) than interacting with V.
5. The independent variables in 5.5 explain 81 percent (R^2 adjusted for degrees of freedom) of the variation in TI and all coefficients display acceptable signs and statistical significance. NW may be a proxy for other correlated financial variables.
6. Autoregressive least squares equation 5.5A did not change signs or in other ways give results materially different from equation 5.5.
7. Each 1 percent increase in direct payments is predicted to directly reduce tractor investment .11 percent in the short run and .12

TABLE 13.5   Statistical Results of Least Squares Regression of Tractor Investment (TI) on Selected Independent Variables with U.S. Annual Data from 1950 to 1987

Equation	Constant	Independent Variables[a]										R^{2b}	
		PF	PMPL	ATFY	DTFY	DTFD	NWV	NW	E	G	TI(t-1)		DW
5.1													
Coef.	16133	243.20	-16000								0.2392	0.258	
s.e.	9399	76.64	6477								0.0711		0.776
pr>|t|	0.1	<.01	-0.58								<.01		
SR Elast.		0.81	-0.58										
5.2													
Coef.	34869								-282.4	-0.707	0.0704	0.442	
s.e.	1821								200.3	0.1645	0.0523		0.501
pr>|t|	<.01								-0.17	<.01	0.19		
SR Elast.									-0.05	-0.16			
5.3													
Coef.	8376	191.90	-6002	0.0508	-0.0822	-0.1519				-0.7806	0.3214	0.870	
s.e.	8579	62.2	6106	0.0337	0.0992	0.1800				0.1558	0.0967		1.420
pr>|t|	0.34	<.01	0.33	0.14	0.41	0.41				<.01	<.01		
SR Elast.		0.64	-0.22	0.14	-0.003	0.03				-0.18			
5.4													
Coef.	27020	138.26	-12764				16.003			-0.6253	0.1653	0.567	
s.e.	8445	71.65	5550				8.249			0.1560	0.0661		0.933
pr>|t|	<.01	0.06	0.03				0.06			<.01	0.02		
SR Elast.		0.46	-0.46							-0.15			

5.5						
	39023	-20004	16.590	-0.4938	0.0695	0.812
	2908	3748	1.960	0.0967	0.0299	0.985
	<.01	<.01	<.01	<.01	0.03	
		-0.72	0.46	-0.11		
5.5A[b]						
Coef.	24063	-16853	8.927	-0.3000	0.5762	0.895
s.e.	2916	2453	1.660	0.0722	0.0826	
SR Elast.	-0.61	0.25	-0.07			

[a]See Table 13.2 for variable definitions and sources.
[b]Estimated by autoregressive least squares; first-order autoregressive coefficient -0.15.

percent in the long run according to equation 5.5, and .07 in the short run and .17 percent in the long run according to equation 5.5A.

Share of Farm Income from Farm Sources

As noted earlier in the discussion of the conceptual model, greater off-farm income reduces dependence on income from the farm for a farm family, reducing demand for farm employment *ceteris paribus*. The coefficient of the proportion of farm net income from farm sources FY was not significant in equations for FAR perhaps because of multicollinearity problems but the logic of a relationship between FY and FAR is inescapable. The impact of government program and other variables on FY is shown in Table 13.6.

1. Prices alone in equation 6.1 account for a considerable proportion of the variation in FY. The principal impact comes through the machinery-labor price ratio. A 10 percent increase in the relative price of machinery is predicted to raise the proportion of farm income from farm sources .1 percent in the short run and .7 percent in the long run. It may be noted that this long-run elasticity is comparable to the elasticity found in equation 6.3 without a distributed lag. Thus 6.3 may be regarded as a long-run equation.
2. Farm commodity program variables alone account for a modest share of variation in FY in equation 6.2. The specification is incomplete without other variables. None of the program variables has a significant coefficient.
3. The significant negative coefficient on G in equation 6.3 indicates that increased government program payments reduce the share of all farm income coming from farming. Because payments are counted as income from "farm sources" and hence raise FY, the relationship is unexpected and is probably a spurious result. The coefficient is negative probably because government payments are large when net cash receipts from farming are low.
4. Equation 6.3A estimated by autoregressive least squares shows no significant impact on FY from farm programs. Results from Table 13.6 are not used subsequently in the summary to calculate the impact of government programs on farm numbers.

Aggregate Farm Resource Productivity

Aggregate farm productivity SX was not included in equations of Table 13.3 because the objective is to measure farm numbers and size in terms of resource volume or area rather than output or sales. Also the

Table 13.6 Statistical Results of Least Squares Regression of Share of Farmers' Income from Farm Sources (FY) on Selected Independent Variables with U.S. Annual Data from 1950 to 1987

Equation	Constant	PF(t-1)	PMPL(t-1)	TIM	E	G	FY(t-1)	R^{2b}	DW
				Independent Variables[a]					
6.1									
Coef.	17.260	-0.1239	0.3115				0.8801	0.966	
s.e.	9.139	0.0753	0.0323				0.0522		
pr>|t|	0.07	0.10	<.01				<.01		2.097
SR Elast.		-0.29	0.01						
6.2									
Coef.	20.567				-0.8737	-0.00049	0.7778	0.412	
s.e.	13.034				0.7088	0.00068	0.2058		
pr>|t|	0.12				0.23	0.47	<.01		0.833
SR Elast.					-0.12	-0.08			
6.3									
Coef.	140.19		3.1855	-0.4470		-0.00545		0.869	
s.e.	10.83		0.5208	0.0822		0.00396			
pr>|t|	<.01		<.01	<.01		<.01			2.339
SR Elast.			0.08	-0.89		-0.03			
6.3A[b]									
Coef.	102.96		5.5459	-0.8192		0.00037		0.899	
s.e.	17.78		10.392	0.2506		0.00032			
SR Elast.	<.01		0.14	-1.64		0.06			

[a]See Table 13.2 for variable definitions and sources.
[b]Estimated by autoregressive least squares; first-order autoregressive coefficient 0.99.

theoretical justification for including SX as an intermediate variable measuring the impact of greater economic stability on farm numbers was not supported empirically. Selected results of various specifications of an SX equation are included in Table 13.7.

1. Price variables included in this study account alone for 78 (adjusted R^2) percent of the variation in productivity SX in equation 7.1. It is notable that higher factor terms of trade, a measure of incentives to expand overall resource use, do not increase productivity. Based on the sign of coefficients, higher real prices PF may reduce productivity. However, significance levels indicated no strong statistical association between terms of trade and productivity, a conclusion consistent with that found much earlier by Heady and Tweeten (1963, 447).
2. Commodity program variables alone account for only 25 percent of the variation in SX (equation 7.2).
3. With prices included with program variables to more completely specify the equation for productivity, no program variable coefficients were significant.
4. Equation 7.3A estimated by autoregressive least squares also gives no support to the proposition that commodity programs enhance farming productivity. The sign of the PMPL coefficient changes between equations 7.3 and 7.3A, but the elasticities are so low in absolute value that the overall impact of the variable can be ignored without much injustice to reality.

Commodity programs reduce output by removing land from production but diverted acres are included in calculation of the SX denominator, all production resources. Hence, by construction, programs reduce productivity. That conclusion could be inferred from the negative coefficient of G for government programs in equation 7.3.

Results in Table 13.7 provide no support for the "treadmill hypothesis" that commodity programs increase productivity either directly through E and G or indirectly by improving price terms of trade PF. Hence SX is ignored in subsequent calculation of responses of farm numbers to commodity program variables.

Conclusions

I conclude by responding to each of the six hypotheses listed earlier and quantify the impact of commodity programs in the short and long run. The last equation in each table is the basis for estimates and conclusions.

TABLE 13.7 Statistical Results of Least Squares Regression of Aggregate Farm Productivity (SX) on Selected Independent Variables with U.S. Annual Data from 1950 to 1987

Equation	Constant	PF(t-1)	PMPL(t-1)	TIM	E	G	SX(t-1)	R^{2b}	DW
7.1									
Coef.	49.44	-0.5817	0.5507				1.093	0.784	
s.e.	19.79	0.2614	0.0869				0.114		
pr>|t|	0.02	0.03	<.01				<.01		1.590
SR Elast.		-0.49	0.01						
7.2									
	61.12				0.1687	0.00225	0.2066	0.248	
	8.72				1.0480	0.00084	0.1063		
	<.01				0.87	0.01	0.06		0.364
					0.01	0.13			
7.3									
	21.21	-0.3316	-1.8040	0.3568		-0.00036	0.9037	0.928	
	13.42	0.1740	0.7125	0.1087		0.00034	0.1046		
	0.12	0.07	0.02	<.01		0.30	<.01		2.452
		-0.28	-0.02	0.26		-0.002			
7.3A^b									
Coef.	-14.16	-0.2051	6.1256	1.4848		-0.00014	0.1883	0.950	
s.e.	15.69	0.1241	12.2970	0.2909		0.00031	0.1956		
SR Elast.		-0.17	0.06	1.07		-0.01			

aSee Table 13.2 for variable definitions and sources.
bEstimated by autoregressive least squares; first-order autoregressive coefficient 0.99.

Hypothesis 1: An increase in diverted acres A and excess capacity E has a very minor direct impact on farm numbers (and on farm size measured in acres if aggregate acreage is fixed so numbers and size are perfectly and inversely correlated). The elasticity of FAR with respect to E is only -.06 according to equation 3.3 and only -.04 according to equation 3.3A. However, excess capacity E may influence FAR through PF.

Hypothesis 2: An increase in direct payments G (the strongest measure of program direct impacts in nearly all equations and much superior to the overall government outlays variable G´, excess capacity E, or diverted acres A) does indeed seem to significantly influence farm numbers. Results are consistent with the hypothesis that commodity programs encourage farms to consolidate and grow in size and decline in numbers. However, the direct impact of a 10 percent increase in G is to decrease farm numbers by only .8 to .9 percent according to equations 3.3 and 3.3A.

Hypothesis 3: The general hypothesis that commodity programs increase farm size and decrease numbers because reduced variance of income raises labor productivity or aggregate productivity is not supported by this study.

Hypothesis 4: The Cochrane (1985) hypothesis that farm commodity programs decrease farm numbers and increase farm size is doubly rejected. That is, no statistical evidence indicated (a) that farm commodity programs increased farm resource productivity SX or (b) that SX influenced farm numbers FAR.

Hypothesis 5: This study does not support the hypothesis that programs influence income through the permanent income affect. The propensity to invest out of transitory positive income was not found to exceed the propensity from permanent income. Programs may induce stability but this study did not indicate that stability of income influenced investment or productivity.

Hypothesis 6: Results of this study indicate that commodity programs do not change farm numbers much in the short run or in the long run. Each 1 percent increase in excess capacity reduces numbers .04 to .06 percent and each 1 percent increase in government payments directly reduces numbers .08 to .09 percent (equations 3.3 nd 3.3A). But each 1 percent increase in excess capacity raises real farm prices at least 2 percent, causing farm numbers to increase .2 to .4 percent -- more than offsetting the direct impact of E and G in the short run. Thus programs in net retain farms in the short run according to the results of this analysis.

Based on the formula shown in the footnote, elasticities of farm numbers with respect to program variables in the long run are estimated to be

$$\frac{dFAR}{dE} \frac{E}{FAR} = -.04 \text{ in the short and long run from equation 3.3A.}[6]$$

$$\frac{dFAR}{dG} \frac{G}{FAR} = -.01 \text{ in the long run from equation 3.3A, 4.4A, and 5.5.}[7]$$

The result is consistent with hypothesis 6 but the magnitude of elasticities is too small to make a case the coefficients differ from zero. Furthermore, equations 3.3, 4.4, and 5.5 give a positive long-term elasticity of farm numbers with respect to more government intervention. However, the autoregressive least square result is preferred because of autocorrelation in ordinary least squares.

In short, statistical inference suggests farm commodity programs as measured herein modestly increase farm numbers in the short run and slightly decrease farm numbers in the long run. The strong impact of payments G dominates all other impacts of programs through excess capacity, acreage diversion, or total government outlays. One interpretation is that income effects from G overshadow diversion effects reflected in A and E because of slippage in controls, targeting of direct payments to farms most vulnerable to failure, or for other reasons. Another interpretation is that the variable G may be reflecting the impacts of acreage diversion because program variables are correlated. The simple correlation coefficient of G with A is .83, with E is .33, and with total government outlays G´ is .59 for the 1950-87 period.

The model was formulated to favor identification of program impacts. If anything, the small impacts found would be expected to have upward bias. Analysis could be improved by more accurate data and by disaggregation by farm commodity type, size, and region. Simultaneous equation estimation techniques also might improve results.

Notes

Comments of Marvin Batte, Lindon Robison, and Carl Zulauf are much appreciated. Shortcomings of this study are the sole responsibility of the author.

1. Whether S and X are viewed on a per farm basis as in S or on an aggregate basis makes no difference for S/X because number of farms cancels in the numerator and denominator.

2. The simple correlation coefficient between TI and SX is only -.03 for the 1950-86 period so collinearity is not a problem between these variables.

3. Off-farm earnings L' was included in the empirical model but the coefficient was insignificant. The variable was dropped to reduce multicollinearity.

4. A special effort was made to deal with the changing definition of a farm. The "old" definition ($250 of sales rather than $1,000) was retained to 1982 and the new definition used thereafter in measuring farm population. A dummy equal to 1.0 for years 1982-87, zeros elsewhere, was included as a independent variable. The coefficient was highly insignificant. One interpretation is that the tax rate reduction beginning in 1982 offset. Another is that a farm with sales of $1,000 in the late 1980s was equivalent in real size to one with $250 in sales in earlier years.

5. Total real production assets replacing TI in the model performed less well.

6. No statistical basis was found to support a different short-run than long-run response of FAR to E in the specification.

7. Long run elasticity where FAR=F is

$(dF/dG)(G/F) =$

$(\partial F/\partial G)(G/F)+[(\partial F/\partial XL)(XL/F)(\partial XL/\partial G)(G/XL)]+$
(　　3.3　　) (　　3.3　　)(　　4.4A　　)

$[(\partial F/\partial XL)(XL/F)(\partial XL/\partial TI)(TI/XL)(\partial TI/\partial G)(G/TI)]+$
(　　3.3　　)(　　4.4A　　)(　　5.5　　)

$[(\partial F/\partial TI)(TI/F)(\partial TI/\partial G)(G/TI)]$
(　　3.3　　)(　　5.5　　)

(Numbers in parentheses reflect the equation from which they come.)

References

Cochrane, Willard. 1985. "The Need to Rethink Agricultural Policy." *American Journal of Agricultural Economics* 67: 1002-1009.

Council of Economic Advisors. 1989. "Economic Report of the President." Washington, DC: U.S. Government Printing Office.

Dvoskin, Dan. 1988. "Excess Capacity in U.S. Agriculture." Agricultural Economic Report 580. Washington, DC: ERS, U.S. Department of Agriculture.

Gale, H. Frederick, Jr. 1989. "Effects of Prices and Income Variables on Number of Farms." Washington, DC: ERS, U.S. Department of Agriculture.

Gardner, Bruce. 1978. "Public Policy and the Control of Agricultural Production." *American Journal of Agricultural Economics* 60: 836-843.

George, P. S. and G. A. King. 1971. "Consumer Demand for Food Commodities in the United States." Giannini Foundation Monograph No. 26. Berkeley, CA: Division of Agricultural Sciences, University of California.

Headley, J. C. 1988. "The Effect of Government Policy on Farm Size and Structure," in Lindon Robison, ed., *Determinants of Farm Size and Structure.* Pp. 157-164.

Heady, Earl and Luther Tweeten. 1963. *Resource Demand and Structure of the Agricultural Industry.* Ames: Iowa State University Press.

Quance, Leroy and Luther Tweeten. 1972. "Policies, 1930-1970." Ch. 2 in Gordon Ball and Earl Heady, eds., *Size, Structure, and Future of Farms.* Ames: Iowa State University Press.

Richardson, James, Edward Smith, and Ronald Knutson. 1988. "Who Benefits from Farm Programs: Size and Structure Issues?" In Lindon Robison, ed., *Determinants of Farm Size and Structure.* Pp. 143-156.

Robison, Lindon, ed. 1988. *Determinants of Farm Size and Structure.* Proceedings of NC-181 Committee on Determinants of Farm Size and Structure in North Central Areas of the United States, held January 16-19, 1988, in San Antonio, TX. East Lansing: Michigan Agricultural Experiment Station.

Shepard, Lawrence and Robert Collins. 1982. "Why Do Farmers Fail? Farm Bankruptcies 1910-78." *American Journal of Agricultural Economics* 64: 609-615.

Spitze, Robert, Daryll Ray, Allen Walter, and Jerry West. 1980. "Public Agricultural Food Policies and Small Farms." Paper I of NRC Small Farms Project. Washington, DC: National Rural Center.

Stanton, Bernard F. 1985. "Commentary," in Bruce Gardner, ed., *U.S. Agricultural Policy: The 1985 Farm Legislation.* Pp. 321-328. Washington, DC: American Enterprise Institute.

Sumner, Daniel. 1985. "Farm Programs and Structural Issues," in Bruce Gardner, ed., *U.S. Agricultural Policy.* Pp. 283-320. Washington, DC: American Enterprise Institute.

Swanson, Louis. Summer 1989. "The Rural Development Dilemma." Publication No. 96. Washington, DC: Resources for the Future.

Tweeten, Luther. 1984. "Causes and Consequences of Structural Change in the Farming Industry." NPA Report No. 207. Washington, DC: National Planning Association.

Tweeten, Luther, Tom Barclay, David Pyles, and Stanley Ralstin. 1984. "Simulated Farm Firm Growth and Survivability Under Alternative Federal Fiscal-Monetary Policies, Initial Size, Tenure, and Uncertainty Conditions." Research report P-848. Stillwater: Agricultural Experiment Station, Oklahoma State University.

Tweeten, Luther and Leroy Quance. 1969. "Positivistic Measures of Aggregate Supply Elasticities." *American Journal of Agricultural Economics* 51: 342-352.

U.S. Department of Agriculture. 1989. "Agricultural Prices." Pr 1-3(89). Washington, DC: Economic Research Service, USDA.

_____. 1989. "Economic Indicators of the Farm Sector: Production and Efficiency Statistics, 1987." ECIFS 7-5. Washington, DC: Economic Research Service, USDA.

_____. 1989. "Economic Indicators of the Farm Sector: National Financial Summary, 1988." ECIFS 8-1. Washington, DC: Economic Research Service, USDA.

_____. 1989. *Agricultural Statistics.* Washington, DC: U.S. Government Printing Office.

TABLE 13.8 Statistical Results of Least Squares Regression of Farm Population (POP) on Selected Independent Variables with U.S. Annual Data from 1950 to 1987

Equation	Constant	Independent Variables[a]								R^{2b}	
		PF(t-1)	PMPL	lnXL	lnTI	lnSX	E	G	POP(t-1)		DW
A.1											
Coef.	13831	-98.084	-2967.98						0.8285	0.985	
s.e.	1326.3	8.513	1413.32						0.0202		1.951
pr>|t|	<.01	<.01	0.04						<.01		
SR Elast.		-1.66	-0.54								
A.2											
Coef.	3691						-118.24	-0.1618	0.7991	0.919	
s.e.	836						93.682	0.0844	0.0448		0.816
pr>|t|	<.01						0.22	0.06	<.01		
SR Elast.							-0.11	-0.19			
A.3											
Coef.	12473	3289.45[b]		13827	-3155.77	-5484.92		-0.1494		0.994	
s.e.	23171	993.50		1424	1024.15	1889.01		0.0364			2.01
pr>|t|	0.54	<.01		<.01	<.01	0.01		<.01			
SR Elast.		0.56		2.36	-0.54	-0.94		-0.18			

[a]See Table 13.2 for variable definitions and sources.
[b]lnPF for year t.

14

Implications of Tax Policy
for Farm Structure

Joseph A. Atwood, Vincent H. Smith, Myles J. Watts,
Glenn A. Helmers, and Boris E. Bravo-Ureta

The policies embodied in federal farm programs are often viewed as a major source of federal government influence on the structure of U.S. agriculture (Tweeten, chapter 13, this volume). Monetary and fiscal policy may also have had important effects on the size and number of firms in the farm sector and the degree of capitalization within each farm (see, for example, Chambers and Just 1982; Rausser et al. 1986). This chapter examines the potential implications of tax policy for the U.S. agricultural sector. Historically, the U.S. tax code has included many incentives that attempt to stimulate investment by lowering the effective cost of capital. These incentives include tax credits, accelerated depreciation, and special treatment of capital gains. If the relative cost of labor, capital, and land inputs is altered by such provisions, the levels and mixes of labor, capital, and land involved in U.S. agriculture are also likely to be affected. However, the effects of other tax provisions as well as market adjustments in asset prices and interest rates often (at least partially) offset the ability of tax policies to reduce the relative costs of capital equipment.

The potential importance of relative input prices for U.S. agriculture has been emphasized by some recent results provided by Kislev and Peterson (1982). These results indicate that almost all of the change in observed farm size (as measured in terms of land per farm unit) can be explained by changes in the relative prices of capital and labor. A competing hypothesis is that changes in cost-size relationships explain part or all of the tendency for farm size to increase over time. For a summary of these issues see Castle (1989). Several statistical and engineering studies have examined the cost-size relationship in U.S. agriculture and the resulting evidence does not seem to support the hypothesis of significant size economies (Hall and LeVeen 1978 or Hallam, chapter 8, this volume). Given these findings, we focus our discussion on the potential effects of tax policies on the relative costs of capital, land, and labor. The effects of tax policy on technological innovation are not examined as it is not clear whether technological change induces increases in farm size, or whether economic pressures to increase farm size induce the development and adoption of new technologies (Kislev and Peterson 1982).

Costs of Capital

Tax policies have both direct and indirect effects on the cost of capital. The direct effects of investment tax credits are clearly to lower the cost of capital if the firm has a positive tax liability. Some of these effects have been shown to be quite large. For example, changes in the level of the investment tax credit have had substantial effects on investment both in agriculture and the aggregate economy (Bernanke 1983; LeBlanc and Hrubovcak 1986). More accelerated depreciation allowances also have the direct effect of reducing the cost of capital, although evidence from replacement investment studies suggests that actual changes in depreciation schedules have had only modest impacts on actual investment (for example, Reid and Bradford 1983; Chisholm 1974; Bates, Rayner and Custance 1979). Samuelson (1964) has pointed out that these types of policies also have second round or indirect effects through interest rates. The consequences of changes in marginal tax rates (shifts in tax brackets) for the user cost of capital are complicated.

Samuelson (1964) also showed that setting the tax depreciation allowance for a given period equal to the change in the before-tax discounted present value of the asset over the period results in asset values that are invariant to changes in the tax bracket. An implication of Samuelson's (1964) result is that if tax depreciation allowances are accelerated relative to actual changes in economic value, an investor could afford to pay more for an asset. Alternatively, at the same acquisition cost, the after-tax cost of the asset would be lowered. Samuelson's (1964) results imply that government tax policies can affect the after-tax costs of depreciable assets and therefore result in the substitution of depreciable capital for labor. However, tax policies with respect to capital inputs are implemented as parts of comprehensive tax programs. Thus tax policies affecting other aspects of the firm's operations may mitigate or magnify the direct effects of a particular policy because of multiple impacts on the opportunity costs of different inputs to the firm. This issue will be examined below in more detail. An additional complication is that usually tax policy has been implemented without directly considering interactions with inflation.

The fact that the interaction between inflation and tax policy has consequences for the cost of capital and input use decisions has been widely recognized (see, for example, Bates, Rayner, and Custance 1979). Feldstein and Summers (1979) reported that between 1954 and 1977 the U.S. practice of determining depreciation allowances using historical costs rather than economic replacement costs greatly overstated total corporate profits during the nineteen seventies, and argued that as a result historical cost depreciation allowances greatly underestimated the true economic cost of capital. They also recognized that these effects were partially offset by the fact that firms were allowed to deduct nominal interest rather than real interest payments. However, they concluded that because interest expenses of firms were interest income to creditors the aggregate effects of nominal interest deductions could be largely ignored at the macro level. The individual firm, however, may find that relative returns and factor rental prices are affected by tax-inflation interactions. Thus, the interactive effects of different tax provisions may alter the costs of capital in unexpected ways.

The potential for these types of effects is demonstrated in the example presented below. The manner in which tax depreciation allowances, discount rates and inflation interact is illustrated through the examination of the before-tax annualized real cost of a machinery investment. The before-tax annualized real cost of an investment can be viewed as the amount the firm has to recover annually if the investment is to break even. Alternatively, the before-tax annualized real cost can be viewed as the maximum amount that a producer is willing to pay to lease the services provided by the asset rather than acquiring the asset directly. Since labor costs are usually secured annually and are fully deductible for agricultural producers, the annualized before-tax cost of the capital asset can be contrasted directly to before-tax labor costs.

In the example, the capital asset is assumed to cost $50,000. Assuming that actual economic depreciation follows a double declining balance schedule, the asset's salvage value is $5,370 at the end of ten years. The real before-tax discount rate is six percent. In each inflationary scenario, the inflation rate is ten percent and the nominal before-tax discount rate is 16.6 percent (assuming that the Fisher effect holds).[1]

For any depreciation method the net present cost of the investment is found by discounting (1) benefits of reduced taxes arising from depreciation (depreciation multiplied by the marginal tax rate) for each year of the depreciation period, (2) salvage value benefit at period's end, (3) losses from depreciation recapture at period's end (the difference between sale value less depreciated value multiplied by the marginal tax rate). These three are added to the initial cost in reaching the net present cost. An after-tax nominal discount rate (nominal before-tax discount rate multiplied by one minus the marginal tax rate) is used in the discounting process. Tax regulations are nominally constructed hence a nominal discount rate is appropriate.

The net present cost is then amortized for the asset life using a real after-tax discount rate. A real discount rate is useful because it results in a cost expression which does not embed a changing monetary value. Nominal amortization involves changing monetary values, and while expressions resulting from nominal amortization are sometimes useful, only expressions based on a constant monetary base are valid for point-in-time expressions where there is an implicit or explicit comparison to a real return. This annualized after-tax cost expression can be converted to a before-tax basis by dividing by one minus the marginal tax rate. Tax law changes relating to investment credit, capital gains exclusions or depreciation allowances different from those examined here can be analyzed using this process.

Table 14.1 shows the real annualized before-tax costs under different tax bracket, depreciation allowance, and inflation scenarios. Three tax depreciation methods are examined: 200 percent and 150 percent declining balances, and straight line depreciation over the life of the asset. Finally, the analysis is carried out for two tax brackets, 15 percent and 28 percent. Table 14.1 shows that for any given inflation and tax rate scenario, annualized costs are slightly lower when more accelerated depreciation methods are used, and individuals in higher tax brackets have lower costs than those in lower tax brackets. When the inflation rate is zero, these differences are not large. However, when inflation and taxes interact, the annualized costs of the machine are from 3 to 5 percent lower for the individual in the higher tax bracket. The differences vary slightly depending on the assumed pattern of depreciation allowances, but in all inflationary settings annualized costs decrease with increases in tax brackets.

Feedback Effects Upon the Cost of Capital

The effects of tax policy upon the costs of a given unit of capital for an individual farm could be based on variants of the above type of analysis. However, from a policy perspective, the analyst must also consider that both the markets for outputs and capital inputs will be

TABLE 14.1 Real Annualized Before-Tax Costs for a $50,000 Depreciable Investment for Before- and After-Tax Analyses.[a]

| | Real Annualized Costs | | |
| | Depreciation | | |
	200%	150%	Straight Line
No Inflation			
15% Tax Bracket	6300	6339	6368
28% Tax Bracket	6219	6297	6351
Inflation			
15% Tax Bracket	6064	6078	6178
28% Tax Bracket	5812	5892	5987

[a] A 10 year period is analyzed, and the real salvage value is assumed to be $5370 (that derived from 200 percent declining balance depreciation). Present values are amortized by the assumed real after-tax discount rate and then converted to before-tax expressions. The real no-tax annualized cost is $6386.

influenced by producer decisions. Tax provisions that reduce the cost of capital may induce substitution of capital for labor but also they may induce all producers to increase production. As a result, output prices may fall and firms may fail. Moreover, the substitution of capital for labor causes an increase in demand for capital goods that may raise the prices of capital inputs (at least in the short run). The net effect on capital prices depends upon the relative elasticities of demand and supply for capital goods (which themselves are not time invariant). Thus, the effects of any cost reducing tax provision may be partially offset by increased capital prices due to increased input demand.

During periods of inflation additional feedback effects occur with respect to interest rates. The example presented in Table 14.1 assumes that the nominal interest rates adjust to expected inflation in accordance with the "Fisher effect." Thus,

$$i_{BT} = r_{BT} + f + r_{BT}f \tag{1}$$

where i_{BT} is the nominal before-tax interest or discount rate; r_{BT} is the real before-tax discount rate; and f is the expected rate of inflation. Since current tax law requires taxes on interest receipts and allows deductions for nominal interest payments, the nominal after-tax discount rate is $i_{AT} = i_{BT}(1-T)$. The real after-tax discount rate is then $r_{AT} = (1+i_{AT})/(1+f) - 1$.

Darby (1975) and Gandolfi (1982) have argued that, as investors are concerned with real after-tax returns and taxes on capital gain are delayed, the equilibrium adjustment of the nominal before-tax interest rate is larger than that predicted by the Fisher effect. The exact level of the adjustment is subject to debate (Tanzi 1980). However, if nominal market rates do increase by more than the Fisher effect, the investor's discount rate would be higher than that used in the preceding examples.

An additional feedback effect with respect to the ability of tax policy to reduce after-tax capital costs is that the tax rate of the firm will often change as a result of the investment itself due to changes in effective tax brackets. This is especially true if interest rates include compensation for expected inflation. Under current law, the full amount of nominal interest is tax deductible. If the nominal revenue stream associated with an asset is growing, the level of nominal interest paid will often exceed the nominal receipts derived from the asset in the early years of its working life. In such cases, the excess deductions will shelter other income of the firm, thereby lowering the effective tax bracket of the firm. While the tax sheltering aspects of the investment might lead firms to acquire more "tax sheltering" investments, the fact that the acquisition itself lowers average (and hence marginal) tax brackets will tend to limit the firm's incentives to acquire additional tax shelters.

The above discussion of feedback effects of tax policy is not meant to be exhaustive. The discussion is intended to demonstrate that feedback effects will often complicate the analysis and prediction of the effects of tax policies upon investment and capital costs.

Costs of Labor

Income tax policy has occasionally attempted to lower the effective costs of hiring certain disadvantaged laborers. However, these policies would seem to have little affect upon the majority of agricultural producers. Any effects that tax policy would have on agricultural labor costs would appear to be indirect. For example, if tax incentives outside agriculture result in increased nonagricultural labor productivity and wages, the opportunity costs of using agricultural labor would rise, increasing the cost of labor relative to capital. Kislev and Peterson (1982) have pointed out that much of the increase in the capital to labor ratio in agriculture can be attributed to changes in the relative costs of capital and labor. During the period of their analysis, the opportunity cost of labor increased significantly. To what degree the increase in nonagricultural wages has been induced by tax policy is a question that lies beyond the scope of this chapter.

Value and Cost of Land

In this section we examine the value of land to investors in differing tax brackets. We assume an individual's tax bracket is sufficiently large to prevent an investment's current income from altering the individual's tax rate. To simplify the analysis it is assumed that two investment alternatives exist, both of which generate income streams over an infinite planning horizon. The first investment generates an income stream $Y(x)$ which may vary with time. The second investment generates a constant before-tax rate of return i which is considered to be the opportunity cost of the first investment. The investment decision is based on the imputed present value of the alternatives. The imputed value of the investment with income stream, $Y(x)$, evaluated at tax rate t is

$$V_n = \int_0^n Y(x)(1-t)e^{-i(1-t)x}dx \qquad (2)$$

where n is the useful life of the asset, x notes time, and i is the interest rate.

We are interested in how changes in $Y(x)$ interact with t to affect V. The change in $Y(x)$, $\delta Y(x)/\delta x = Y'(x)$, is introduced by integrating (2) by parts to obtain:

$$V_n = \frac{1}{i}\left[Y(0) - Y(n)e^{-i(1-t)n} + \int_0^n Y'(x)e^{-i(1-t)x}dx\right] \qquad (3)$$

If, for large values of x, $Y(x)$ increases at a rate proportionately less than $i(1-t)$, then, as n approaches infinity, the assets value V is:

$$\lim_{n \to \infty} V_n = \frac{1}{i}\left[Y(0) + \int_0^\infty Y'(x)e^{-i(1-t)x}dx\right] \equiv V \qquad (4)$$

because, as $n \to \infty$, $Y(n)e^{-i(1-t)n} \to 0$.

The effect of a change in the tax rate on the value of the investment can be obtained by differentiating (3) to obtain:

$$\delta V / \delta t = \int_0^n Y'(x) x e^{-i(1-t)x} dx \qquad (5)$$

From equation (5), it follows that if the income stream is constant through time, then Y'(x) equals 0, and, as expected, changes in tax rates have no effect on the value of the investment. If the income stream is increasing through time, then the value of the investment increases with the tax rate (i.e., if $Y'(x) > 0$ then $\delta V / \delta t > 0$).

The effect of the interaction of the tax rate and the time path of the income stream is illustrated in the following two cases.

Case I: A bond produces an annual constant income stream of $1000 for an infinite period. The market interest rate is 10 percent. The imputed value of the bond can be evaluated by the familiar capitalization formula as

$$V = \frac{\$1000(1-t)}{.10(1-t)} = \$10,000$$

Note that, as predicted by the above results, the value of the bond is independent of the tax rate.

Case II: A land investment is expected to generate an income stream of $1000 initially which increases by 4 percent annually. Y(x) is thus $1000e^{.04x}$. Assuming the interest rate is 10 percent, using (2) the land's value is

$$V = \frac{\$1000(1-t)}{.10(1-t) - .04}$$

In this case, the imputed value of the land is affected by the tax rate. For example, at a zero tax rate its imputed value is $16,666 while at a 50 percent tax rate its imputed value rises to $50,000.

These results indicate that high tax bracket investors will be attracted to investments whose income streams have the greatest growth rates, leaving investments with little income growth to those in lower tax brackets. The implications of these tax induced investment incentives will be discussed later. Next we examine a tax prescription which will reduce the tax impact on imputed values.

A tax adjustment exists that makes the value of investments independent of tax rates. Such an adjustment may be desirable for both equity and efficiency reasons. The adjustment would result in investors in different tax brackets valuing investments equally, *ceteris paribus*, and the bias in relative values introduced by the tax system would be ameliorated. Following the approach offered by Samuelson, this adjustment can be identified. To facilitate the derivation of the tax adjustment which leaves the value of the investment invariant to changes in tax rates, the beginning of the investment period is defined as an arbitrary point in time m. The taxable income adjustment that leaves the asset value invariant with regard to taxes is denoted as g(x). The resulting imputed value function for the asset may be written as

$$V(m) = \int_{m}^{n} \left\{ Y(x) - t[Y(x) + g(x)] \right\} e^{-i(1-t)(x-m)} dx \qquad (6)$$

where V(m) is the value of the asset at time m, g(x) is the tax adjustment at time m and the other variables have been previously defined. Differentiating (6) with respect to m yields,

$$\frac{\delta V(m)}{\delta m} = V'(m) = -Y(m)(1-t) + g(m)t + i(1-t)V(m) \qquad (7)$$

The objective is to find a tax adjustment, g(m), such that the value of the asset is invariant to the tax rate; that is, to define g(m) such that $\delta V(m)/\delta t = \delta V'(m)/\delta t = 0$. Rearranging (7) gives

$$V(m) = \frac{Y(m)(1-t) + V'(m) - g(m)t}{i(1-t)} \qquad (8)$$

Differentiating (8) with respect to the tax rate, t, yields

$$\frac{\delta V(m)}{\delta t} = \frac{\left[-Y(m) + \frac{\delta V'(m)}{\delta t} - g(m) \right] i(1-t)}{[i(1-t)]^2}$$

$$+ \frac{\left[Y(m)(1-t) + V'(m) - g(m)t \right] i}{[i(1-t)]^2} \qquad (9)$$

Setting $\delta V(m)/\delta t = \delta V'(m)/\delta t = 0$ and solving for g(m), it follows that

$$g(m) = V'(m) \qquad (10)$$

Equation (10) implies that change in an asset's value should be taxed as it occurs if the value of the asset is to be independent of the tax rate.

Similar results have been used to justify tax deductions for depreciation. However, taxation of capital gains as they *accrue* and at the same rate as ordinary income could be justified analogously. Capital gains currently are taxed when they are realized, not as they accrue, and so capital gains taxes may be deferred (or never paid). The relationship between valuation, taxes, and income paths coupled with the manner in which capital gains are currently taxed, may have implications for the control of productive assets by different classes of owners.

Suppose, to some degree, that investments can be classified as providing either constant nominal or constant real income streams. For instance, in moderately inflationary economies nominal income streams from real estate and common stocks are likely to increase over time. On the other hand, many financial investments such as saving accounts, money market accounts, and most bonds yield constant nominal returns. The results presented above indicate that those assets with constant real returns (increasing nominal returns under inflation), such as common stocks and real estate, would be valued more highly and therefore acquired by investors in higher tax brackets. Individuals with lower tax brackets would be more likely to acquire assets with constant nominal income streams. This suggests that farmers with relatively high incomes in relatively high tax brackets are able to pay higher prices for production assets such as land than farmers with low incomes in low tax brackets. Thus farmers with high tax brackets are likely to operate relatively large farms. To the extent that the tax structure is progressive with respect to marginal tax rates, it encourages the acquisition of additional acreage by large farms rather than new entrants or small farms. A caveat has to be added. If new entrants have high taxable incomes they also will be willing to pay higher prices than low tax bracket individuals for land and other fixed assets. Thus, progressive tax structures may also encourage "gentleman farming." These types of operations may be quite small and relatively inefficient as farming operations both because of their size and, as discussed below, because of tax sheltering considerations.

Taxes and Farm Structure:
A Review of the Empirical Evidence

The theoretical results presented above suggest that tax policy can have structural implications in U.S. agriculture. Several empirical studies have examined the effects of tax policy on size economies in U.S. agriculture. Using panel data for the 1975-1979 period, Batte and Sonka

(1985) compared before-tax and after-tax cost/size relationships. Given the panel structure of the data, analysis of covariance was used to estimate before- and after-tax average cost curves. The tax rules prevailing in 1979 were applied to each of the five years included in the study. The results reveal statistically significant size economies while there is no evidence of diseconomies. Most of the available size economies are captured by farms of 500 to 750 tillable acres. When federal income tax liabilities were added to other costs of production there was a small reduction in the level of size economies which, according to the authors, is evidence that over the sample period the prevailing tax structure was progressive. The analysis also showed that the investment tax credit tends to reinforce the presence of size economies, which suggests that this credit is of greater value to larger farmers compared to small farmers. The authors note, however, that the impact of the tax credit is not sufficiently large to completely offset the progressivity of the tax structure.

Batte (1985) used the same data set to examine the potential effects of tax provisions prevailing in 1980, 1981, 1982, and 1983 on farmers' tax liabilities. A simulation model was constructed to compute various tax related variables for each farm, for each year, and for each set of tax provisions. The analysis showed that the Economic Recovery Tax Act of 1981 (ERTA) resulted in a reduction of tax liabilities for the average producer compared to the 1980 provisions primarily because of more accelerated depreciation. In addition, marginal tax rates were less progressive under the 1981 law. The introduction of lower marginal tax rates in 1982 and again in 1983 led to a decrease in average tax liabilities. The progressivity of the tax system changed little between 1982 and 1983. Regression analysis was used to examine the relationship between tax liabilities and gross income. The results of the statistical analysis supported the hypothesis that income tax liabilities, marginal tax rates, and the progressivity of the tax system decreased steadily from 1980 to 1983.

Rossi and Durst (1989, 85) examined the impact of the 1986 Tax Reform Act (TRA) on the progressivity of personal federal income and self employment taxes for sole proprietors. Their approach was to construct an agricultural tax simulation model to compute federal income tax liabilities for a sample of 15,000 taxpayers reporting farm income for 1982. The authors defined expanded income as "adjusted gross income plus the deduction for married couples when both work, accelerated depreciation in excess of straight line, and the capital gains exclusion". They argued that using expanded income instead of adjusted gross income is preferable because the former income definition includes items that provide tax deductions without actual expenditures. Hence it is a better reflection of the ability to pay.

The analysis by Rossi and Durst (1989) showed that, for the various income classes, average tax rates ranged from 4 to 23 percent prior to the 1986 TRA. They found a positive association between expanded income and the average tax rate for all but one income class. The exception is for farmers with expanded incomes in excess of $500,000 whose average tax rates declined. Subsequent to the 1986 TRA, average tax rates ranged from 5 to 24 percent and in all cases were positively associated with expanded income. Using Lorenz curve analysis, the authors found that the TRA of 1986 increased the progressivity of federal income and self employment taxes even though marginal tax rates were reduced sharply due to broadening of the tax base. Rossi and Durst (1989) also evaluated the effects of reintroducing preferential treatment on capital gains, a provision that was eliminated in the 1986 TRA. Eliminating preferential treatment on capital gains increased the maximum rate for capital gains from 20 to 33 percent. Reintroducing preferential treatment under the tax code established by the 1986 TRA would yield a maximum capital gains tax rate of 13.2 percent. Average tax rates would then range from 7 to 17 percent with a reduction in these rates of between 25 and 40 percent for the three top expanded income groups. However, tax rates for the lowest three expanded income groups would not change. Consequently, this reintroduction would have a major impact in average tax rates and on the progressivity of the tax structure and potentially on the distribution of asset ownership in agriculture.

Grisley et al. (1989) examined the effective progressivity of the federal personal income tax system using a sample of Pennsylvania dairy farms. Data from 590 sole proprietors filing tax returns in 1984 with taxable income in excess of the maximum zero bracket amount of $3,400 were used to estimate a series of single equation econometric models. The dependent variable was either the nominal average tax rate (income tax liability divided by adjusted gross income) or the effective average tax rate (income tax liability net of all individual tax credits divided by adjusted gross income). The authors found a statistically significant positive relationship between nominal tax rates and farm size as measured by adjusted farm cash operating receipts. However, when the investment tax credit was included, no significant relationship between the average effective tax rate and farm size was found. These results suggest that the tax system prevailing in 1984 gave larger farmers a greater incentive to invest and hence to further increase the size of their operations. Grisley et al. (1989) used the econometric results to perform a simulation analysis to assess the likely impacts of four provisions: (1) the elimination of favorable treatment of capital gains; (2) the use of economic life for depreciation; (3) the replacement of accrual accounting for cash accounting; and (4) the exclusion of interest expense as a deduction when computing taxable personal income. The first two provisions were enacted in the Tax Reform Act of 1986, while the latter two were given serious consideration but were not implemented. The simulation results indicate that each of these provisions provided more benefits to large firms than to small firms. The results also showed that, despite the fact the farm income can be sheltered from taxation more easily than non-farm income, part-time and intermediate farmers had average effective tax rates comparable to farmers obtaining over 80 percent of their total net income from the farm business operation.

Hanson and Eidman (1985, 271) used panel data from 1967 to 1978 for a sample of 76 Minnesota farmers to evaluate the magnitude of tax expenditures by type and size of farm, and to analyze the impact of tax expenditures on tax progressivity. Tax expenditures were defined as ". . . deviations from the underlying tax rate burden for one group relative to another". The computer model used to estimate tax expenditures included the following provisions: (1) business interest expense; (2) investment credit; (3) cash basis tax accounting; (4) capital gains on qualifying livestock; and (5) accelerated depreciation. The results showed that average annual tax expenditures for the farms in the sample rose by 203 percent from the first half of the period (1967-72) to the second half of the period (1973-78) while average sales only increased by 88 percent. Regression analysis results showed that tax expenditures increased at a decreasing rate with respect to farm sales. A classification of farms into crop, hog, dairy, or feeder cattle showed the same relationship between tax expenditures and sales for all farm types. The analysis also showed that between 1967 and 1972 the effective tax rate was generally progressive but there was a clear reduction in progressivity during the period 1973-78 primarily because of a marked drop in effective federal tax rates for large farms and large increases in state income taxes for medium and small farms. In a more recent study, Hanson and Eidman (1986) updated their results by adding data for the period 1979-82. The authors concluded ". . . that tax expenditures continued to be of comparable importance (in real terms) to the sample during the more financially depressed period 1979-82".

Long investigated the use of agricultural investments as a tax shelter or avoidance. Two models were formulated in which alternative measures of tax shelter were used as dependent variables. The first tax shelter measure, INCIDENCE, is a binary variable set to one if a taxpayer reports a net loss from farming and zero otherwise. The second variable, LOSSES, is a measure of the dollar amount of reported farm losses. The data used to estimate the model was obtained from the 1983 Individual

Tax Model File which contains a stratified random sample of individual income tax return prepared by the Internal Revenue Service (IRS). Taxpayers with adjusted gross income below -$200,000 or above $200,000 were excluded because, given confidentiality rules, their state of residence (data required for estimation purposes) could not be obtained. Long (1990) also excluded taxpayer tax returns that used income averaging or that reported negative nonfarm income. A major objective of the paper was to evaluate the impact of marginal tax rates on tax avoidance. Long (1990) found a significant positive relationship between the probability of reporting a farm loss and the individual's marginal tax rate. A similar relationship was found for the amount of disposable nonfarm income. When the incidence model was estimated only for taxpayers reporting over $100,000 in nonfarm income, these positive associations were considerably more pronounced. Two separate equations were estimated in which LOSSES was the dependent variable; the first for taxpayers with nonfarm income less than $100,000 and the second for those with nonfarm income exceeding $100,000. The first model showed that a unit rise in the marginal tax rate led to an increase in farm losses of between $500 and $600, while the second showed that a unit rise in the marginal tax rate lead to an increase in farm losses ranging from $4,700 and $5,700. Long (1990) concluded that investing in farming to avoid taxes increases with income, but that this practice was likely to be discouraged by the reductions in federal tax rates introduced in the 1986 TRA. An important implication of Long's (1990, 11) study for farm structure is that ". . . tax-loss farming may create some troubling side effects, such as biding up the prices of farmland and other resources and subjecting genuine farm operators to competition with investors who consider farm profit in the economic sense as unnecessary for their purposes".

LeBlanc and Hrubovcak (1986) developed a dynamic optimization model to examine the impact of tax policies on aggregate investment in agriculture. In their analysis, agricultural capital was separated into four asset groups: (1) short-lived equipment (tractors and trucks); (2) long-lived equipment (plows, planters, and harvesters); (3) structures; and (4) land. Aggregate time series data for the period 1955-1978 were used to estimate a model consisting of four investment equations, one for each asset category. The estimated model was used to assess the impact of investment tax credits, interest deductibility, accelerated depreciation, and the complete set of tax changes adopted since 1954 on net investment and optimal capital stock for each of the four asset groups. The effects of tax policies were incorporated via the rental rates on capital. Simulations indicated that accelerated depreciation had almost no impact on net investment. In contrast, over the 1962-1978 period, the investment tax credit accounted for over $3 billion or 12 percent of total net investment in agricultural equipment, and for $500 million or 5 percent of total net investment in structures. The interest deduction was also found to have important effects, accounting for investment of over $1.3 billion in equipment, $701 million in structures and $3.5 billion in land. The results of a simulation incorporating all the tax changes that occurred between 1954 and 1978 provided additional evidence that tax policy (in particular the investment credit) is of major importance in shaping the structure of the capital stock in agriculture. The greatest effect of the provisions included in the latter simulation is on short-lived equipment. The authors concluded that about 20 percent of net investment in agricultural equipment can be attributed to tax policy during the period analyzed. This translates into an excess of $5 billion, in real dollars, for equipment and over $1 billion for structures. The authors conclude that tax policy, by encouraging investment in agriculture, counteracts farm policy efforts designed to curtail output. It is worth noting that LeBlanc and Hrubovcak's (1986) findings with respect to net investment are consistent with those obtained in several studies of replacement investment (see, for example, Reid and Bradford 1983; Bates, Rayner, and Custance 1979; Chisholm 1974).

Conclusions

This chapter has emphasized that the structure of U.S. agriculture, as it is reflected by farm size and the mix of inputs used by farm enterprises, is influenced by federal tax policy in important ways. The most significant consequences of changes in tax policy probably result from their effects on relative input prices. Kislev and Peterson (1982) suggest that more than 90 percent of the increase in farm size that has occurred during the twentieth century is the result of changes in relative opportunity input costs. Tax policy initiatives, such as, changes in marginal tax rates, investment tax credits, and accelerated depreciation allowances, alter relative input prices and therefore change optimal input mixes.

Among others, Leblanc and Hrubovcak (1986) have shown that, at least in the case of capital inputs, some of these direct effects have the potential to be quite large. In particular, adjustments in marginal tax rates and changes in investment tax credit provisions appear to have had substantial effects on investment. On the other hand, modifications of the provisions for accelerated depreciation allowances appear to be relatively unimportant. However, these types of changes in the federal tax structure result in economy wide responses that also have indirect implications for agriculture. For example, the introduction of large investment tax credits stimulates the demand for capital goods in all sectors of the economy, increasing the demand for loanable funds and *ceteris paribus* raising real interest rates. In addition, the economy wide increase in the demand for investment goods is likely to raise asset prices. These types of general equilibrium or second round effects mitigate the initial or first round consequences of tax policy changes. Most empirical analyses of the impacts of tax policy on agricultural input use and structure have not taken account of such feed back effects.

Tax policy has also been shown to have some subtle implications for the ownership and control of productive assets such as land. In inflationary economies, individuals with higher marginal tax rates have relatively large incentives to acquire access to investments such as land that provide income streams that increase in nominal terms over time. Thus, if the tax structure becomes more progressive, individuals with high incomes (and high marginal tax rates) are more likely to acquire land. This suggests that under a progressive tax structure owners of large farms are more likely to acquire land than new entrants even if the new entrants have equivalent levels of human capital. The reason is that at any given market price for the land, its after-tax opportunity cost to the large scale (high income) farmer is lower than for the small scale (low income) farmer. Moderation of the progressivity of marginal tax rates as a result of the 1986 Tax Reform Act may therefore have reduced tax incentives for mergers of farm enterprises resulting from differences in income levels.

Several other aspects of the 1986 TRA also appear to have implications for farm size. The reduction in the progressivity of the structure of marginal tax rates has, according to Long (1990), reduced incentives for the ownership of agricultural assets as tax shelters. On the other hand, the TRA has resulted in more progressivity in the structure of the average tax rate in the farm sector because of the elimination of preferential treatment for capital gains and the abolition of the investment tax credit (Rossi and Durst 1989). Thus it is not clear whether small farmers have been made more or less competitive by the 1986 Tax Recovery Act.

Finally, one other important issue with respect to tax policy deserves serious consideration. It is clear that if changes in tax policy are permanent, they lead to changes in relative input prices and that these changes have important ramifications for farm size and structure. However, over the past 40 years federal tax policy has been extremely volatile. For example, since 1980 there have been three major revisions in the tax treatment of depreciation allowances and marginal income tax rate schedules have also been adjusted on at least five occasions. Moreover, "[S]ince 1950, the investment tax credit has appeared, been expanded and disappeared so frequently that now it almost seems to exhibit the properties of the Cheshire cat—once observed it immediately begins to fade away" (Smith 1990). In this type of unpredictable environment, tax policy adjustments may have more implications for the timing of capital acquisitions and replacement decisions than for levels and mixes of input use. The role of uncertain tax policy requires special attention in future studies of U.S. agricultural structure.

Notes

1. The Fisher effect assumes that, given continuous compounding, the nominal interest rate is equal to the sum of the real interest rate and inflation rate. As is demonstrated below, under discrete compounding the nominal interest rate includes an additional interaction term.

References

Bates, J. M., A. L. Rayner, and P. K. Custance. 1979. "Inflation and Farm Tractor Replacement in the U.S.: A Simulation Model." *American Journal of Agricultural Economics* 61: 355-358.

Batte, M. T. 1985. "An Evaluation of the 1981 and 1982 Federal Income Tax Laws: Implications for Farm Size Structure." *North Central Journal of Agricultural Economics* 7: 9-19.

Batte, M. T., and S. T. Sonka. 1985. "Before and After-Tax Size Economies: An Example for Cash Grain Production in Illinois." *American Journal of Agricultural Economics* 67: 600-608.

Bernanke, B. S. 1983. "The Determinants of Investment: Another Look." *American Economic Review, Papers and Proceedings* 73: 71-75.

Castle, E. N. 1989. "Is Farming a Constant Cost Industry." *American Journal of Agricultural Economics* 71: 574-582.

Chambers, R. G., and R. E. Just. 1982. "An Investigation of the Effects of Monetary Factors on U.S. Agriculture." *Journal of Monetary Economics* 9: 235-247.

Chisholm, A. H. 1974. "Effects of Tax Depreciation Policy and Investment Incentives on Optional Replacement Decisions." *American Journal of Agricultural Economics* 58: 776-783.

Darby, M. R. 1975. "The Financial Effects of Monetary Policy on Interest Rates." *Economic Inquiry* 13: 266-276.

Dean, W., and H. O. Carter. 1962. "Some Effects of Income Taxes on Large-Scale Agriculture." *Journal of Farm Economics* 44: 754-768.

Feldstein, M., and L. Summers. 1979. "Inflation and the Taxation of Capital Income in the Corporate Sector." *National Tax Journal* 32: 445-470.

Gandolfi, A. R. 1982. "Inflation, Taxation and Interest Rates." *The Journal of Finance* 37: 797-807.

Grisley, W., T. G. Fox, L. Jenkins, and I. Hoover. 1989. "Dairy Farm Size and Federal Income Tax Progressivity." *Agricultural Finance Review* 49: 46-56.

Hall, B., and E. P. LeVeen. 1978. "Farm Size and Economic Efficiency: A Case of California." *American Journal of Agricultural Economics* 60: 589-600.

Hanson, G. D., and V. R. Eidman. 1985. "Agricultural Income Tax Expenditures - A Microeconomic Analysis." *American Journal of Agricultural Economics* 67: 270-278.

_____ . 1986. "Evidence of the Stability of Income Tax Expenditures to Farmers." *Agricultural Finance Review* 46: 69-83.

Kislev, Y., and W. Peterson. 1982. "Prices, Technology, and Farm Size." *Journal of Political Economy* 90: 578-595.

LeBlanc, M., and J. Hrubovcak. 1986. "The Effects of Tax Policy on Aggregate Agricultural Investments." *American Journal of Agricultural Economics* 68: 767-777.

Long, J. E. 1990. "Farming the Tax Code: The Impact of High Marginal Tax Rates on Agricultural Tax Shelters." *American Journal of Agricultural Economics* 72: 1-12.

Rausser, G. C., J. A. Chalfant, H. A. Love, and K. G. Stamoulis. 1986. "Macroeconomic Linkages, Taxes and Subsidies in the U.S. Agricultural Sector." *American Journal of Agricultural Economics* 68: 854-860.

Reid, D. W., and G. L. Bradford. 1983. "On Optimal Equipment Replacement of Farm Tractors." *American Journal of Agricultural Economics* 65: 326-331.

Rossi, C., and R. Durst. 1989. "Estimates of Farm Tax Progressivity After Tax Reform." *Agricultural Finance Review* 49: 82-91.

Samuelson, P. A. 1964. "Tax Deductibility of Economic Depreciation to Insure Invariant Valuations." *Journal of Political Economy* 72: 604-606.

Smith, V. H. 1990. "The Effects of Changes in the Tax Structure on Agricultural Asset Replacement." *Southern Journal of Agricultural Economics* 22: 113-121.

Tanzi, V. 1980. "Inflationary Expectations, Economic Activity, Taxes, and Interest Rates." *American Economic Review* 70: 12-21.

15

Structural Implications of Agricultural Finance

Cole R. Gustafson and Peter J. Barry

A major reason for studying farm structure, as outlined in Chapter 2, is to understand how and why the agricultural sector is changing and what such changes may mean for the future. Agricultural finance is one factor that is often linked with changes in farm structure. Past practices of farm lending, which include liberal lending in favorable times and capital rationing in less favorable times, have strongly affected the size, profitability, and welfare of family farms. However, this linkage does not necessarily imply causation. Rather, financial markets in agriculture can be viewed as accommodating in the sense that they only indirectly respond to underlying economic and political shifts in the sector. That is, financial markets provide the financial capital needed to achieve structural changes in response to economic incentives created by changes in underlying supply and demand conditions, and/or by changes in agricultural policy. For example, public credit programs frequently facilitate structural change by reducing costs of transition.

In addition to debt financing, agricultural finance also encompasses farmers' use of equity capital and financial leasing of farmland and other types of assets.[1] Taken together, debt, equity, and leasing obligations comprise the range of financial claims on the assets a farm business controls. Relative changes in accumulated retained earnings, capital gains, and the interests of outside investors have differentiated farms in terms of capital mix, profit potential, management form, and risk-bearing capacity. The tenure and ownership structure of agriculture have also been altered by a variety of operating and financial leases.

This chapter describes and explains recent and historical changes in agricultural finance markets and institutions, and identifies the structural effects of these changes on the farm sector. The first two sections discuss the role of finance in U.S. agriculture and review the concepts of financial leverage, liquidity and control. Farmers' historical use of debt and equity capital then is described, differentiating between financing for growth and financing for survival. Following sections of the chapter describe the financial institutions and public credit programs for agriculture. In both cases, structural implications for the agricultural sector are explicitly considered. Finally, future structural implications of agricultural finance are discussed.

Role of Finance in Agriculture

Capital assets in U.S. agriculture are primarily financed by debt and equity, although considerable leasing also occurs. Major sources of credit include the cooperative Farm Credit System (FCS), commercial banks, life insurance companies, merchants and dealers, and government agencies, including the Farmers Home Administration (FmHA) and Commodity Credit Corporation (CCC). Individuals, primarily sellers of farmland, also offer considerable credit through contracts for deed and installment sales. Equity capital is derived mainly through retained earnings and unrealized capital gains, although specialized segments of the industry can attract outside equity capital.

Unlike most corporate nonfarm businesses where ownership, management, and labor functions are separated, individual farmers and immediate family members typically own and operate agricultural production units. Although this structure is most prominent, the corporate form of organization is being adopted in an increasing number of agricultural enterprises and commodities (e.g., broiler operations, cattle feeding, citrus and vineyards, and certain types of hog production) where agriculture is becoming more industrialized and vertically integrated. In addition, a small but growing number of family-oriented farm businesses have incorporated to facilitate estate transfers and tax management.[2]

Given these organizational structures at the production level, financial markets have several key functions. To establish viable economic units, farmers require significant levels of capital for the purchase (or leasing) of land, buildings, machinery and livestock. Thus, one function of agricultural financial markets is to aid beginning farmers in resource acquisition. Financial markets also provide capital to farms that require additional financing for operating purposes (seed, feeder animals, etc.) because of lengthy and seasonal production processes. When new investment is necessary, financial markets facilitate the replacement of worn-out assets, permit changes in business organization as economic conditions warrant, and allow growth and modernization when profitable opportunities arise. Similarly, financial markets aid disinvestment by those farmers who seek to exit the sector, retire, and/or transfer assets to succeeding generations.

Perhaps just as important, though, financial markets provide credit, which is defined as a farmer's unused borrowing capacity (Barry, Baker, and Sanint 1981). Well-functioning financial markets allow farmers to convert credit into cash with minimal transaction costs whenever liquidity needs arise. Agricultural prices and production are highly volatile, leading to fluctuating incomes, cash flows, and liquidity needs. Credit is important to farmers because it allows them to make optimal decisions, even to the point of selecting business plans that leave them with relatively illiquid assets. If unforeseen liquidity needs do occur, farmers are then able to convert their credit to loanable funds at minimum cost. If credit were not available, farmers would have to either maintain larger reserves of financial assets, forego more profitable investment alternatives, or incur the transaction costs associated with untimely liquidation of assets.

Although financial markets in agriculture function quite well, they are not perfect. Lenders do not always provide sufficient capital to meet farmers' legitimate credit needs. Credit availability typically declines when financial institutions are regulated more closely, have limits placed on their lending or funding activities, or when the agricultural sector's financial health temporarily declines.

Farmers' need for debt and equity capital is a derived demand based upon expected product prices, input costs, rates of productivity, management capabilities and tolerances for risk. The quantity, mix, and cost of financial capital employed affect the organization, size, and system of production each farm unit chooses. The aggregate results of these choices define farm structure. For example, farmers with limited capital often lease rather than own assets, thus influencing the tenure structure of their farm businesses. Such choices also depend on supplies of debt and equity capital. Financing alternatives differ considerably in terms of maturity, down payment and collateral requirements, repayment patterns, maximum loan size, and reporting and monitoring demands. When the provisions of a specific method of financing or the range of financing alternatives change, the structure of agriculture often changes.

Key Financial Concepts

To more thoroughly understand the impact of finance on farmers' decisions, the key concepts of financial leverage, liquidity, and control are reviewed in this section.

Farm operators are assumed to maximize the expected rate of return on equity capital:

$$r_e = (r_a P_a - i P_d) \qquad (1)$$

where r_e is the rate of return to equity, r_a is the rate of return on assets, P_a is the ratio of assets to equity capital employed by the farm business, i is the interest rate on debt capital, and P_d is the firm's leverage ratio, debt/equity. This simplistic analysis does not acknowledge the effects of income taxation, family consumption or risk on equity growth (Barry, Hopkin, and Baker 1988, Chapter 6).

By substituting $P_a = 1 - P_d$ into (1) and rearranging, the direct impact of leverage on equity growth is expressed as:

$$r_e = P_d(r_a - i) + r_a \qquad (2)$$

Increasing leverage (a higher P_d) magnifies equity growth as long as marginal asset returns exceed borrowing costs ($r_a - i > 0$) when additional debt capital is used. Therefore, in the absence of risk, increased financial leverage appears attractive because it accelerates the expected rate of growth of the firm's equity capital.

However, financial risk from additional leverage constrains most farmers' use of debt capital. Higher leverage leads to greater financial risk because interest payments on debt represent a fixed obligation and lead to increased variability of equity returns. The increases in financial risk associated with higher leverage can be illustrated using the mean-variance (EV) approach (Robison and Barry 1987, Chapter 16). In this approach, risk is quantified in the context of a portfolio model where leverage implies negative holdings of a risk-free or risky asset. Farm managers are assumed to maximize expected utility, which in turn is a function of the desired trade-off in utility terms (λ) between the expected level (r_e) and variance of equity returns (σ_{re}^2):

$$E[U(r_e)] = f(\lambda, r_e, \sigma_{re}^2) \qquad (3)$$

Since i is not random, the variability of equity returns can be derived from equation (1) as follows:

$$\sigma_{re} = \sigma_{ra} P_a \qquad (4)$$

Even if interest rates on debt are fixed, leverage heightens the variability of equity returns because lower proportions of equity in a firm's capital structure increase the value of P_a, leading to linear increases in equity risk. Thus, leverage magnifies the variability of asset returns, r_a.

For most farmers though, interest rates are not fixed. On variable rate loans, i is a stochastic variable because interest rates on debt capital vary with market conditions. When both asset returns and interest rates vary, equity return variability can be expressed as:

$$\sigma_{re}^2 = \sigma_{ra}^2 P_a^2 + \sigma_i^2 P_d^2 - 2P_a P_d \sigma_{ra,i} \tag{5}$$

As expected, interest rate variability ($\sigma_i^2 P_d^2$) contributes directly to equity return variability. However, the appealing aspect of this portfolio formulation is the explicit recognition of the correlation ($\sigma_{ra,i}$) between movements in asset returns and interest rates. In the past decade, these two variables have been negatively correlated, compounding equity return variability (Leatham and Baker 1988). If they were positively correlated, rising interest costs would be offset by increased asset returns and vice versa, leading to greater stability in equity returns. Of course, farmers with fixed rate loans are immune from these sources of risk.

A firm's choice of leverage directly affects its liquidity position and ability to meet financial obligations when unfavorable events occur. When borrowing increases, liquidity is reduced because interest payments command a larger proportion of residual cash flow. Moreover, interest payments are fixed obligations, even though variable rate loans may be used. Since asset returns are variable, sufficient funds may not be available for debt service, leading to possible default and bankruptcy of the farm operation.

Although a firm's optimal leverage position is a function of the variables described above, the choice is ultimately based on individual preference because it depends critically on each investor's risk attitude. (The chapter by Robison discusses risk and risk preference in a more general setting.) Since higher leverage frequently results in greater variation of equity returns and increased returns to equity, highly risk-averse individuals tend to prefer positions of lower leverage. In doing so, they derive greater expected utility from the lower probability of equity losses (and possible bankruptcy) than they would from marginal increases in equity returns.

In addition to altering the level and variability of equity returns, financial leverage also affects the degree of control individual farm operators have over management decisions that influence farm performance. Increased debt financing and loss of control are often highly correlated. Even though lenders cannot breach their fiduciary responsibility and directly control a farming unit, most farmers seriously consider lender's suggestions as debt financing increases. In the extreme case of bankruptcy when a firm's debt/asset ratio exceeds 1.0, farmers ultimately convey their control to court-appointed trustees. Moreover, business control can be shared in the case of leases as well, especially share leasing of farm real estate.

In general, the preferences of lenders and the implications for the cost (i) and availability (P_d) of credit play an important role in farm financial planning (Baker 1968). More conservative lenders tend to ration credit more strongly through higher interest rates and/or lower loan limits, greater collateral requirements, more extensive loan documentation, and other nonprice responses. Similarly, differences in lending preferences (i.e., loan safety and repayment for lenders versus business profitability and liquidity for borrowers) and risk attitudes between lenders and farm borrowers may influence the cost and availability of financial capital and, thus, the direction and rate of farm firm growth. Lenders typically prefer to finance loan programs that are self-liquidating and asset generating, because such purposes have stronger repayment expectations and greater loan security (Barry, Hopkin, and Baker 1988). Similarly, effective risk management by farmers may generate additional credit at lower interest rates, thus building liquidity and providing for more rapid firm growth and structural change (Barry and Willmann 1976; Pflueger and Barry 1986). In a dynamic sense, lenders respond to a farmer's choice of lender, sequence and source of borrowing and repayment, financing instrument, asset structure, enterprise mix, and tenure position (Barry and Baker 1984). Finally, the growing use of risk-adjusted interest rates by agricultural lenders in the 1980s, often based on formal credit scoring and risk assessment models, has shifted the balance of lender responses away from nonprice methods toward price responses, thus providing clearer signals to borrowers about factors affecting the cost of borrowing (Ellinger, Barry, and Mazzocco 1990).

Financing Agriculture

In theory, financial capital is allocated to sectors of the economy according to rates of return that are commensurate with potential investment risks. Although the real value of the agricultural sector's assets has increased less than 10 percent from 1945 to 1988 (Table 15.1), the volatility of asset values during intervening years has greatly affected the expectations and willingness of farmers and outside investors to provide financial capital. Increasing demand for agricultural products, rising inflationary expectations, and favorable taxation policies resulted

TABLE 15.1 Farm Financial Trends, 1945-88

Year	Farm Assets	Farm Debt	Farm Equity	Net Farm Income	Off-Farm Income	Cash to Total Expenses	Equity to Assets
	Billion Dollars, 1982=100					Percent	
1945	649.7	48.4	601.3	78.3	n/a	87.3	92.6
1950	639.7	51.5	588.7	56.9	n/a	84.5	92.0
1955	634.2	62.5	572.1	41.5	n/a	82.1	90.2
1960	682.5	80.3	602.3	36.2	27.5	83.0	88.2
1965	770.1	116.3	653.8	38.2	37.6	83.9	84.9
1970	773.3	125.7	647.6	34.3	41.9	83.7	83.7
1975	970.5	154.3	816.4	43.0	40.3	82.9	84.1
1980	1,285.4	208.5	1,076.9	18.8	40.5	83.4	83.8
1985	759.8	169.4	590.4	29.2	38.4	84.1	77.7
1988	709.6	122.4	587.1	37.7	42.6	86.4	82.7

Source: U.S. Department of Agriculture.

in increasing capital flows and steadily rising farm asset values prior to 1980. When each of these trends reversed in early 1980, asset values and capital demands rapidly declined.

Many structural differences among farms, including size, asset and liability composition, enterprise combinations and off-farm employment levels, can be traced, in part, to individual farmers' decisions regarding their mix of capital financing. This section describes the trends and diversity in farm financing methods.

Equity Financing

Historically, the vast majority of agricultural assets has been financed with equity capital, which represents the ownership interests of farm operators and outside investors. In 1988, 83 percent of all farm assets in the U.S. were equity financed (Table 15.1). The distribution of equity financing across different farm sizes varies (Figure 15.1). While the majority of small farms (sales less than $20,000) rely entirely on equity capital, equity constitutes a smaller share of asset financing as farm size increases. Moreover, equity financing varies by farm type. Cotton, livestock, nursery, tobacco and vegetable farms rely more on equity financing because production is more variable (Morehart et al. 1989).

In agriculture, returns to equity arise from both current earnings (i.e., residual income from farm production) and nominal capital gains. The latter is comprised of both the effects of inflation and real capital gains (i.e., capital asset appreciation net of inflation). The contribution of both current earnings and real capital gains to total equity returns has varied considerably over the past 40 years (Figure 15.2).

Except for the early 1980s, current equity returns for the sector as a whole have been positive, similar to the profit experience of other domestic industries (U.S. Council of Economic Advisors). This generalization, however, masks variation across regions, enterprises, and individual farming situations. Periods of low farm income have a mixed impact on current agricultural equity returns. Most U.S. farms are predominately organized as sole proprietorships. Consequently, the distribution of residual income between labor earnings and capital returns is frequently an arbitrary, personal decision based on life-cycle and life-style choices of the operator.

The variability of real capital gains has been far greater than the variability of current rates of equity return, often by a factor of five (Figure 15.2). When current rates of return declined from 1973 to 1980, real capital gains rose above 10 percent and provided a collateral base against which farmers could borrow for family living and expansion. Thereafter, real capital gains turned to real capital losses and farmers' equity positions quickly eroded. Although many factors influence real

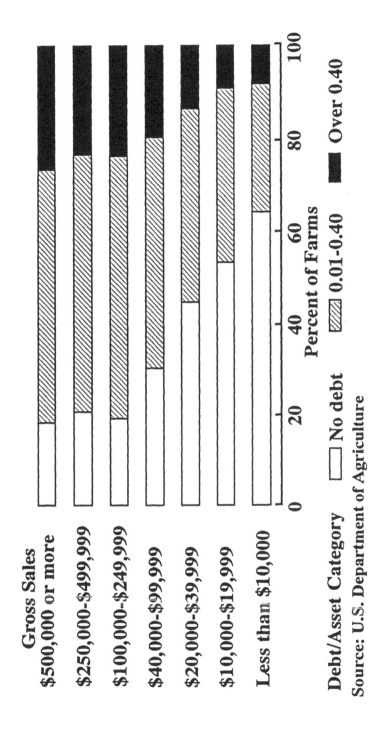

FIGURE 15.1   Distribution of Farms by Debt/Asset Ratio and Gross Sales, 1988

FIGURE 15.2 Composition of Equity Returns

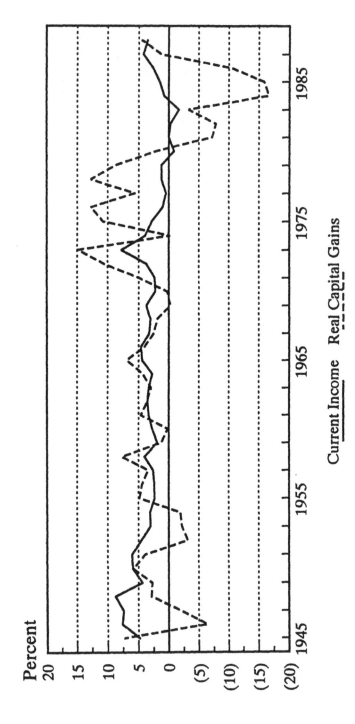

Source: U.S. Department of Agriculture

capital gain levels, they are primarily determined by farmers' expectations about future returns to assets.

In addition to expectations of current returns and real capital gains, several factors limit the extent of owner-equity financing in agriculture. First, larger farm sizes make it difficult for individual, beginning farmers to amass sufficient equity capital for initial investment. Even if such quantities of equity capital were available, prudent farmers could diversify and reduce unsystematic portfolio risks by investing a portion of their funds in nonagricultural assets (Barry 1980; Kaplan 1985; Young and Barry 1987).

Nonfarm equity funds are an alternative source of equity capital. Certain farming operations, primarily cattle feeding, citrus groves, tree nut and vineyards, have been able to attract nonfarm equity capital for speculative and tax-shelter opportunities. Scofield (1972) and Moore (1979) estimated that 15 to 20 percent of the capital in these industries in the 1970s was from nonfarm sources. Rising costs of debt financing and the difficulties of financially distressed operators heightened farmer interest in nonfarm equity financing during the 1980s. Barry (1984) states nonfarm equity capital could be an important method of financing new agricultural technologies.

However, several unique characteristics of the agricultural production sector limit the potential for nonfarm equity capital (Lowenberg-DeBoer, Featherstone, and Leatham 1989). Historically, the ownership and management functions of farming have been difficult to separate. Nonfarm investors are confronted with significant search costs, have difficulty monitoring farm manager's actions (principal agent problem), and transferring their claims to other equity holders. In addition, many states have enacted legislation discouraging corporate forms of farm ownership.

Nonfarm equity capital in agriculture has increased in several ways though. One example involves the use of shared appreciation mortgages whereby lenders holding delinquent farm loans reduce outstanding principal in exchange for a portion of future capital gains earned by the encumbered assets. In addition, nonfarm investors own substantial amounts of farmland available to farm operators through lease arrangements, thus providing another source of outside capital to agriculture.

Debt Financing

Although equity financing dominates, debt capital is a critical element in most agricultural investments. In terms of the distribution of debt among farmers, the 1988 Agricultural Economics and Land Ownership

Survey (Bureau of the Census) indicated that 52.8 percent of U.S. farmers were utilizing debt capital at the time of the survey. Among states, the percentages of farmers using debt ranged from 34.9 percent in West Virginia to 74.6 percent in North Dakota. Moreover, the debt was concentrated in larger farming operations. For example, 86.0 percent of the farms with a value of agricultural products sold exceeding $500,000 utilized debt capital while fewer than 40 percent of the farms with less than $5,000 of agricultural products sold utilized debt. Among tenure structures, 69.4 percent of part owners utilized debt, while only 45.7 percent of full owners and 48.2 percent of tenants utilized debt.

In 1988, 17 percent of agricultural assets were financed with debt capital (Table 15.1). On a percentage basis, growth in debt financing exceeded relative increases in equity capital before 1980. This increase occurred because of farmers' conservative use of debt following the Great Depression and because of the heavy financing requirements needed to facilitate the substantial increases in farm size and reductions in farm numbers that occurred from post World War II until the 1970s.

High real net farm incomes after WW II facilitated the return of many servicemen to agriculture. They could rejoin the family farm and earn a reasonable debt-free income without either diminishing other family members earnings or significantly expanding farm sizes. In 1945, less than 7.5 percent of the agricultural sector's assets were financed with debt. Leverage increased modestly from 1945-60 as debt fears eased, in spite of declining real asset values.

When real asset values began appreciating in the 1960s and 1970s, farmers became even more interested in financing for growth purposes. Rising inflation rates lowered real costs of borrowing, increased the demand for land as an investment hedge against rising prices, and depreciated the value of the U.S. dollar--which in turn led to increased agricultural exports and stronger agricultural commodity prices. These favorable events lowered the financial risks of farm borrowing and significantly increased the equity of farmers who chose to leverage. Moreover, lenders for the most part were willing to accommodate the growing demands for farm debt, and expanded lending limits for Federal Land Banks authorized by the 1971 Farm Credit Act contributed significantly to the rapid growth in farm real estate debt.

Other phenomena also motivated farmers' desire for growth and need for financing, especially operators of larger-sized farms. The availability of confinement livestock production systems, larger farm machinery, and operators with greater management skills meant that farm sizes had to grow to achieve size economies and maintain income-generating capacity. Capital growth and changes in debt structure facilitated many of these structural adjustments (Baker 1977, 1968; Baker and Irwin 1961). Debt financing was a low-cost way to expand and change business organization.

However, rapid growth in debt financing became a burden to many farmers in the 1980s when the Federal Reserve System allowed interest rates to rise and vary more in an effort to control inflation. Declining real net farm incomes compounded farmers' debt servicing problems (Table 15.1). In aggregate real terms, farmers were servicing three times their 1945 debt level with half as much income. The sector remained highly solvent in aggregate terms, but many individual farmers were financially distressed. Although farm debt was concentrated among the nation's largest farms (Figure 15.1), mid-size, commercial-scale farms experienced the greatest financial problems (Jolly et al. 1989; Morehart, Johnson, and Banker 1989). Very large farms were better able to cope with high leverage due to the benefits of economies of size and perhaps greater skills in finance and risk management, and small farms had nonfarm income to fall back on.

Increased farm mechanization, less reliance on family sources of labor, and rising use of purchased inputs are generally believed to have increased the cash requirements for farming and reduced the debt-servicing ability of farmers. But as Table 15.1 illustrates, the proportion of cash to total production expenses remained relatively stable since 1945. In fact, the interest portion of cash expenses rose from 3.4 to 13.4 percent in 1988–implying the share of other purchased inputs actually declined. However, this additional cash flow available for debt service was not enough to mitigate financial pressures in the 1980s, and farmers had to revise their growth oriented strategies.

Instead of financing for growth, farmers of the 1980s sought financing to survive. Financially stressed farmers responded to the new environment of greater business risk (lower farm profits and asset values) and higher financial risk (higher, more volatile interest rates and higher leverage) by living more frugally, reducing farm expenses, lowering capital expenditures, liquidating capital assets, and seeking off-farm employment. Even with these intra-firm adjustments, many farms still had difficulty servicing debts.

As a consequence, delinquent borrowers petitioned public and private lenders for assistance and special consideration. When financing for survival, debtors requested and obtained interest rate reductions, lender forbearance, debt deferrals, loan restructuring, a specialized bankruptcy chapter (Chapter 12), mediation of their grievances, and other programs of education and financial assistance to aid their plight. Many of these lending practices reflected the effects of new credit programs at federal and state levels. With these adjustments and farmer's renewed interest in debt conservation, aggregate real farm debt was reduced over 41 percent from peak 1980 levels. These financial constraints directly affect the size and growth potential of U.S. farms.

Private Financial Markets in Agriculture

The farm commodity program, special provisions of the U.S. tax code, and a separate farm bankruptcy chapter are all evidence of society's preference for maintaining Jefferson's agrarian model of pluralistic family farms. This preference even transcends the regulation of private financial markets through establishment of the FCS, geographical bounds on financial institutions, and creation of FmHA programs that provide direct loan programs to farmers unable to qualify for financing from commercial lenders and guarantee marginal farm loans originated by private lenders (Lins and Barry 1980). These programs ensure that adequate supplies of reasonably priced debt capital are available to farm borrowers with a host of differing structural characteristics.

However, this multi-source, decentralized financial market structure with some institutions specializing in financing agriculture has been costly to maintain, both in terms of public and private resources. Farmer defaults, interest rate concessions and debt write-offs of the 1980s were a significant hardship for agricultural lenders. The public cost of these activities is typified by the losses experienced by FmHA and the significant financial assistance provided to FCS under the Agricultural Credit Act of 1987.

But these are only part of the many challenges in the current operating environment of lenders. Increased competition from nondepository financial institutions, broader geographic bounds on financial markets, greater interest rate volatility, rapid development of electronic funds transfer, and weakened support for public credit programs have greatly affected the decision environment of traditional agricultural lenders. These challenges have impacted on their loan volumes and ability to supply agricultural credit (Figures 15.3 and 15.4). While market shares of the major agricultural lenders exhibited stability prior to 1970, competitive opportunities motivated some lenders to aggressively increase loan volumes in the 1970s while others remained cautious. The aggressive lenders experienced significant financial losses and lost most of their gains when conditions reversed in the 1980s.

Agricultural lending is a dynamic activity. As indicated earlier in the chapter, financial market changes are signaled to farmers through lender credit policies that involve loan pricing decisions, collateral requirements, minimum and maximum loan size restrictions, financial eligibility criteria, and forbearance procedures. Borrowers respond to these credit factors

FIGURE 15.3   Real Estate Loan Volume by Lender

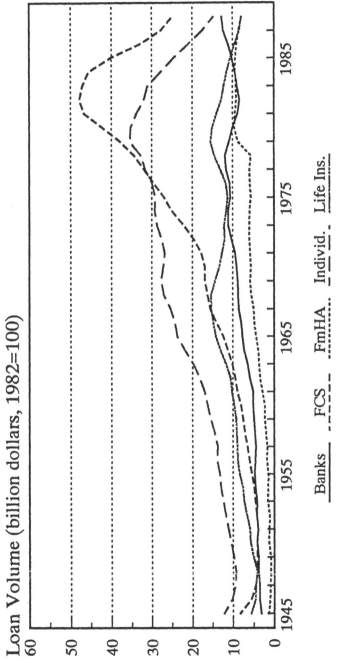

Loan Volume (billion dollars, 1982=100)

Banks ___   FCS _._._   FmHA ....   Individ. _ _ _   Life Ins. ....

Source: U.S. Department of Agriculture

FIGURE 15.4   Non-Real Estate Loan Volume by Lender

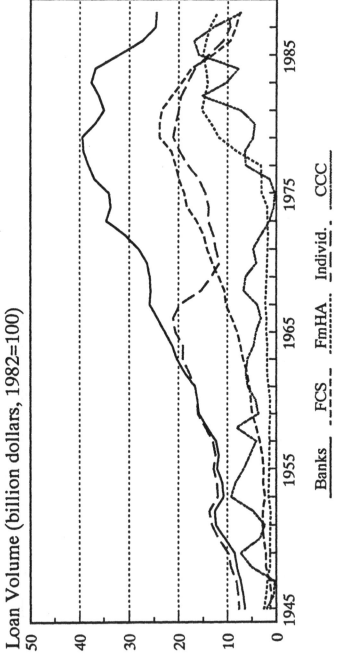

Loan Volume (billion dollars, 1982=100)

Banks   FCS   FmHA   Individ.   CCC

Source: U.S. Department of Agriculture

and reorganize their firms accordingly, leading to structural change in the sector (Baker and Irwin 1961). The diversity of financial institutions serving agriculture, and their unique methods of responding to financial market changes, enable farms with differing resources to exist. If farm credit was administered by a single lender, farm structures would be far more uniform.

Commercial Banks

In 1989, commercial banks once again achieved the largest share of agricultural loans, through increases in both real estate and non-real estate lending activities. Much of the agricultural lending by banks is provided by small and intermediate-sized rural banks that rely on local deposits for loanable funds. At the discretion of state authorities, banks may be permitted to (1) branch statewide, (2) branch on a limited basis, or (3) maintain unit operations. Banks in many states have been able to circumvent branching restrictions through formation of bank holding companies. The development of holding companies has not only liberalized geographic restrictions within states, it also has enabled operations in adjacent states that permit reciprocal arrangements, regional banking markets, and even some arrangements that are national in scope. New banking legislation under consideration in the early 1990s may further liberalize the geographic scope of banking. Still, sizable numbers of unit banks remain. Due to limited capital and reserve requirements at federal and state levels, these banks have correspondent relationships with large money center banks that enable them to service the credit needs of larger farm borrowers and agribusinesses through loan participations.

After the 1920s banking crisis, commercial banks were highly regulated through usury laws, interest rate ceilings on major sources of funds (Regulation Q), limits on the range of products and services banks could offer (Glass-Steagall), prohibition of interstate activity, and rigorous examinations by regulatory authorities to ensure safety, soundness, and fair competition among themselves and with other financial institutions; maintenance of adequate capital reserves and liquidity; and responsiveness to the diverse credit needs of their service areas. These rigid regulations severely limited financial intermediation by banks in the 1970s. Financial innovations allowed some banks to adjust to or circumvent these regulations. Regulatory discretion also mitigated the severely unfavorable effects of some regulations. Consequently, regulations were liberalized with passage of the Depository Institutions Deregulation and Monetary Control Act of 1980 and the Garn-St. Germain Depository Institutions Act of 1982. These acts decontrolled interest rates, unified reserve requirements and aided the ailing thrift industry.

However, the deregulation pendulum swung too fast as many of the nation's thrifts adopted unsafe lending practices and eventually failed when optimistic projections did not materialize. Various types of commercial banks experienced higher failure rates as well, due in part to the financial problems of borrowers. In August 1989, the Financial Institutions Reform, Recovery and Enforcement Act restricted the risky activities of thrifts, strengthened procedures for liquidating insolvent thrifts, and imposed more stringent capital standards. Although the legislation applies directly to thrifts, similar policies are being adopted by other financial regulators.

Continued liberalization is subject to substantial debate. Nevertheless, market and technology pressures will force banks to become more competitive and cost-efficient, offer a greater variety of financial products and services, and diversify their portfolios across broader geographic regions. Agricultural banks are responding through adoption of modern spread and gap management techniques, organizing secondary markets for agricultural real estate loans, developing uniform methods of analyzing farmers' financial statements, evaluating borrowers' creditworthiness more thoroughly, improving loan documentation, and targeting services toward market segments offering greatest profit potential.

As these changes are implemented, farm borrowers will face greater differentiation. Farms with diverse enterprises, credit factors, geographic locations, etc., will no longer borrow at identical interest rates, receive equal quantities of credit, or be expected to repay loans in an identical manner. Instead, more competitively priced credit will flow to the sector's most creditworthy borrowers—greatly affecting the profitability, investment capabilities, and structure of the sector.

When small rural bankers place increasing levels of farm loans on secondary markets or wholesale them to large money center banks, farm borrowers will be subjected to national credit standards rather than local lending criteria. Instead of allocating capital within small trade areas, lending pools associated with broad regions of the country with unique types of agriculture may enable farmers in these areas to receive credit on more favorable terms than other areas. A requirement for secondary market loan pools is that they must be diversified geographically. Loans from farming areas that reduce portfolio risk will command premiums and lead to regional capital inflows.

Farm Credit System

Like commercial banks, the Farm Credit System (FCS) has responded to changing financial markets, significant agricultural loan losses, and additional regulatory oversight through internal reorganization, adoption of new management practices, and development of innovative products.

Three unique characteristics of the FCS, however, directly affect the structure of U.S. farms.

First, the FCS is a national system of federally chartered, cooperatively organized, privately owned banks and associations. The district banks benefit from joint and several liability when farm credit securities are issued in national financial markets. However, this benefit became a hindrance in the financial stress periods of 1980 when banks had to develop loss-share arrangements, without the power or inclinations to monitor or control the activities of peer institutions. The recently developed Farm Credit System Insurance Corporation may resolve some of these problems and, perhaps, even replace joint and several liability.

By operating together, Farm Credit banks and lending associations have the potential to realize significant size economies and are able to balance credit risks and loan losses geographically. The significant restructurings since passage of the 1987 Agricultural Credit Act suggest that further cost savings (i.e., size economies) are still possible. These features may assist the system in providing commercial-scale farmers with stable supplies of debt capital even when great volatility exists in both financial and agricultural markets. Access to dependable credit is critical to optimal farm management decisions. At present, however, FCS does not have the legislative authority to collect deposits, sell investment securities, or diversify its lending portfolio beyond agriculture. Although these constraints ensure that the system will remain a specialized agricultural lender, they limit the system's ability to adjust to new financial market and regulatory environments.

Second, FCS was initially capitalized with federal funds in 1916 to provide farmers with long-term farm loans at more favorable rates and terms than were available from other sources.[3] Although the government's original capital was soon repaid, FCS received an additional $4 billion line of financial assistance in 1987 under the Agricultural Credit Act. The assistance was in the form of debt capital, which must be repaid. FCS's actual use of the assistance has (so far) been less than anticipated. Hence, FCS remains a government-sponsored enterprise (GSE). FCS securities are traded in financial markets with agency status, where special regulatory exemptions and preferences are granted, and government support is implied if the institution experiences financial difficulties (as occurred in the 1980s). The agency status of FCS securities enables the system to raise funds below the costs of corporate bonds of comparable maturities.[4] As a cooperative, these benefits are passed on to member-borrowers as reduced borrowing costs (i.e., lower levels of interest rates on farm loans, net of the cooperative capital requirement), which in turn directly affects farm structure by lowering the marginal costs of land acquisition, machinery investment and the purchase of technologically improved inputs.

FCS's loan pricing mechanism is the third characteristic that affects farm structure. While other commercial lenders and FCS both determine interest rates by the average cost of their loan funds, the liabilities of other commercial lenders are typically of shorter maturity than those of FCS, allowing more rapid adjustment in average costs of funds and, thus, in loan rates. When interest rates were relatively stable prior to 1979, this practice created few market distortions. However, when interest rates began to increase, FCS's method of loan pricing insulated farm borrowers from rising capital costs and encouraged excessive use of debt. Hence, more capital was channeled to the agricultural sector than market fundamentals warranted. This situation was reversed when interest rates began to fall in the mid-1980s, leading to overpriced debt capital. As a result, many FCS districts are now experimenting with loan pricing based on the marginal cost of loan funds.

Individuals

Though often overlooked, private individuals also extend significant credit to the farm sector, primarily through seller financing of farm real estate. Before 1976, individuals held the majority of farm real estate loans. Unique characteristics of seller financing include informal negotiations, below market loan rates, low down payment requirements, flexible repayment terms, liberal forbearance, and tax advantages for sellers who spread receipts from sales over more than one year. Because many of these advantages are capitalized in farmland prices (Thompson and Kaiser 1985), seller financing has only a minimal structural impact on the rent versus farmland ownership decisions of farm operators. If seller financing reduced farmland ownership costs, farmers would be more inclined to own than rent, which in turn would affect a key variable describing the structure of agriculture--the proportion of tenant farmers. In addition, the significant availability of seller financing may have helped to foster a smaller scale of farm operations relative to other industries, facilitating the entry of new and/or young farmers into agriculture through ownership. However, little empirical evidence exists to document this conjecture.

A closely related means of finance is the "sweat equity" farm laborers earn, especially the children of family farmers. Cash farm wages are typically less than the economic value of labor contributed by farm workers. When purchasing the capital assets of an employer, workers are frequently able to monetize these noncash earnings. By agreeing to have their wages withheld, farm laborers are in effect financing farm operations.

Insurance Companies

From 1948 to 1967, insurance companies were the second largest farm real estate lender. Except for a slight rise in the late 1970s, their share has declined as other lenders have adopted more liberal and aggressive lending policies. With no legislative mandate or regulator requiring them to service the capital needs of marginal farmers, insurance companies are able to choose lending territories offering greatest profit potential and stability. By acquiring more lucrative and large farm loans, insurance companies sometimes jeopardize the financial health of competing lenders in a given trade area. Although more capital is available to farmers with greater creditworthiness, marginal borrowers experience greater difficulty obtaining credit because competing lenders are unable to balance their risks against loans of higher quality. Moreover, the capacity of competing lenders to offer innovative financial services to both creditworthy and marginal farmers is reduced because lending profits often subsidize other bank activities.

Trade Credit

Merchant and dealer financing options are becoming more diverse and important to farm operators. Trade credit has traditionally been used as a marketing technique to increase sales. In the 1980s, trade credit essentially became a requirement for input suppliers--particularly in areas experiencing significant financial stress. Trade credit was the only source of financing for many delinquent farmers when they exceeded established credit lines with traditional lenders. More recently, several farm-supply cooperatives have organized credit subsidiaries to extend operating credit to farmers under more favorable terms than other commercial sources.

Trade credit provides farmers with an additional source of capital but often limits their total credit capacity. Lenders generally prefer to avoid split financing because of the difficulties in establishing security claims and in evaluating cash management by farm borrowers over time.

Although trade credit generally has only minor structural impacts, it sometimes can affect farmers' decisions. Perhaps one of the largest structural impacts of trade credit occurred in the mid-1980s when several major farm equipment manufacturers merged. To liquidate excess inventories in a period of great financial stress, manufactures heavily subsidized interest rates on new equipment financing. Structural implications of this action include the possibility for increased farm sizes, debt, and productivity (to the extent this machinery embodied technical innovations).

Financial Operating and Real Estate Leases

Tenure is a major structural characteristic of U.S. farms. Chapter 18 reviews changing tenure patterns, focusing primarily on land rental and leasing issues. However, leasing of non-real estate assets also occurs through operating leases (asset rental on a monthly basis or less), financial leases (asset rental on an annual basis) and custom hire (joint rental of an asset and labor). Discontinuation of the investment tax credit and less favorable depreciation schedules have renewed farmer interest in leasing. However, these two tax changes have also reduced the attractiveness of rental terms that leasing firms can offer.

These forms of leasing affect farm financial structure in several ways. Leasing a previously owned asset that is considered strictly as off-balance sheet financing simultaneously reduces both assets and liabilities on the firm's balance sheet. The magnitude of each depends on the operator's financial position and terms of the leasing arrangement. Higher lease payments may exceed reductions in interest and principal payments, thereby leading to cash flow shortages, although lease payments are fully tax deductible.

Another structural effect of a farm's tenure position involves the effects of real estate leasing on a farm's profitability and debt carrying capacity. Ellinger and Barry (1987) show that a higher ratio of leased to owned land is associated with higher accounting rates of return to farm assets, higher financial leverage (i.e., higher debt-to-asset ratios), and greater liquidity. These relationships occur because of the nondepreciability of farmland and because part of the return to land occurs as unrealized capital gains. Thus, land investments exhibit relatively low debt carrying capacity and lower accounting rates of return compared with other types of assets (Barry and Robison 1986). As tenancy increases, higher leverage and higher accounting rates of return are anticipated because land is a lower proportion of total assets.

Public Financial Markets in Agriculture

Government lending provides capital to borrowers who do not meet commercial credit standards, but still have the potential to establish financially viable farming operations. Eligible borrowers generally must have inadequate resources or limited experience or have been adversely affected by physical, economic or social events, and have the potential to "graduate" to commercial financing within a reasonable period. Other government lending programs facilitate the storage and marketing of agricultural commodities.

Government lending programs affect the sector's structure through loan guarantees, subsidized interest rates, stabilization of farmers' incomes, and other more favorable financing terms than commercial lenders can provide. Consequently, participating farmers can acquire additional debt for farm expansion or reorganization. Public sector lenders consist of FmHA, CCC, and various state-sponsored credit programs, many of which assist young or beginning farmers.

The U.S. Department of Agriculture administers both FmHA and CCC programs. FmHA programs, which essentially began in the 1930s, involve both direct lending and guarantees of agricultural loans originated by other commercial lenders, although the emphasis in the 1990s is on guaranteed loans. FmHA also administers a variety of emergency loans to farmers who have experienced the effects of a natural disaster. A local office and committee of peer farmers supervise borrowers using FmHA direct loan funds, although recently high delinquency rates on this portfolio cast doubt on this supervision's effectiveness.

FmHA lending policies greatly affect farm structure because participation is limited to family sized farms that do not meet commercial credit standards. Large-scale corporate farms, commercially viable mid-size farms and small part-time hobby operations are ineligible. Participating farms benefit from credit access that was previously unavailable, loan interest rates priced at the government's cost of funds, and management guidance provided by the supervisory committee. Some of the benefits continue as FmHA shifts to loan guarantees, especially in terms of generating credit from lenders who otherwise would not be involved. A recent General Accounting Office review (1990) finds that the majority of FmHA loan funds are used for farm operating expenses (50 percent) and to refinance existing farm debts (30 percent). Only 6 percent of FmHA funds were used to purchase property.

Overall, FmHA programs have slowed the decline in farm numbers and the concentration of resources and control among larger sized farms. In addition, these programs have the effect of retaining unproductive excess human capital resources in agriculture (Lee and Gabriel 1980). Participating farmers are motivated to increase farm sizes and efficiency, but only to the point where they still qualify as a family sized farm. FmHA also may experience moral hazard problems when borrowers have incentives to take actions that avoid graduation to commercial credit (LaDue 1990). In fact, FmHA has been criticized for not graduating borrowers to other lenders when their farm operations become commercially viable (General Accounting Office 1988). The bimodal distribution of farm sizes would accelerate if the mid-sized farms FmHA serves ceased to exist.

The rising market share and loan volume of FmHA financing in the 1980s indicated that the reliance on federal credit programs to cope with financial problems in agriculture has been at a high level for a long time. A major concern was that such credit would not solve the agricultural problems existing at that time. Moreover, excessive public credit tends to exceed and erode the fundamental mission of "credit" as a means of financing commercial activity in agriculture. Under adversity, the liquidity adjustments permitted by the credit markets may involve debt restructuring, for example. But these markets cannot withstand chronic use of this response. Moreover, public credit programs are not considered to be effective means of transferring income or wealth among sectors, or in serving as policy instruments for structural change. Thus, the downsizing of FmHA loan programs and the shift toward loan guarantees in the early 1990s are consistent with the need to avoid excessive reliance on credit as a means of solving farm financial problems and as an instrument of structural change.

The CCC has traditionally used deficiency payments to enhance agricultural prices and incomes of U.S. farmers. Most recently though, a portion of these payments has been available to farmers before planting, thus reducing their need for commercial operating loans. As expected, interest-free capital can have sizable structural impacts if leverage increases. Other CCC programs involve storage loans and crop disaster payments (as a supplement to crop insurance), which influence farmers' storage, marketing, and inventory decisions and modestly affect structure by reducing the business risks of farming.

Numerous states have enacted credit programs to assist beginning farmers, although the loan volumes are small. Methods include direct loans funded by revenue bonds for farmland purchase, loan guarantees to buy land, and tax incentives to encourage farmland sales and leasing to young farmers. Like FmHA credit programs, beginning farmer programs affect structure because participants face various farm size and net worth restrictions. Lowenberg-Deboer and Boehlje (1983) note that these programs are relatively expensive and do not enhance the viability of beginning farm operations. Mane and Gustafson (1990) compared the economic performance of beginning farmers who did and did not participate in North Dakota's beginning farmer program. Participating farmers owned a significantly larger proportion of the acreage operated by their farm unit. Total farm size, machinery investment, liquidity, gross farm income, and net farm income were not statistically different.

Thus, beginning farmer programs affect structure mostly through tenure differences, not farm size. In general, the limited resources of state programs result in only negligible aggregate effects.

Structural Implications: A Future Perspective

The development of secondary farm loan markets, structural change among financial institutions, continued integration of U.S. agricultural finance markets, and greater internationalization of U.S. capital markets will likely influence the future structure of agriculture.

The Agricultural Credit Act of 1987 established the Federal Agricultural Mortgage Corporation (FAMC) to create a secondary market for farm real estate and rural housing loans. Although this market has been slow to evolve, it has the potential to significantly alter the delivery of credit to farmers.[5] Lemieux (1989) finds that overall credit availability in the farm sector will increase, although it is unclear whether depository institutions or the Farm Credit System will experience greater market share gains. In terms of structural implications, she expects stringent, inflexible, and more uniform loan standards to accelerate the trend to fewer, larger farms.

In 1990 the Food, Agriculture, Conservation, and Trade Act facilitated the development of Farmer Mac II--a secondary market for the guaranteed portion of FmHA loans. If secondary markets in agriculture are successful, one can expect an extension of the concept to intermediate and operating loans with similar structural effects.

Continued integration of U.S. agricultural financial markets will result from the development of secondary loan markets, lenders' adoption of uniform financial analysis methods, increased use of FmHA guaranteed loans, and more consistent actions by financial institution regulators. The stringent and inflexible loan standards of the secondary market were discussed above. Widespread adoption of the Farm Financial Standards Task Force recommendations will yield consistent benchmarks for gauging the financial performance of farms. FmHA guaranteed loans lead to greater uniformity in lending practices because participating lenders must abide by FmHA's national loan approval process. Greater uniformity of financial institution regulation will limit the diversity in lender actions. Thus, all four developments either standardize loan officers' methods of compiling, analyzing, and interpreting financial statement information or set forth criteria for loan acceptance--leading to more uniform credit decisions by loan officers.

Farmers and lenders will respond to these developments through business reorganization. More uniform credit decisions could translate into more specialized farm structures. Agricultural lending institutions will become more centralized and specialized. Agricultural banks are quite specialized at present by virtue of their small size and rural locations. As lending authority becomes more regionally and nationally concentrated, through FCS restructuring and continued geographic liberalization of banking, financial institutions specializing in specific crop or livestock enterprises could eventually emerge. Larger, more commercially oriented farm borrowers will be preferred, and more sophisticated practices in farm business management will be expected.

U.S. agricultural financial markets will become more susceptible to international capital market fluctuations in the future. Increased reliance on foreign capital to manage U.S. federal budget deficits, reduced insulation from external monetary and fiscal policy changes, greater capital demands by East European and Soviet countries, and fewer restraints on international lending/investment activity imply greater volatility in interest rates and availability of financial capital. These trends will make it more difficult for U.S. farmers to finance their operations with debt capital.

Conclusions

A historic preference has been expressed for a pluralistic farm structure in the U.S. with the control of farm business largely held by farm families. Moreover, financial institutions, credit programs, and regulations were designed to achieve these structural characteristics. For example, FmHA policies have perpetuated family sized farms. However, economic forces have often pushed toward larger farms and fewer farmers, thus creating pressures between public preferences and economic forces. Moreover, with significant public intervention, agricultural financial markets probably have provided farmers with greater quantities of debt capital (relative to nonfarm businesses) than economic fundamentals have warranted. Thus, overall structural impacts of agricultural finance are mixed. Nevertheless, sizable federal budget deficits, large private and public losses associated with the farm financial crisis of the 1980s, and rapidly changing capital markets will, in all likelihood, place an upper bound on future public intervention.

New methods of credit delivery will evolve to fill these voids. Greater privatization of agricultural financial markets would normally provide farm borrowers with more flexible financing opportunities, yielding more heterogeneous firm structures. However, increased standardization of financial information, credit analyses, and financial institution regulation will probably lead to more uniform farming units.

Notes

1. Leasing is a form of financial leveraging because it creates fixed-obligation financial claims on the assets controlled by the business and on the income generated by the business.

2. Data from the Agricultural Economics and Land Ownership Survey (1988) indicate the following forms of business organizations in agriculture: proprietorships (87.7 percent), partnerships (9.1 percent), family corporations (2.5 percent), other corporations (0.2 percent), and other (0.5 percent). The average size of partnerships (781 acres) is more than double that of individual or family units (369 acres), while family held and other corporations averaged 1,938 acres and 3,370 acres, respectively.

3. Congress originally authorized a cooperative system of 12 Federal Land banks. In 1923 and 1933, Congress authorized the establishment of the Federal Intermediate Credit Banks and Production Credit Associations, respectively.

4. Various proposals have occurred during the 1980s to reduce or remove the agency status of the FCS and other government sponsored enterprises. At this time, it remains unclear whether any such changes will be enacted.

5. Reasons for the slow evolution include disputes among organizations responsible for regulating the market, a regulation that requires lenders to set aside reserves against the total value of the loan instead of the 10 percent loan portion retained, excessive credit pre-existing in the sector, inadequate loan volume to support a viable market, and uncompetitive pricing of loans.

References

Baker, C. B. 1968. "Credit in the Production Organization of the Firm." *American Journal of Agricultural Economics* 50: 507-520.

_____ . 1977. "Introduction to Economic Growth of the Agricultural Firm." *Economic Growth of the Agricultural Firm*. Technical Bulletin 86, College of Agriculture Research Center, Washington State University, Pullman.

Baker, C. B., and G. D. Irwin. 1961. *Effects of Borrowing from Commercial Lenders on Farm Organization*. Illinois Agr. Exp. Sta. Bul. 671.

Barry, P. J. 1980. "Capital Asset Pricing and Farm Real Estate." *American Journal of Agricultural Economics* 62: 549-53.

_____ . 1984. *Regulatory and Performance Issues for Financial Institutions: Their Effects on Technology Adoption and Structural Change in Agriculture*. Office of Technology Assessment, U.S. Congress.

_____ . 1985. "The Farmers Home Administration: Current Issues and Policy Directions." *Looking Ahead*. The National Planning Association, VIII(4): 4-11.

Barry, P. J., and C. B. Baker. 1984. "Financial Responses to Risk," in Peter J. Barry, ed., *Risk Management in Agriculture*. Ames: Iowa State University Press.

Barry, P. J., and L. J. Robison. 1986. "Economic Versus Accounting Rates of Return on Farmland." *Land Economics* 62: 388-401.

410

Barry, P. J., and D. R. Willmann. 1976. "A Risk Programming Analysis of Forward Contrasting with Credit Constraints." *American Journal of Agricultural Economics* 58: 62-70.

Barry, P. J., C. B. Baker, and L. R. Sanint. 1981. "Farmers' Credit Risks and Liquidity Management." *American Journal of Agricultural Economics* 63: 216-27.

Barry, P. J., J. A. Hopkin, and C. B. Baker. 1988. *Financial Management in Agriculture*. The Interstate Printers and Publishers, Inc., Danville, IL.

Bureau of the Census. 1991. *Agricultural Economics and Land Ownership Survey (1988)*. Vol. 3, 1987, Census of Agriculture, AG87-R5-2.

Ellinger, P. M., and P. J. Barry. 1987. "The Effects of Tenure Position on Farm Profitability and Solvency: An Application to Illinois Farms." *Agricultural Finance Review* 47: 106-118.

Ellinger, P. M., P. J. Barry, and M. Mazzocco. 1990. "Farm Real Estate Lending by Commercial Banks." *Agricultural Finance Review* 50: 1-15.

General Accounting Office. *Farmers Home Administration: Farm Loan Programs Have Become a Continuous Source of Subsidized Credit*. GAO/RCED-89-3, Washington, DC, Nov. 22, 1988.

_____ . *Farmers Home Administration: Use of Loan Funds by Farm Program Borrowers*. GAO/RCED-90-95BR, Washington, DC, Feb. 8, 1990.

Jolly, R. W., A. Paulson, J. D. Johnson, K. A. Baum, and R. Prescott. 1989. "Incidence, Intensity, and Duration of Financial Stress Among Farm Firms." *American Journal of Agricultural Economics* 67: 118-15.

Kaplan, H. M. 1985. "Farmland as a Portfolio Investment." *Journal of Portfolio Management* 11: 73-8.

LaDue, E. L. 1990. "Moral Hazard in Farm Lending." *American Journal of Agricultural Economics* 72: 774-780.

Langemeier, M. R., and G. F. Patrick. 1990. "Farmers' Marginal Propensity to Consume: An Application to Illinois Grain Farms." *American Journal of Agricultural Economics* 72: 308-16.

Leatham, D. J., and T. G. Baker. 1988. "Farmers' Choice of Fixed and Adjustable Interest Rates Loans." *American Journal Agricultural Economics* 70: 803-12.

Lee, J., and S. C. Gabriel. 1980. "Public Policy Toward Agricultural Credit." *Future Sources of Loanable Funds for Agricultural Bank*. Federal Reserve Bank of Kansas City.

Lemieux, C. M. 1989. "Farmer Mac: How Will it Affect Agricultural Lending?" *Agricultural Finance Review* 49: 37-45.

Lins, D. A., and P. J. Barry. 1980. "Availability of Financial Capital as a Factor of Structural Changes in the U.S. Farm Production Sector." Farm Structure U.S. Senate, Committee on Agriculture, Nutrition and Forestry, Washington, DC.

Lowenberg-Deboer, J., and M. Boehlje. 1983. "Evaluation of State Legislative Programs to Assist Beginning Farmers." *Agricultural Finance Review* 43: 9-20.

Lowenberg-Deboer, J., A. M. Featherstone, and D. J. Leatham. 1989. "Nonfarm Equity Capital Financing of Production Agriculture." *Agricultural Finance Review* 49: 92-104.

Mane, D., and C. R. Gustafson. 1990. *An Economic Evaluation of the Bank of North Dakota's Beginning Farmer Loan Program*. Ag. Econ. Misc. Rpt. No. 127, North Dakota State Univ., Fargo.

Moore, C. 1979. "External Equity Capital in Production Agriculture." *Agricultural Finance Review* 39: 72-82.

Morehart, M. J., J. D. Johnson, and D. E. Banker. 1989. *Financial Characteristics of U.S. Farms, January 1, 1989*. Ag. Info. Bull. No. 579, Agriculture and Rural Economy Division, Economic Research Service.

Pflueger, B. W., and P. J. Barry. 1986. "Crop Insurance and Credit: A Farm Level Simulation Analysis." *Agricultural Finance Review* 46: 1-14.

Robison, L. J., and P. J. Barry. 1987. *The Competitive Firm's Response to Risk*. New York: Macmillan Publishing Co.

Scofield, W. H. 1972. "Nonfarm Equity Capital in Agriculture." *Agricultural Finance Review* 33: 36-41.

Thompson, C. S., and E. H. Kaiser. 1985. "Effects of Seller Financing on Prices Paid for Farmland." *Agricultural Finance Review* 45: 40-44.

U.S. Council of Economic Advisors. 1990. *Economic Report of the President*. House Document No. 101-121, U.S. Government Printing Office, Washington, DC.

U.S. Department of Agriculture. 1989. *Economic Indicators of the Farm Sector: National Finance Summary*. ECIFS 8-1, Agriculture and Rural Economy Division, Economic Research Service.

Young, R. P., and P. J. Barry. 1987. "Holding Financial Assets as a Risk Response: A Portfolio Analysis of Illinois Grain Farms." *North Central Journal of Agricultural Economics* 9: 77-84.

16

A Changing Food and Agribusiness Sector: Its Impacts on Farm Structure

Michael A. Hudson, Bruce J. Sherrick,
and Michael A. Mazzocco

Changes in participants' perceptions of the environment and appropriate behavior follow from changes in the environment . . . the sequence continues and the environment evolves.

Shaffer 1980

Introduction

Change has been perhaps the one constant present in agriculture over the past several decades. Technological improvements, changes in size and scope of operations, evolving consumer preferences, shifting production locations, increased capital requirements, and increased exposure to risk are illustrative of the changes which have impacted farm structure. These and other changes have brought about an evolution in agriculture -- away from being production driven toward being market driven.

A recent speech by Purina Mills Chief Executive Officer Ed McMillan reveals this shift in noting that "for the first time in the history of American Agriculture, the end-consumer -- not the producers or packers -- now drives the entire food chain." McMillan (1990) further suggests that "everyone is going to have to change the focus of their business, shifting from being production-driven to being consumer driven." This new focus leads to the emergence of new relationships along the production-marketing continuum for food and fiber products; relationships which have direct impacts on the structure of agriculture.

Professor Stanton identifies a key impact noting that to control both production and marketing decisions farmers "must establish a specialized marketing niche in local markets, utilize roadside markets or pick-your-own-operations, or some other specialized marketing relationship." This, Stanton suggests, follows from the fact that:

> Input suppliers on one side of farm producers and processors and marketing firms on the other side have become fewer in number [and to some degree more specialized] with the power to impose substantial requirements on the production and marketing decisions of producers. Complete independence in deciding what to plant or produce and when and how much to sell has become less and less possible . . . (Chapter 2, this volume).

As the sector has shifted its focus to the consumer, the role of the agricultural producer has changed. Once the driving force of the system, the producer is now driven by the system. New linkages and relationships among levels in the marketing system have evolved which dramatically change the structure of agriculture. But as suggested in the opening quote from Shaffer (1980) -- it has been an evolutionary process, occurring largely in response to participants' perceptions of the environment. This view recognizes the influence of technology on costs of production. Costs, in turn, influence size and organization of production units. The size and organization then influence profits which in part determine the investment capacity for new technologies which change the cost relationships. Vertical and cross-function relationships continue to evolve as the input supply and output processing sectors work together with the agricultural production sector in seeking new arrangements to satisfy the consumer. Farm structure both impacts these changes and is impacted by them due to implicit linkages along the production-marketing continuum.

This chapter explores some of these relationships. The first section discusses evidence of changes in some of the input supply and output processing industries. This is followed by a discussion of changes in the entire food and agribusiness sector. The changing structure of vertical relationships within the sector is then examined, followed by some concluding remarks.

A Traditional View of Changes in Agribusiness

Changes in the structure of U.S. agribusiness have been studied to varying degrees for a number of years. This section documents the aggregate data for a small sampling of industries to illustrate the types of changes occurring. The first part describes changes in some farm input industries, namely 1) feed and feed milling, 2) credit, 3) agricultural chemicals, and 4) farm machinery. The second part illustrates changes in some output processing industries: 1) meat packing, 2) grain milling, and 3) dairy processing.

Input Supply

Farm input suppliers have been adjusting to new technology and changing demographics of agricultural producers. The importance of purchased inputs, both direct production factors and services, continue to challenge producers' ability to perform as purchasing agents.

Feed. The feed and feed milling industries have been historically among the most important in terms of providing purchased inputs to the production sector. Feed purchases account for 13.5 percent of farmers' total outlay for purchased inputs and surpass farmers' expenses for rent, interest, energy, or fertilizer (USDA/ERS 1990).

The two most evident changes the industry has undergone are a slight increase in concentration and a largely increased reliance on vertical integration to lessen market uncertainties. In addition to the increase in market share held by the top 20 companies, the industry has witnessed sharp swings in the total numbers of feed mills that produce 1,000 tons or more per year. In total, the U.S. declined from 7,917 mills in 1969 to 6,340 in 1975 and increased back to 6,723 by 1984. Further details are provided in Table 16.1 which also gives total mills and tonnages produced within regions.

Increased vertical integration in the poultry industry has been accelerated by the feed manufacturers. Many of the large integrators have developed feed milling capacity in response to producers' requirements for highly specialized ingredients and forms. For example, in the Delta and Southeast, 70 percent of formula feed in 1985 was fed to animals owned by the mills, and this percentage appears to have increased (USDA/ERS 1990).

Credit. The agricultural credit industry has undergone as many significant changes as any of the input industries. For example, the Farmers Home Administration has significantly changed its role by shifting its activities away from lending directly to producers toward their loan guarantee program. Their volume in 1990 was down by $200 million, dropping below $1 billion for the first time since 1972 (USDA/ERS 1991). This fact contributed substantially to the estimated decline of 1.3 percent in total farm debt which was estimated to be $133.9 billion at the end of 1990. This total figure represents a 31 percent

TABLE 16.1 Number of Feed Mills and Formula Feed Production by Region

Region	Feed Mills (number)			Formula Feed Production (million tons)		
	1969	1975	1984	1969	1975	1984
Northeast	707	611	600	10.1	11.9	9.3
Lake States	1,149	870	1,163	9.6	10.4	13.5
Corn Belt	2,459	1,900	1,983	21.9	19.3	20.4
Northern Plains	978	873	918	9.6	9.6	10.9
Appalachian	608	482	542	8.0	7.5	7.2
Southeast	469	391	328	9.2	9.5	9.8
Delta States	239	196	184	6.1	6.2	6.1
Southern Plains	583	473	531	10.8	12.5	14.3
Mountain	423	304	302	7.0	8.0	5.8
Pacific	302	239	172	8.7	9.7	12.4
Total U.S.[a]	7,917	6,340	6,723	101.0	104.5	109.6

[a]Excludes Alaska and Rhode Island.

Source: USDA/ERS, 1990.

decline from the peak of $192 billion in total farm debt in 1983. Other lender market shares are more fully described in Table 16.2.

The early and mid 1980s were difficult for the Farm Credit System (FCS), which has undergone significant organizational and structural change as a result. FCS has experienced a declining market share (since 1982) in a contracting market, compounding its loan volume problems. However, this trend subsided by the beginning of the 1990s, with a stabilization in both market share and loan volumes. The composition of FCS loans has begun to shift away from long term loans toward short and intermediate term loans.

Commercial banks hold the largest share of the total farm debt at nearly 35 percent in 1990, up from 33 percent a year earlier (USDA/ERS 1991). In the aggregate, agricultural banks (as defined by the Board of Governors of the Federal Reserve System) have maintained a strong level of performance and of market share. However, banking reform has led to increases in affiliations of small banks with larger banks or bank holding companies, raising concerns by some that long term prospects for service to agriculture by rural banks could be jeopardized.

Although the organized secondary market for agricultural loans (Farmer Mac) had a slow beginning, the prospects for its success have

TABLE 16.2 Percentage Distribution of Debt by Lender and Total Farm Debt (December 31)[a]

Year	FCS	Commercial Banks	FmHA	Insurance Companies	Individuals & Others[b]	Total ($ mil.)
1974	28.6	29.7	5.0	7.6	29.0	75,853
1975	29.7	29.0	5.4	7.3	28.6	85,012
1976	30.2	29.2	5.2	7.1	28.3	96,067
1977	29.8	28.2	5.8	7.4	28.9	110,857
1978	29.5	27.0	6.9	7.6	28.9	127,402
1979	29.9	24.5	9.5	7.4	28.6	151,550
1980	31.8	22.6	10.5	7.2	28.0	166,779
1981	33.8	21.3	11.4	6.7	26.9	182,276
1982	34.0	22.1	11.3	6.3	26.3	189,501
1983	33.4	23.6	11.2	6.1	25.6	192,695
1984	33.2	24.7	12.1	6.1	24.0	189,871
1985	31.4	25.3	13.9	6.3	23.1	173,950
1986	29.1	26.7	15.5	6.5	22.1	154,244
1987	27.6	28.7	16.4	6.4	20.9	141,951
1988	26.5	30.8	15.7	6.4	20.5	137,575
1989	26.3	32.8	13.9	6.5	20.5	135,609

[a]Excludes operator households.
[b]Includes individuals and others (land for contract, merchants, and dealers credit, etc.) and CCC storage and drying facilities loans.

Source: USDA/ERS, 1991.

improved. In 1990, two firms began attempts to pool loans under Farmer Mac authority and a third firm is in the process of becoming certified as a pooler. If Farmer Mac becomes highly successful, the availability of capital for prospects for long term loans to agricultural producers could increase dramatically.

The lender category labelled "Individuals and Others" in Table 16.2 contains merchant and dealer credit which is becoming increasingly important to farmers and ranchers. Originally, many of these suppliers developed small credit programs to facilitate and encourage sales of their own products. They are increasingly willing and able to finance whole farms and input purchases from other sources as well. Among the more prominent of these types of lenders are John Deere, Farmland Industries, J.I. Case, and Pioneer Hi-Bred International. Anecdotal evidence suggests that the informational data bases maintained by these dealers and merchants in the course of non-credit business makes the information gathering and processing for their credit business rather innocuous. This type of cross-function information sharing will likely assist these firms to become even more prominent in the credit arena in the future. For further discussion of these issues see the chapter by Gustafson and Barry.

Agricultural Chemicals

The most significant change within the agricultural chemical industry during the past several years has been consolidation. In 1977, 12 companies held 66 percent of the $8 billion world crop protection chemicals market. By 1987 eight companies held 70 percent of a $17 billion market (Thomas 1989).

A major force driving consolidation of chemical producers is the high cost of research and development (R&D). The industry's average cost to bring a new product to the marketplace can currently range to $80 million (nominal) and take about 8 years. In the early 1980s the industry average was about half that amount or $40 million. In 1987, average R&D expenditures were approximately 10 percent of sales compared to only 6 percent in the mid-70s. Most companies believe that $70 to $100 million are required to operate an effective R&D program. Considerable efficiencies have been gained in some cases by combining complementary R&D programs from different companies.

Another force for consolidation of this industry is the increasingly international nature of the business. The U.S. share of the world market decreased from 39 percent to 26 percent from 1978 to 1987 (Thomas 1989). With markets elsewhere, namely China, Brazil, and India, growing at the fastest rates in the world, a global presence is necessary to increase market share. This presence is often achieved most quickly through acquisitions of existing, foreign companies. Information technologies have also significantly reduced the complexities of operating companies in international locations.

Farm Machinery. The profitability of this industry ebbs and flows with the levels of farm income, interest rates, land values, crop prices, and the amount of land in production. The 1970s were a period of prosperity for many farmers and in turn for input suppliers such as farm machinery manufacturers. There was high demand for large tractors and other farm equipment, both domestically and abroad. Primary reasons for this high demand were rapidly increasing land values, low real interest rates, widely available credit, strong investment tax incentives, and rising commodity prices. As a result, farm machinery dealers greatly increased both their short-term and long-term production capacity. Through the 1970s expenditures on tractors and other farm equipment increased by 6 percent per year. From 1950-1979 expenditures on farm machinery increased from $2.2 billion to a record $11.7 billion (USDA/ERS 1987).

The downturn in the farm economy of the early 1980s created perhaps even more severe hardships in the farm machinery sector. The economic conditions occurring in the U.S. after 1979 caused a major downward shift in the demand for tractors and other farm equipment, which translated to a reversal in farm machinery sales. The value of annual farm machinery purchases which peaked in 1979 ($11.2 billion) declined to only $4.6 billion by the end of 1986.

A considerable downsizing of the capacity of the sector, and a reorganization of assets followed this decline in the financial condition of the farm machinery industry. Consolidation has tended to occur through direct buyouts, mergers, and joint ventures. The most notable of these was the 1985 purchase of International Harvester's farm machinery division by J.I. Case. Other major consolidation activities of this industry include: the purchase of Allis-Chalmers (farm equipment division and financial subsidiary) by Klochner-Humbold-Deutz A.G. of West Germany; Allied Product Corporation's purchases of White Farm Equipment's U.S. tractor and implement assets and Avco's New Idea combine assets, as well as Bush Hog and Kewanee; and Ford-New Holland's purchase of Versatile, a low-volume Canadian manufacturer of four-wheel drive tractors.

Table 16.3 estimates market share for the North American wheel tractor market for 1983 and 1986. The 1986 market share estimates are based on 1983 sales taking into account any recent mergers or direct buyouts. The top four firms accounted for 68 percent of total tractor sales in 1983. By 1986 this figure had increased to 80 percent. By 1990, some

TABLE 16.3 North American Wheel-Tractor Market Shares[a]

Manufacturer	1983	Post-mergers (by 1986)
Deere and Co.	25.0	26.0
Case-IH	--	21.5
IH	12.0	--
J.I. Case	9.0	--
Steiger	0.5	--
Ford-New Holland	--	18.5
Ford Tractor Co.	17.5	--
Versatile	1.0	--
Kubota (Japan)	13.5	13.5
Massey-Ferguson	8.5	8.0
Deutz-Allis	--	7.0
Allis-Chalmers	7.0	--
Allied Products	--	2.0
White	2.5	--
Others	3.5	3.5
Total	100.0	100.0
Top 4 Firms	68.0	80.0

-- = Not applicable

[a]Market shares are based on North American wheel tractor sales for 1983, prior to mergers completed by 1986. Thus, the 1986 shares do not reflect 1986 sales, but only the effect of the mergers of 1983 shares.

Source: USDA/ERS, 1990.

industry sources placed John Deere's market share at nearly 42 percent and J.I. Case at 37 percent.

The farm machinery distribution system in the U.S. is based on independent franchised dealerships. The distribution system also underwent similar and severe reorganization along with the manufacturing industries. In the 1940s there were approximately 35,000 of these dealerships but by 1979 this figure had decreased to 11,342 and by 1987, there were only about 8,000 dealerships remaining to serve U.S. agricultural producers.

Output Processing

Output processing continues to be the dominant portion of value added in the food and agribusiness sector. Although many other activities and industries contribute to this portion of the economy, three industries are chosen for illustration: meat processing, grain milling and dairy processing. A more thorough documentation of other processing industries may be found in Connor (1988).

Meat Processing. The structure of the U.S. meat processing industry and its impact on the structure of production agriculture has been a source of concern since the late 1800s. Ward (1988) and Purcell (1990) each offer a historical perspective on the early development of related market concentration issues and the development consolidation and mergers through the late 1980s. Data in Table 16.4 show the movement toward fewer, larger meat packing plants with increasing market share during the period 1972-1985. Four firm concentration ratios identify the market share (percent) controlled by the four largest firms. Evidence suggests that regional four-firm concentration ratios in major beef producing areas are strongly higher. For example, the comparable 1985 four firm ratio in Colorado was 99.9 percent; for Arizona, 99.7 percent; and for the thirteen major beef producing states was 89.6 percent (Kahl, Hudson, and Ward 1989).

TABLE 16.4 Comparative Structure of the Meat Packing Industry

	1972	1985
Beef		
Number of Beef Packing Plants	807	435
Beef Packing Plants > 500,000 head capacity		
Number	3	17
Volume (million head)	2.0	14.4
Share of slaughter (percent)	7.5	53.4
Four Firm Concentration - slaughter (percent)	29.0	52.3
Four Firm Concentration - boxed beef (percent)		64.0
Pork		
Number of Pork Packing Plants	597	403
Pork Packing Plants > 1 mil. head capacity		
Number	23	34
Volume (million head)	30.1	54.0
Share of slaughter (percent)	37.0	67.1
Four Firm Concentration - slaughter (percent)	32.0	32.6

Source: Ward 1988, Chapter 1.

This increase in market concentration has led to numerous studies of potential oligopsony power being exerted on cattle producers. (See GAO, 1990, for a brief review of ten such studies.) According to the GAO, evidence is not sufficiently strong to conclude that beef packer concentration has led to lower cattle prices, although conditions could change to allow packers to exert downward price pressure.

Nevertheless, farm structure has been impacted by changes in the packing industry. With the increase in large-capacity plants, packers are more focused on securing dependable sources for large volumes of fed animals. This has led to vertical coordination arrangements with large cattle and hog feeders, including direct contract arrangements and purchase of large feeding operations. As reported in GAO (1990, 5), Census of Agriculture data indicate a 40 percent drop in the number of cattle feeding operations in the thirteen major beef producing states during the 1980s, with most of the reduction in small operations. Changes in federal income tax regulations have eliminated many incentives supporting custom feed operations. Changes in federal income tax regulations have eliminated many incentives supporting custom feed operations. Some large pork integrators in the Southeast offer price collars to help reduce producers' market risk. When taken together with the additional consideration of economies of size, the trend toward fewer, larger beef and hog feeders is likely to continue, although at a reduced pace.

Grain Milling

U.S. grain milling operations cover a vast array of farm products and primary outputs. Two-sub-industries are chosen to exemplify changes in grain milling industries, wheat flour milling and corn wet milling. Further information is available in selected references.

Wheat Flour Milling. The flour milling industry has undergone changes similar to the meat packing industry: fewer firms with larger plants and increasing four-firm concentration ratios. However, the industry remains much more diffuse than meat packing, as illustrated in Table 16.5. Major features of the change in wheat flour milling are the location of plants and use of products. As noted by Harwood, Leath and Heid (1989, 67), the growth in the U.S. flour grind has been a response to consumer demand for value-added products. As a result, 85 percent of the flour grind is now shipped to commercial bakers. With simultaneous pressure on transportation costs and relative changes in transportation rates between flour and wheat, there has been a shift in location of flour grinding from rural to more metropolitan areas.

TABLE 16.5 Comparative Structure of the U.S. Flour Milling Industry

	1963	1972	1982	1987
Number of Companies	510	340	251	240
Number of Plants	618	457	360	358
Plants with more than 10,000 cwt daily capacity:				
Number	n.a.	25[a]	42[b]	54
Percent of industry capacity	n.a.	34.7[a]	49.8[b]	52.6

n.a. = not available
[a] = 1973
[b] = 1983

Source: Harwood, Leath and Heid; U.S. Department of Commerce.

Wet Corn Milling. The development of technology in the wet corn milling industry has impacted the corn processing market during the last decade, while also affecting markets for other agricultural commodities. As noted in Table 16.6, there has been a contraction in the number of commercial wet corn milling firms over the past 25 years, but that trend is reversing. Expanding markets for bulk by-products have led to investment in expanded plant numbers and size. Furthermore, not all wet corn milling concerns are represented in U.S. Census of Manufactures data because of industry definitions. The dramatic increase from 1972 to 1987 in the percent of the U.S. corn crop being wet milled is indicative of the growing demand for wet milling by-products.

The main by-products of wet corn milling are corn oil, corn gluten feed, corn gluten meal, dry starch, high fructose corn syrup, ethanol and carbon dioxide. Markets for these products have competed with animal protein feeds (e.g. soybean meal, cotton seed meal), and sugar and other sweeteners (GAO 1990). Continued pressure for a more environmentally acceptable fuel policy will likely lead to expanding ethanol capacity, which will require more wet (or dry) corn milling capacity and lead to more production of the other by-products listed above. Wasson (1990) estimates that ethanol production to meet motor fuel demand in the nine cities with the most severe pollution problems would require an additional 550 million bushels of corn annually.

Dairy Processing. The dairy processing industry is conventionally divided into five product categories: creamery butter; cheese; dry, condensed, or evaporated milk; ice cream and frozen desserts; and fluid

TABLE 16.6 Comparative Structure of the U.S. Wet Corn Milling Industry

	1963	1972	1982	1987
Number of Companies	49	26	25	32
Number of Plants	60	41	42	60
Million bushels wet ground	n.a.	261.8	480.6	796.7
Percent of annual corn crop wet ground	n.a.	4.7	5.8	11.23

n.a. = not available

Source: U.S. Department of Commerce; U.S. Department of Agriculture; Wasson.

milk. Table 16.7 illustrates the relative change in the use of whole milk as an ingredient in these product groups. It is readily apparent that during the twenty years from 1967 to 1987 there occurred a substantial decline in the amount of whole milk being processed into butter. It is also evident that fluid milk production remains the dominant use of whole milk while increased cheese production accounts for 77 percent of the increased processing of whole milk during this period.

Table 16.8 provides a closer look at the structure of the fluid milk and cheese processing industries during the fifteen year period 1972-1987. Many plants produce both products, but they are not counted in both categories. The number of cheese producing firms and plants declined somewhat during this period, while larger plants were being emphasized. In fluid milk production there were less than one-third as many firms involved at the end of the period. The number of plants fell by more than 60 percent, although the decline in large plants is much less dramatic. Although Table 16.8 points toward increasing concentration in dairy manufacturing, there remains a relatively large number of firms in the industry. Market concentration is better measured on a regional basis because the level of competition and its impact on farm structure varies by region.

The Changing Food and Agribusiness Sector

Sonka and Hudson (1989, 305) describe the food and agribusiness sector as "a number of interrelated subsectors which work together formally and informally to produce goods and services." In a traditional view of the sector, the authors suggest, agribusiness is viewed as the set of activities "beyond the farm gate" – the input supply and output

TABLE 16.7 Consumption of Whole Milk by Dairy Processing Firms (by Product Class)

Product	1967		1972		1977		1982		1987	
	mil. cwt.	% of total	mil. cwt.	% of total	mil. cwt.	% of total	mil. cwt.	% of total	mil. cwt.	% of total
Creamery Butter	110.5	11.5	53.1	5.2	21.4	2.0	32.1	2.8	19.9	1.5
Cheese	133.2	13.9	192.6	19.0	257.4	24.2	332.0	29.3	403.1	30.8
Dry, Cond., Evapor.	97.1	10.1	90.2	8.9	140.6	13.2	126.2	11.1	107.7	8.2
Ice Cream & Frozen	5.7	0.6	6.9	0.7	11.9	1.1	9.7	0.9	16.6	1.3
Fluid Milk	614.4	63.9	671.9	66.2	632.0	59.4	633.2	55.9	763.2	58.2
Total	960.9	100.0	1014.7	100.0	1063.3	100.0	1133.2	100.0	1310.5	100.0

Source: U.S. Department of Commerce.

TABLE 16.8 Comparative Structure of Cheese and Fluid Milk Industries

	1972	1977	1987
Cheese			
Number of firms	739	660	507
Number of plants	872	791	643
Number of plants with more than 100 employees	48	52	81
Fluid milk			
Number of firms	2,025	1,156	654
Number of plants	2,507	1,924	946
Number of plants with more than 100 employees	373	318	264

Source: U.S. Department of Commerce.

marketing functions referred to by Stanton. Specifically, Sonka and Hudson (1989, 305-306) suggest that the traditional view of the agribusiness sector includes:

> the production of genetics and seedstock for crops and livestock, the production and distribution of inputs which are combined with seedstocks in commodity production, financing and other services required by agricultural producers, and procurement and processing of commodities produced by farms and ranches.

The changes in the structure of agricultural production noted earlier, coupled with an increased incidence of vertical and horizontal coordination lead the authors to suggest the need for an alternative characterization of the sector. As depicted in Figure 16.1, "the sector can be thought of as a sequence of interrelated activities made up of: genetics and seedstock firms, input suppliers, agricultural producers, merchandisers or first handlers, processors, retailers, [and] consumers" (Sonka and Hudson 1989, 306), supported by firms providing various services, financing, and research and development as shown in the diagram.

Three important differences are identified between the "traditional view" of the agribusiness sector and this "emerging view" of the food and agribusiness sector. First, the definition "implicitly includes agricultural production as part of agribusiness, thus eliminating the artificial and

426

FIGURE 16.1   The Food and Agribusiness Sector

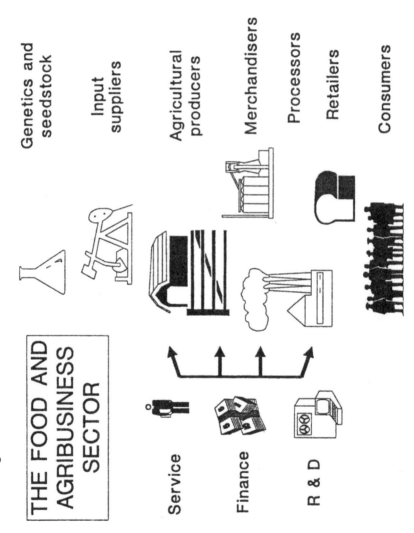

arbitrary exclusion of major commercial production enterprises" (Sonka and Hudson 1989, 307). Second, consumers are included to reflect the consumer driven nature of the system discussed above. Finally, the word food is used in describing the system, recognizing the fact that "the ultimate goal is the health, well being, and satisfaction of consumers of food" (Sonka and Hudson 1989, 307).

Hudson (1990) uses this perspective as a point of departure in considering agribusiness competitiveness, suggesting that changes within the sector have fundamentally altered the perspective within which firms view the vertical continuum between the producer and the final consumer through the marketing channels. Considering the perspective presented above, Hudson (1990) suggests that traditionally firms have operated either in the production subsector of the system or in the processing subsector; subsectors divided by the slaughter/milling line shown in Figure 16.2.

The slaughter/milling demarcation, Hudson (1990) suggests, has led firms in the production subsector to focus primarily on serving producers, attempting to surround them with services as shown in Figure 16.3. Services provided are either inputs into the production process, such as feed, seed, or fertilizer, or activities which facilitate the movement of products into the marketing channel, such as transportation. The strategic focus of the firm operating in this environment is the producer, shown at the center of the diagram. Competitive advantage for agribusiness firms in this production-driven system is achieved by assembling a set of activities which serves the production subsector.

The changes noted in the introduction, a move away from a production driven system towards a consumer driven system, alter the perspective presented in Figures 16.2 and 16.3. As the consumer becomes the focal point, competitive advantage is no longer achieved solely by surrounding the producer with goods and services.

Figure 16.4 depicts this view of competitive position in which firms surround the consumer and attempt to satisfy both the individual and societal needs shown. To establish competitive positions in the world depicted in Figure 16.4, new vertical relationships emerge, blurring the traditional lines between competitors, buyers, and suppliers. This new perspective is presented in Figure 16.5 which depicts the food and agribusiness system in which we currently operate. Note that the lines which were solid in Figure 16.2, indicating the clear demarcation between the various levels of the continuum are now broken -- indicating a high degree of interaction across the levels. Also, note that the slaughter/milling line is no longer present as a demarcation between the production and processing subsectors. A value-added processing subsector has now emerged to satisfy the new demands of consumers for

FIGURE 16.2   A Traditional Perspective on the Food and Agribusiness Sector

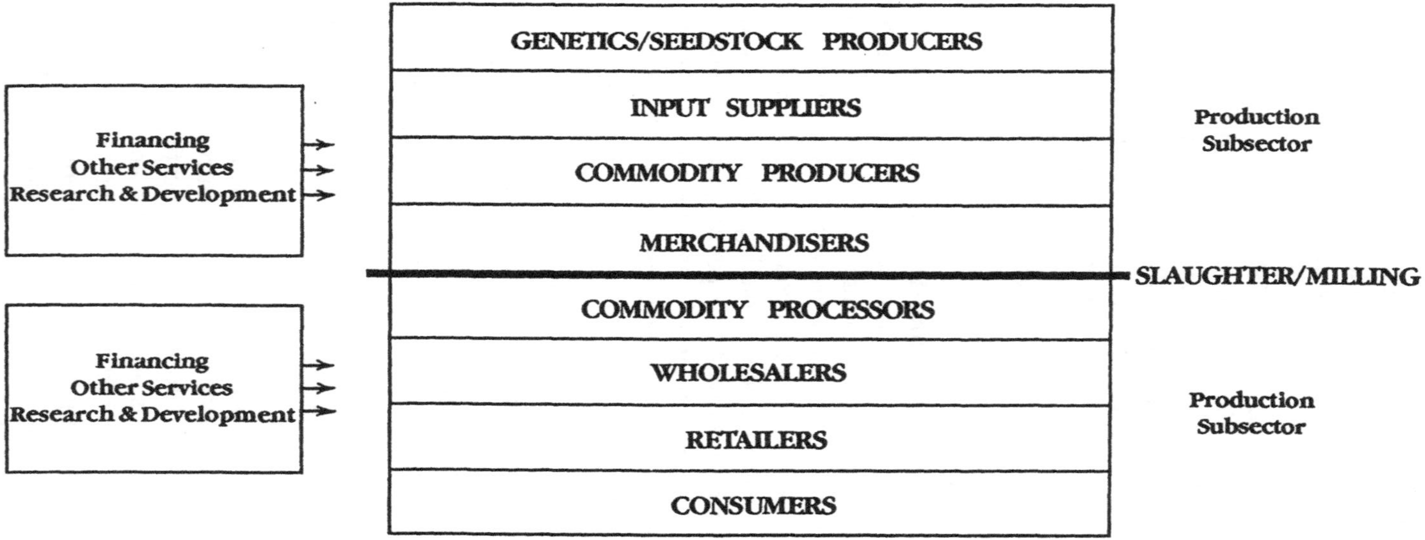

FIGURE 16.3 A Traditional View of Competitiveness Focusing on the Production Sector

Input Oriented Goods and Services

Machinery and Equipment

Financing

Feed, Seed and Fertilizer

Production Information

Commodity Producers

Marketing Services

Transportation

Commodity Processing

Market Information

Output Oriented Goods and Services

FIGURE 16.4   An Emerging View of Competitiveness Focusing on the Consumer

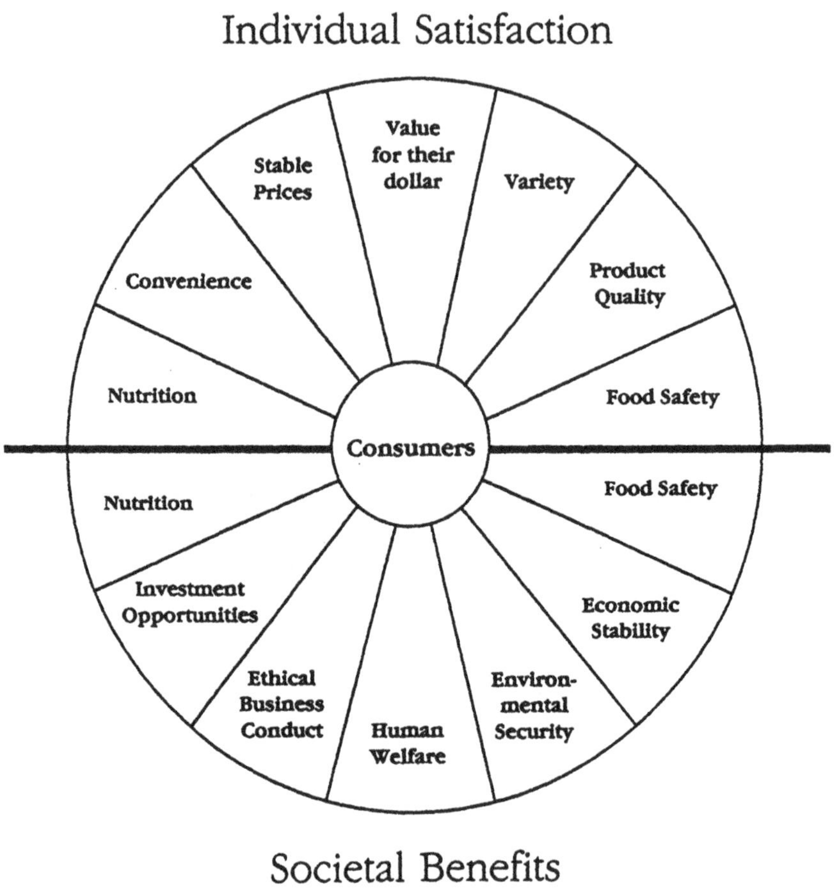

Individual Satisfaction

Value for their dollar · Stable Prices · Variety · Convenience · Product Quality · Nutrition · Food Safety · Consumers · Nutrition · Food Safety · Investment Opportunities · Economic Stability · Ethical Business Conduct · Human Welfare · Environmental Security

Societal Benefits

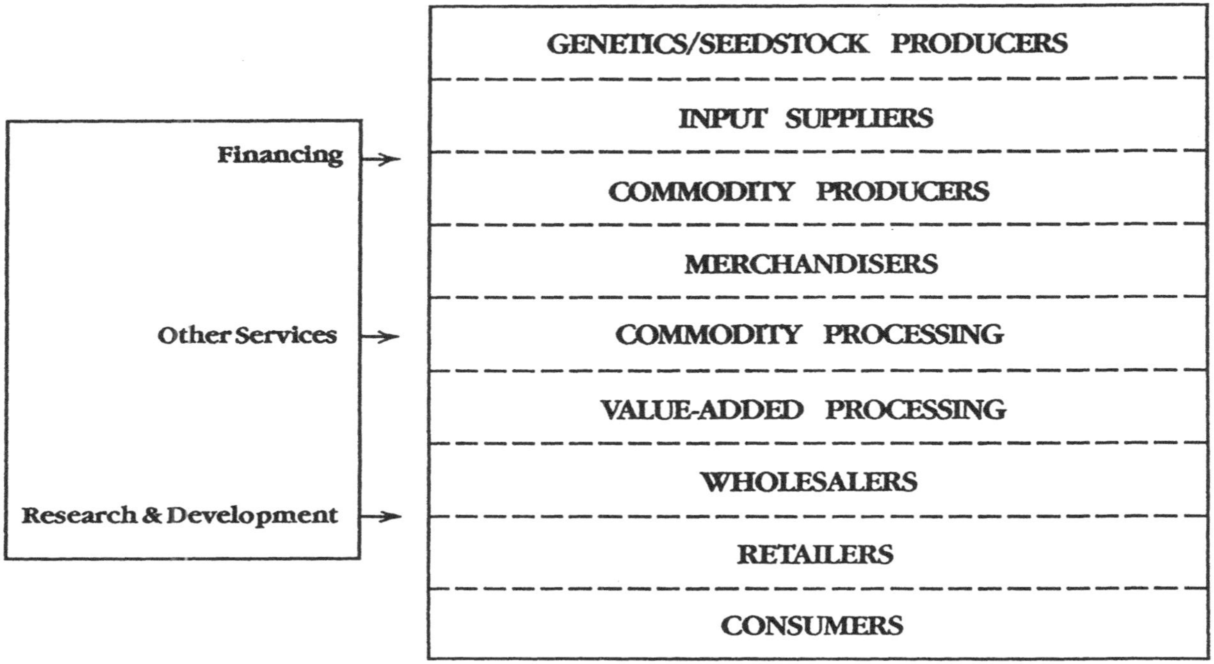

FIGURE 16.5   An Emerging Perspective on the Food and Agribusiness Sector

convenience, variety, quality, and many of the other factors shown in Figure 16.4.

The perspective[1] discussed by Sonka and Hudson (1989), and expanded upon by Hudson (1990) has significant implications for operations within the sector and for the structure of agriculture. The mindset of firms has changed in response to the new consumer driven system. Firms historically viewed as an input suppliers serving the production subsector, such as Purina Mills, now talk of a "mindset [toward becoming] a major player in the food chain" (McMillan 1990). As such, their role changes as unforeseen levels of concentration emerge within the production-marketing channel for food and fiber products and "vertical coordination -- through contractual arrangements, vertical integration, and sharing of information technology -- increasingly characterizes [the sector]" (Hudson 1990, 188).

Vertical Relationships

The production-marketing continuum for food and fiber products shown in Figure 16.5 is a vertical channel through which inputs are transformed into commodities, which are processed into products which are distributed to consumers. As the system becomes more consumer driven the potential benefits from involvement in multiple levels of the system increase. Further, as the demands of the consumer expand to include quality, variety, food safety, and other attributes, coordination of activities within the continuum becomes a necessity.

There are a number of ways in which vertical coordination of the production-marketing continuum can be achieved. The price system traditionally has been viewed as the primary mechanism for coordinating the system. Price signals are sent from one level to the other throughout the marketplace and economic incentives or disincentives perceived by decision makers result in necessary reallocations of effort to address the needs of the system. In an increasingly complex world with significant market segmentation, price signals or the responses to them can become distorted and misallocation may occur. This distortion occurs primarily because of a lack of information sharing between the various levels of the system described above -- the producer does not receive an accurate signal to stimulate production of the desired product either because the attributes which demand a premium price are added after production of the commodity or because the prices which consumers pay for specialized products are not incorporated into the commodity price. Specific examples of such distortions can be found in livestock markets where leaner products and certified origin products command higher premiums, but localized markets continue to reflect the general price level for the "average" commodity, i.e., that which is wanted by the largest market segment. These situations lead to the search for new coordinating mechanisms and strategies.

Vertical integration is perhaps the most commonly considered form of vertical market coordination to replace price. Vertical integration involves the ownership or contractual control of multiple levels of the production-marketing continuum in order to coordinate activities. Vertical integration is often driven more by a need to acquire information and technologies than physical production facilities. The poultry industry is perhaps the most popular example of vertical integration in agriculture; control over the entire production and marketing process is maintained in this industry by integrators who contract with farm producers to grow-out broiler chicks. Through involvement in production of genetics and seedstock, production of feed and other inputs, and in processing and marketing the final product, the integrators are able to coordinate activities along the production-marketing continuum to allow them to serve specific market niches.

Vertical coordination also can be achieved through contractual arrangements and through various nonmarket means, such as information sharing by participants at different levels of the system. As Schrader (1986, 1166) notes, "improved information technologies facilitate nonmarket coordination because of greater ability to monitor and control production processes." Information technology may in fact become the key to vertical coordination within the system in the 1990s, as retail scanner data and other forms of electronic information gathering impact the sector.

Vertical relationships are clearly important within the food and agribusiness sector described above. The consumer driven system which we have frequently alluded to here mandates interaction at many levels of the system in order to satisfy the diverse demands of various consumer segments. For example, if the consumer demands leaner cuts of meat, and the technology develops to track the identity of carcasses through the entire marketing-production channel, a wide range of participants could be called upon to alter their behavior or product in response to changing consumer demands. The key question which emerges is how this increased reliance on vertical coordination will impact the structure of agriculture and the role which farms play within the food and agribusiness sector.

Research Results

Research related to vertical coordination is mixed. The seminal work published in 1963 by Mighell and Jones provides an informative and thorough conceptual framework, including suggested research approaches. Progress, however, has been "with glacial slowness" (Godwin and Jones 1971, 813). However, more work has been done on firm concentration ratios within industries and the relationships to profits in the course of industrial organization and antitrust investigations.

Despite repeated calls for research which examines interactions along the production-marketing continuum, with an eye on identification of the motivations for and benefits of coordination, few efforts have addressed the issue.[2] Two exceptions should be noted. First is the work of Reimund, Martin, and Moore (1981) which addressed structural change in agriculture by considering vertical relationships and the motivations for and process of vertical integration. This paper provides the prototype framework below.

A second notable exception is the work of the regional project NC-117 which generated a large amount of research related to markets and pricing problems within the sector, specifically with regard to the role of price as a coordinating mechanism. The works of Connor et al. (1985), and of Marion and the NC-117 Committee (1986) are also outputs of this project which address the vertical system. It should be noted, however, that limited attention is given to the motivations for specific vertical relationships within these works or their impacts on farm structure -- emphasis instead is on the structure of the system and its performance, primarily at the output processing level discussed in the preceding section.

Schrader (1986, 1166) highlights an important limiting factor in vertical coordination research, noting that the "data system is not providing the data needed to assess the vertical structure of commodity systems or to test many hypotheses regarding it." The author further suggests that despite the usefulness of the conceptual framework offered by Mighell and Jones (1963), "little or no progress (has) been made to incorporate these vertical concepts into the data system." Schrader (1986, 1166) expects this situation to improve in the future as the task becomes "less formidable as the number of farm, supply, and marketing firms declines."

The degree of vertical integration, both forward and backward, from various levels of the system is considered by Kilmer (1986). Presenting percentages of integration and comparing across the past several years, Kilmer (1986) concludes that vertical integration in the system will increase over the next decade. The factors which will have the most influence on integration within the vertical system are identified by Kilmer (1986, 1160) as: "(a) concentration, (b) capital intensity, (c) flow economics, (d) number of inputs and outputs per firm, (e) economies of scope, (f) firm size, and (g) future demand."

Before leaving this topic, it should be noted that a number of studies have been made of the vertical relationships in specific industries and locations. The majority of these works are descriptive in nature, with a focus on explaining the behavior of the system, rather than the motivations for integration or coordination, or the impacts on farm structure.

A Prototype Model

The work of Reimund, Martin, and Moore (1981) provides a useful model for considering the motivations for and the process of structural change within the vertical continuum. By examining the evolution of the broiler, fed cattle, and processing vegetables industries from the early 1950s to the 1970s, the authors were able to identify key factors involved in the structural change process. Examination of these factors will reveal the important role that various levels of the vertical system play in structural changes and will allow inferences about the future impacts of vertical coordination efforts on farm structure.

According to Reimund, Martin, and Moore (1981, 41), prior to the initiation of the structural change process:

> each of the three subsectors [broilers, fed cattle, and processing vegetables] was organized traditionally. That is, the farm production stages had an atomistic structure. [In this environment,] coordination with nonfarm stages of the subsectors was achieved through open market transactions, with the pricing mechanism serving as the primary basis for producer decisions relating to levels of output [and] [m]arkets were generally accessible to all producers.

Another common thread in the three subsectors was that the production processes were all tied to the production of other commodities. For example, broilers were produced from male chicks of heavy layer chicken breeds. Similarly, cattle feeding was done on grain farms as a means of utilizing employees during the off-season help and as a way to market grain. In fact, the authors note that the "degree of association of these [production] activities with other commodities [was often] so close prior to the structural changes that they could not be considered as distinct subsectors" (Reimund, Martin, and Moore 1981, 41).

Based on these observations and on a detailed review of data related to structural change in these three industries, Reimund, Martin, and

Moore (1981) posit a prototype model of structural change which provides insight into the process and may be useful in considering possible future changes in other subsectors. The model consists of four stages and suggests specific impacts in each stage.[3]

Stage One - Technological Change: New technology is developed and adopted by early adopters which increases the capital requirements of the industry. As output per unit of labor and land increase, economies of size develop and the value of resources in new producing areas increases.

Stage Two - Shift in Location of Production: New producers, capital, and other resources enter in new producing areas. As the interregional competitive balances change in favor of the new producing areas, production begins to concentrate in the new areas.

Stage Three - Growth and Development: New production technology is extended as innovative and aggressive entrepreneurs develop larger operations, and specialization and concentration of production occurs. Market economies develop in new growth areas and output per farm rapidly increases. New information systems develop as rapid growth and expansion alters risks – risks which are accepted by the new aggressive, innovative, risk-taking producers. High risks are associated with inefficient operations and coordination procedures and risks are magnified by periods of overproduction.

Stage Four - Adjustment to Risks: New risk aversion strategies are developed and new types of coordinating procedures established. The use of forward sales and production contracts increases and more coordination develops. Control of product flows and characteristics shifts from producers to other stages of the sector closer to the final consumer, and all stages become more industrialized.

The authors apply this prototype model to the hog, dairy, feed grain, and fruit subsectors to show its applicability to other subsectors. They conclude that the causal factors identified in the model are not activated with the intent of changing the structure of agriculture, but to "satisfy rather specific perceived private or social needs" (Reimund, Martin, and Moore 1981, 64). Indeed the authors identify several actions taken which contributed to the structural change which were generally intended to improve the performance of the vertical system, either by improving marketing efficiency, by better meeting the needs of consumers, or by attempting to reduce uncompensated risk to producers and others within the system.

Reimund, Martin, and Moore (1981) thus provide evidence of a number of factors which contribute to structural change, many of which can be viewed in the context of attempts to enhance the effectiveness of the production-marketing continuum in delivering goods and services to consumers. In light of the discussion in the preceding sections related to the emerging perspective of a consumer-driven food and agribusiness system, there are clear implications for farm structure.

As the system is driven more and more by consumers, it is quite likely that additional structural change will occur -- as Kilmer (1986) noted in suggesting that the level of vertical integration would increase within the sector during the last half of the '80s and the first half of the '90s. However, as Schrader (1986) suggests, changes in information technology offer new means of nonmarket coordination -- thus, structural changes in the future may be less dramatic than in the past. It could be argued, for example, that information technology will be the key to coordination of the food and agribusiness system within the next decade (see Batte and Johnson). If this is the case, then the responsibility for coordination will rest with the firm controlling the information, but changes in farm ownership will not be as likely because the ownership is then more separable from the "information" and production technologies that were previously embodied in ownership of the assets. Thus the concern over the "concentration of power in the hands of few decision makers [which] continues to undergird the discussion of farm structure issues" (Stanton) may become less important as the system is coordinated not through ownership of physical resources but through the control of coordinating information. The extent to which consolidation and vertical coordination are driven by needs to access and control information will likely increase. Increasingly, information technologies are permitting more specialized and specific information control across levels in the production-consumer chain.[4]

Illustrative Industry Examples

The possible role of information in coordinating the vertical system can be examined through specific industry examples. By considering the traditional role of firms at various levels of the system and then positing their role in a system where information technology is used as a means of nonmarket coordination, implications for farm and firm structure will emerge.

The Feed Supply Industry. Historically the feed supply industry has provided a key input into the production process for livestock, particularly cattle and hogs. As the sector has evolved over the past fifteen years livestock producers have become fewer in number and larger in size. Feed firms have consolidated and increased the level of services provided to producers, especially the small to medium-sized producers. Services including financial and management advice, veterinary assistance, and marketing consulting assistance are illustrative. All of these services are information services.

Thus, as a feed firm begins to view itself as a player in the food chain, it is possible, perhaps likely, that their role as a provider of information will expand. In fact, this is exactly the strategy which Purina Mills has announced, suggesting that they have "carved out a new position as the informational link to a growing network of breeders, suppliers, producers, packer/processors, marketing nation-wide serving end consumers" (McMillan 1990).

The implication of this strategy -- which although only attributed to a single firm is illustrative of what can be expected from the industry -- for farm structure is that independent operations will continue to be important. The strategy of being an information provider or coordinator explicitly excludes involvement in ownership of the assets related to the production of feed animals. In fact, the risk position of the firm in providing information is reduced, as the individual entrepreneur -- the farm operator -- bears all of the production related risks.

The Seed Industry. A similar case for information as a coordinating mechanism can be built in the seed industry. Consider, for example, Pioneer Hi-Bred International, the largest seed corn firm in the nation. The company recently announced a program called *Better-Life Grains* which serves as a certification procedure for grains grown under specified conditions -- with a symbol being placed on the final product noting that it "contains *Better-Life Grains*." Thus Pioneer uses information to coordinate the efforts of the producer with those of the processor and marketer of the final product.

Numerous other examples could be developed to illustrate the possible roles of a firm like Pioneer in using information as a coordinating mechanism. For example, suppose a processor of corn seeks a certain mix of amino acids in the grain. Pioneer can use its information on how to breed for that gene and generate seed to fit the needs of the processor, then sell the seed to the producer who grows it and delivers it to the processor for a premium price. Pioneer's role in the process is one of information handler and the information serves a key coordinating function.

The implications of the role of the seed company in the examples given here are similar to those noted for the feed supply industry. The independent farm operator is relied upon to bear much of the risk -- as in an uncoordinated system -- and the firm uses information to coordinate the vertical system.

The Livestock and Meat Industry. The role of information in the livestock and meat industry is likely to be slower to evolve than in either of the previous examples. From the production side of the industry, the role of the feed supply firm can occur as noted above. The more exciting and challenging role of information as a coordinating mechanism derives from the other end of the vertical continuum -- from the consumer. As scanner data allow the industry to gather more specific information with regard to the demands of consumers for various meat products, the information can be fed back into the system to coordinate the activities of value-added processors, packers, and even initial producers. Linkages with feedlots and feed suppliers will be important, but the first step will be for someone to step forward and exploit the opportunity. An important role of information technology in this industry also exists in tracking the performance of various types of animals through the packing processing levels to provide feedback for improving the production process.

Initially, farm structure impacts are likely to be minimal here. The industry has already undergone a major structural change and excess capacity in the packing industry continues to decline. The key structural impact may be the alignment of groups of producers working with specific packers to schedule delivery of the particular quality and quantity of animals needed to fit the needs of the retail marketplace. Once the potential of using information technology to track the animal from conception to consumption can be realized, a major structural change could begin to occur in this industry.

Other examples could be considered, but these are illustrative. The role of information technology as the key coordinating mechanism of the 1990s suggests minimal changes in farm structure. But, the reorientation of the sector towards the consumer indicates that farm structure will be impacted by changes which affect the vertical system. Perhaps, the role of information coordination will be to align more closely, compensation for the economic risks borne at the particular level in the system. Specific impacts on structure will depend upon the scope, focus, and success of the innovations which evolve.

Conclusions

Evidence of changes in agricultural input supply and output processing industries was presented to document the different types of dynamics occurring within various industries. A changing perspective of the food and agribusiness sector has been used throughout the chapter to illustrate the inherent linkages between farm structure and the input supply and output processing subsector. It has been suggested that the system, once production driven and focused around the farm, has changed to a consumer driven system -- much like other sectors of the economy. This change in focus raises questions related to coordination of activities within the system.

Research on vertical relationships was examined to illustrate the factors which have led to changes in farm structure in the past. Based on the forces identified in previous work, it was argued that information (and not vertical integration by ownership) will be the primary coordinating mechanism for the 1990s. Three examples were offered to suggest possible impacts on farm structure. The net impacts appear to be minimal, assuming information technology evolves and is used by firms as a means of nonmarket coordination. If information technology does not emerge as a coordinating mechanism, then it is likely that market mechanisms will continue, which may lead to a further decline in the number of farms or to increased reliance on vertical integration.

The future, as always is uncertain. Leadership within the food and agribusiness sector seems to be emerging from a number of firms seeing the potential to use information in bettering serving the consumer -- the new driving force of the system. The examples given here show leadership coming from major agribusiness firms and they suggest a possible expanded role for food firms. It is also entirely likely that farm operations will provide similar leadership, either individually or working together. The key to future success and a strong agricultural infrastructure will be serving the people who consume the products of the system -- coordination of effort would seem to be the key for the decade ahead.

Notes

The authors thank Corey Waldinger for research assistance.

1. It should be noted that this perspective is not necessarily new, indeed it may be a blinding flash of the obvious. Davis and Goldberg (1957) suggested the agribusiness sector as a series of interrelated activities over 30 years ago. Also, Goldberg (1988) has noted the impacts of mergers, acquisitions, and restructuring on the sector during the 80s -- a key factor leading to the perspective suggested in Hudson (1990). Nonetheless, the discussion presented here provides a basis for considering vertical relationships in the context of a consumer-driven system and examining the possible impacts on farm structure.

2. Examples addressing this issue include Shaffer (1968) and Purcell (1973). Shaffer (1968) called for research to systematically examine and contrast the performance of what is with what might be in the way of organizing the food and fiber sector. Purcell (1973) provided a summary of the criticisms leveled toward

marketing systems research and suggested possible barriers to change, concluding that the inherent goal conflict between players at different levels was a key obstacle to coordination; calling for systems research which helped to identify the benefits of coordination. The cry for action has been repeated on occasion, yet much of the work that exists fails to consider the interrelationships discussed above, opting instead for examining the system at one level or the other and attempting to draw inferences with regard to system performance based on activities at a single stage in the process.

3. The discussion which follows draws heavily upon Reimund, Martin, and Moore (1981, 54-55) which present the prototype model in some detail.

4. Recent discussions regarding the role of information technology as a coordinating mechanism within the food and agribusiness system are provided in Streeter, Sonka, and Hudson (1991) and in Streeter and Hudson (1991).

References

Batte, M. J., and R. Johnson. "Technology and its Impact on American Agriculture." Chapter 12, this volume.

Connor, J. M. 1988. *Food Processing: An Industrial Powerhouse in Transition.* Lexington Books, Lexington, MA.

Connor, J. M., R. T. Rogers, B. W. Marion, and W. F. Mueller. 1985. *The Food Manufacturing Industries: Structure, Strategies, Performance, and Policies.* Lexington, MA: Lexington Books.

Davis, J. H., and R. A. Goldberg. 1957. *A Concept of Agribusiness.* Division of Research, Graduate School of Business, Harvard University, Cambridge, MA.

GAO. 1990. *Beef Industry: Packer Market Concentration and Cattle Prices.* RCED-91-28.

Godwin, M. R., and L. L. Jones. 1971. "The Emerging Food and Fiber System: Implications for Agriculture." *American Journal of Agricultural Economics* 53: 806-815.

Goldberg, R. A. 1988. "A Global Agribusiness Market Revolution: The Restructuring of Agribusiness," in R. A. Goldberg, ed., *Research in Domestic and International Agribusiness Management: A Research Annual* 9: 145-156. Greenwich, CT: JAI Press.

Gustafson, C. R., and P. J. Barry. "Structural Implications of Agricultural Finance." Chapter 15, this volume.

Harwood, J. L., M. N. Leath, and W. G. Heid, Jr. 1989. *The U.S. Milling and Baking Industries.* USDA/ERS, AER-611.

Hudson, M. A. 1990. "Toward a Framework for Examining Agribusiness Competitiveness." *Agribusiness: An International Journal* 6: 181-189.

Kahl, Kandice, M. A. Hudson, and C. E. Ward. 1989. "Cash Settlement for Live Cattle Futures Contracts." *Journal of Futures Markets* 9: 237-248.

Kilmer, R. L. 1986. "Vertical Integration in Agricultural and Food Marketing." *American Journal of Agricultural Economics* 68: 1155-1160.

442

Marion, B. W., and NC 117 Committee. 1986. *The Organization and Performance of the U.S. Food System*. Lexington, MA: Lexington Books.

McMillan, E. L. 1990. "Changing Strategies to Meet Consumer Needs." Keynote Address to the National Agri-Marketing Association Convention, St. Louis, MO.

Mighell, R. L., and L. A. Jones. 1963. *Vertical Coordination in Agriculture*. Agricultural Economic Report No.19, Economic Research Service, U.S. Department of Agriculture.

Purcell, W. D. 1973. "An Approach to Research on Vertical Coordination: The Beef System in Oklahoma." *American Journal of Agricultural Economics* 55: 65-68.

_____. 1990. "Economics of Consolidation in the Beef Sector: Research Challenges." *American Journal of Agricultural Economics* 72: 1210-1218.

Reimund, D. A., J. R. Martin, and C. V. Moore. 1981. *Structural Change in Agriculture: The Experience for Broilers, Fed Cattle, and Processing Vegetables*. Technical Bulletin No. 1648, Economic Research Service, U.S. Department of Agriculture.

Schrader, L. F. 1986. "Responses to Forces Shaping Agricultural Marketing: Contracting." *American Journal of Agricultural Economics* 68: 1161-1166.

Shaffer, J. D. 1968. "Changing Orientations of Marketing Research." *American Journal of Agricultural Economics* 50: 1437-1449.

_____. 1980. "Food System Organization and Performance: Toward a Conceptual Framework." *American Journal of Agricultural Economics* 62: 310-318.

Sonka, S. T., and M. A. Hudson. 1989. "Why Agribusiness Anyway?" *Agribusiness: An International Journal* 5: 305-314.

Stanton, B. F. "Farm Structure: Concept and Definition." Chapter 2, this volume.

Streeter, D. H., S. T. Sonka, and M. A. Hudson. 1991. "Information Technology, Coordination, and Competitiveness." *American Journal of Agricultural Economics* 73: in press.

Streeter, D. H., and M. A. Hudson. 1991. "Information Partnerships in the Food and Agribusiness Sector: An Alternative Coordination Strategy." *Journal of Food Distribution Research* 27: in press.

Thomas, T. W. 1989. "Agricultural Chemicals: An Industry of Change." *Farm Chemicals*. Pp. 38-45.

U.S. Department of Agriculture. *Agriculture Statistics*, various issues.

_____. Economic Research Service. 1990. *Seven Farm Input Industries*. AER-635.

_____. Economic Research Service. 1991. *Agricultural Income and Finance: Situation and Outlook Report*. AFO-40.

_____. Economic Research Service. 1987. *Agricultural Input Industry Indicators in 1974-85*. Agricultural Information Bulletin Number 534.

_____. Economic Research Service. *Agricultural Resources: Situation and Outlook*. August 1987 and earlier issues.

U.S. Department of Commerce. *Census of Manufactures; Volume II, Industry Statistics*. Various issues.

Ward, C. E. 1988. *Meatpacking Competition and Pricing.* Blacksburg, VA: Research Institute on Livestock Pricing.

Wasson, L. S. 1990. "Impact of Demand-Enhancing Farm Policy in the Agricultural Sector: A Firm Level Simulation of Ethanol Production Subsides." Unpublished M.S. Thesis, Texas A&M University.

17

Structural Diversity Under Risk: Choice of Durable Assets

Lindon J. Robison

Introduction

Earlier chapters of this book have explained and described changes in the structure of U.S. agriculture. Part of the structural changes discussed have been related to the quantity of inputs--acres per farm, livestock numbers per farm, workers per farm, income per farm, etc. Another view of farm structure deals with the design of agricultural firms. This view focuses on the combinations, construction, sizes, and durability of assets that provide services to the farm firm.

Little has been written about the design of U.S. agricultural firms. In part, this is attributable to agricultural economists' focuses on production functions and interest in choosing quantities of inputs. To explain the differences in the design of agricultural firms, a different approach is needed. The purpose of this chapter is to develop a different approach to the structural question in agriculture, and to provide a theoretical basis for the observed structural diversity in U.S. agriculture by focusing attention on the design of durables.

Background

The competitive theory of the firm, popularized by Samuelson (1970), holds output price constant and implies a horizontal long-run supply curve with identical firms producing identical quantities. Clearly, this view of the agricultural industry doesn't match reality. Gibrat observed that the distribution of industrial firms is log normally distributed, suggesting firm size is a cumulative stochastic process. Based on a sample of Ontario dairy firms, Fox and Dickson (1989) tested and rejected this explanation of firm size in agriculture. Shapiro et al. (1987) performed a similar study and obtained similar results. Chesher (1979), on the other hand, found support for Gibrat's law. Still, the point is, agricultural firms do not look alike, and changes in firm size cannot be represented as converging to some standard size with identically designed operation.

An alternative to the perfectly competitive industry view is that firm size is a reflection of the distribution of cost functions. This view follows the Viner (1952) approach. Oi (1983) and Panzar and Willig (1983) continue in this tradition.

There are, of course, many other explanations why farm firms do not look alike. One explanation is risk. A second reason is that production can occur with farm firms of several different designs, each efficient for a given size. In addition, in agriculture there are resources that are fixed--with salvage values much less than their value in use. Thus adjustment costs are high.

Finally, it may *not* be possible for firms to duplicate one another. For example, humans who are unique cannot be duplicated to provide identical management services. Services from a single manager are likely to change over time as age and experience alter his or her abilities (Sumner and Leiby 1987). Farm units are not equal distance from suppliers and farm product purchasers. No two tracts of land are identical. Farm firms are likely to have begun farming at different times and to have experienced different prices, real interest rates, and policy cycles. The consequence of these differences will be that balance sheets will differ--altering the kinds, cost, and amount of credit firms can expect to receive from financial intermediaries. Finally, with these differences even in the most alike firms, each firm will find it has different comparative advantages resulting in distinctiveness in firms, even among those producing the same product.

There are many other factors that would lead us to declare that firms should be distinct. This chapter focuses on three: differences in durable assets, preferences for risk, and asset fixity. This chapter is organized as follows. First, it identifies important differences among durables. Then it uses the differences among durables to classify them into 12 categories. Next, this chapter shows how differences among durables lead to distinct average cost curves. Finally, this chapter shows how risk, durable characteristics, and asset fixity lead firms to choose distinct types of operations that result in distinct average cost curves even under

446

equilibrium conditions. Consequently, there is no reason to expect farm firms to look alike.

Classifying Durable Assets

A durable factor of production is a stock of potential services available in more than one time period. In addition, a durable retains a unique identity over time apart from other factors of production. If the durable were a service (a flow) such as heating, lighting, pulling, growing, painting, thinking, etc., it would not be capable of providing services in more than one period or even existing for more than one period. The durable's potential to supply services over time is called its lifetime capacity (LTC). Any attempt to classify durables must focus on their service potential.

Such a focus on services might prompt the question: what alters the LTC of each durable? For example, does the durable's capacity to provide services depend on time and use, or is it endurable? Then one might ask, at what rate does the durable provide services? A single rate of service available from the durable leaves the decision maker with few, if any, choices. On the other hand, if a durable provides services at multiple rates, in each period a use decision must be made.

Finally, one might ask: what are the acquisition (sale) possibilities associated with the durable? Does its service potential come in divisible quantities, like seed and gasoline? Or is it available in lumpy units, like barns, bulls, combines, and cars? All of these factors will be considered as a durable classification scheme is developed.

Altering a Durable's LTC

Most durables have their service potential altered by the passage of time. This aging process may, in some cases, improve a durable's service potential as in the case of aging cheese and growing crops. In other cases, time may reduce a durable's service potential as in the case of a barn roof. Thus the study of durables must include the effect of time on the durable's LTC.

The durable's LTC may also be changed by use. For example, a car with 100,000 miles registered on its odometer has a different LTC than the same car with 50,000 miles on its odometer. Durable use rates, then, along with time, are needed in durable analysis.[1]

Most durables leave a residual product after the service extraction is completed. That is, because durables are not often completely consumed on transformed in the process of providing services, it exists in some form beyond its economic life. Typical of this durable type are barns, breeding stock, and glass bottles.

For completeness, we suggest that some durables are very nearly unaffected by either time or use. We refer to this type of durable as an endurable. Land can be described as endurable. Land, however, may be lost or destroyed. It retains its properties only if it receives sufficient service from other durables to replenish its service capacity lost through use and time. When the analyst implicitly assumes this regeneration process, it is appropriate to label land as endurable.

A painting could also be considered an endurable. It provides sight services for art connoisseurs. In the process of providing these services, its service potential is not reduced or altered regardless of how many people view the painting. But imperceptibly to all but the trained observer, the painting is decaying. Time, temperature, and moisture all extract their toll. Only if they are of no interest to the research or if their effect is negligible over the time period of interest, can paintings be called endurables. Despite these considerations, endurables represent a useful simplification in some studies because they allow us to ignore decay/growth in durables' LTCs.[2]

Thus durables can be classified according to ways in which their LTC changes: by use, by time, or not at all. Many durables, however, have their capacity altered by both time and use. Thus any one durable may be included in more than one category depending on which decay method dominates.

Durables and the Services They Provide

So far this chapter has emphasized that a durable's most important characteristic is its potential to supply services. But in this regard, there are important differences in the supply of services durables provide.

An important question to be answered about the durable's service supply is: what is the cost of varying service? Does the durable have more than one service rate? Consider the durable we call breeding stock. The services they provide are reproduction services that produce offsprings; a calf, a colt, a lamb, etc. The only control the durable owner has in this process is a breeding decision and a nutrition level. If the animal is bred, the service extraction rate begins and under most circumstances cannot be postponed, or even terminated. "On" or "off" appear to be the only alternative choices for service extraction.

When the "on" decision cannot be reversed, except at a very high cost, the durable's service flow is described as irreversible. For example, reversing the reproductive service of bred livestock is the cost of aborting the offspring or killing the animal. Other durables whose services once turned "on" cannot be reversed without incurring significant costs are clocks, body organs, and fertilized seeds. A clock has time keeping value only when it doesn't stop (unless, of course, it's a stop watch). A transplanted heart is a durable that its recipient and donor would agree would incur high reversibility costs if stopped. Finally, sprouted seeds are durables that cannot have their service flow interrupted without incurring high costs. These durables all have the characteristic in common that once they are turned on, it is very costly to turn them off.

It will often occur that single use durables called expendibles have their service rate fixed. Seeds, fertilizer, fuel, and medicines, for example, are single use durables. They can only be used once, and in the process are converted to services or to durables of another form. Thus, if they are used only once, their service extraction rate is fixed.

An electric light fixture is also a durable. It is also either on or off. But the action that turns on its service flow is easily interrupted or stopped. So during any given time period, the light may be turned on or off several times. This interruptibility in the service flow provided by an on and off switch makes the durable services controllable. Control allows services to be extracted from the durable at varying rates depending on the need for light.

Many durables have a fixed capacity to provide services in any given period, such as the storage space in a barn. This does not, however, mean that the durable has a fixed service supply rate. The service supply rate can be controlled. A person may decide on how much hay or other durables will receive the storage services available from the barn. So the barn has a variable service supply rate, limited by total capacity.

Consider the service supply rate of some selected durables. Cars can be driven at various speeds or not at all, altering the rate at which they supply transportation services. A well (and pump) also provides transportation services. It moves water from below ground to above ground at variable rates. Finally, stored fertilizer, fuel, and seed are all durables if the time period of analysis is appropriately defined. However, once they are applied and their services turned on, the service flow level provided is fixed.

Some durables may provide multiple services. Multiple use potentials also add divisibility to service extraction. A tractor can pull many different kinds of farm implements increasing the divisibility of its services. It can also serve as a stationary power source to power a mill or a water pump, or it can provide transportation services.

Thus durables can be classified according to the reversibility or the cost of turning off their service flow. Durables with reversible service flows are described in our classification scheme as having a "variable" service extraction rate. Durables with irreversible or a high cost of turning off service flows are described in our classification scheme as having a "fixed" service extraction rate.

Acquisition Characteristics of Durables

Thus far our discussion has described how the durable's LTC, its capacity to provide services, can be altered by time and use. Then this chapter described the service potential and the control afforded the decision maker in determining the supply of services provided by the durable. The cost of reversing service levels was a key consideration in determining the control of the durable's service flow. Now is described how the service potential of the durable can be altered through purchase.

In some cases, a durable's LTC may be increased through the purchase of repairs. Other times, the most efficient means of increasing LTC is to purchase a new and sometimes a differently designed durable. These decisions require careful attention to acquisition characteristics of durables; characteristics which will enable us to complete our durable classification scheme.

We now describe durables as either lumpy or divisible in acquisition. This classification, however, requires a careful description of what is acquired. We already agreed that we acquire LTC or the inventory of potential services when we gain control of a durable. But the form of the inventory is of interest. To increase the inventory of gasoline services, we simply buy more gasoline. There is no economy of scale in the purchase of gasoline except perhaps in the transaction cost: one gallon of gasoline provides one-half the amount of services available from two gallons of gasoline. Another way to describe this relationship would be to say that the LTC of gasoline is linearly related to the amount of gasoline purchased.

When the LTC of the durable is linearly related to the stock of the durable, we call the durable "divisible in acquisition." Examples of this type of durable include fertilizer, seeds, gasoline, feed, bonds of the same company, stocks of the same company, paint, and most animals used in commercial production.

Another kind of durable, one we call lumpy in acquisition, exhibits economies of size in acquisition. Consider a fence around one square mile. This would require four miles of fence. Now consider a fence around four square miles. This would require eight miles of fence. Doubling the fence length quadruples the area enclosed. Thus, the area enclosed (a service potential) is not linearly related to miles of fence. This is an example of economy of scale.

Another example of the same phenomenon is the supply of horsepower. One tractor design may be preferred for small amounts of horsepower. But the ideal way to supply twice as much horsepower is not to purchase two tractors of the same size and use them pulling in tandem like engines on a train. Instead, the ideal way to supply twice as much horsepower is to design a different tractor with the capacity of supplying the necessary power. This is because there is economy of scale. For one thing, a redesigned tractor only requires one driver as opposed to two drivers needed to extract horsepower from two smaller tractors.

Durables may also be considered to be "lumpy" in acquisition if they are unique in some important characteristics. Earlier this chapter described livestock used in commercial production as divisible in acquisition. There are, however, animals that can be considered exceptions. These are the animals that possess some unique design characteristic. For example, Secretariat, the race horse, possessed a unique phenotype and genotype that simply cannot be replaced or duplicated. Thus, Secretariat and the acquisition of highly selected breeding stock, in general, are likely to be lumpy in acquisition.

Thus, durables that achieve economies of scale in the acquisition of LTC must be redesigned for each required level of LTC. There will be no one optimal design. The optimal design will depend on the level of services required from the durable. So this type of durable is considered as lumpy in acquisition. This is because the optimal way to increase its LTC is not simply to increase the quantity of the existing durable. It requires a change in design.

Thus durables can be classified according to their acquisition characteristics. Those durables in which there exist economies of scale provided by design changes are called lumpy in acquisition. Those durables exhibiting no economy of scale in design are called divisible in acquisition.

Twelve Types of Durables

These distinct durable characteristics described so far result in 12 durable types. Table 17.1 summarizes the 12 different durable types resulting from our classification scheme and provides examples of each type.

Miscellaneous Problems Encountered
When Classifying Durables

One problem encountered when attempting to classify durables is the following. Durables may have more than one use; and different uses may result in the durable being classified differently. Land is an example of this difficulty.

TABLE 17.1 A Classification of Durable Assets According to Acquisition, Use, and Changes in Lifetime Capacity (LTC) of the Durable to Provide Services

	Acquisition Characteristics		Use Characteristics		How LTC is Changed			Example of Durable (Services)
	Lumpy	Divisible	Fixed (irreversible)	Variable (reversible)	Time	Use	Enduring	
1.	x		x		x			Water dams (erosion inhibiting)
2.	x		x			x		Breeding stock (producing offspring)
3.	x		x				x	Cert. of Dep. (earning interest)
4.	x			x	x			Barn (storing)
5.	x			x		x		Tractor (pulling)
6.	x			x			x	Painting (sighting)
7.		x	x		x			Roof (barrier services)
8.		x	x			x		Milk cows (producing milk)
9.		x	x				x	Land (providing loc. to support buildings, etc.)
10.		x		x	x			Highway road sign (giving information)
11.		x		x		x		Light bulb (turning darkness to light)
12.		x		x			x	Paper clips (holding papers together)

Land used to grow crops is a durable divisible in acquisition and considered indestructible in most studies. Land devoted to growing crops has a high reversibility cost so we consider its supply of services fixed. But land used for recreation or for roads has its services supplied determined by visitors to the recreation site or by the number of travelers using the road. Finally, some studies of land consider erosion as a problem which takes land out of the enduring class and puts it in the time or use decay class. The point is that the unique classification of the durable often depends on its application when it is a durable capable of providing many different services.

Another classification problem is that many durables change form or location or both. In the process, they become different types of durables. For example, paint in cans is a durable divisible in acquisition. Moreover, since it can be applied a drop at a time, it is also variable in use. In this case, variable in use means that at any time the decision maker may stop the extraction of services by closing the lid on the paint can and refusing to paint any more. On the other hand, paint applied to the exterior of a building is different than paint in the can. Now the acquisition of applied paint is lumpy, not divisible. In addition, it is nearly impossible or at least very costly to stop the service flow from the applied paint. Thus, we now classify the paint on a surface as lumpy in acquisition with a fixed service extraction rate, and a decay determined by time.

Another classification problem is that durable decay often depends on both time and use. An unused car wears out by time. Cars driven frequently wear out through use and time. Our decay categories simply list extremes--with most durables wearing out through time and use.

The other problem encountered in the classification of durables is that our definition is not capable of excluding anything that lasts for more than one period and supplies a service. Laws, wills, riparian rights, and right of ways would all be durables under our classification. Thus institutions are considered durables since: ". . . institutions are sets of ordered relationships among people that define their rights, their exposure to the rights of others, their privileges, and their responsibilities" (Schmid 1989, 6).

Having proposed a classification scheme to organize our understanding of durables, this chapter now turns to an evaluation of the use of durables as topics for research in the literature of agricultural economists.

Durable Asset Types and Average Cost Curves

Twelve different types of durables have been identified. What is deduced next is that differences in durable characteristics result in different costs of extracting services. Thus the average cost curve of firms, even ones that produce the same product, may be quite distinct if they use different durables.

Average cost curves for durables divisible in acquisition, variable in use, and decay through use most likely meet the requirements of our perfectly competitive models in which all firms look alike.[3] For this durable type (durable type 11), the average cost of extracting services is a constant (Figure 17.1). On the other hand, for durables whose service capacity is divisible in acquisition, has variable service flows, but whose capacity is altered by time alone (durable type 10), the average cost curve decreases with the level of services extracted (Figure 17.2). And for durables whose capacity is altered by use and time, the average cost curve may be "U" shaped (Figure 17.3). Of course, each of the other durables may produce still different shaped cost curves.

Risk and the Choice of Durables

To say that different durables are associated with different average cost curves does not alone explain why firms may be distinct. The complete explanation requires the introduction of uncertainty. Indeed, with certainty and differences in durables, firms in an industry should look alike (assuming it were possible to duplicate firms) because once an output level is established, the cost minimizing plant would be preferred. But once uncertainty enters, and not one but a set of service levels are required of the durable, it becomes less clear which of a family of durables is preferred assuming not all decision makers have the same risk preferences.

To illustrate this latter point, consider the two average cost curves graphed in Figure 17.4. Durable 1 has a lower average cost of providing services between service level extraction rates BC. Consequently, durable 1 is to be preferred as long as the service extraction rate is between those levels. Outside of those levels of production, however, durable 2 is preferred. Now consider the two durables, durable 3 and durable 4, described in Figure 17.5. If the single service extraction rate were A, we would be indifferent between durables 3 and 4. And if the service extraction rates required were A-ΔA or A+ΔA, with equal likelihood we would also be indifferent between the two durables. Thus it is possible to be indifferent between different durables even under uncertainty. On the other hand, if one is uncertain about the range of services required, the preferred durable likely depends on individual risk preferences.

FIGURE 17.1   The Total and Average Cost of Services from Durable Type 11, Divisible in Acquisition, Variable Use Rate, and Decay Through Use

Panel a

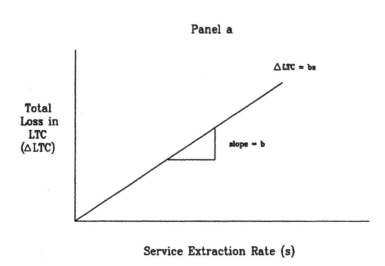

Total
Loss in
LTC
(△LTC)

△LTC = bs

slope = b

Service Extraction Rate (s)

Panel b

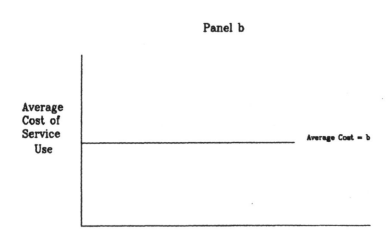

Average
Cost of
Service
Use

Average Cost = b

Service Extraction Rate (s)

FIGURE 17.2   The Total and Average Cost of Services from Durable Type 10,
               Divisible in Acquisition, Variable Use Rate, and Decay by the Passage
               of Time

Panel a

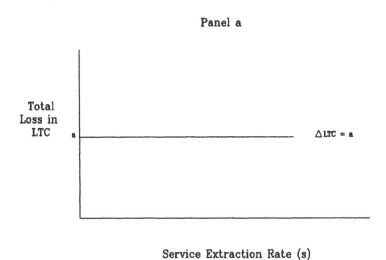

Service Extraction Rate (s)

Panel b

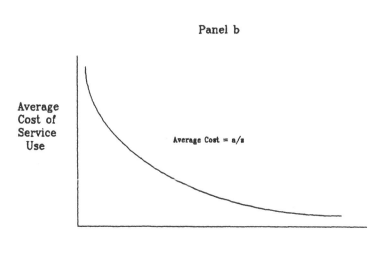

Service Extraction Rate (s)

FIGURE 17.3 The Total and Average Cost of Services from a Combination of Durable Types 10 and 11, Divisible in Acquisition, Variable in Use, and Decay from Use and the Passage of Time

Panel a

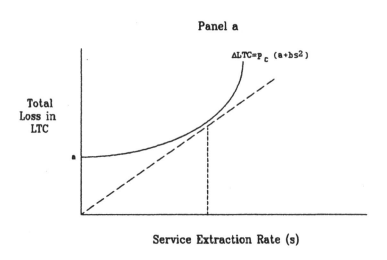

Panel b

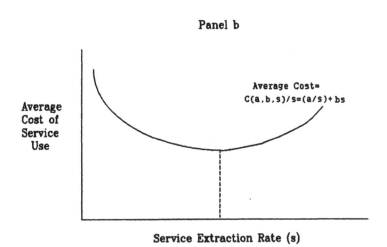

FIGURE 17.4   Average Cost of Services Provided by Durables 1 and 2

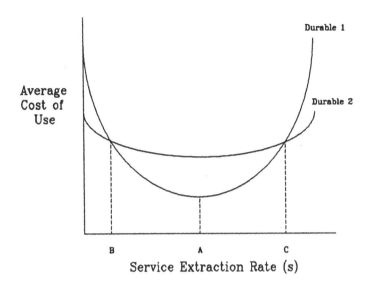

FIGURE 17.5   Average Cost of Services Provided by Durables 3 and 4

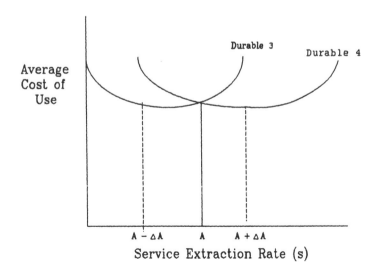

Risk and Structural Diversity

An important article in the risk literature is Sandmo's. He showed that a firm's output sold at a risky price depends on the decision maker's risk attitudes. Thus firms facing the same random output price and the same certain costs would choose to produce at different levels. Others, including Cass and Stiglitz (1972); Levy (1973); and Robison and Barry (1977), deduced income and substitution effects for risky attitudes corresponding to differences in risk preferences. Robison and Barry (1987, Chapter 11) reviewed how returns to scale can lead alternatively to diversification or specialization under risk. White and Irwin (1972) concluded from Census data that specialization is related to firm size. Meanwhile, from a survey of 1,000 crop farmers in California, Pope and Prescott (1980) concluded that as firm wealth increases, firms place less emphasis on diversification. But as firm size (wealth) increases, they found that firms diversify.

Firm behavior, in response to a host of risks, has been studied. For example, Robison, Barry, and Burghardt (1986) studied the effects of financial stress. Jones-Lee (1976); Ahsan et al. (1982) and Trechter (1984) studied the effects of alternative insurance plans for the firm. Hartman (1976); Turnovsky (1973); and Robison and Barry (1987, Chapter 17) studied the role of flexibility. And Fleisher (1985); Gardner (1978); and Gardner, Just, Kramer, and Pope (1984) studied the influence of public policy on firm behavior.

A common characteristic of the studies of firm behavior under risk just cited is their focus on the firm. Recent studies by Appelbaum and Katz (1986) and by Meyer and Robison (1991) deduce aggregate risk models. Appelbaum and Katz (1986) analyze equilibrium for competitive firms facing output price risk when operating in a constant cost industry. In the Meyer and Robison (1991) study, equilibrium conditions are found for a competitive industry when an input's price adjusts to maintain industry equilibrium. In this latter paper, the authors find that many policies designed to reduce risk in the farm sector simply increase land prices and intensify production.

In what follows, we will show an important result using the industry equilibrium framework of Meyer and Robison (1991). This result is that industry and firm equilibrium conditions under risk can be satisfied and still permit diversity--even in the long run.

Firm and Industry Equilibrium Levels Under Risk

One of the most widely used models in risk is the portfolio model. Popularized in the finance literature by Markowitz (1952) and extensively applied in agricultural economics (Robison and Brake 1979), it characterizes a particular class of durables: those divisible in acquisition and use, and whose decay may be attributed to use (durable type 11).[4] For this class of durables, however, the average cost curve is a constant. As a result, separation properties associated with this model arrive at the conclusion that all firms hold identical combinations of risky assets although the relative amounts of risky and safe assets may be different. The reason for this conclusion is that design changes are not needed to obtain efficient combinations.

What is shown next is that for a durable type with the combined decay characteristics of durable types 10 and 11, not only will the firms invest different amounts in risky and safe assets, but will purchase different durables and produce at different levels.

To illustrate, suppose a durable is required which has the combined characteristics of durable types 10 and 11. If s is the service level required and p_c is a cost per unit of durable capacity, then capacity costs $C(s)$ are expressed as:

$$C(s) = p_c\,(a + bs^2) \tag{1}$$

where $p_c a$ is the value of lost capacity as a result of the passage of time and $p_c s_2$ is the value of the service capacity lost in the process of supplying s units of service. The cost of the service flow is the sum of time and use costs described in Figure 17.3. A quadratic form of the cost function was chosen because it corresponds to accepted wisdom that the loss in capacity increases at an increasing rate as the service level per period increases. Thus $C'(s)>0$ and $C''(s)>0$.

Next, suppose that the engineers have determined that the acquisition cost of the durable $C(a,b)$ depends on design paramaters or a and b, and can be expressed as:

$$C(a,b) = \frac{p_a}{a} + \frac{p_b}{b} \tag{2}$$

where p_a and p_b are exogenously determined and a,b>0. The form of $C(a,b)$ is chosen arbitrarily for this example. In practice, it is determined by physical laws, material costs, and estimated by engineers.

The total cost curve $C(a,b,s)$ for this durable in the first period of use can be written as the opportunity cost of the acquisition price, $rC(a,b)$, where r is the opportunity cost of capital, plus the decline in value of the durable as a result of time and use, $C(s)$. The sum of these two costs equals:[5]

$$C(a,b,s) = rC(a,b) + C(s)$$

$$= r\left(\frac{p_a}{a} + \frac{p_b}{b}\right) + p_c(a + bs^2)$$

The average total cost curve is obtained by dividing $C(a,b,s)$ by s. The short-run average cost curve for this particular cost curve is the familiar "U" shaped curve graphed in panel b of Figure 17.3. To find the long-run average cost curve we differentiate $C(a,b,s)$ with respect to "a" and "b" and find their values that minimize cost for a given service level s to equal:

$$a = (rp_a/p_c)^{1/2}$$

and

$$b = (rp_b/p_c)^{1/2} / s$$

Substituting these optimal values for a and b into $C(a,b,s)$ and dividing the result by s, we obtain the long-run average cost curve (LRAC) or the envelope of short-run average cost curves. This curve has the shape of the "lazy L" which some authors (e.g., Madden) have claimed characterizes average cost curves for many U.S. agricultural firms. It is described graphically in Figure 17.6.

$$C(a,b,s) / s = 2(rp_c p_b)^{1/2} + 2(rp_c p_a)^{1/2} / s \quad \text{for } s > 0$$

Short run average cost curve (SRAC) is tangent to the LRAC at a point for a particular fixed set of values for design parameters a and b. The only difficulty with this characterization of average costs in the agricultural sector is that it would predict firms to be all very large since average costs continue to decline with size. Thus the certainty model alone, even with a more general description of costs, does not predict the diversified agricultural structure that we know exists. However, adding risk provides the result consistent with our observation of a diversified agricultural structure.

To introduce uncertainty into this problem, we assume the services s from the durable are sold at a stochastic price $\hat{p}$ with expected value p and standard deviation σ_p. Then expected profits, μ, equal:[6]

$$\mu = ps - p_c(a + bs^2) - \left(\frac{p_a}{a} + \frac{p_b}{b}\right)r$$

where the standard deviation of profits σ equals:

$$\sigma = s\sigma_p$$

The mean-standard deviation (MS) frontier of combinations of (μ,σ) can be found by solving for s in the expression for σ and substituting the result into the expression for μ. The result is:

$$\mu = p\frac{\sigma}{\sigma_p} - p_c\left[a + b\left(\frac{\sigma}{\sigma_p}\right)^2\right] + \left(\frac{p_a}{a} + \frac{p_b}{b}\right)r \tag{3}$$

However, the result in (3) is equivalent to a short-run MS frontier since parameters a and b are fixed (see Figure 17.6). Also notice that the mean-standard deviation (MS) set described is not linear but concave with a maximum μ at the point $\sigma = p\sigma_p/2bp_c$. To find the long-run MS frontier μ is maximized with respect to a and b and the resulting expression substituted into equation (3).

Maximizing (3) with respect to a and b produce the results:

$$a* = (rp_a/p_c)^{1/2}$$

and

$$b* = (\sigma_p/\sigma)(rp_b/p_c)^{1/2}$$

The resulting long-run MS frontier obtained by substituting a* and b* into (3) for a and b equals:

$$\mu = p\frac{\sigma}{\sigma_p} - 2\left[(rp_cp_a)^{1/2} + (rp_cp_b)^{1/2}\left(\frac{\sigma}{\sigma_p}\right)\right]$$

$$\tag{4}$$

$$= -2(rp_cp_a)^{1/2} + \left[\frac{p}{\sigma_p} - \frac{2(rp_cp_b)^{1/2}}{\sigma_p}\right]\sigma$$

a linear MS frontier. Figure 17.7 is a graph of alternative long-run and short-run MS frontiers. Since all the points on the long-run frontier are

FIGURE 17.6   Short-Run Average Cost Curves (SRAC) and Long-Run Average Cost
Curves (LRAC)

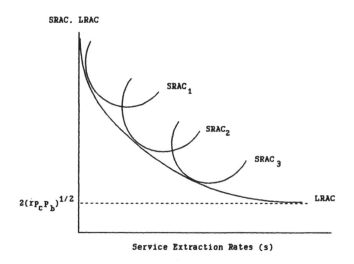

Service Extraction Rates (s)

FIGURE 17.7   A Long-Run MS Frontier and Alternative Short-Run MS Frontiers
MS_1, MS_2, and MS_3

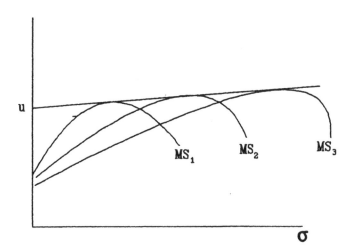

MS efficient, no single point is preferred to the others. Only different attitudes toward risk would identify firms along the frontier--with each different MS point representing a firm with a distinct durable.

Industry Equilibrium Requirements Under Risk

In another paper, Meyer and Robison (1991) imposed industry equilibrium on firms by allowing them to invest in a market with MS efficient choices of

$$\mu* = \phi + \eta\sigma$$

where ϕ is a risk-free rate of return, η is a parameter measuring the diversification potential of the asset in question and where σ is the standard deviation of profits per unit of investment. If the firm could trade in such a market, then increases in risk occasioned by producing at higher service levels must also be compensated at rate η. That is, in equilibrium the firm is expected to earn η on its increases in risk. Thus, differentiating (4) with respect to σ and setting the result equal to η, we obtain the equilibrium requirement:

$$\frac{d\mu}{d\sigma} = \frac{p}{\sigma_p} - \frac{2(rp_cp_b)^{1/2}}{\sigma_p} = \eta \qquad (5a)$$

or

$$p = \eta\sigma_p + 2(rp_cp_b)^{1/2} \qquad (5b)$$

The firm's opportunity set subject to equilibrium conditions can be found by substituting the right-hand side of (5b) for p in equation (3). The result is:[7]

$$\mu = \eta\sigma - 2(rp_cp_a)^{1/2}$$

This description of equilibrium also requires the assumption of asset fixity. Namely, that resources (such as land) are fixed in agriculture because their use value is bounded by their acquisition and salvage values. However, service extraction rates can be varied and, at the margin, earn returns available in the capital market.

This equilibrium condition has some important implications for agricultural policy. Some perceive that the well-being of the agricultural sector can be improved by stabilizing prices for its services. This model suggests, however, that efforts to reduce the variance of agricultural prices σ_p will simply cause prices to adjust so that the rate of return on agricultural inputs equals that available in the market. Thus, a reduction in the standard deviation of prices, σ_p, simply reduces prices:

$$\frac{-dp}{d\sigma_p} = -\eta$$

Moreover, any changes in opportunities in the market, reflected by changes in η, will end up increasing p as follows:

$$\frac{dp}{d\eta} = \sigma_p$$

Moreover, a recession affecting the value of services in the nonagricultural sector will eventually influence the price of agricultural services. Alternatively, if the adjustment is obtained through changes in the capacity price p_c, then in equilibrium the following is true:

$$p_c = \frac{1}{rp_b}\left(\frac{p - \eta\sigma_p}{2}\right)^2$$

Then any policy that increases p or decreases σ_p will increase or decrease p_c. In that case, all firms' choices of a and b, and hence their structural preferences, will be changed. For example, suppose p increases causing equilibrium p_c to rise. As a consequence:

$$\frac{da}{dp_c} < 0$$

and

$$\frac{db}{dp_c} < 0$$

An Example

An example will illustrate the solution. Suppose a firm faces the lazy "L" average cost functions described in this chapter. Moreover, let the firm's preferences be characterized by the linear mean-variance function:

$$y_{CE} = \mu - \frac{\lambda}{2}\sigma^2$$

where $\lambda = \lambda(p_a, p_b, p_c, p, \sigma_p)$. Then assume the firm maximizes its certainty- equivalent income over its choices of s, a, and b. Thus the firm maximizes:

$$\max_{s,a,b} y_{CE} = ps - p_c(a+bs^2) - \left(\frac{rp_a}{a} + \frac{rp_b}{b}\right) - \frac{\lambda}{2}s^2\sigma_p^2$$

The first-order conditions are:

$$\frac{\partial y_{CE}}{\partial s} = p - 2p_c bs - \lambda\sigma_p^2 s = 0$$

$$\frac{\partial y_{CE}}{\partial a} = -p_c + \frac{rp_a}{a^2} = 0$$

$$\frac{\partial y_{CE}}{\partial b} = -p_c s^2 + \frac{rp_b}{b^2} = 0$$

It can be easily shown that second-order conditions for a maximum are satisfied and the solutions for s, a, and b are, after substituting industry equilibrium conditions for p:

$$s = \frac{p - 2(rp_c p_b)^{1/2}}{\lambda\sigma_p^2} = \frac{\eta}{\lambda\sigma_p}$$

$$a = \left(\frac{rp_a}{p_c}\right)^{1/2}$$

and

$$b = \frac{\lambda\left(\frac{p_b}{p_c}\right)^{1/2}\sigma_p^2}{p - 2(rp_b p_c)^{1/2}} = \frac{\lambda\sigma_p\left(\frac{rp_b}{p_c}\right)^{1/2}}{\eta}$$

Notice that the firm's choice of b depends on λ, the firm's risk preference. That is, the firm selects from the efficient set of mean-variance solutions, a structure that is in long-run equilibrium and unique to its own preferences.

Thus we should not be surprised to find firms whose average costs are located all along the long-run average cost curve and at the same time are in equilibrium. Moreover, we should not be surprised to learn that agricultural policies affecting the probability distribution of prices paid for agricultural services will alter the structural characteristics of agricultural firms. In particular, reducing σ_p^2 will cause them to increase in size:

$$\frac{ds}{d\sigma_p^2} < 0$$

Conclusions

The chapter relaxed the assumption that the designs of production units are fixed. It then provided a description of the different durables and design opportunities available to firms as they make their production and marketing plans and determine their response to risk. Then, an efficient mean-standard deviation set of designs was derived when durables decay as a function of both time and use. The conclusion reached was that design choices in the face of risk provide one explanation for the distinctiveness observed between agricultural firms.

A conclusion of this chapter is that governmental policies affecting the probability distributions of returns facing firms will also influence the designs that firms select. For the simple design problem presented in this chapter, lowering output price risk would lead to designs their firms for increasing levels of output.

Clearly, this chapter has only introduced the subject of design selection and how risk affects the preferred design. Much more work is needed to better characterize the relevant design choices facing firms and how design choices differ between farm types. Perhaps research in this area will help to answer some of the yet unresolved questions relating to firm size and structure.

Notes

The author thanks Jack Meyer and Glenn Fox for helpful comments received on an earlier version of this chapter. The author also thanks Kent Gwilliam for helpful suggestions on the topic of classifying durables.

1. Expendables, such as feed, seed, fuel and like durables, are affected by use. Unlike durables, expendables are consumed or transformed into a new stock in the process of providing services. The consequence of this transformation is the absence of any disposable inventory that may require sale or disposal. Moreover, because disposal is not a concern, expendables are often introduced into production functions without any distinction between it and the service it

provides. In the remainder of this chapter no distinction is made between durables (stocks that last more than one period) and expendables (stocks that last until they are used. The important characteristic they share is their stock characteristics.

2. This category of durables includes catalysts—those durables that provide services but are not themselves changed. An example would be water in a car radiator.

3. Usually cost functions are defined over outputs. In this study, the output is an intermediate product, namely the services provided by durables.

4. Decay may be attributed to use since each unit of a financial input becomes an output at the end of the period; a bond or stock is both an input at the beginning of the period and an output at the end of the period.

5. Obviously, design considerations should include multiperiod use and decay considerations. But for our purpose, a one period model is sufficient to demonstrate that durable design may be distinct even under equilibrium considerations.

6. Notice that since $\hat{p}$ is multiplied by s, this problem satisfies location/scale conditions and is capable of being described in mean-standard deviation units.

7. Silberberg uses a similar approach, allowing product price to change, to obtain equilibrium.

References

Ahsan, S. M., A. A. G. Ali, and N. J. Kurian. 1982. "Towards A Theory of Agricultural Insurance." *American Journal of Agricultural Economics* 64: 520-529.

Appelbaum, E., and E. Katz. 1986. "Measures of Risk Aversion and Comparative Statics of Industry Equilibrium." *The American Economic Review* 76: 524-529.

Cass, D., and J. E. Stiglitz. 1972. "Risk Aversion and Wealth Effects on Portfolios With Many Assets." *Review of Economic Studies* 39: 331-354.

Chesher, A. 1979. "Testing the Law of Proportionate Effect." *Journal of Industrial Economics* 27: 403-411.

Debertin, D. L. 1986. *Agricultural Production Economics*. Macmillan Publishing Co.

Fleisher, B. 1985. "A New Measure of Attitude Toward Risk." Unpublished Ph.D. thesis, Michigan State University.

Fox, G., and E. Dickson. 1989. "Farm Growth in the Ontario Dairy Industry: A Skeptical Look at Gibrat's Law," in *Determinants of Farm Size and Structure*. Iowa State University, Ames, IA.

Gardner, B. L. 1978. "Public Policy and the Control of Agricultural Production." *American Journal of Agricultural Economics* 60: 836-843.

Gardner, B. L., R. E. Just, R. A. Kramer, and R. D. Pope. 1984. "Agricultural Policy and Risk," in P. J. Barry, ed., *Risk Management in Agriculture*. Chapter 16. Ames, IA: Iowa State University Press.

Hartman, R. 1976. "Factor Demand With Output Price Uncertainty." *American Economic Review* 66: 675-681.

Jones-Lee, M. W. 1976. *The Value of Life*. London: Martin Robertson.

468

Levy, H. 1973. "The Demand for Assets Under Conditions of Risk." *Journal of Finance* 38: 79-96.

Madden, J. P. 1967. *Economies of Size in Farming.* AER-107. Washington, DC: Economic Research Service, U.S. Department of Agriculture.

Markowitz, H. 1952. "Portfolio Selection." *Journal of Finance* 7: 77-91.

Meyer, J. 1987. "Two-Moment Decision Models and Expected Utility Maximization." *American Economic Review* 77: 421-430.

Meyer, J., and L. J. Robison. 1991. "Aggregate Effects of Risk in the Agricultural Sector." *American Journal of Economics* 73: 18-24.

Oi, W. Y. 1983. "Heterogeneous Firms and the Organization of Production." *Economic Inquiry* 21:147-155.

Panzar, J. C., and R. D. Willig. 1983. "On the Comparative Statics of a Competitive Industry With Inframarginal Firms." *American Economic Review* 68 (3).

Pope, R., and R. Prescott. 1980. "Diversification in Relation to Farm Size and Other Socioeconomic Characteristics." *American Journal of Agricultural Economics* 62: 554-559.

Robison, L. J., and P. J. Barry. 1977. "Portfolio Adjustments: An Application to Rural Banking." *American Journal of Agricultural Economics* 59: 311-320.

_____. 1987. *The Competitive Firm's Response to Risk.* New York and London: Macmillan Publishing Company.

Robison, L. J., P. J. Barry, and W. G. Burghardt. 1987. "Borrowing Behavior Under Financial Stress by the Proprietary Firm: A Theoretical Analysis." *Western Journal of Agricultural Economics* 12: 144-151.

Robison, L. J., and J. R. Brake. 1979. "Application of Portfolio Theory to Farmer and Lender Behavior." *American Journal of Agricultural Economics* 61: 158-164.

Samuelson, P. A. 1970. *Foundations of Economic Analysis.* New York: Atheneum.

Sandmo, A. 1971. "On the Theory of the Competitive Firm Under Price Uncertainty." *American Economic Review* 61: 65-73.

Schmid, A. A. 1989. *Benefit-Cost Analysis - A Political Economy Approach.* Boulder, San Francisco, and London: Westview Press.

Shapiro, D., R. D. Bollman, and P. Ehrensaft. 1987. "Farm Size and Growth in Canada." *American Journal of Agricultural Economics* 69: 477-483.

Silberberg, E. 1974. "The Theory of the Firm in Long Run Equilibrium." *American Economic Review* 64:734-741.

Sumner, D. A., and J. D. Leiby. 1987. "An Econometric Analysis of the Effects of Human Capital on Size and Growth Among Dairy Farms." *American Journal of Agricultural Economics* 69:465-470.

Trechter, D. 1984. "An Economic Analysis of Farm Revenue Insurance." Unpublished Ph.D. thesis, Michigan State University.

Turnovsky, S. J. 1973. "Production Flexibility, Price Uncertainty, and the Behavior of the Competitive Firm." *International Economic Review* 14: 395-412.

Viner, J. 1952. "Cost Curves and Supply Curves," in K. E. Boulding and G. J. Stigler, ed., American Economic Association, *Readings in Price Theory.* Homewood, IL: Richard D. Irwin.

White, T. and G. Irwin. 1972. "Farm Size and Specialization," in G. Ball and E. Heady, ed., *Size, Structure, and Future of Farms.* Ames, IA: Iowa State University Press.

18

Empirical Analysis of Tenure Patterns and Farm Structure

Larry Janssen

Introduction

In this chapter, the importance of various farmland tenure arrangements and their relationships to structural changes in U.S. agriculture are examined. Farm real estate represents about two-thirds of the physical asset wealth of the U.S. farming sector (USDA 1991). Claims on current income flows and potential or realized capital appreciation are major sources of income and wealth for farmers and landowners. The major economic policy concern about farm real estate is whether existing or prospective land tenure arrangements facilitate or impede efficient resource allocation, including capacity for technological change. The major social policy concerns are related to the continued access of various social classes, such as beginning farmers without substantial family assistance or expanding farmers, to ownership and/or use of farmland.

Land tenure is "concerned with the many relationships that govern access to and use of land resources and claims on goods and services that flow from them" (Moyer et al. 1969, 1). Land tenure is an important issue in the contemporary agricultural development of many nations throughout the world. Historically, land tenure has been a major issue in U.S. agricultural development. Many private interest groups and public leaders continue to express concern about various contemporary land tenure trends (Strange 1988; Paarlberg 1980, 184-203; Raup 1980).

Land tenure is an important component of farm structure because it is concerned with the extent of ownership and control of the farmland resource. Land tenure also has a considerable influence on:

1. Who controls production decisions at the farm level;
2. The organization of resources into farm business units;
3. The degree of freedom to make business decisions and the degree of risks assumed by the owner; and
4. The ease of entry into farming and the transfer of farmland to the next generation.

Major institutional changes in farmland tenure can lead to other structural changes in agriculture. However, the causality usually flows the other way. Most incremental changes in land tenure are induced by technological changes in production, marketing and finance or in institutional changes "upstream" or "downstream" from production agriculture.

A key issue concerning U.S. farmland tenure is the extent of farm operator *control* of the farmland resource by leasing or ownership. For each operator, the relative advantages and disadvantages of farmland leasing or ownership depends on their resource situation and their social/economic values.

Some key advantages of land ownership by farm operators are: (1) greater security of tenure, (2) farmland may be a source of loan collateral, (3) greater managerial freedom and independence, and (4) farmland may serve as a possible hedge against inflation. Also, land ownership is a source of prestige and personal satisfaction to the owner, and a source of tangible wealth that can be passed on to heirs. The major disadvantages of farmland ownership by farm operators are usually associated with a limited capital position. These disadvantages are: (1) cash flow problems associated with farm mortgage payments exceeding net returns from the purchased land, (2) lower current rate of return on invested capital compared to investment in farm machinery, livestock or operating inputs, (3) reduced amount of working capital due to farmland debt servicing requirements, and (4) farmers with limited capital and sole reliance on farmland ownership often find it difficult to expand their business to an adequate size (adapted from Kay 1981, 252).

Leasing provides farm operators the right to use farmland without obtaining ownership or title. Common types of farmland leasing arrangements are cash leases, cropshare leases, and livestock share leases. Selection of a specific lease arrangement involves economic considerations of distributing income, expenses, and risk/uncertainty between the landowner and renter.

Key advantages for farm operators leasing farmland include: (1) greater flexibility in selecting the size of the farm operation, (2) more flexible financial obligations, compared to typical land purchase financing arrangements (mortgage or installment land contract), and (3) more working capital for purchasing machinery, livestock, or operating inputs. The major disadvantages of farmland leasing are: (1) uncertainty involved with lease renewal can result in a fairly rapid, unwanted reduction in farm size, (2) slow equity accumulation during times of rising farmland prices, and (3) poor facilities and reluctance of landlords (and renters) to invest in improvements (adapted from Kay 1981, 253).

Following a brief discussion of the historical development of U.S. farmland tenure systems, twentieth century trends in U.S. farmland tenure are presented. Next, a more detailed discussion of farmland tenure trends from 1950-1987 in the North Central states is presented. Third, farmland ownership trends in the U.S. and in the North Central region are discussed based on USDA land ownership surveys conducted in 1946, 1978 and 1988. Key socio-economic indicators are compared to the farm ownership and tenure status of farm operators and landlords. Fourth, contemporary information on farmland leasing patterns based on recent farm-level survey data from South Dakota and Nebraska, are presented. Major changes in farmland leasing and tenure arrangements that occurred in these two North Central states between 1951 and 1986 are evaluated.

Historical Development of U.S. Farmland Tenure Systems

U.S. farmland ownership and tenure systems are based on our English common law heritage along with Spanish influence in some Southwestern states. Legislation favoring private ownership and inheritance was in response to dissatisfaction with feudal land tenure systems in Europe. The constitutional and legislative framework for land tenure in the U.S. was largely developed by the early 1800s. Key characteristics of the U.S. farmland tenure system are:

1. Widespread private ownership of farmland was a major public policy objective in the 19th and early 20th Century. Federal and state lands were usually transferred to private owners via grants, sale or homesteads;

2. Private ownership rights are conditional, not absolute, and these rights can be revised over time. Society retains rights of taxation, eminent domain, and police power. Society can acquire private lands for "public purposes" and must provide "adequate compensation;

3. Individuals, partnerships, corporations and other legal "persons" may hold private rights (surface, subsurface and above surface rights) in land for defined or indefinite periods of time. These

rights include the right to enter, use, sell, buy, lease, or mortgage real estate;

4. Rights in private land can be maintained across generations by use of written wills;

5. Public regulations of private land ownership and use are minimal (compared to land regulations in many European nations) but are increasing over time as population and development pressures increase in many areas of the United States; and

6. All citizens have some rights in public lands, with some rights to public lands only available by lease or permits. (Adapted from Moyer et al., 1969, 3 and from Halcrow 1977, 334-342.)

The major Federal policies that have had or continue to have a substantial impact on U.S. farmland tenure include land distribution, agricultural credit, and farm price/income support programs. From 1840 to the mid-1920s, Federal land programs distributed land to settlers at nominal costs. These programs included the Preemption Act (1841), Homestead Act (1862) and subsequent land settlement acts accounting for different conditions in the Plains and Western states.

The most important result of the Homestead Act of 1862 was symbolic -- it represented the triumph of the "family farm" philosophy of land ownership by farm families. The land tenure ideals expressed in these Acts continue to have political and social importance, and are a major reason why U.S. land tenure systems are periodically re-examined.

Federal agricultural credit legislation of 1916 and 1933 established the cooperative Farm Credit System (FCS). The Farm Credit System pioneered the development of long-term mortgage financing of farm real estate and was charged with providing a "dependable" source of credit to farmers, regardless of general economic conditions. According to (Halcrow 1977, 401): "the introduction of long-term amortized loans at uniform rates of interest revolutionized the entire farm mortgage market and expanded the flow of long-term capital into agriculture".

Other Federal credit legislation in the 1930s and 1940s was specifically designed to improve farmland tenure conditions. Federal legislation from 1932-1937 provided for considerable amounts of emergency refinancing of farm mortgage loans, a temporary mortgage foreclosure moratorium, and established procedures for transferral of acquired farmland back to farmers as soon as feasible. Finally, Congress established the Farmers Home Administration (FmHA) in 1946 to provide credit to low-resource and beginning farmers (Halcrow 1977, 399-400).

Federal farm price and income support programs have considerable indirect impacts on land tenure, because farm program benefits are usually tied to the amount of eligible (base) acres. The capitalization of

U.S. farm program benefits into farmland values has provided incentives for landownership (Tweeten 1979, 486-487; Reinsel and Krenz 1972). Commodity farm program effects are especially important in many North Central and Southern states where dependence on Federal farm program benefits is greater than in other states.

U.S. Farmland Tenure Trends in the Twentieth Century

Land tenure statistics, compiled by the U.S. Census of Agriculture, classify farm operators into three main categories:

1. *Full owners* operate only land that they own. They may also rent out land to other farmers;
2. *Part owners* operate land that they own and also rent additional land from others; and
3. *Tenants* operate only land they rent in from others or work on shares for others.

Until the 1950s, full ownership of farmland by farm families was often considered the "ideal" system of land tenure in the United States. Full owners were the dominant tenure class from 1900 to 1950, consisting of 47 percent to 58 percent of total U.S. farm operators. The number of full owner operators varied from 2.96 million to 3.44 million farm operators. From 1900 to 1920, full owners operated 515 - 519 million acres, which was a majority of land in farms. Full ownership of farmland declined to 435 million acres by 1930, a direct consequence of the agricultural depression of the 1920s. The extent of full ownership did not increase until farm economic conditions improved during World War II and in the late 1940s. By 1950, full owner operators controlled 526 million acres (Table 18.1).

From 1950 to 1987, the number of full owners declined 60 percent, a percentage similar to the rate of decline in total farm numbers. Consequently, full owners remained a majority (57 percent - 59 percent) of farm operators during this period. However, the total amount of farmland operated by full owners sharply declined from 526 million acres in 1950 to 318 million acres in 1987. Average size of farm operated per full owner increased from 169 acres in 1950 to about 256 acres in 1987. By comparison, average U.S. farm size increased from 215 acres in 1950 to 462 acres in 1987.

Full ownership, once the "ideal" and dominant farm tenure system, has more recently been associated with smaller farm size. Despite their relative decline, full owners remain an important component of U.S.

TABLE 18.1   Farmland Tenure of U.S. Farm Operators, 1900-1987, by Number of Farm Operators and Land in Farms

Year	Thousands of Farm Operators by Tenure Class			
	Full Owners	Part Owners	Tenants	Total
1900	3261	451	2025	5737
1910	3413	594	2355	6362
1920	3435	558	2455	6448
1930	2968	657	2664	6289
1935	3258	689	2865	6812
1940	3121	615	2361	6097
1945	3340	661	1858	5859
1950	3113	825	1444	5382
1954	2757	857	1168	4782
1959	2140	811	760	3711
1964	1836	782	540	3158
1969	1706	671	353	2730
1974	1424	628	262	2314
1978	1298	681	279	2258
1982	1326	656	259	2241
1987	1238	609	240	2087

Year	Millions of Acres by Tenure Class			
	Full Owners	Part Owners	Tenants	Total
1900	519	125	195	839
1910	519	133	227	879
1920	515	176	265	956
1930	435	246	306	987
1935	452	266	337	1055
1940	449	300	312	1061
1945	519	371	252	1142
1950	526	423	212	1161
1954	495	470	193	1158
1959	459	498	167	1124
1964	432	533	145	1110
1969	375	550	138	1063
1974	360	535	122	1017
1978	332	561	122	1015
1982	342	531	114	987
1987	318	520	127	964

Source: U.S. Dept. of Commerce, U.S. Census of Agriculture, Vol. I, various years.

agriculture and are concentrated among small, part-time and residential farms. In 1987, only 36 percent of full ownership farms had annual gross farm sales exceeding $10,000, compared to 67 percent of full tenant farms and 75 percent of part ownership farms. Full owners are often older than part owner or tenant operators. In 1987, the average age of full owners was 55 years, compared to 49.8 years for part owners and 41 years for tenants (USDC, 1987 Census of Agriculture, Table 48).

Tenancy was an important public policy issue before World War II. The number of tenant farmers increased from 2.02 million in 1900 to 2.87 million farm operators in 1935 (Table 18.1). An important part of this number were sharecroppers primarily located in the Southeastern states.

The decline in tenant operated farms began in 1935 and continued until 1974. Between 1935 and 1950, tenant farms decreased 50 percent in numbers from 2.87 million to 1.44 million operators. From 1950 to 1974, tenant operated farms decreased another 81 percent to 262,000 farms. Since 1974, tenant farms have varied from 240,000 to 279,000 farm operations and account for 11 - 13 percent of U.S. farm numbers. By 1982, tenant farms were no longer located primarily in the Southeastern states.

The relative importance of tenancy is indicated by the proportion of all land in farms operated by tenants in different time periods. Land operated by tenants increased steadily from 195 million acres in 1900 to 337 million acres in 1935. At its peak, one-third of total farmland was tenant operated. By 1950, less than 20 percent of farmland was tenant operated. Since 1974, full tenants have operated 11 percent - 13 percent of U.S. farmland. Prior to 1950, the average size of tenant operated farms was less than three-fourths of U.S. average farm size. Since 1969, the average size of tenant operated farms has equalled or exceeded the average size of all U.S. farms.

The steady decline of full tenancy from the late 1930s to the early 1980s and the steady decline of full ownership since 1950 are important structural changes in American agriculture. Part ownership has become the dominant land tenure arrangement of U.S. farm operations. In this tenure class, a farmer owns part of the land operated and rents in the rest. The rented land may be one parcel of cropland or pasture; it may also be 80 - 95 percent of total acres operated.

The number of part owners increased from 451,000 in 1900 to a peak of 857,000 farm operations in 1954 and has slowly declined to about 609,000 farms in 1987. However, acres operated by part owners has more than quadrupled from 125 million acres in 1900 to above 520 million acres from 1964 - 1987. Since 1969, part owners have operated more than 50 percent of U.S. farmland. Throughout this century, the average size of part owner farms has been considerably larger than the average size of all U.S. farms. The relative importance of part ownership as a proportion of farmland acres varies considerably by state and region. Most states in the North Central region have a higher proportion of acres operated by part owners than the U.S. average of 54 percent (Figure 18.1).

The overall importance of rented land has increased for part owners. In 1950, 46 percent of farmland acres operated by part owners was rented land. In 1987, the proportion of rented farmland had increased to 53 percent of acres operated by part owners!

The dominant trend to part ownership since 1950 indicates renting part of the land is now a natural part of the commercial sector of U.S. agriculture. In many cases, the most efficient method of expanding commercial farm operations is to rent rather than purchase additional farmland. Leasing often conserves expanding farmer's working capital by reducing financial outlays to acquire farmland. Part ownership also permits these farmers to obtain the advantages of farmland ownership and the advantages of farmland leasing. In an economic environment of farm expansion, part ownership is an important capital management strategy to increase current returns and to reduce business risk.

The social status of renting farmland has changed in this century. Renting farmland is now seen as part of successful operations and tenancy is no longer viewed as a major social problem in most of the United States. Landlords provide an important source of capital to part owners and to full tenants in today's capital intensive agriculture (Stanton 1989, Chapter 4, this volume).

Landlords have continued to provide capital to the U.S. farm sector because rates of return to farmland ownership (current returns + expected capital appreciation) have been competitive with many other long-term investments. Also, many landlords were raised on farms, have some farm management experience and are more familiar with farmland as an investment opportunity. In many cases, landlords are reluctant to sell the "family farm" because it is a major part of their family heritage.

Farmland Tenure Trends in the
North Central Region, 1950 - 1987

Increased part ownership is the dominant farm tenure trend in the United States and in every state in the North Central region. From 1950 to 1987 part owners have increased as a percent of all farm operators, as a percent of cropland operated and as a percent of land in farms. In 1987, part owners operate a majority of farmland (and cropland) in every North Central state, except Missouri where part owners operate 46 percent of the state's farmland. More than 60 percent of farmland in

FIGURE 18.1 Percent of Farmland Operated by Part Owner Operators, by State and Region, 1987

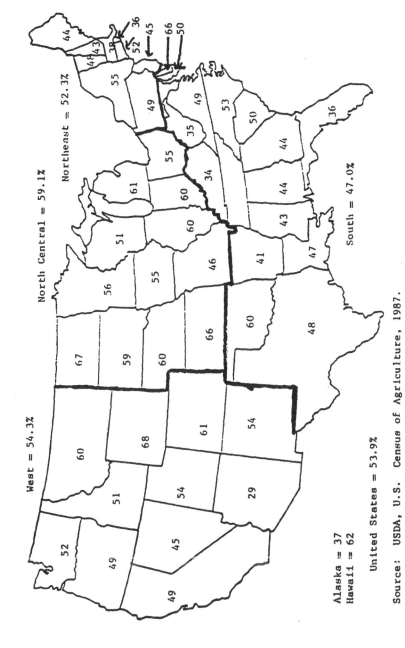

West = 54.3%

North Central = 59.1%

Northeast = 52.3%

South = 47.0%

Alaska = 37
Hawaii = 62

United States = 53.9%

Source: USDA, U.S. Census of Agriculture, 1987.

North Dakota, Nebraska, Kansas, Indiana, Illinois and Michigan is operated by part owners. Average farm size is largest for part owners and smallest for full owners in every North Central State and in the United States (Figure 18.1, Tables 18.2 and 18.3).

Nearly three-fifths of U.S. farm operators are full owners and this proportion has been fairly steady since 1950, even though the amount of land in farms owned by farmers in this tenure class has substantially declined. In the North Central region, full owners were 56 percent - 57 percent of farm numbers from 1950 - 1969, but have declined to 49.4 percent of farm numbers in 1987 (Table 18.2).

There are substantial variations in full ownership trends among North Central states. From 1950 to 1987, the proportion of full owners has increased 5-10 percentage points in South Dakota, Nebraska and Kansas and has decreased 5 - 13 percentage points in North Dakota, Minnesota, Wisconsin, Michigan, Indiana, and Ohio. The proportion of full owners has remained fairly constant in Illinois, Iowa and Missouri.

In 1987, the proportion of full owners is lowest in the Northern Plains states (North Dakota, South Dakota, Nebraska and Kansas). These are the only states in the North Central region where the number of part owners is nearly equal to or exceeds the number of full owners. These states are dominated by commercial family farms and have the lowest percentage of part-time farms or residential farms.

The proportion of full owners are highest in Missouri, Michigan, Ohio and Wisconsin (58 percent - 66 percent of farm numbers). A high proportion of farmland in Ohio and Michigan are in metropolitan counties and much of the rural landscape is characterized by small, part-time and residential farms. Missouri and Wisconsin have a moderate rural population density and many small farms are located in marginal farming regions.

Full tenancy has declined as a percent of farm numbers and land in farms in every state in the North Central region. Since 1969, the proportion of full tenant farms has been *higher* in the North Central region than in the United States! In 1987, North Central tenants are 13.9 percent of farm numbers and operate 13.8 percent of land in farms. The extent of full tenancy is highest in the Cornbelt states of Iowa and Illinois (>20 percent of land in farms) and lowest in Missouri and Wisconsin (<10 percent of land in farms).

Structural Changes in U.S. Farmland Ownership, 1946 - 1988

The U.S. Census of Agriculture provides useful socio-economic information about farmland tenure. However, the focus of Census

TABLE 18.2 Farmland Tenure Changes in the U.S and North Central Region, 1950-1987

Tenure Class	Number of Farms (percent)			Land in Farms (percent)		
	1950	1969	1987	1950	1969	1987
UNITED STATES						
Full owner	57.4	62.5	59.3	36.1	35.6	32.9
Part owner	15.3	24.6	29.2	36.4	51.4	53.9
Manager[a]	0.4	–	–	9.2	–	–
Tenant	26.9	12.9	11.5	18.3	13.0	13.2
Total	100.0	100.0	100.0	100.0	100.0	100.0
Thousands of Farms	5388.4	2730.0	2087.8			
Millions of Acres	1161.4	1151.9		1161.4	1062.9	964.5
NORTH CENTRAL REGION[b]						
Full owner	55.9	56.7	49.4	36.2	34.9	27.1
Part owner	19.7	27.9	36.7	36.6	49.6	59.1
Manager	0.3	–	–	1.8	–	–
Tenant	24.1	15.4	13.9	25.4	15.5	13.8
Total	100.0	100.0	100.0	100.0	100.0	100.0
Thousands of Farms	1868.1	1151.9	862.0			
Millions of Acres	396.4			396.4	372.4	350.5

[a]The manager tenure classification was discontinued after the 1964 Census of Agriculture and, in subsequent Census periods, these farms were redistributed across the other tenure categories. The manager classification consisted of farm units where a manager operated the land for others and was paid a wage, salary or commission.
[b]States included in the North Central region are shown in Figure 18.1.
Source: U.S. Dept. of Commerce, U.S. Census of Agriculture, Vol. I, 1987, 1978, 1964, and 1950.

TABLE 18.3 Farmland Tenure Changes in the North Central States, 1950-1987[a]

Tenure Class	Number of Farms (percent)			Land in Farms (percent)		
	1950	1969	1987	1950	1969	1987
ILLINOIS						
Full owner	44.9	45.4	44.0	28.5	27.1	19.3
Part owner	20.5	29.5	36.6	29.0	44.4	60.4
Tenant	34.6	25.1	19.4	42.5	28.5	20.3
Total	100.0	100.0	100.0	100.0	100.0	100.0
Thousands of Farms	194.5	123.6	88.8			
Millions of Acres				30.6	29.9	28.5
INDIANA						
Full owner	63.4	63.5	57.8	44.5	40.1	27.7
Part owner	17.3	24.3	31.3	27.7	43.9	60.2
Tenant	19.3	12.2	10.9	27.8	16.0	12.1
Total	100.0	100.0	100.0	100.0	100.0	100.0
Thousands of Farms	166.1	101.5	70.5			
Millions of Acres				19.4	17.6	16.2
IOWA						
Full owner	46.8	50.8	45.9	36.6	36.0	25.2
Part owner	15.0	25.2	33.5	20.6	38.3	54.5
Tenant	38.2	24.0	20.6	42.8	25.7	20.3
Total	100.0	100.0	100.0	100.0	100.0	100.0
Thousands of Farms	202.5	140.4	105.2			
Millions of Acres				34.0	33.6	31.6

MISSOURI

Full owner	63.8	70.8	66.5	54.9	54.3	43.8
Part owner	15.9	19.9	24.6	25.2	35.6	46.4
Tenant	20.3	9.3	8.9	19.9	10.1	9.8
Total	100.0	100.0	100.0	100.0	100.0	100.0
Thousands of Farms	229.5	137.1	106.2			
Millions of Acres				34.7	32.4	29.2

NEBRASKA

Full owner	35.6	39.3	41.8	27.2	30.1	23.1
Part owner	25.9	36.1	37.0	43.6	58.6	60.3
Tenant	38.5	24.6	21.2	29.2	18.4	16.6
Total	100.0	100.0	100.0	100.0	100.0	100.0
Thousands of Farms	106.9	72.3	60.5			
Millions of Acres				45.5	45.8	45.3

NORTH DAKOTA

Full owner	39.1	39.0	32.4	29.0	25.4	19.2
Part owner	39.1	46.6	49.7	52.9	61.7	67.3
Tenant	21.8	14.4	17.9	18.1	12.9	13.5
Total	100.0	100.0	100.0	100.0	100.0	100.0
Thousands of Farms	65.3	46.4	35.3			
Millions of Acres				40.9	41.5	40.3

OHIO

Full owner	68.0	66.5	59.5	52.5	47.2	33.5
Part owner	14.1	21.6	29.5	22.5	38.2	54.9
Tenant	17.9	11.9	11.0	25.0	14.6	11.6
Total	100.0	100.0	100.0	100.0	100.0	100.0
Thousands of Farms	198.6	111.3	79.3			
Millions of Acres				20.7	17.1	15

(continued)

TABLE 18.3 (continued)

Tenure Class	Number of Farms (percent)			Land in Farms (percent)		
	1950	1969	1987	1950	1969	1987
SOUTH DAKOTA						
Full owner	31.2	38.3	40.8	17.6	28.4	29.8
Part owner	38.3	44.5	42.8	63.8	60.8	59.1
Tenant	30.5	17.2	16.4	18.6	10.8	11.1
Total	100.0	100.0	100.0	100.0	100.0	100.0
Thousands of Farms	66.3	45.7	36.4			
Millions of Acres				42.5	45.6	44.8
WISCONSIN						
Full owner	71.9	73.3	58.1	65.3	63.9	41.4
Part owner	12.4	19.6	33.4	17.0	29.0	51.2
Tenant	15.7	7.1	8.5	17.7	7.1	7.4
Total	100.0	100.0	100.0	100.0	100.0	100.0
Thousands of Farms	168.0	99.0	75.1			
Millions of Acres				23.0	18.1	16.6

[a]Farm numbers for 1950 exclude all farms in the "manager" tenure class. This excluded 0.1% - 0.4% of farms by state. Farmland acres for 1950 exclude acres operated by the "manager" tenure class in each state.

Source: U.S. Dept. of Commerce, U.S. Census of Agriculture, 1987, 1978, 1964, and 1950, various states.

inquiry is about the operator and the farm operation, and does not usually include information about nonoperator landlords. Compared to land tenure information, relatively little is known about farm landlord characteristics or about trends in farmer and nonfarmer ownership of farmland. Congress authorized nationwide surveys of landowners in 1946 and 1978 to obtain some information about land ownership patterns (Inman and Fippin 1949; Lewis 1980; Daugherty and Otte 1983). In 1988, a special survey of the U.S. Census of Agriculture (Agricultural Economics and Land Ownership Survey) obtained land ownership information from a sample of Census farm operator respondents and from their landlords.

In this section, regional information on U.S. farmland ownership in 1988 is presented, followed by detailed comparisons of the 1946 and 1978 farmland ownership data (Harris and Gilbert 1985). Ownership data presented from each survey are limited to privately owned farmland in a farm operation.[1]

Ownership of U.S. Farmland, 1988

In 1988, farm owner operators comprised 56 percent of the estimated 2.95 million private owners of U.S. farmland and owned 60 percent of privately owned farmland acres. The proportion of farmland acres owned by farm operators is highest in the Northeast (72 percent) and Western (67 percent) regions, and lowest in the South (58 percent) and North Central (55 percent) regions (Table 18.4).

The North Central region is the largest agricultural region in the U.S. and contains over two-fifths of privately owned farmland acreage and nearly half of farmland owners. It is also the only region where the number of nonoperator owners is about equal to the number of farm owner operators. In the East North Central division, nonoperator owners own half of the farmland and greatly exceed the number of farm operator owners (Table 18.4).

An estimated 44.5 percent (371 million acres) of privately owned farmland acres in the United States are leased to farmers and ranchers. The percentage of leased farmland (49 percent) is highest in the North Central region. In all regions, farmers and ranchers obtain most of their leased land from nonoperator owners.

Major differences in sampling methods and classification systems makes it very difficult to directly compare results from the 1978 and 1988 Land Ownership surveys. The principal differences concern the total number of farmland owners and the relative importance of nonoperators as owners of farmland.[2] In the following section, key results from two

TABLE 18.4 U.S. Farmland Ownership Characteristics, National and Regional Summary, 1988[a]

	Type of Owner			Farmland Owned By:			
				Farm Operators		Nonoperators	Total
	Owner Operators	Nonoperator Owners	Total Owner	Owned and Operated	Rented to Others	Rented to Others	Privately Owned Farmland
	Thousands of Owners			Millions of Acres			
United States	1656	1296	2952	462.1	32.6	338.4	833.2
Region[b]							
Northeast	97	83	180	14.0	0.6	5.7	20.3
South	684	443	1127	139.3	10.0	108.9	258.2
West	215	109	324	134.8	7.5	69.7	212.0
North Central	662	660	1322	174.1	14.5	154.1	342.7
East North Central	281	332	613	41.9	2.2	43.4	87.5
West North Central	379	330	709	132.3	12.3	110.7	255.3

[a]Acreage data reported in this table is based on privately owned acres of farmland reported by farm operator owners and nonoperator owners. It excludes all agricultural land leased from public agency landlords. The owner operator class includes full owners, part owners, and owner-operator landlords. The estimated nonoperator owners of farmland (1,296,000) is less than the estimated number of nonoperator landlords (1,590,000) reported in the 1988 Land Ownership Survey. The main reason for the difference is that data reported in this table are landholder (ownership) units which is less than the number of farm owner operators and their nonoperator landlords.

[b]U.S. Census regions (Northeast, South, West and North Central) are shown in Figure 18.1. Data from the East North Central and West North Central divisions are shown to highlight differences within the North Central region. East North Central division includes states of Ohio, Indiana, Illinois, Michigan and Wisconsin. West North Central division includes states of Minnesota, Iowa, Missouri, North Dakota, South Dakota, Nebraska, and Kansas.

Source: U.S. Department of Commerce, Bureau of Census. 1987 Census of Agriculture-Agricultural Economics and Land Ownership Survey (1988), Vol. 3, Part 2. Table 67.

studies that compare farmland ownership in 1988 (1978) with farmland ownership in 1946 are presented.

Changes in U.S. Farmland Ownership Patterns

Wunderlich (1991) focused on the changing number (and concentration) of farmland owners from 1946 to 1948 and on selected differences in characteristics of operator/nonoperator owners responding to the 1988 survey. Key findings from comparing farm ownership data in 1946 to 1988 are:

1. Fewer people are directly involved in agriculture as farmland owners or as farm operators. From 1946 to 1988, the number of farmland owners declined nearly 40 percent (5.0 million to 2.95 million owners), the number of farmland operators declined by 62 percent (5.8 million to 2.2 million farm operators), while the U.S. population has increased by 75 percent to 245 million people. Thus, fewer people are making decisions on agricultural land use, conservation and investments in improvements.

2. Concentration of farmland ownership has increased only in the sense that fewer people own farmland. Among those owning farmland, the proportionate distribution of farmland ownership has not changed from 1946 to 1988. The concentration of farmland ownership measured by value is similar to the concentration of U.S. household income.[3]

3. Nonoperator owners are an increasingly important component of farm tenure and farm structure. The number of nonoperators owners increased from 0.9 million in 1946 to 1.3 million in 1988; the proportion of nonoperator owners increased from 18 percent in 1946 to 44 percent in 1988. A majority of nonoperator owners are women, are more than 65 years old and are retired. Furthermore, a majority of nonoperator owners and farm landlords are not engaged in or retired from farming.

Harris and Gilbert (1985) made detailed comparisons of U.S. farmland ownership patterns in 1946 and 1978. They caution that some differences in the construction of the 1946 and 1978 land ownership surveys present some difficulties in making direct comparisons. However, selected key comparisons and inferences can be made:

1. The proportion of part owner operators and nonoperator landlords has increased while the proportion of full owner operators and owner-operator landlords has declined. Part

owner operators have greatly increased their percent of total acres owned, while the percent of total farmland acres owned by owner-operator landlords has greatly declined (Table 18.5);

2. Nonoperator landlords, including retired farmers and investors without farm backgrounds, have become an increasing component of the farmland rental market. In 1978, nonoperator landlords were 80 percent of all farm landlords, compared to about 55 percent of all farm landlords in 1946 (Table 18.5). In 1978, nonoperator landlords rented out 87 percent of all privately owned agricultural land leased in the U.S. (Lewis 1980; Baron 1983). Most landlords (85 percent) are individuals and usually lease to only one renter (Wunderlich 1983);

3. In both time periods, nonoperator landlords were a much higher percentage of farmland owners and owned a higher proportion of farmland acres in the North Central region than in other regions. This result corresponds with the 1988 findings that farmland leasing is a more important economic activity in the North Central region than in other regions.

4. Farmland owners tend to be older people. Nearly half of farmland owners in both time periods were 55 years of age or older and they owned 52 percent-56 percent of private farmland acreage. The age distribution of farmland ownership has slowly increased as average lifespan has increased (Table 18.5);

5. Land tenure and ownership classes follow a systematic pattern by age of operator or owner. Nonoperator landlords are the oldest group, followed by owner-operator landlords, full owner operators and part owner operators. Full tenants are the youngest operators;

6. In 1946 and 1978, a majority of male landowners were full owner operators and a majority of female landowners were nonoperator landlords. In 1978, women were sole owners of 14 percent of U.S. farmland, up from 9 percent in 1946;

7. The occupational status of U.S. farmland owners has drastically changed from predominantly farmers (and retired farmers) to nonfarm workers, managers and professionals. In 1946, 65 percent of U.S. farmland owners listed "farming" as their primary occupation, compared to only 34 percent in 1978 (Table 18.4). In 1946, most full owner operators listed "farmer" or "retired farmer" as their primary occupation, but only 43 percent reported "farmer" or "retired farmer" as their primary occupation in 1978;

8. The proportion of farmland acres owned by "farmers" has declined from 70 percent in 1946 to 59 percent in 1978. However, the average number of farmland acres owned by each "farmer"

TABLE 18.5   Ownership and Distribution of U.S. Farmland by Tenure, Primary Occupation, Gender, and Age, 1946 and 1978[a]

	Percent of Owners[b]		Percent of Acres[c]	
Tenure	1946	1978	1946	1978
Full owner operators	56	47	33	37
Part owner operator	11	18	16	28
Owner-operator landlord	15	7	29	10
Nonoperator landlord	18	29	22	25
Primary Occupation[d]				
Farmer	65	34	70	59
Retired farmer	8	10	9	12
Business professional	10	21	14	14
Clerical-laborer	14	30	5	11
Housewife	3	5	2	3[e]
Gender				
Male	89	83	91	86
Female	11	29	9	14
Age of Owner				
Under 35	8	8	6	6
35 - 44	18	17	16	14
45 - 54	25	23	26	24
55 - 64	24	23	25	27
65 - 74	18	18	18	19
75 and over	7	7	9	10

[a]Farmland ownership data for 1978 were based on a subset of the 1978 USDA Land Ownership Survey responses which could be compared to the 1946 farmland ownership survey responses as determined by Harris and Gilbert. The 1946 farmland ownership data source is Inman and Fippen, 1949.

[b]Percent of owners is based on an estimated total of 5,025,000 farmland owners, in 1946 and an estimated 4,402,000 farmland owners in 1978.

[c]Percent of acres are based on an estimated total of 975 million acres in the 1946 survey and 783 million acres in the 1978 survey.

[d]Due to nonresponses to some questions in the 1978 survey, percent of owners by primary occupation and by age are based on 4,133,000 owners.

[e]Totals may not add up due to rounding.

Source: Harris and Gilbert, 1985. Tables 4, 5, 10, 11, 12, 17, 18, 19, 20, 21, and 22.

has increased, while the average number of farmland acres owned by persons in other occupations has decreased; and

9. Corporations and partnerships owned about 23 percent of U.S. privately owned farmland in 1978. Most of this land was held by family partnerships and family farm corporations with less than 10 members. In most instances, family farm corporations have incorporated for tax purposes and for credit financing purposes and are not substantially different in operation than other family partnerships (Harris and Gilbert 1985, 52; Raup 1973).

Based on the above findings, the most important changes in farmland ownership and tenure since World War II are: (1) increased separation of ownership and control of farmland; (2) reduced agricultural orientation of farmland owners and landlords; and (3) emergence of part owner operators as the dominant farm tenure class.

Farmland ownership remains widespread, but a lower proportion of farmland is owned by current farm operators. The tremendous outmigration of farm people to urban centers has also changed the occupational status of many farmland owners. Farmland generated current income and substantial capital appreciation (during the 1946-1978 period). For these and other reasons many families elected to retain ownership of the former "family farm" unit and have become landlords.

Part owners have replaced full tenants as the major renters of farmland. An increasing proportion of renters have to rent land from nonoperator landlords instead of farm operator landlords. Many of these landlords no longer live in the local area and a majority of nonoperator landlords are women. In the next section, some implications of these changes in farmland tenure and farmland leasing markets are explored in a 1986 case study of Nebraska and South Dakota farmer renters, farmer landlords, and nonoperator landlords.

An Empirical Study of Farmland Leasing and Land Tenure Arrangements in South Dakota and Nebraska

As discussed previously, farmland leasing is a widely used method for transferring use rights of farmland. The proportion of U.S. farmland leased since 1930 has been in the range of 35-45 percent each year. The proportion of 1987 farmland acres rented is 38 percent in South Dakota and 46 percent in Nebraska. Nearly two-thirds of farmers in both states are involved in farmland rental markets as renters or as landlords. Part owners lease 70 percent of rented farmland in South Dakota, and 64 percent of rented farmland in Nebraska (USDC 1989).

Prior to 1986, the last comprehensive statewide study of the farmland rental market in Nebraska and South Dakota was conducted in 1951 (Hurlburt 1954). The 1951 survey was a regional project of the North Central Land Tenure Committee and was conducted in six North Central states (Iowa, Kansas, Minnesota, Nebraska, South Dakota, and Wisconsin).

In 1986, a two-state study of farmland leasing practices and performance of farmland rental markets was conducted in Nebraska and South Dakota with the assistance of a grant from the USDA Economics Research Service. A common survey questionnaire was developed and mailed to a random sample of 5 percent of farm landlords and renters in both states -- 5800 Nebraska and 4100 South Dakota landlords and renters. This farmland leasing survey was completed by 1615 Nebraska and 1155 South Dakota farm landlord and renter respondents (Johnson et al. 1987; Peterson and Janssen 1988; Lundeen and Johnson 1987).

The principal findings and implications from the 1986 Nebraska-South Dakota farmland leasing study are summarized by key characteristics of: (1) farmland rental market participants, and (2) agricultural lease agreements.

Characteristics of Farmland Rental Market Participants

The rental market for agricultural land remains predominantly local in nature, with 95 percent of farm operators and 55 percent of landlords residing in the same county or a county adjacent to their leased land. However, absentee land ownership is also common with 27 percent of landlords residing in another State. If absentee ownership increases in the future, there will likely be greater use of professional farm management services, and a greater trend to cash leases and more formal leasing arrangements.

Farm rental income for most landlords was a modest proportion (<30 percent) of total household income, while most farm renters (75 percent) were highly dependent on net farm income. This suggests landlords and renters may not experience or perceive changing agricultural economic conditions in the same manner or to the same degree. It also suggests that many landlords are not traditional (farm oriented) landlords, as was the case in the past.

Respondents' age systematically varied by tenure status. Full tenants were the youngest group while nonoperator landlords were often near or past retirement age. Between these extremes were part owner operators and owner operator landlords. This age continuum by tenure status illustrates the importance of the farmland rental market in transferring control of agricultural production from aging farmers or landlords to younger farmers.

Most landlords managed their own leases, but 15 percent of South Dakota landlords and 20 percent of Nebraska landlords reported leases managed fully or in part by someone else. Most outside managers were relatives of the landowner, suggesting the importance of family relationships in land ownership and control. However professional farm management services were used by 6 percent (2 percent) of Nebraska (South Dakota) landlords, especially those living in another state.

Most women respondents (84 percent) were nonoperator landlords and a majority were over 65 years of age. Women were 40 percent of nonoperator landlords and only 10 percent of farm operator respondents. Women landlords were much more likely than male landlords to have someone else manage their farm leases.

The majority of leased land involved a contract between unrelated individuals. However, a majority of renters and about three-eighths of landlords reported one or more leases with family members or relatives. Lease terms between family members did not significantly differ from lease terms between unrelated individuals, except that fewer written leases occurred between family members.

Multiple leasing by farm operators (leasing land from more than one landlord) was the rule rather than the exception. Also, a majority of farm operators with multiple leases used a combination of cash leases and share leases. Thus, today's renter often uses a sophisticated process of land resource control via farmland rental. In so doing, the renter's risks associated with losing any one parcel are reduced. It also suggests many renters may have more knowledge of and experience with farmland rental agreements than many landlords.

Characteristics of Farmland Lease Agreements

Despite the degree of landlord absentee ownership and multiple leasing among renters, most leasing agreements tend to be comparatively informal (verbal, year to year agreements). This suggests that patterns and terms of typical leasing agreements are well established within localities. In many rental relationships, there are few incentives for more formal agreements.

Cropshare leases were the dominant form of cropland leasing in Nebraska and in South Dakota. Almost all cropshare leases were one of the following tenant-landlord shares of output: 2/3-1/3, 3/5-2/5, or 1/2-1/2. The dominant share agreement varied by region and crops grown. In both states, the dominant tenant-landlord share is 2/3-1/3 in the wheat and small grain regions and 3/5-2/5 in the dryland corn and soybean regions. The 1/2-1/2 cropshare lease was predominant on sprinkler irrigated land and also occurred on the most productive nonirrigated cropland.

Most (75 percent) cropshare lease respondents reported the landlord and tenant sharing expenses for one or more variable inputs, but less than 10 percent reported all variable input expenses shared. In almost all cropshare leases, shared inputs were shared in the same proportion as crop output was shared. Fertilizer was the most commonly shared input expense, followed by insecticide or herbicide expenses. Input costs were more likely to be shared on tracts with relatively high per acre input costs including corn, soybean and irrigated crop leases where the tenant's share was 1/2 or 3/5 of crop output.

From an economic efficiency viewpoint, the output shares and inputs shared in a cropshare lease should reflect the relative contributions of the renter and landlord (Heady 1952; Hurlburt 1962). Crop enterprise budgets were used to estimate the renter's and landlord's relative cost contributions for typical crop share leases in different regions of South Dakota and Nebraska. Overall, the dominant output and input shares reported in the survey reflects a reasonable degree of economic efficiency. In most cases, participation in the Federal farm program improved the relative cost contribution of renter and landlord in relation to their output share. This suggests the share rental market for cropland in both states has adjusted to the growing importance of Federal farm programs in the 1980s.

Cash leases completely dominated rangeland lease agreements and were an important type of cropland lease in both states. Almost all rangeland and pasture leases involved cash payments per acre or per animal unit month. Cash leases were used for 40 percent of cropland acres leased in South Dakota and 28 percent of cropland acres leased in Nebraska.

Cash rental rates per acre varied greatly by region and land use. For example, average 1986 nonirrigated cropland cash rents varied from $10-$11 per acre in western South Dakota to $63-66 per acre in east central Nebraska. Within each region, average per acre cash rental rates were highest for irrigated cropland, followed by nonirrigated cropland, alfalfa hayland, native hayland, and native pasture. However, cash rental rates as a percentage of reported farmland values did not significantly (p=.05) vary by region.

Except for changes in annual cash rental rates, the incidence of change in the details of cash and share agreements were infrequent. Moreover, the typical lease agreement had been in effect for more than a decade and most respondents reported considerable satisfaction with their leasing agreements. These favorable perceptions and low incidence of reported changes in lease provisions suggests relatively slow, but deliberate, adaptation of farmland market participants to changes in agricultural or economic conditions.

Changing Characteristics of Farmland Rental Markets
Between 1951 and 1986

In this section, selected comparisons are made between findings from the 1986 survey and findings from the 1951 survey of farmland rental markets in Nebraska and South Dakota (Janssen and Johnson 1989; Hurlburt 1954).[4] Emphasis is placed on changes in leasing practices and changes in characteristics of farmland rental market participants between the two time periods. Categorical modeling procedures (SAS 1985, CATMOD procedure) were used to test for statistically significant differences (p=.01) in sets of coefficients between time periods.

Comparisons between the two time periods indicates gradual, evolutionary changes are occurring in the Nebraska and South Dakota farmland rental markets. Based on the statistical analyses of the 1951 and 1986 surveys, key changes occurring in the farmland rental market in both states between 1951 and 1986 are:

1. From 1951 to 1986, the proportion of part owners, relative to full tenants, has significantly increased in all regions of both states. Overall, part owner operators have increased from 42 percent to 78 percent of farmland renters. Also, part owner operators tend to rent in more acres than full tenants;

2. The incidence of female landlords has significantly increased from 1951 to 1986. The proportion of female landlords has increased from 22 percent to 36 percent of all landlords in South Dakota, and 26 percent to 41 percent of all landlords in Nebraska;

3. In both states, the distribution of renters and landlords has shifted toward older age groups and the median age of renters and landlords has increased. The principal change for landlords is an increase in the proportion of landlords 65 years or older (due to increased incidence of older female landlords) and a decrease in the proportion of landlords 45-64 years of age. The increased proportion of renters 45 years or older (30 percent of renters in 1951 and 48 percent in 1986) is directly related to the pronounced shift toward part owner operators that has occurred since the early 1950s;

4. Multiple landlords per renter are much more prevalent in 1986 than in 1951. In 1951, 55 percent of respondent renters leased land from only one landlord, compared to only 34 percent in 1986. The proportion of respondent renters leasing from 5 or more landlords has increased from about 2 percent in 1951 to 12-13 percent in 1986. These phenomena have occurred at similar rates in most regions of both states. These findings provide

additional evidence that today's commercial farm operators are involved in fairly sophisticated land acquisition behavior that likely involves a combination of farm expansion and risk reduction motivations. Multiple tracts, typically owned by different landlords, are acquired to obtain adequate acreage, and the business risk associated with losing any single parcel is reduced; and

5. The incidence of cash leases for cropland has dramatically increased from 1951 to 1986, while combination share-cash leases have drastically declined. The proportion of share leases (with no cash payments) has remained about the same. For share leases, the incidence of landlords and renters sharing input cost has greatly increased in both states. For example, input cost sharing occurred in 20 percent of South Dakota cropshare leases in 1951 and 70 percent of cropshare leases in 1986.

Conclusions and Implications

Land tenure is an important component of farm structure because it is concerned with ownership and control of the farmland resource. Land tenure and ownership patterns can facilitate or impede other structural changes occurring in agriculture.

A review of U.S. agricultural development indicates the historical importance of land tenure and its impact on other structural characteristics of U.S. agriculture. The basic system of U.S. agricultural land tenure was developed by the late 1800s and public policies concerning land distribution, agricultural credit and farm price/income support programs have had considerable impact on present land tenure and ownership patterns.

Until the 1950s, full ownership of farmland by farm families was the dominant system of land tenure in terms of acres operated and proportion of farms. By 1987, a majority of farmers were full owners, but they controlled less than one-third of U.S. farmland.

The steady decline of full tenancy since the late 1930s and the steady decline of full ownership since 1950 are important structural changes in American agriculture. Part ownership has been the dominant form of land tenure since 1950, and has increased in relative importance throughout the twentieth century. The dominant trend to part ownership indicates renting part of the land is now a natural feature of U.S. commercial agriculture. Renting is now seen as part of successful farm operations and tenancy is no longer viewed as a major social problem.

A review of U.S. farmland ownership trends from 1946 to 1988 indicates farmland ownership remains widely dispersed, but there are some major changes in characteristics of farmland owners and renters. Part owners have replaced full tenants as the major renters of farmland. Part owners and nonoperator landlords have replaced full owners as the major owners of farmland. Nonoperator landlords have replaced farm operator landlords as the major owners of rented farmland. Corporations and partnerships own an increasing percentage of privately owned farmland acreage (23 percent in 1978), but most of this land is owned and operated by family farm corporations and partnerships.

A majority of U.S. farmland owners are 55 years of age or older and a majority of farm landlords are 65 years of age or older. The occupational status of U.S. farmland owners has drastically changed from predominantly farmers (and retired farmers) to nonfarm workers, managers, professionals, and persons retired from nonfarm occupations. The incidence of female landlords has greatly increased and females are sole owners of a relatively small, but growing, percentage of farmland.

Farmland ownership and tenure trends in the North Central states are generally consistent with national trends. The relative importance of part owners, nonoperator owners, and farmland leasing are substantially greater in the North Central region than in other regions. In 1987, a majority of farmland in all North Central states, except Missouri, is operated by part owners. The extent of part ownership is greatest in the Northern Plains states where commercial agriculture is the dominant economic activity.

The combination of technological changes in agriculture and structural changes in farmland ownership and tenure also have important impacts on farmland rental markets. The author's comparison of Nebraska and South Dakota farmland rental market characteristics in 1951 and 1986 indicates changes in market participants and leasing practices have been evolutionary in nature. There has been some increase in the complexity of farmland leasing as part ownership has increased and a greater proportion of landlords are nontraditional landlords. Multiple landlords and multiple leases are the norm for most renters, compared to past typical practice of renting from only one landlord. The greater use of purchased inputs, increased complexity of farm programs and other management decisions in cropland agriculture has led to increased use of cash leases, but also to greater incidence of shared input costs on share leases. The combination cash-share leases have greatly declined in usage. Increased use of cash leases for cropland is likely to reduce security of tenure for many renters, unless a long-term written lease agreement is made (Scott 1985).

Farmland rental markets in Nebraska and South Dakota appear to be functioning in a reasonably efficient and equitable manner. Examination of returns to share leases and cash leases indicates farmland rental markets are efficient in adjusting to geographic differences and to Federal farm program changes. Regional differences in crop output shares and in the array of inputs shared reflect differences in cropping patterns, yield risk, and cultural practices. The examination of relative cost contributions indicates landlords and renters usually negotiate leases within a reasonable degree of economic efficiency and equity.

Based on the above findings, the most important changes in farmland tenure, ownership and leasing since World War II are: (1) increased separation of ownership and control of farmland; (2) emergence of part owners as the dominant tenure class; (3) reduced agricultural orientation of many farmland owners and landlords; and (4) evolutionary changes in farmland leasing agreements. These interrelated changes have important future implications.

Farmland rental markets have become and will remain as a "permanent" part of the organization of production agriculture in the North Central region and in the United States. Landlords provide a major source of capital to most commercial farm operators. Their relative importance will continue to gradually increase because: (1) commercial farmers are usually able to achieve higher current rates of return by investing in other production assets; (2) farmland ownership is a source of current returns and potential capital appreciation with risk-return characteristics that are attractive to many investors (farmers and landlords); and (3) farmland remains a major source of "consumption income" (utility) for many owners, even though their primary income may be obtained from nonagricultural pursuits.

Evolutionary changes have occurred and will continue to occur in farmland leasing agreements. The long term duration of most rental agreements, the largely local nature of leasing markets, and the renter-landlord relationship are major sources of relative stability in leasing agreements. At the same time, incremental changes occur in response to major changes in production technology and economic conditions. Consequently, farmland leasing remains an effective means of production control for farm operators (especially part owners) and ownership control for landlords.

The overall conclusion is that land tenure, ownership, and rental market changes are continuing to occur and generally facilitate other structural changes in U.S. agriculture. Farm management, resource and policy economists should continue: (1) to monitor ongoing changes in land tenure, ownership and rental market; (2) to examine probable socio-economic consequences of alternative changes in tenure and ownership patterns; and (3) to recommend specific changes which improve efficiency and equity of leasing agreements. Performance of these tasks would be enhanced by a regional or national effort to systematically collect detailed information on leasing agreements and land tenure/ownership patterns.

Notes

1. Privately owned farmland in a farm operation excludes: (1) agricultural land leased by farmers and ranchers from Federal, state and tribal agencies, and (2) rural tracts, containing small amounts of agricultural land, that were not operated as a farm by the owner or leased to other farmers. These rural tracts are primarily used as rural residences or vacation (second home) units and are not included as farm operations in the U.S. Census of Agriculture.

Almost all public (Federal and state) and tribal trust lands used for agricultural purposes are located in the Great Plains and Western states and are used for livestock grazing. In 1988, farmer and ranchers leased an estimated 54 million acres from these "public landlords" (USDC 1990). Additional Federal lands (primarily Forest Service and Bureau of Land Management administered lands) were used by ranchers with long term grazing permits. In 1978, an estimated 158 million acres of Federal lands were used for agricultural purposes on a grazing permit or lease basis (Frey 1982).

2. The 1988 Agricultural Economics and Land Ownership Survey (AELOS) is an integrated survey of farm finance and land ownership (USDC 1990). It is a direct successor to the 1979 Farm Finance Survey, but has much more information on land ownership. In both surveys, a sample of farm operators included in the Census of Agriculture were contacted for additional information, including information about their landlord(s). These landlords were also contacted and asked to complete the survey instrument. Survey results are expanded to state and national estimates from responses of the farm operator and their landlord(s).

The 1978 USDA Land Ownership Survey results are based on extrapolation of responses from ownership units located within specific land areas included in the sample. Because of major differences in sampling methods, the 1988 Land Ownership survey obtained an estimated 2.95 million private owners of U.S. farmland, and an estimated 3.25 million farmer owners and landlords. The estimated number of farmland owners was 4.4 million in the 1978 Land Ownership Survey.

3. Concentration (Gini) ratios are often used to measure the distribution of income, wealth, or ownership in a region or nation. Concentration (Gini) ratios vary from 0 - 1, with values near 1.0, indicating a highly uneven concentrated distribution of ownership, while a value near zero indicates a nearly even distribution of ownership and minimal concentration. The concentration (Gini) ratio for all farmland owners is 0.69 based on farmland acreage and 0.44 based on farmland value. For all farmland owners the Gini ratio has barely changed from 1946 to 1988 (Wunderlich 1991). Furthermore, the concentration ratio of 0.44 for farmland ownership based on land value is similar to the concentration ratio of 0.43 for U.S. household income. Thus, concentration of farmland ownership wealth is similar to concentration of income among U.S. households.

4. The 1986 and 1952 farmland rental market surveys obtained considerable amount of comparable data on farmland rental practices and characteristics of farmland rental market participants in Nebraska and in South Dakota. Farm

renters were respondents to both surveys, but landlords were also included as respondents to the 1986 survey. In the 1951 survey, objective characteristics of landlords were obtained from the farm renter respondent. Since there were few differences in responses of landlords and renters, the 1986 survey responses of landlords and renters were pooled and direct comparisons were made, where possible, to responses from the 1951 survey.

References

Baron, Donald. 1983. "The Status of Farmland Leasing in the United States." *Rents and Rental Practices in U.S. Agriculture.* Pp. 73-92. Oak Brook, IL: Farm Foundation.

Daugherty, Arthur B., and Robert C. Otte. *Farmland Ownership in the United States.* U.S. Department of Agriculture, Economics and Statistics Service, ESS Staff Report No. AGES811208.

Frey, H. Thomas. 1982. *Major Land Uses in the United States: 1978.* Agricultural Economic Report 487. Washington, DC: U.S. Department of Agriculture.

Halcrow, Harold. 1977. *Food Policy for America.* New York, NY: McGraw Hill.

Harris, Craig K., and Jess Gilbert. 1985. "Measuring the Social Dimensions of Landownership and Control." *Transfer of Land Rights.* Pp. 31-98. Oak Brook, IL: Farm Foundation.

Heady, Earl. 1952. *Economics of Agricultural Production and Resource Use.* New York, NY: Prentice Hall, Inc.

Hurlburt, Virgil L. 1954. *Farm Rental Practices and Problems in the Midwest.* North Central Regional Publication No. 50, Ames, IA: Iowa State University.

_____. 1962. "Rent Determination Within the Farm Firm." *Rent Theory, Problems and Practices.* Pp. 5-26. North Central Regional Research Publication No. 139, Columbia, MO: University of Missouri.

Inman, Buis T., and William H. Fippin. 1949. *Farmland Ownership in the United States.* Bureau of Agricultural Economics, Miscellaneous Publication 699, Washington, DC: U.S. Department of Agriculture.

Janssen, Larry, and Bruce B. Johnson. 1989. "Farmland Leasing and Land Tenure in South Dakota and Nebraska, NC-181 Proceedings." in Arne Hallam, ed., *Determinants of Farm Size and Structure,* Ames, IA: Iowa State University.

Johnson, Bruce B., Larry Janssen, Michael Lundeen, and J. David Aiken. 1987. *Agricultural Land Leasing and Rental Market Characteristics: A Case Study of South Dakota and Nebraska.* Completion report to U.S. Department of Agriculture, Washington, DC.

Kay, Ronald D. 1981. *Farm Management: Planning, Control and Implementation.* New York, NY: McGraw Hill.

Lewis, James A. 1980. *Landownership in the United States, 1978.* U.S. Department of Agriculture, Economics, Statistics, and Cooperative Service, Agricultural Information Bulletin No. 435.

Lundeen, Michael, and Bruce Johnson. 1987. *Farmland Leasing in Nebraska.* Agricultural Economics Report 152, Lincoln, NE: University of Nebraska.

Moyer, D. David, Marshall Harris, and Marie B. Harmon. 1969. *Land Tenure in the United States: Development and Status.* Agricultural Information Bulletin No. 338, Economics Research Service, Washington, DC: U.S. Department of Agriculture.

Paarlberg, Don. 1980. *Farm and Food Policy - Issues of the 1980s.* Lincoln, NE: University of Nebraska Press.

Penson, J. B., and Marvin Duncan. 1981. "Farmers' Alternatives to Debt Financing." *Agricultural Finance Review* 41: 83-91.

Peterson, Scott. 1987. *Analysis of Farmland Rental Markets in South Dakota.* M.S. thesis (unpublished), Brookings, SD: South Dakota State University.

Peterson, Scott, and Larry Janssen. 1988. *Farmland Leasing in South Dakota.* Agricultural Experiment Station B-704, Brookings, SD: South Dakota State University.

Raup, Philip M. 1973. "Corporate Farming in the United States." *Journal of Economic History* 33: 274-290.

Raup, Philip M. 1980. "Some Issues in Land Tenure, Ownership and Control in Dispersed vs. Concentrated Agriculture." *Increasing Understanding of Public Problems and Policies.* Pp. 153-159. Oak Brook, IL: Farm Foundation.

Reid, Joseph D., Jr. 1979. "Tenancy in American History." in J.A. Roumasset, ed., *Risk and Uncertainty in Agriculture.* College, Laguna, Philippines: SEARCA.

Reinsel, Robert, and Ronald Krenz. 1972. "Capitalization of Farm Program Benefits into Land Values." U.S. Department of Agriculture, Economics Research Service, ERS - 506.

SAS Institute Inc. 1985. *SAS User's Guide: Statistics, Version 5 Edition.* Cary, NC: SAS Institute.

Scott, John T., Jr. 1985. *Implications of Current Conditions on Land Rent and Tenure Patterns in the Years Ahead.* Agricultural Economics Publication No. 4604, Urbana, IL: University of Illinois.

Stanton, B. F. 1989. "Changes in Farm Size and Structure in American Agriculture in the Twentieth Century." (Chapter 4, this volume).

Strange, Marty. 1988. *Family Farming - A New Economic Vision.* Lincoln, NE: University of Nebraska Press.

Tweeten, Luther. 1979. *Foundations of Farm Policy.* 2d. ed. Lincoln, NE: University of Nebraska Press.

U.S. Department of Agriculture. 1991. *Economic Indicators of the Farm Sector: National Financial Summary, 1989.* Economics Research Service, Washington, DC.

U.S. Department of Commerce, Bureau of the Census. 1989. *1987 Census of Agriculture, U.S. and State Summaries.* Vol. 1.

_____ . 1980. *1978 Census of Agriculture, U.S. and State Summaries.* Vol. 1.

_____ . 1967. *1964 Census of Agriculture, U.S. and State Summaries.* Vol. 1.

_____ . 1952. *1950 Census of Agriculture, U.S. and State Summaries.* Vol. 1.

_____ . *1979 Farm Finance Survey, 1978 Census of Agriculture - Special Reports.* Vol. 5, part 6.

_____ . *Agricultural Economics and Landownership Survey (1988), 1987 Census of Agriculture.* Vol. 3, part 2.

Weisberger, Pat. 1979. *U.S. Cropland Rental Practices.* ESCS Report 46, Washington, DC: U.S. Department of Agriculture.

Wunderlich, Gene. 1983. "The Facts of Agricultural Leasing." *Rent and Rental Practices in U.S. Agriculture.* Pp. 39-54. Oak Brook, IL: Farm Foundation.

_____ . 1991. *Owning Farmland in the United States.* Agricultural Information Bulletin 637, Washington, DC: U.S. Department of Agriculture.

19

The Structure of Families and Changes in Farm Organization and Structure

Larry Janssen, Ron Stover, and Virginia L. Clark

Introduction

The interrelationships between farm business and farm household decisions and activities are another important element of farm structure. Stanton indicates that "understanding the nature of change in (farm) structure requires conscious recognition of the role of the household in actions of family members" (Chapter 2). In a family farm setting, the decisions and actions of family members influence the course of the farm business and the welfare of the farm household. Therefore, it is important to investigate interrelated decisions and actions of farm families and their farm business.

Recent review papers in economics and in family life studies have emphasized the importance of identifying factors influencing successful farm management and successful families (G. Johnson 1988; Defrain and Stinnett 1988). Our thesis is that the quality of farm management and the quality of family management are two key endogenous factors that influence farm firm growth and survival.

Glenn Johnson (1988) urged farm management researchers: (a) to redevelop a multidisciplinary, holistic approach to management issues, (b) to summarize and integrate management concepts from several disciplines and focus on variables that are controllable by managers, and (c) to conduct a combination of case studies and large scale empirical studies of farm management behavior.

Defrain and Stinnet (1988, 138) indicated family researchers and agricultural economists have each spent about 70 years developing methods for respectively measuring: (a) family strengths and weaknesses, and (b) financial success of farm operations. They suggest that "it's about time the two research traditions started talking with each other".

This chapter is written within the spirit of these authors' suggestions. The next section contains a selected literature review of: (a) structural changes in American families with implications for farm families, and (b) characteristics of "successful" families. Extensive amounts of agricultural economics literature are available on characteristics of "successful" farm managers and is reviewed elsewhere (Fox, Berger, and Dickson, this volume). The remaining sections area report of selected empirical findings from a 1989 multidisciplinary study[1] (economics, sociology and home economics) of 549 South Dakota farm families and their family farm operations. Appropriate comparisons to other recent empirical studies of farm family behavior are also presented. Empirical results are discussed for the following topics: (a) work roles of farm couples; (b) decision making roles of farm couples; (c) farm management and farm financial position; (d) family functioning (satisfaction, coherence, stress and agreement) and farm financial position; and (e) farm couple goals concerning continuation of the farm operation and farming lifestyle.

Review of Literature on U.S. Family Life

Like farming, the American family has experienced many changes in the past 50 years. Key structural changes in the American family are identified and the effects of these changes on farm families are discussed.

Structural Changes in U.S. Families

Current American marriages and families are characterized by diversity and changing family design in the number of family members and relationships between members. For example, there are fewer children per family today than in the past. An examination of fertility rates suggests white females had an average of 2.1 children in the 1930s, 3.6 children during the 1950s Baby Boom era, and 1.7 children in the early 1980s (Thornton and Freedman 1983).

There is considerable evidence that young people are delaying marriage and having children. The age of first marriage for both males (25.9) and females (23.6) is the highest in this century. Furthermore, the average age of women having their first child has increased from 22 to 25 years.

Divorce rates are another way of looking at the design of the American family. Divorce rates in the United States increased gradually from the mid-1800s to the 1950s, more than doubled in the 1960s and 1970s and stabilized in the 1980s! If current trends continue, the probability is about 50 percent that a marriage started in the late 1980s will end in divorce (Norton and Moorman 1987). Between 70-75 percent of divorced people remarry.

A fourth change in the design of the American family is in the number of adults per household. In 1984, 25 percent of all U.S. families were single parent families compared to only 13 percent of families in 1970. The proportion of single parent families is continuing to increase rapidly. "The impact on society is only now being measured, but the trend is already redefining our concept of the all-American family" (*Newsweek*, July 15, 1985, p. 42).

Only 74 percent of U.S. households involve a "family". The remaining households are single people (23 percent) or two or more unrelated people sharing quarters (about 3 percent). The number of single person households increased by more than 60 percent from 1970 to 1980. Combining the number of households in which someone lives alone with the number of single parent households, we find more than one-third of all households contain only one adult (Hacker 1983).

Another method for examining change in design is to identify the distribution of roles (parent, provider, homemaker etc.) within the marriage. A profound change involving the distribution of roles has been the movement of married women into the work force. In the late 1940s, less than 20 percent of all wives worked (part-time or full-time) in the paid labor force (Bianchi and Spain 1986). By 1987 more than 60 percent of all married women worked outside of their home and the percentage is still climbing (Otten 1988). While the proportion of wives working full-time has always lagged the proportion that worked part-time, the gap has narrowed to 8 percentage points, 54 percent part-time workers and 46 percent full-time workers (Jorgensen 1986). These trends indicate that the U.S. family is most often a two-wage earner family, and many are dual career families.

The movement of married women into the labor force has led to substantial research effort to determine how this change effects the internal functioning of the family. Szinovacz's (1984) thorough review of contemporary empirical and theoretical work on this topic indicates that most women are assigned a helping, and not co-provider, role in the economic support of their families. Secondly, the wife maintains primary responsibility for performance of household work and child care tasks, although there has been a modest increase in the husband's performance of these tasks.

Each of the changes in U.S. family design (structure) has affected farm families, but there are also some substantial differences in the degree of changes. For example, married farm women have more children than nonfarm married women, but the gap has narrowed over time. Secondly, marriages of farm couples are far more stable than are marriages between nonfarm couples. In 1980, over 87 percent of all farm women who had been married were still living with their first husband compared to only 71 percent of nonfarm women (Hacker 1983). Third, a considerably lower proportion of farm households are single parent households, a factor directly related to lower divorce rates.

In other respects, family trends occurring in farming communities are producing a U.S. farm family with characteristics similar to those of nonfarm families. For example, farmers are also delaying marriage (Sanders 1985). Secondly, off-farm labor force participation rates for U.S. farm women have doubled from 22 percent in 1960 to 44 percent in 1980. Third, educational levels of farm men and women have increased substantially during the past 50 years.

Characteristics of Successful Families

Changing trends in American families provides many challenges for contemporary behavioral research on characteristics of "successful families." Research approaches and findings with strong potential application to analysis of farm families are briefly reviewed herein.

Economics of the Family. Economic analysis, emphasizing recent developments in human capital theory provides a unique and important perspective in studying family life (Schultz 1974). First, investment in human capital improves the potential of "increased future earnings and of consumer services that accrue to the individual as satisfactions over his (her) lifetime" (Schultz 1974, 7). Second, individuals and families are decisionmaking units that maximize their utility (satisfaction) by allocating their time between market and nonmarket (household production and consumption) activities in response to monetary prices and "time costs" associated with each activity. Household production includes many nonmarket activities such as: "the quality and quantity of children, prestige, recreation, companionship, love and health status . . (which) cannot be identified as consumption or output as usually measured" (Becker 1974). The above concepts suggest that attention to time allocation activities of farm families and the factors that explain differences in allocation decisions are a useful line of inquiry.

Becker (1974) extends the above concepts to economic analysis of marriage behavior patterns, and concludes that a persons gain from marriage depends on their income, human capital, and relative differences in their own wage rate compare to the wage rate of their spouse. Becker also examines traits of marriage partners as substitutes or complements, concludes that most traits are complements, and indicates that marriage partners will usually have similar preference functions. Furthermore, marriages exhibiting a greater degree of loving and caring, will include a high level of "sharing". This implies that the level of satisfaction of spouses and extent of shared decisionmaking is higher among more "successful marriages".

While economists have usually focused on allocative decisions of families, other behavioral and social scientists have debated various objective and subjective approaches to measuring "quality of life" in families. The emerging consensus among these disciplines is that subjective measures, using personal perceptions of family members, should be combined with objective measures (socio-economic indicators) to properly assess quality of family life (Campbell, Converse, and Rodgers 1976; Jurich, Schumm, and Bollman 1986; Stoeckeler and Gage 1978).

Family Quality of Life and Satisfaction. Research findings have consistently indicated that a strong relationship exists between quality of life measures and reported level of satisfaction with family life (Andrews and Withey 1976; Campbell, Converse, and Rodgers 1976; Bubolz, et al. 1980; Schumm, et al. 1986). In most instances, satisfaction with family life was one of the most important predictors of quality of life, especially among women. Farm wives also reported more satisfaction with life overall than was reported by other women (Knaub et al. 1988).

Joint decisionmaking, similar views about gender role orientation, and a feeling of spousal support were highly correlated with marital satisfaction and quality of life in farm couples (Schumm and Bollman 1981, Bokemier and Maurer 1987; Keating 1987). Working off the farm decreases wive's life satisfaction in part due to handling most housework and child rearing tasks.

Quality of life research has also shown family income, financial position and education levels are positively associated with life satisfaction for farm and nonfarm couples (Light, Hertsgaard, and Martin 1985). Other factors including occupation and job satisfaction, residence, health, and home management are also important to overall life satisfaction.

Family Cohesion and Adaptation. The quality of life approach to identifying successful families has yielded useful measures of satisfaction, but it fails to explain how families develop a sense of satisfaction with their quality of life. The family stress and resources approach explains processes leading to satisfaction. Angell (1936), in studies of Depression families, first identified family integration and adaptability as fundamental resources of crisis resistant families. A considerable volume of research (Olson et al. 1983) has confirmed the importance of these characteristics to effective family functioning.

Recent studies have combined the advantages of the two approaches, providing a basis for identifying successful farm families. A pertinent study by Antonovsky and Sourani (1988), combining quality of life and family stress concepts used a current definition of family success as adaptation or fit -- fit between family members and fit between the family and the outside community. These adaptation definitions assume successful resolution of problems associated with stressors should provide one with a sense of satisfaction about family life (Antonovsky and Sourani 1988, 89). Using couple data, they found that a family's sense of coherence showed a strong relationship to family adaptation.

Prior research on family success (whether it emphasizes family satisfaction or family cohesion) has determined there are several intervening factors which have an important bearing on interrelationships among family members. Four factors important to farm families are: (1) stress, (2) decisionmaking styles, (3) extent of couple agreement or disagreement, and (4) work roles of farm men and women, especially off-farm employment.

Level of Stress. While farm families often report greater satisfaction than urban families, they also report more stress symptoms and higher levels of stress than urban families (Walker and Walker 1987). The greater stress reported by farm families has led to development of measures (Walker and Walker 1987; Weigel, Weigel, and Blundall 1987) to identify the most stressful features of farm family living. In the farm population, as well as the general population, death of a spouse or family member ranks as the most severe type of stressor; divorce also is a major stressor. Beyond these major disruptions, farm family members experience considerable stress from farm crises such as machinery breakdown, production loss, or weather-caused delays.

Family Decisionmaking and Couple Agreement. Family decision-making styles have been identified as crucial in family success. For example, three basic styles of family decisionmaking have been identified by Retting (1987) -- individualistic/competitive, group accommodative, and group collaborative. Families using the individualistic/competitive style do not usually reach consensus on goals and standards, nor do they establish ways they all can work together (Constantine 1983).

Harmony, loyalty, unity, and solidarity are characteristic of families that use the group accommodative style of decisionmaking. If a conflict of interest occurs, the group preference takes priority over the individual, and a high emphasis on cooperation and concern for others is present.

Families using a collaborative decisionmaking style consider both individual and group needs. Family members work creatively to find solutions which will maximize the goals of all (Hocker and Wilmot 1985). Research indicates that more successful families will use a collaborative decision style (Constantine 1983; Olson, Russell, and Sprenkle 1983; Hocker and Wilmot 1985). Stress-oriented studies (Wilkening 1981) indicate that this type of shared decisionmaking is an important characteristic of "crisis-proof" families.

Off-farm Employment. In response to social and economic changes in U.S. society and within U.S. agriculture, a majority of U.S. farm families rely on off-farm work by at least one adult. Although off-farm work has functioned as an adjustment to change, it also is important to note it has potential negative consequences. Walker and Walker (1987) found work overload was one of the major stressors affecting farming families, and concluded that off-farm work accounted for much of the stress.

A recent study of 933 North Dakota farm families indicates that financial stress was an important factor influencing farm family members to obtain off-farm employment. Off-farm earnings were primarily used to generate adequate cash flow for debt payments and for family living expenses. Other factors significantly related to the off-farm employment decision were: age (-), years of education (+), years of previous job experience (+), presence of and number of children (-), farm size (-), and beef farm/ranch (+) (Leistritz et al. 1986).

Empirical Study of South Dakota
Farm Families and Family Farms

The remaining sections of this chapter are a discussion of selected empirical results from a winter 1989 survey of married couples operating family farms in South Dakota. The South Dakota Family Farm multidisciplinary project, funded by the Midwest Technology Development Institute, was developed to identify key characteristics which have enabled many farm families and family farms to succeed in the current economic and social environment. Faculty members from home economics, sociology, and agricultural economics participated as members of the multidisciplinary team.

Two basic assumptions were used by team members to develop and execute the project. First, two components were needed to identify successful farm families: (1) financial/business viability and (2) quality of family life. Second, information should be obtained from a large scale representative sample of farm families and from the farm couple (both spouses) instead of just the farm operator. We did not assume that all farm operators and spouses would necessarily share similar perceptions of farm and family life.

Data Collection Procedures

The mail survey was sent to a random sample of 2000 farm households (6 percent of farm households in each county) in South Dakota.[2] Two separate questionnaires were sent to each farm household -- one addressed to the farm operator and one addressed to their spouse. Each contained a core set of questions to be answered by both parties and another set of questions to be answered only by the operator or by their spouse. The data gathered includes: basic demographics, farm resources and enterprises, financial and income information, work roles and tasks performed, farm and family decision making, family relationships, and satisfaction with family life and farm life.

Approximately 750 of the 2000 farm families contacted returned the surveys. A total of 626 farm operators and 566 spouses completed their respective questionnaires (a survey was considered useable if most questions, excluding farm financial data, were completed). A total of 549 married farm couples completed both questionnaires and the empirical results are based on their responses. In all 549 cases, the husband was the farm operator.

Respondent Family and Farm Characteristics

Respondent farm operators are the same average age and operate somewhat larger farms (1577 acres vs. 1215 acres) than is the case with all farms included in the 1987 South Dakota Census of Agriculture (Table 19.1). South Dakota families operating small farms with annual gross farm sales of less than $40,000 are underrepresented among respondents. These small farms are 53 percent of all South Dakota farm numbers (USDC, 1987) but only 21 percent of respondent farm operations.[3] Otherwise, respondent characteristics appear to be representative of the farm population in South Dakota.

All 549 South Dakota respondent farm operations are classified as: (1) *part-time commercial farms* providing at least one-half of full-time equivalent employment for family members, or (2) *full-time commercial family farms* providing income and employment for one household or, at most, extended family of 3-4 households. No respondents operated smaller *residential farms* or very large *industrial farms*[4] (Stanton and Bills 1988).

TABLE 19.1 Selected Respondent Family and Farm Characteristics, South Dakota, 1989

Characteristic	Operator	Spouse
Family Items		
Age-range (years)	25-86	21-82
Average age	49.2	46.3
Education Level		
Percent completing:		
High school degree	82.3	91.5
Four year college degree	17.0	17.7
Lived on farm as child (percent)	94.5	73.6
Percent of families with:		
Any children living at home	64.0	
Children under 5 years	19.3	
Children 5-12 years	32.1	
Children 13-18 years	23.3	
Adult children	62.5	
Adult children farming with parents	12.6	
Farm Items		
Partnership/corporations (percent of farms)	17.0	
Land tenure: (percent of farms)		
Full owner	20.8	
Part owner	64.4	
Full tenant	14.8	
Average number of		
Acres-operated	1577	
Acres-owned	838	
Gross farm sales (percent of farms)		
Less than $10,000	5.0	
$10,000 - $39,999	16.3	
$40,000 - $99,999	39.0	
$100,000 - 249,999	29.9	
$250,000 or more	7.8	

Source: South Dakota Family Farm Survey, 1989.

Work Roles of Farm Couples

The intermix of technological, economic and gender role changes in American society has also influenced work roles of farm couples. Farm families increasingly rely on off-farm employment and greater participation of farm women in the farm operation.

Farm Couple Labor Force Participation. The role of off-farm employment on farm structure is examined by classifying each farm couple by their incidence of off-farm vs. farm employment: (1) *traditional* couples are those farm couples where neither spouse works off-farm; (2) *husband, off-farm* are farm couples where only the husband works off-farm; (3) *wife, off-farm* are farm couples where only the wife works off-farm; and (4) *dual career* farm couples are couples where both spouses work off-farm (Deseran, Falks, and Jenkins 1984).

Nearly half of South Dakota respondent farm families (47.7 percent) have one or both spouses working off-farm. Compared to the United States, South Dakota has a much higher proportion of farm couples that are *traditional* or *wife, off-farm* households and a much lower proportion of *husband, off-farm* or *dual career* households (Table 19.2). These major differences by labor force category are largely accounted by differences between the economic structure of South Dakota and U.S. agriculture. South Dakota has few residential farms, which tend to be *husband, off-farm* or *dual career* farm households, while as many as 40 percent of U.S. households may be residential farms. Conversely, South Dakota has a much higher proportion of full-time commercial farms which would be classified as either *traditional* or *wife working off-farm* households. It is important to note that a similar proportion of farm women in South Dakota and in the United States have off-farm employment. This suggests that the impacts of the wife's employment on farm family functioning and economic well-being may be similar in South Dakota and in the United States. The major difference between South Dakota and U.S. farm families is in the incidence and extent of operator off-farm employment.

Extent of Farm Work and Off-farm Work. The above classification system does not measure the extent of time involvement of operator or spouse in farm work or off-farm work. Based on respondent's work data, we classified their annual farm and off-farm labor as: (1) minimal/none (2) part-time/seasonal, or (3) full-time (Table 19.2).

Nearly two-thirds (65.7 percent) of South Dakota farm operators worked full-time, year around and 34.3 percent worked part-time/seasonal on their farm. About 30.6 percent of farm operators reported working 60 hours or more per week in all but the winter season! Farmers with part-time, off-farm employment usually worked full-time

TABLE 19.2 Employment Structure, Farm Labor Profile and Off-farm Labor Profile of South Dakota Farm Couples, 1989

A. *Employment Structure*

Employment Structure	Off-farm Employment		S.D. Farm Couples, 1989[a]	U.S. Farm Couples, 1977[b]
	Husband	Wife	Percent of Farm Couples	
Traditional	No	No	52.3	28.8
Husband, off-farm	Yes	No	5.4	26.5
Wife, off-farm	No	Yes	28.2	13.3
Dual career	Yes	Yes	14.1	31.4
			100.0	100.0

B. *Farm Labor and Off-farm Labor Profile*

Extent of Farm Work[c]	Operators	Spouses	Extent of Off Farm Work[d]	Operators	Spouses
	Percent of Respondents[a]			Percent of Responses[a]	
Minimal	0.0	43.6	None	80.5	57.7
Part-time	34.3	43.6	Part-time	10.4	17.0
Full-time I	35.1	7.3	Full-time I	1.4	9.8
Full-time II	30.6	5.5	Full-time II	7.7	15.5
	100.0	100.0		100.0	100.0

[a]Based on responses from 520 of 549 South Dakota farm couples completing the South Dakota Family Farm Survey, 1989.
[b]Based on 1,772 farm households in the 1977 U.S. Current Population Survey as reported in Deseran, Falk, and Jenkins (1984).
[c]Farm work classification is based on the average number of hours worked in each of four seasons (Spring, Summer, Autumn and Winter).

Minimal	=	Farm work is less than 20 hours per week in each season.
Part-time	=	Farm work exceeds 20 hours per week in 2-4 seasons, but is less than 40 hours per week in at least one season.
Full-time I	=	Farm work is 40 hours or more per week in all seasons, but is less than the amount reported as Full-time II.
Full-time II	=	Farm work is 60 hours or more per week in three seasons and 40 hours or more per week in the fourth season.

[d]Off-farm work classification is based on number of hours and months worked per year.

Part-time	=	100-1499 hours of off-farm work per year
Full-time I	=	1500 hours or more of off-farm work per year including
		(1) 9-10 months of full-time work and
		(2) 12 months of work, 30-39 hours per week
Full-time II	=	12 months of off-farm work, 40 hours or more per week

Source: South Dakota Family Farm Survey, 1989.

hours on their farm, while farmers with full-time, off-farm employment usually worked part-time hours on their farm operation.

Nearly one-eighth (12.8 percent) of farm women reported working full-time, year around on farm-related tasks, 43.6 percent reported part-time/seasonal farm work and 43.6 percent reported minimal involvement with farm-related tasks. The extent of farm work was inversely related to incidence of off-farm work. Three-fifths of the farm women with off-farm employment reported minimal involvement in farm work. Nearly 70 percent of farm women with no off-farm work, also reported working part-time or full-time on the farm.

Performance of farm tasks and family/household tasks followed distinct gender roles. Certain farm tasks (tillage operations, chemical applications) were mostly performed by men. Several farm tasks (harvesting crops and hay, taking care of livestock, running farm errands, and keeping farm records) were shared by a majority of South Dakota respondent farm couples. Farm women regardless of their extent of off-farm employment or farm work, assumed most household tasks with occasional or no help from their husband (Clark, Janssen, and Stover 1990).

Decisionmaking Roles of Farm Couples

Family life research indicates that successful families are much more likely to use shared decisionmaking (group collaborative) styles than other family decisionmaking styles. Respondent operators and spouses identified how decisions were made in their family in areas related to both farm and family management (12 questions). Each person indicated if he/she made the decision alone; if the spouse made the decision; if he/she made the decision with his/her spouse; or if the decision had never come up. Responses to all twelve questions (Table 19.3) indicate that many of the families used a collaborative style (made the decision with their spouse). Percentages ranged from 19.5 percent in response to the decision to try a new agricultural practice to an 84 percent collaborative response to the selection of family goals. A majority (84.0 percent - 54.5 percent) of families (both spouse and operator) indicated they used the collaborative style in making the following types of decisions: (a) selecting family goals, (b) selecting leisure activities, (c) attending church, (d) buying major household appliances, (e) buying or selling land, and (f) renting more or less land (Table 19.3). In almost all cases where less than 50 percent of respondent couples used a collaborative decisionmaking style, the decisions related specifically to farm management practices and were designated as individual decisions

TABLE 19.3 Couple Decision Styles Indicated in Responses to Farm and Family Decisions by Percent, South Dakota, 1989.

Decisions:	Individual Operator	Individual Spouse	Collaborative	Disagree[a]	Never[b]
Buy or sell land	7.8	0.2	64.1	23.9	3.9
Rent more or less land	15.3	0.2	54.4	27.5	2.6
Buy major farm/ranch equipment	23.1	0.0	48.0	28.7	0.2
Produce a new crop or type of livestock	34.0	0.0	30.5	33.5	2.0
Try a new agricultural practice	50.6	0.0	19.5	28.4	1.5
When to sell agricultural products	44.7	0.2	27.9	27.2	–
Work on non-farm job or family business	1.9	0.2	41.1	45.9	10.9
Buy major household appliance	0.6	3.3	75.1	20.8	0.2
Attend church	0.7	2.8	76.8	17.7	2.0
Select family goals	0.8	0.4	84.0	13.8	1.0
Delegation of household tasks	0.2	5.4	80.3	13.2	0.6
Selection of leisure activities	0.6	0.9	83.1	14.7	0.7

The columns Individual Operator, Individual Spouse, Collaborative, Disagree[a], and Never[b] fall under the heading *Decision Style*.

[a]Includes styles such as competitive (each says they decide by themselves), operator or spouse says they make decision and the other person says they make it together.
[b]Both respondents reported the decision has never been made.

Source: South Dakota Family Farm Survey, 1989 completed by 549 farm couples.

made by the operator. Specific examples include buying major farm equipment, trying a new agricultural practice, etc.

Farm Management and Farm Financial Position

Farm management is the application of planning, implementation and control concepts to the activities of agricultural production, marketing, and finance. Managerial activities include psychological, sociological, administrative, and economic dimensions and are best understood in a multidisciplinary framework (Boehlje and Eidman 1983; Johnson 1988).

Excellent examples of the multidisciplinary approach can be seen in numerous research reports from the North Central region's Interstate Managerial Study conducted in the late 1950s and successive studies from the Management Factor in Farming project conducted in the 1960s (Johnson et al. 1961; Justus and Headley 1968). In brief, *successful farm managers* were found: (a) to maximize both monetary and nonmonetary values defined by the multiple sets of goals established by the family, (b) to acquire accurate knowledge and information quicker and at lower cost than other managers, and (c) *to act to ensure and control outcomes instead of passively reacting to the economic and social environment that they faced.*

In response to the farm finance crisis of the 1980s, many farm managers changed their management practices to ensure survival of their family farm and to position their firm for future growth (Barry, Ellinger, and Eidman 1987; Ekstrom, Hardie, and Leistritz 1987). We examined farm management practices and other adjustments made by respondent farm families by their reported financial position. We hypothesized that the incidence of off-farm employment, the use of farm management records for decisionmaking purposes, and numerous changes in farm management practices were strongly related to the financial leverage and net farm income position of the farm operation. Also, farmers that have made many management changes are more likely to be "successful" from a farm income viewpoint and have greater financial viability. In other words, they have been "acting to ensure and control outcomes".

Two key financial indicators – net farm income and farm debt/asset ratio -- were used to classify respondents by farm financial position. These measures of farm financial viability are proxy measures of farm business success. A similar approach is used by USDA in their national studies of farm financial stress and farm financial viability (Johnson et al. 1987; Reimund, Brooks, and Velde 1986; Harrington and Carlin 1987).

Financial Position of Respondent Farm Operations. A total of 420 South Dakota respondent farm firms were classified by the following financial positions: (a) favorable, (b) marginal income, (c) marginal solvency, and (d) vulnerable (Table 19.4). An additional 129 farm couples

TABLE 19.4 Farm Financial and Income Indicators by Farm Financial Position, South Dakota, 1989

	Financial Position[a]				All Classified Farms[b]
	Favorable	Marginal Income	Marginal Solvency	Vulnerable	
Number of responses	210	70	92	48	420
		Thousands of Dollars			
Total assets	$436.9	$230.8	$424.3	$233.6	$377.3
Current assets	199.5	105.4	230.0	125.5	182.5
Real estate assets	237.4	125.4	194.3	108.1	194.8
Total debt	61.6	30.1	272.5	152.7	113.6
Current debt	25.1	15.0	123.9	86.7	52.4
Real estate	36.5	15.1	148.6	66.0	61.2
Net worth	$375.3	$200.7	$151.8	$ 80.9	$263.7
Gross farm income	131.1	57.2	168.2	70.1	120.1
Gross farm expense	91.5	57.8	134.7	71.9	93.1
Net farm income	+ 39.8	(-0.6)	+ 33.5	(-1.8)	+ 27.0
Government payments	+ 19.5	+ 9.5	+ 27.5	+ 10.6	+ 18.6

(continued)

TABLE 19.4 (continued)

| | Financial Position[a] | | | | |
	Favorable	Marginal Income	Marginal Solvency	Vulnerable	All Classified Farms[b]
Debt/asset ratio (percent)					
Total	14.1	13.0	64.2	65.4	30.1
Current	12.6	14.2	53.9	69.1	28.7
Long term	15.4	12.0	76.5	61.0	31.4
Asset turnover (percent)	30.0	24.8	39.6	30.0	31.8
Net farm income/gross farm income (percent)	30.3	(-1.0)	19.9	(-2.6)	13.4
Return on equity (percent)[c]	10.6	(-0.3)	22.1	(-2.2)	10.2

[a]Definition of farm financial position:
Favorable: Total debt/asset ratio is 0.0–0.40 and 1988 net farm income exceeds $10,000.
Marginal income: Total debt/asset ratio is 0.0–0.40 and 1988 net farm income is less than $10,000.
Marginal solvency: Total debt/asset ratio exceeds 0.40 and 1988 net farm income exceeds $10,000.
Vulnerable: Total debt/asset ratio exceeds 0.40 and 1988 net farm income is less than $10,000.
Not Classified: Insufficient financial and farm income data to classify farm.
[b]Totals are only for those farms that are classified and do not include those in the not classified category.
[c]Return on equity is defined as net farm income/net worth.

Source: South Dakota Family Farm Survey, 1989.

did not provide sufficient data on farm assets, debts, income and/or expenses to classify their farm by financial position.

Financial stress is still evident on many South Dakota farms. Overall, only 50 percent of the 420 classified farm firms were in a favorable financial position, 16.8 percent were in a marginal income position, 21.8 percent were in a marginal solvency position, and 11.4 percent remained in a vulnerable financial position (Table 19.4). One fifth of respondent farm couples reported no farm debt.

Key findings from analysis of farms by financial position are:

(a) Average total assets of farms in a favorable and marginal solvency position are nearly double the amount of total assets of farms in a marginal income or vulnerable financial position. These larger farms (based on total assets) have substantially higher average net farm incomes and higher average rates of return to equity;

(b) Government farm payments were about 15 percent of gross farm income and a high percent of net farm income in all farm finance classes. Overall, Federal farm payments were about 68 percent of net farm income;

(c) Higher asset turnover rates, higher net margin percentages and subsequent higher rates of return on equity were the key differences between favorable (>$10,000) and less favorable (<$10,000) net farm income levels. These financial indicators indicate producers achieving higher net farm incomes are not only larger in average size (based on total assets and gross farm income) but also achieve higher unit production levels and lower unit costs.

(d) Nearly 47 percent of farm operators in a favorable or marginal solvency position reported gross farm sales (excluding government payments) of $100,000 or more, compared to less than 23 percent of farmers in a marginal income or vulnerable position.

Selected Respondent Characteristics by Farm Financial Position

Key demographic characteristics of respondents are also related to farm financial position and help explain farm and family management differences by financial position. For example, farm couples on highly leveraged farms (marginal solvency and vulnerable financial position) are an average of 8.5-9.7 years younger than other farm couples (Table 19.5). These same couples also have a higher average number of children living

TABLE 19.5 Selected Respondent Characteristics by Farm Financial Position, South Dakota, 1989

		Financial Position[a]				
	All Farms	Favorable	Marginal Income	Marginal Solvency	Vulnerable	Not Classified
N[b] =	549	210	70	92	48	129
			Average (Mean) Number of Years			
Age-years						
Operator*d	49.2	50.2	53.1	41.7c	43.4c	53.0
Spouse*	46.3	47.6	59.0	49.0c	41.4c	47.4
Number of years married	24.8	25.7	28.4	19.7c	19.5c	26.8
			Average Number of People			
Household size	3.3	3.1	3.0	4.0	3.7	3.3
			Percent of Respondents Reporting Off-farm Employment			
Off-farm employment						
Operator**	19.4	16.7	19.4	14.4	45.6	18.0
Spouse***	43.1	41.2	34.3	51.7	65.2	36.4
			Percent of Respondents Reporting Limiting Health Condition			
Health condition limiting work						
Operators*	15.0	11.0	20.3	12.0	22.9	17.8
Spouse*	11.0	7.8	14.5	8.9	16.6	13.4

Percent of Families Making Adjustments

Selected adjustments made in past 2 years due to financial need						
Family member has taken off-farm work to help meet expenses****	29.2	19.7	24.6	44.0	55.3	27.0
Postponed medical care to save money****	23.2	12.1	24.6	31.9	51.1	23.8
Reduced/cancelled medical insurance****	21.0	15.3	17.4	28.3	42.6	18.9
Fallen behind in paying bills****	21.5	9.8	11.6	41.3	51.1	20.5
Any adjustment****,e	55.0	39.0	45.7	75.0	91.7	55.0

aSee Table 19.4 for definitions of financial position.
bN = number of responses. A few respondents did not answer questions on health conditions or financial adjustment and are not included in ratio calculations.
cWaller-Duncan K-ratio t-test was used to evaluate significant differences between mean number of years by financial position. A 'c' indicates that the average (mean) number of years is significantly different (p = 0.05) from the average (mean) number of years reported for respondents in a *favorable* financial position.
dChi-square probability level of significance
* = 0.05, ** = 0.01, *** = 0.001
Based on data for farms where financial position is classified.
eThe category "any financial adjustments" includes any of the four most common adjustments listed above and any of the following adjustments: (a) sold possessions or cashed in insurance, (b) borrowed money from friends or relatives, (c) unable to pay property taxes, and (d) used public assistance programs.

Source: South Dakota Family Farm Survey, 1989.

at home and a larger average household size. In general, farm couples operating highly leveraged farms are often in an earlier position of the family life cycle than many farm couples operating farms in a lower leverage position. This result likely occurs because younger families are more likely to be in an expansion phase of their farm business and are more dependent on debt capital.

Off-farm Employment. Farm couples operating highly leveraged farms are much more likely to report off-farm employment than other farm couples (Table 19.5). Operators of farms in a vulnerable financial position are much more likely to have off-farm employment (45.6 percent) than operators of other farms (14.4 percent - 19.4 percent). A majority (56 percent) of spouses living on highly leveraged farms have off-farm employment compared to 39 percent of other farm spouses. Overall, off-farm employment decisions are strongly associated with age and life cycle position of the farm couple, with farm size, and with farm financial position.

Over three-fifths of farm couples reporting off-farm employment indicate that "a family member has taken off-farm work to help meet expenses". Farm couples operating highly leveraged farms are much more likely to report that off-farm work is necessary to make "ends meet".

Family Health and Financial Adjustments. Operators and spouses operating farms with low net farm incomes (marginal income and vulnerable financial position) were twice as likely to report health problems that limit the amount of work that they can perform (Table 19.5). Incidence of health problems are more related to low net farm income than to the farm couples age!

Farm couples with low net farm incomes are almost twice as likely to report "postponing medical care to save money". Furthermore, farm couples operating highly leveraged farms report much higher incidence of "reducing or cancelling medical insurance" in the past two years to reduce outlays. Clearly, the degree of farm financial stress, incidence of health problems, and vulnerability to further health and financial problems are interrelated.

Farm financial position and incidence of family financial adjustments are closely related. A majority (55 percent) of farm couples have made one or more of eight possible family financial adjustments in the past two years (Table 19.5). Over four-fifths of farm couples operating highly leveraged farms made family financial adjustments compared to about two-fifths of other farm couples. Four family financial adjustments were each used by more than 20 percent of respondent farm families: (a) off-farm employment, (b) postpone medical care, (c) reduce or cancel medical or life insurance, or (d) fall behind in paying bills. This data clearly indicates the lingering effects of the farm finance crisis for many farm couples.

Farm Management Characteristics
by Farm Financial Position

Production, marketing and financial management are key ingredients of successful farm management. King and Sonka (1985) suggest that successful management practices differ in response to major changes in the environment faced by farm managers. Managing information and managing business/financial risks are two major farm management issues in today's economic environment.

Farm Operator's Use of Farm Management Records for Making Decisions. Numerous management studies have shown the importance of using farm records for making management decisions. Recent studies indicate farmers regard the preparation and use of management records as very important to their success in the modern "information era" (Carlson 1988; Mu'min and Hepp 1988). However, actual behavior of most farmers indicates that managing information is not a priority use of their time and is ranked low in terms of task enjoyment (Carlson 1988).

More than three-fifths of respondent operators reported using these records for making management decisions: (a) yield or production records, (b) net worth statements, (c) income statements, and (d) annual cash flow statements (Table 19.6). In addition, 42 percent of respondent operators reported using enterprise budgets as a decision tool, while 30 percent reported using multi-year cash flow plans for making decisions. Few producers (10.3 percent) formulate written business goals. Farm couples using farm records in making decisions, regardless of financial leverage position, reported higher average net farm incomes!

Overall, farm operators in a marginal solvency position were *most likely* to report using farm records. Farmers in a marginal income position were *least likely* to report using farm records of any kind. In general, farmers in a higher leverage position were more likely to report using financial records, in part because their lenders were requiring preparation of these statements.

Changes in Farm Management Practices. Farmers, as entrepreneurs, are involved in managing changes in their operation in response to many sources of risk and profit seeking opportunities. In the 1980s, many farmers were forced to make numerous changes in their operation in efforts to insure long-term business survival. Data in Table 19.7 indicates that most respondent farm operators made numerous changes in farm management practices in the previous five years (1984-1988).

Reducing short-term debt and long-term debt were priority management changes for 70 percent of farmer respondents. Purchasing

TABLE 19.6 Farmers' Use of Farm Management Records, Overall and by Farm Financial Position, South Dakota, 1989

	All Farms	Financial Position[a]			
		Favorable	Marginal Income	Marginal Solvency	Vulnerable
N[b] =	549	210	70	92	48
Type of Farm Management Records	Percent of Farms Using Record	Deviation from Percent of All Farms Using Record			
Yield or production records***c	61.8	+2.8	-11.0	+18.2	+2.8
Crop or livestock enterprise budgets***	41.8	-1.9	-15.2	+26.7	+ 4.0
Net worth statement***	62.2	-2.5	-4.6	+16.7	+12.8
Income statement***	62.8	-1.7	-6.1	+17.2	+8.0
Annual cash flow***	63.0	-5.8	-14.5	+24.9	+16.2
Multi-year cash flow plan*	30.0	-3.2	-0.3	+11.8	+5.4
Business goals for this year*	74.7	-0.4	-3.7	+10.1	+8.3
Written business goals	10.3	-3.3	+0.5	+4.0	+0.6

[a]See Table 19.4 for definitions of farm financial position. All farms include responses by 129 farms that were not classified by financial position.
[b]N = number of responses. Some operators did not respond to some specific questions on farm management records.
[c]Chi-square probability level of significance
  * = 0.05, ** = 0.01, *** = 0.001

Based on data where financial position is classified. No * implies chi-square distribution is not significant at 0.05 probability level.

Source: South Dakota Family Farm Survey, 1989.

TABLE 19.7 Changes in Management Practices in Past Five Years by Farm Financial Position, South Dakota, 1989

	All Farms	Financial Position[a]			
		Favorable	Marginal Income	Marginal Solvency	Vulnerable
N[b] =	549	210	70	92	48
	Percent Adopting Practice	Deviation from Percent of All Farms Adopting Practices			
Selected Management Changes					
Raising new crops**c	34.2	-1.4	-10.0	+10.9	-2.3
Raising livestock	30.6	-3.5	+1.1	+4.6	+5.0
Low input farming	23.8	+0.3	-6.1	+7.6	-7.9
No till farming	20.9	-0.8	+4.1	+3.2	-7.9
Forward contracting*	20.1	-0.8	-6.3	+12.1	-11.4
Futures/Options	15.7	-0.7	-0.9	+4.5	-4.6
Crop insurance***	48.6	-0.8	-12.5	+13.6	-6.0
Computer analysis of farm finances	17.4	-0.2	-5.5	+5.1	-2.2

(continued)

TABLE 19.7 (continued)

	All Farms	Financial Position[a]			
		Favorable	Marginal Income	Marginal Solvency	Vulnerable
Reduce long-term debt***	70.4	-2.2	-7.1	+9.2	+5.7
Reduce short-term debt***	70.5	-5.0	-9.2	+12.7	+9.5
Purchase land	26.2	-6.8	-7.6	+3.8	-1.8
Sold land	5.8	+0.7	-0.7	-2.0	+0.7
Transfer land back to lender/seller	3.6	-1.6	-0.1	+1.9	+3.1
Rent less acres	16.8	-1.6	+3.2	-0.3	+2.8
Rent more acres	39.7	-0.9	-4.1	+8.6	-7.1
Reduce machinery	11.9	-3.7	+5.6	+4.8	-1.0

[a]See Table 19.4 for definitions of farm financial position. All farms includes responses by 129 farms that were not classified by financial position.
[b]N = number of responses. Some operators did not respond to some specific questions on management changes.
[c]Chi-square probability level of significance
* = 0.05, ** = 0.01, *** = 0.001

Based on data where financial position is classified. No * implies chi-square distributions not significant at 0.05 probability level.

Sources: South Dakota Family Farm Survey, 1989 and Stover, Clark, and Janssen, 1991, page 120.

crop insurance was a management change for nearly half (48.6 percent) of respondents. Expanding by renting more acres was a management change for nearly 40 percent of the farm operators. All of these management changes reduced their financial risk (Table 19.7).

Many farmers have been forced to reduce the scope of their operations by reducing machinery inventory or renting less land. However, very few respondents (<6 percent) reduced their operation by selling land or transferring land to the seller or lender.

The following changes in management practices were strongly related to farm financial position: (a) raising new crops, (b) forward contracting, (c) purchasing crop insurance, and (d) reducing farm debt. Each of these changes in management practices are methods to reduce adverse consequences of business/financial risk or are profit-seeking opportunities. These results are generally consistent with recent studies indicating that farmers perceive greater risk in marketing and finance, but their management responses emphasize changes in financial and production management (Boggess, Anaman, and Hanson 1985; Branch and Olson 1987; Ekstrom et al. 1987; Mu'min and Hepp 1988; and Carlson, 1988).

Management Profile of Farmers by Financial Position. Farmers in a marginal solvency position are in the forefront of making changes in their production, marketing, and financial management practices. These same farmers are also more likely to use various Federal and state programs, including: (1) Federal crop insurance, (b) farm financial counseling and/or farm mediation, and (3) 1988 drought assistance programs. Also, this group of farmers are more apt to prepare and use various farm management records in making decisions. All of these changes in management practices have enabled them to reduce the adverse consequences of risk in their operation and to engage in additional profit-seeking opportunities.

Farmers in a vulnerable financial position have also made many changes in their operation. However, they are not as likely to use farm management records for decisionmaking purposes and have not been able to make as many changes that simultaneously reduce risk and increase their profit potential.

Farmers in a marginal income position have made the least amount of management changes, are less likely to use Federal and state programs, and are the least likely to use farm management records for making decisions. Farmers in a favorable financial position, while similar in age to those in a marginal income position, have many management characteristics that are similar to those in a marginal solvency position. This group was much more likely to make changes that permitted expansion of their operation, with some attention to reducing financial risk.

The future structural composition of commercial farms will likely be those in a favorable financial position and many in a marginal solvency position that are making considerable financial progress.

Family Functioning and Farm Financial Position

Contemporary investigations of "successful family" life are focused: (a) on the extent that participants are satisfied with their family life; (b) on the extent of cohesion within the family; (c) on the stress endured by family members; and (d) on the amount of agreement on basic issues within the family. Family life satisfaction and the extent that a family operates as a cohesive social unit are generally considered as direct measures of "successful family life", while family stress and couple agreement are usually considered as intervening variables. Based on the literature review, we have developed Likert-scale index measures of family life satisfaction, family cohesion, family life stress, and couple agreement from questions answered by operators and spouses.[5] These measures of family functioning are examined by respondent's farm financial position.

Family Satisfaction Index. Successful, stress-resistant families are assumed to have a strong sense of well-being or satisfaction with life -- successful adaptation to family circumstances. The 10-item Family Adaptation Scale (Antonovsky and Sourani 1988) was used to measure family satisfaction in this study, along with four additional items on satisfaction with their spouse and with their marriage. The Family Adaptation Scale asks respondents to assess their level of satisfaction from "completely satisfied" to "dissatisfied" on a 1-5 scale. Items in this scale reflect family life, for example: "extent to which family members are close to each other" and "how the family fits into the neighborhood".

The responses of each operator and of each spouse are separately summed and divided by the number of items involved. The end result is a mean score across all satisfaction items for each operator and for each spouse. This mean score is the Family Satisfaction Index. This same procedure is used to develop the indices on family coherence, family stress, and couple agreement.[6]

Family Coherence, Family Stress, and Couple Agreement Indices. A shortened version of the Family Sense of Coherence Scale (FSOC) (Antonovsky and Sourani 1988) was used to measure family coherence in this study. The Family Coherence Index is developed from the mean score (using a 1-5 scale) of respondent's answers to 12 questions designed to measure their degree of *stress resistance.* Stress resistance is the ability to cope with problems (stressors) as they occur and is enhanced by our ability to view changing situations (events) as comprehensible, manageable and challenging instead of totally confusing, unmanageable and catastrophic!

Perceived stress was measured by a shortened version of the Family Stress Inventory (FSI), a standardized instrument developed by Walker and Walker (1987) to measure occupational and personal stress associated with stressors of a farming lifestyle. The shortened version of the FSI used in the present study contained 12 items. Six items are stressors specific to farming and a rural lifestyle (examples are "no control over weather or commodity prices" and "traveling long distances for services, health care and shopping") and six are related to family stressors (examples are "conflict with spouse" and "relationships with children"). For each stressor item, respondents indicated their degree of perceived stress on a 1-5 scale. The mean score for each respondent is their Family Stress Index.

Family life research indicates most married couples have disagreements in their relationships. However, family functioning depends on their general agreement on "core issues" important to their family. We identified 12 core issues important to farm families including: making major decisions, child rearing, household finances and nine other items. Each respondent was asked about their extent of agreement (or disagreement) with their spouse on each issue, using a 1-5 scale from "always agree" to "always disagree". Their response to each issue was summed and divided by the appropriate number of items. For each spouse, their mean score is the Couple Agreement Index.

Family Functioning by Farm Financial Position. For each index, paired operator and spouse responses are highly correlated with each other. Farm operators and spouses generally expressed high levels of family life satisfaction with an average operator score of 1.71 and average spouse score of 1.82 (Table 19.8). Only 3 percent of respondent operators and 5 percent of spouses had satisfaction scores of 3.0 or above, which indicates few respondents were generally dissatisfied with their family life.

Family coherence index scores are considerably lower than family life satisfaction scores for both operators and spouses. These results correspond with previous research (Antonovsky and Sourani 1988) which indicates respondent's self-reported satisfaction scores are higher than their stress resistance (family coherence) scores.

A review of respondent family life satisfaction scores and their stress index scores indicates farm couples have generally high satisfaction levels *and* moderate-to-high levels of stress. These results correspond with findings in a previous study (Walker and Walker 1987). The average

TABLE 19.8 Farm Couple Satisfaction, Coherence, Stress and Agreement by Farm Financial Position

| | Farm Financial Position[a] | | | | |
	Favorable	Marginal Income	Marginal Solvency	Vulnerable	All Classified Farms
N[b] =	210	70	92	48	420
	Mean (Average) Index				
Satisfaction Index (1-5)[c,d]					
Operator***	1.65	1.64	1.84*[e]	1.71	1.71
Spouse***	1.76	1.61	1.95*	2.11*	1.82
Coherence Index (1-5)[c,d]					
Operator***	2.17	2.10	2.14	2.36*	2.12
Spouse***	2.14	2.13	2.11	2.51*	2.19

Stress Index (1-5)^{c,d}				
Operator****	2.26	2.39	2.50*	2.75*	2.40
Spouse****	2.31	2.32	2.56*	2.73*	2.42
Agreement Index (1-5)^{c,d}				
Operator**	1.96	1.99	2.19*	2.21*	2.04
Spouse**	1.89	1.93	2.05*	2.25*	1.97

[a]See Table 19.4 for specific definition of farm financial position.

[b]Number of responses.

[c]Satisfaction index, coherence index, stress index and agreement index are Likert-scale indices with a range of 1.0 - 5.0. An index value of 1.0 respectively, the highest level of satisfaction, highest coherence level, lowest amount of stress or highest level of agreement. An index value of 5.0 represents, respectively, the least level of satisfaction, least coherence level, highest level of stress or least amount of agreement.

[d]A one way ANOVA test was conducted for each index where average index score (of operator and spouse) = f (farm financial position). A ** indicates the ANOVA F-value is significant at the 0.01 probability level. A *** indicates the ANOVA F-value is significantly at the 0.001 probability level.

[e]The Waller-Duncan K-ratio T-test was used to test if any specific mean index values are significantly different from each other by farm financial position. A * indicates the mean index value has a statistically significant difference (p = 0.05) from the mean index value shown for those in a favorable financial position.

The Waller-Duncan Duncan option of the PROC GLM program in SAS was used to develop the statistical information presented in this table.

Source: South Dakota Family Farm Survey, 1989.

stress index is almost the same (2.40 vs. 2.42) for respondent operators and for spouses. About 17 percent of respondents report relatively high levels of stress with average stress index scores exceeding 3.0.

Most respondents report fairly high levels of agreement with their spouse on each of the 12 core issues. The average couple agreement index score are 2.04 for operators and 1.97 for spouses. Only 5 percent of respondents reported general disagreement with their spouse on a majority of the 12 items.

Respondent average index scores for family life satisfaction, family coherence, family stress and couple agreement are systematically related to the financial position of their farm operation. Respondents operating highly leveraged farms have significantly lower levels of family life satisfaction than other farm couple respondents. Also, respondents operating highly leveraged farms report significantly higher stress levels and lower levels of agreement with their spouse on core issues. Finally, couples operating farms in a vulnerable financial position have a significantly lower sense of coherence than families operating farms in a less vulnerable financial position. Unfortunately, many couples that are in the most vulnerable financial situation also have lower ability to cope with their situation!

These results provide substantial support for the hypothesis that farm family functioning and farm financial position are strongly and systematically related to each other. However, it is also important to note that family functioning variables are not perfectly correlated with economic measures of "farm business viability". Further theoretical and empirical research is needed to identify the direction of causation and feedback loops between family functioning measures and farm business viability measures.

Farm Couples' Assessment of Their Farm Operation

Discussion of respondent farm couples is incomplete without their assessment of and expectations about their farm operation. Overall, operator's assessment of their farm operation is more favorable than their spouses' assessment. More than 90 percent of farm couples reported that their farm is an "ideal place to raise their family" and is a major source of satisfaction for them. Nearly 90 percent of farm operators, but only 60 percent of spouses, indicated their farm "offers me a good place to put my own ideas into operation". The spouses response is directly related to their extent of involvement in the farm business.

Between 62 percent - 65 percent of farm operators agreed that their farm operation is "financially successful" and "provides us with a good income". However, only 55 percent of spouses agreed with these statements. Nearly 20 percent of spouses and 17 percent of operators disagreed with a favorable economic assessment of their farm operation, while the remainder were uncertain. As expected, farm couples' assessment of the income and financial elements of their farm business is strongly related to their farm financial position. Three-fourths of farm operators in a favorable financial position, but only one-third of those in a vulnerable financial position, perceived that their farm provided a good income and is financially successful.

Expanding farm size is a high priority objective for 30 percent of farm operators. Another 60 percent placed high priority on maintaining the present size of their farm business, while 10 percent wanted to reduce the size of their existing farm operation. These objectives were related to operator age, but unrelated to their farm financial position.

Seventy percent of farm operators fully expected to continue operating their farm or ranch for at least another five years. Another 6 percent expected to retire or quit and 24 percent were uncertain about their future plans. Almost 86 percent of farm operators indicated that improvements in their farm financial position over the past 5 years would permit them to farm another five years or more, if they choose to do so. Nearly 80 percent of farmers in a favorable financial position fully expected to continue operating their farm, compared to only 50 percent of farmers in a high leverage financial position.

Nearly half (48 percent) of the respondents expect their farm operation will eventually be operated by one of their children and one-fourth of these respondents are already farming with their adult children. Another 27 percent expect their farm to remain in family ownership, but will be operated by someone else. Another 25 percent expect their farm to be sold to someone else or are uncertain about the future of their farm.

Summary, Conclusions, and Implications

Interrelationships between farm business and farm household decisions and activities are important elements of farm structure. In a family farm setting, the decisions and actions of family members influence the course of the farm business and the welfare of the household. In this chapter, we examined key household-business interrelationships in a multidisciplinary study (economics, sociology and home economics) of 549 South Dakota farm couples operating "part-time" or "full-time" commercial family farms.

Work Roles of Farm Couples

Nearly half (47.7 percent) of respondent South Dakota farm families have one or more adults engaged in off-farm employment. Compared to

U.S. totals, South Dakota has a similar percentage of farm women and a much lower percentage of farm men employed off-farm. The primary explanation is that South Dakota has very few "residential farms" and a high proportion of "commercial family farms."

A majority of respondent women (56 percent) report active involvement in the farm operation. The extent of their farm work involvement is inversely related to their incidence of full-time off-farm employment. Performance of specific farm tasks and family/household tasks follow distinct gender roles.

Decisionmaking Roles of Farm Couples

Stress-oriented studies indicated that shared decisionmaking is an important characteristic of "crisis-proof" families and is a key characteristic of "successful families". Most South Dakota farm couples (75 percent - 84 percent) use a shared decisionmaking approach to family/household decisions and a majority use that approach to farm business decisions involving farmland rental or purchase. The operator is the principal decisionmaker on most other farm-related decisions.

Farm Management
and Family Management Interrelationships

Many differences in farm management and farm household characteristics of respondents are related to their farm financial position. Farm operations with higher net farm incomes are not only larger in average size (based on total assets), but also have higher asset turnover rates, higher net margin percentages, and higher rates of return on equity than other farms. Couples operating farms in a *marginal solvency* position (high leverage and higher net income) have made the most changes in their management practices, while couples operating farms in a *marginal income* position (low leverage and low net income) have made the fewest changes. Farm couples in a *favorable* financial position have made changes that permitted expansion of their operation, while those in a *vulnerable* financial position emphasized debt reduction and other survival strategies.

South Dakota farm couples operating highly leveraged farms are an average of 8-10 years younger and are much more likely to report off-farm employment than other farm couples. These same farm couples also have family life index scores indicating lower family life satisfaction, lower family cohesion (coherence), higher stress and lower couple agreement on basic issues than respondents operating farms in a favorable financial position.

Conclusions and Implications

High levels of family life satisfaction and coherence are major attributes of "successful families." These families also exhibit high levels of couple agreement on basic issues and have greater ability to handle stress. Production ability and timeliness, financial management, ability to handle change and other stressors, and positive attitudes toward work, family and other key human relationships are major characteristics of "successful farm management".

The findings of this research support the assumption that successful family life and successful farm business management are very much interrelated. Operating a successful farm business relies to a great extent on successfully managing the internal relations with the farm family and the interaction between the family and the farm business. Research which does not recognize this intrinsic link between family life and farm management neglect some crucial factors that lead to success (or failure) of farm businesses.

The future structural composition of commercial family farms will likely be those in a favorable financial position and many of those in a marginal solvency position that are making considerable financial progress. These farm couples have many of the characteristics of "successful" farm managers acting to ensure and control outcomes instead of passively reacting to the economic and social environment that they face (Johnson 1988).

Notes

1. The multidisciplinary study of South Dakota Family Farms was financially supported by the Midwest Technology Development Institute, Farm Enterprise Partnership. Matching Support was provided by the SDSU Agricultural Experiment Station.

2. The random sample of 2000 farm households were selected by SDASS (South Dakota Agricultural Statistics Service). Contractual agreement required the USDA agency to administer the mail survey (including two follow-up mailings), because they are required to maintain confidentiality of all names on their mailing list.

3. The lower response rate of farm families operating farms generating annual gross farm receipts of less than $40,000 is due to: (1) the farm population list maintained by SDASS excludes many small farm unit, and (2) the lengthy questionnaire was oriented to families where one or both spouses have a substantial economic (time and financial involvement) commitment to a farm business operation.

4. According to (Stanton and Bills 1988, 18) a *full-time commercial farm* is "an establishment where agricultural production and marketing is the primary

534

occupation of the operator, and where 12 months or more of operator, family or regular hired labor are employed." A *part-time commercial farm* is an "establishment where agricultural production is an important contributor to family income and where 2 to 12 months of operator, family or hired labor in total are employed." A *residential farm* is an "establishment where agricultural production occurs, but is not an important contributor to family income; where less than 2 months of total labor are employed in agricultural production and marketing."

5. A copy of the specific questions used to develop each index can be obtained by contacting the authors. These questions are listed in Clark, Janssen and Stover (1990).

6. In some cases, respondents did not answer all questions used to develop each index. To handle the issue of missing data, a set of minimum criteria was established. For each index, if information was missing for more than one question for *either spouse*, the case (consisting of both husband and wife) was dropped from the analysis.

7. It should be noted no financial related questions are included in the family life satisfaction and family coherence scales. Only one financial related question is included in the twelve stressor items in the family stress scale and only two financial related questions are included in the twelve item couple agreement scale. The conclusion that the average stress index score and average couple index score is systematically related to farm financial position is not changed if the financial questions are removed from each scale.

References

Andrews, F. M., and Withey, S. B. 1976. *Social Implications of Well-being: American's Perceptions of Life Quality*. New York: Plenum.

Angell, R. O. 1936. *The Family Encounters The Depression*. New York: Charles Scribner.

Antonovsky, A., and Sourani, T. 1988. "Family Sense of Coherence and Family Adaptation." *Journal of Marriage and the Family* 50: 79-92.

Barry, P. J., Ellinger, P. N., and Eidman, V. R. 1987. "Firm Level Adjustments to Financial Stress." *Agricultural Finance Review, 47: Special Issue*. Ithaca, NY: Cornell University.

Becker, Gary S. 1974. "A Theory of Marriage." in T.W. Schultz, ed., *Economics of the Family*. Pp. 299-344. Chicago, IL: University of Chicago Press.

Bharadwaj, L., and Wilkening, E. A. 1977. "The Prediction of Perceived Well Being." *Social Indicators Research* 4: 421-439.

Bianchi, S. M., and Spain, D. 1986. *American Women in Transition*. New York: Russell Sage.

Boehlje, M., and Eidman, V. 1983. "Financial Stress in Agriculture: Implications for Producers." *American Journal of Agricultural Economics* 65(5): 937-944.

Boggess, W. G., Anaman, K. A., and Hanson, G. D. 1985. "Importance, Causes, and Management Responses to Farm Risks: Evidence from Florida and Alabama." *Southern Journal of Agricultural Economics* 17(2): 105-116.

Bokemeier, J., and Maurer, R. 1987. "Marital Quality and Conjugal Labor Involvement of Rural Couples." *Family Relations* 36: 4177-424.

Branch, W. F., and Olson, C. E. 1987. "Ranch Manager Risk Perceptions and Management Responses." *Journal of American Society of Farm Managers and Rural Appraisers* 51(2): 58-63.

Bubolz, M. M., Eicher, J. B., Evers, S. J., and Sontag, M. S. 1980. "A Human Ecological Approach to Quality of Life: Conceptual Framework and Results of a Preliminary Study." *Social Indicators Research* 7: 103-136.

Campbell, A., Converse, P. E., and Rodgers, W. L. 1976. *The Quality of American Life*. New York: Russell Sage Foundation.

Carlson, J. E. 1988. "Farmers Perceptions about the Management of their Farms." *Journal of American Society of Farm Managers and Rural Appraisers* 52(1): 91-96.

Clark, V., Janssen, L., and Stover, R. 1990. *Successful Family Farming in Times of Crisis: Empirical Results from South Dakota.* (Tech. Res. Rep. submitted to Farm Enterprise Partnership, Midwest Technology Development Institutes, St. Paul, MN). Brookings, SD: South Dakota State University.

Constantine, L. L. 1983. "Dysfunction in Open Family Systems I: Application of a Unified Theory." *Journal of Marriage and the Family* 45(4): 7.

Defrain, J., and Stinnett, N. 1988. "Strong Families and Strong Farming Organizations. Is This a Connection." in L. Robison, ed., *Determinants of Farm Size and Structures.* Pp. 129-142. Proceeding of NC-181 Committee. Michigan Ag. Expt. Sta. Journal No. 1289a.

Deseran, F. A., Falk, W. W., and Jenkins, P. 1984. "Determinants of Earnings of Farm Families." *Rural Sociology* 49(2): 210-229.

Ekstrom, B. L., Hardie, W., and Leistritz, F. L. 1987. "Management Adjustments in the Face of Farm Financial Stress." *North Dakota Farm Research* 45(2): 3-6.

Hacker, A. 1983. *US: A Statistical Portrait of the American People.* New York: The Viking Press.

Harrington, D., and Carlin, T. A. 1987. *The U.S. Farm Sector: How is it Weathering the 1980s?* (Agricultural Information Bulletin 506). Washington, DC: U.S. Department of Agriculture.

Hocker, J. L., and Wilmot, W. W. 1985. *Interpersonal Conflict.* Dubuque, IA: William C. Brown.

Johnson, G. L. 1988. "Farm Managerial Inquiry: Past and Present Status and Implications for the Future." in L. Robison, ed., *Determinants of Farm Size and Structures.* Pp. 7-24. Proceedings of NC-181 Committee, Michigan Ag. Expt. Stat. Journal No. 12899.

Johnson, G. L., Halter, A., Jensen, H., and Thomas. D. W. 1961. *A Study of Managerial Processes of Midwestern Farmers.* Ames: Iowa State University Press.

Johnson, J., Morehart, M., Neilsen, L., Banker, D., and Ryan, J. 1987. *Financial Characteristics of U.S. Farms.* (Ag. Info. Bul. 525) Washington, DC: U.S. Dept. of Agriculture.

Jorgensen, S. R. 1986. *Marriage and the Family: Development and Change.* New York: Macmillan.

Jurich, A. P., Schumm, W. R., and Bollman, S. R. 1986. "Place of Residence and Quality of Life Conceptual Framework." in J.L Hafstrom, ed., *Compendium*

536

of Quality of Life Research. Pp. 1-15. Urbana, Illinois: University of Illinois, Illinois Agricultural Experiment Station.

Justus, F., and Headley, J. C. 1968. *The Management Factor in Farming: An Evaluation and Summary of Research*. St. Paul: University of Minnesota, Agricultural Experiment Station.

Keating, N. C. 1987. "Reducing Stress of Farm Men and Women." *Family Relations* 36(4): 358-363.

King, R. P., and Sonka, S. T. 1985. "Management Problems of Farms and Agricultural Firms." Paper presented at AAEA Conference on "Agriculture and Rural Areas Approaching the 21st Century: Challenges for Agricultural Economics," Iowa State University, Ames, IA.

Knaub, P. K., Draughn, P. S., Wozniak, P., Little, L. F., Smith, C., Weeks, O. 1988. "Wives Employed Off the Farm: Impact on Lifestyle Satisfaction." *Home Economics Research Journal* 17: 36-46.

Leistritz, F. L., Leholm, A. G., Vreugdenhil, H. G., and Ekstrom, B. L. 1986. "Effect of Farm Financial Stress on Off-Farm Work Behavior of Farm Operators and Spouses in North Dakota." *North Central Journal of Agricultural Economics* 8(2): 269-281.

Light, H., Hertsgaard, D., and Martin, R. E. 1985. "Education and Income: Significant Factors in Life Satisfaction of Farm Men and Women." *Research in Rural Education* 3: 7-12.

Mu'min, R. A., and Hepp, R. E. 1988. *Evaluating Managerial Effectiveness*. (Agricultural Economics Report No. 507). East Lansing: Michigan State University, Department of Agricultural Economics.

Newsweek. (1985, July 15). "Playing Both Mother and Father." Pp. 42-43.

Norton, A. J., and Moorman, J. D. 1987. "Current Trends in Marriage and Divorce among American Women." *Journal of Marriage and the Family* 49: 3-14.

Olson, D. H., Russell, C. S., and Sprenkle, D. H. 1983. "Circumplex Model VI: Theoretical Update." *Family Process* 22: 69-83.

Olson, D., McCubbin, H. I., Barnes, H., Larsen, A., Muxem, M., and Wilson, M. 1983. *Families: What Makes Them Work?* Beverly Hill, CA: Sage.

Reimund, D. A., Brooks, N. L., and Velde, P. D. 1986. *The U.S. Farm Sector in the Mid-1980s*. (Agricultural Economics Report 548). Washington, DC: U.S. Department of Agriculture.

Retting, K. D. 1987. "A Cognitive Conceptual Family Decision Making Framework." Presentation for NCR 116 Family Resource Management Research Reporting Technical Group, St. Louis, MO.

Sanders, W. 1985. "Farm Women and Marriage." *Oxford Agrarian Studies* 14: 114-127.

Schultz, T. W. 1974. *Economics of the Family*. Chicago, IL: University of Chicago Press.

Schumm, W. R., and Bollman, S. R. 1981. "Interpersonal Processes in Rural Families." in R. T. Coward and W. M. Smith, Jr., eds., *The Family in Rural Society*. Pp. 129-146. Boulder, CO: Westview Press.

Schumm, W. R., Bugaighis, M. A., Bollman, S. A., and Jurich, A. P. 1986. "Meaning and Impact of Family Satisfaction upon the Subjective Quality of

Life: Perception of Fathers, Mothers, and Adolescents." in J.L. Hafstrom, ed., *Compendium of Quality of Life Research*. Pp. 57-70. Urbana, IL: University of Illinois, Illinois Agricultural Experiment Station.

Stanton, B. F., and Bills, N. L. 1988. *Approaches to Defining and Classifying Farms*. (Cornell Ag Econ Staff Paper 88-17). Ithaca: Cornell University.

Stoeckeler, H. S., and Gage, H. G. 1978. *Quality of Life*. (Agricultural Experiment Station Miscellaneous Report 154). Minneapolis: University of Minnesota.

Stover, R. G., Clark, V. L., and Janssen, L. L. 1991. "Successful Family Farming: The Intersection of Economics and Family Life." *Research in Rural Sociology and Development* 5: 113-129. Greenwich, CT: JAI Press.

Szinovacz, M. E. 1984. "Changing Family Roles and Interactions." *Journal of Family Issues* 5(2): 163-201.

Thornton, A., and Freedman, D. 1983. "The Changing American Family." *Population Bulletin* 38(4): 3-37.

United States Department of Commerce. 1987. U.S. Census of Agriculture - South Dakota. (Part 41). Washington, DC: Bureau of the Census.

Walker, L. S., and Walker, J. L. 1987. "Stressors and Symptoms Predictive of Distress in Farmers." *Family Relations* 36: 374-378.

Weigel, R. R., Weigel, D. J., and Blundall, J. 1987a. "Stress, Coping, and Satisfaction: Generational Differences in Farm Families." *Family Relations* 36: 45-48.

Wilkening, E. A. 1981. "Farm Families and Family Farming." in R.T. Coward and W.M. Smith, Jr., eds., *The Family in Rural Society*. Pp. 127-38. Boulder, CO: Westview Press.

20

Structural Change in Farming and Its Relationship to Rural Communities

Thomas A. Carlin and William E. Saupe

This chapter continues the theme of the preceeding section on "the factors that induce structural change in agriculture", but also pertains to the topic of this final set of chapters on "the impacts of structural change". That is because there is a two-way relationship between communities and farms. The structure of farming and changes in structure affect development in the communities in their midst, but communities also affect the structure of the farms that surround them.

Regional Development and Economic Concepts

We have witnessed a major transformation of rural America away from farming as the dominant industry. Over the last 40 years, the contribution of farming to the personal income of rural people has declined substantially. In 1950, at the beginning of the rapid decline in farm numbers, over 2,000 nonmetropolitan counties in the 48 contiguous states were "farming dependent", i.e. at least 20 percent of total earnings came from farming. By the early 1980s, only 505 nonmetropolitan counties could be so designated (see Figure 20.1). Manufacturing, government, recreation, and retirement are among the industries now dominating most rural economies (Bender et al. 1985).

This change in rural America's economic base reflects the major transformation that has occurred in the structure of the U.S. farming sector. Since 1950, the number of farms has declined over 60 percent and average farm size has more than doubled. The farm resident population

FIGURE 20.1 Farming Dependent Counties*, 1950 and early 1980s

1950

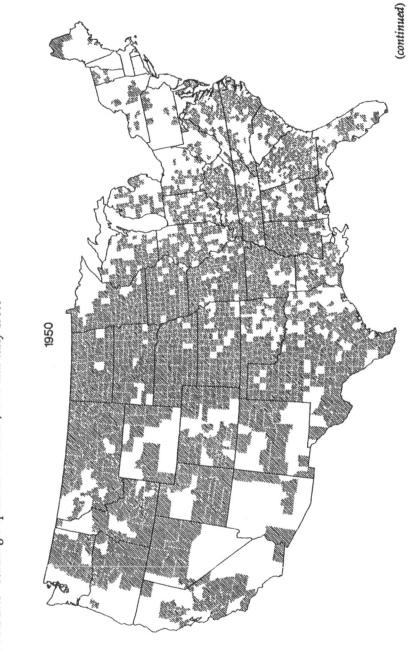

(continued)

540

FIGURE 20.1 *(continued)*

Early 1980's

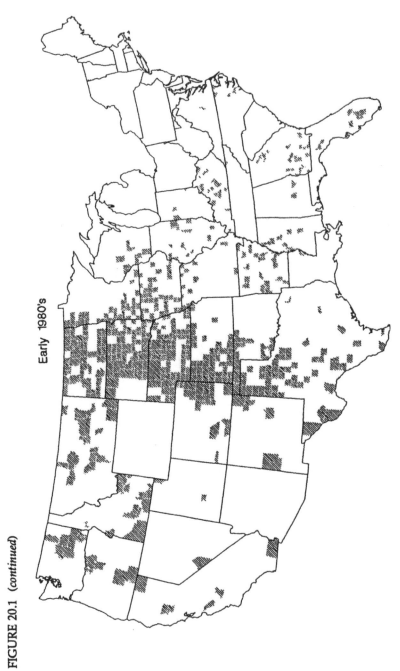

*At least 20 percent of total earnings in the county were from farming. Data for the early 1980s reflects a 5-year period ending in 1986.

declined from over 23 million persons in 1950 to less than 5 million today.

The synergistic nature of the relationships between farm structure and rural communities is emphasized in this chapter. It is often tempting to treat farm structure and community development as if they were mutually exclusive topics when in reality they are highly related. We discuss how farm structure affects the communities in which the farms are located, and how in turn community attributes affect the organization of farming. The relationship has a two-way effect.

The Community and Development

As used here, "community" involves a geographic area, business and social linkages or interactions among persons, and a commonality of mutual interests. We often think of communities as villages, towns, counties or other political jurisdictions. "Community growth" is an increase in the economic activity in a community. Growth might contribute to "community development", which involves the enhancement of human well-being, improved quality of life, and increased equality of opportunity (Shaffer 1989). Development is a more subjective concept than growth, but it is more relevant as we consider the synergistic relationships between the farming sector and rural communities. The theoretical and conceptual base for our discussion draws from various aspects of regional economics paradigms and economic theory (Shaffer et al. 1986).

Supply-Oriented Theories. Supply-oriented community development theories focus on the importance of labor, capital, technology, land, and other natural resources to increase community output and income. Economists apply production function analysis to the community as the unit of interest, rather than (say) to a farm firm. But the same familiar assumptions, marginal conditions, equilibrium conditions, etc. apply in the analysis. The limited availability of private, public, and human capital inputs constrain community development. Community development policies that follow from this paradigm often focus on the need to increase productivity, usually through increased capital investment.

Resource endowment theory, a supply-oriented theory, focuses on the stock of natural resources, saying that they are the basis for producing externally desired goods and services. The ability of a community to develop depends on its ability to utilize these natural resources by combining other inputs (backward linkages) with the natural resources to produce an output that is processed (forward linkages) into the desired good or service. If suppliers of inputs and processors of output can be provided within the community, the community can expand its total output (or valued added) and income.

Farming serves as an example of resource endowment theory, making use of land and climate as natural resources. Farmers purchase fertilizer, machinery, and other inputs and combine them with natural resources to produce food and fiber. Food and fiber products are further processed into consumer products. But farming provides a development dilemma as it is an unstable contributor to community economic development. This is because of its vulnerability over time due to the narrowness of the product line, the income inelasticity of demand for the product, and potential changes in the derived demand for the resource because of changes in technology or consumer preferences. Investments in farming to improve efficiency have resulted in a substantial reduction in the need for farm labor. Those communities that have not captured the forward linked value-added agricultural industries or that have not diversified their economies face increasing levels of unemployment and population outmigration. This situation characterizes much of the western Corn Belt and Northern Plains.

Demand-Oriented Theory. Demand oriented development theories focus on the external demand for a community's goods and services, and the community's comparative advantage in producing them. The most explicit of these is "export base theory". It suggests that the economic vitality of a community depends on the production of goods and services that can be exported to an external market. Generally there is a basic industry that produces products or services for export. This industry essentially "drives" the local economy. There are secondary industries present that support either the basic industry or provide goods and services to people residing in the community. Agriculture, mining, forestry and manufacturing were initially put forth as examples of export based industries, but more recently any activity that brings money into the community is considered an export activity. These include such diverse activities as tourism, and the receipt of social security benefits and other transfers.

In the case of farming communities, the local economy is developed around activities associated with producing, transporting, processing and marketing farm products. Sometimes only a limited number of these activities are performed in the local rural community. The food processing industry, for example, is predominantly an urban industry with two-thirds of the establishments and three-fourths of the employment located in metropolitan areas (Francis and Petrulis 1988).

If an export industry in a community is experiencing employment declines (e.g. farming) and there are limited employment opportunities in other local industries, then population outmigration likely occurs, threatening the viability of consumer based services. This in turn can lead to a decline in the community's business district. If the process proceeds long enough, it becomes difficult for the community to maintain adequate public services. Soon the community can no longer maintain itself as a viable entity. In this way the role of various communities in the region change as economic activity adjusts itself spatially to accommodate the new economic and social environment.

Other Theories. By its very nature, community development deals with the spatial location of economic activity, what we sometimes refer to as economic geography. "Location theory" deals with the attributes of space as it relates to markets, transportation, resources, and production. In location decisions, a firm probably first considers the location of the market for its product, and then to the minimization of transportation and labor costs, and the possibility of capturing economies of scale in the industry. More recently, the location's desirable attributes from the point of view of the employees have become more important, e.g. educational facilities, climate, recreation. Location theory can probably be extended from determining the most desirable location for a firm as the unit of interest, to the optimal locations of communities.

"Central place theory" provides tools to understand where retail and service functions are clustered in the region. Community trade and service activities depend on the distance people will travel to purchase goods and services, the costs of providing goods and services, and the size of market needed to earn minimum profits. Some trade and service activities depend more on the volume of the export commodity produced (e.g. some farm inputs), whereas others depend on population size (e.g. retail trade).

Finally, "welfare theory" draws attention to the distributional impacts of changes in the national economy, the farming economy, and the regional economy. For example, development and adoption of cost reducing and output increasing farm technology can lead to increased agricultural production, lower farm prices and incomes, accelerated farm exit, reduced trade and income in some sectors of the local rural economy, but lower cost of food for consumers. Public intervention in this process is a political decision, but social scientists can articulate the alternatives and their differential impacts.

The Synergistic Relationship

The study of farm structure usually focuses on how land, labor, and capital are organized into farming units to produce food and fiber and the distribution of income and wealth that results from that activity.

Discussions about the structure of U.S. farming typically feature national level statistics about the number and distribution of farms by various variables such as sales, tenure, operator age, etc. Yet, the structure of agriculture at the national level is the summation of a diverse set of regional and local farming sectors, all of which are influenced by their local economic environment. For example, farming in the vast, sparsely settled western Corn Belt and Northern Plains is different from farming undertaken at the urban fringe or in the Appalachian region (Ahearn and Banker 1988). As a result, the small farm component of U.S. farming is dominated by farms in the South whereas the large farm component is dominated by farms in the western Corn Belt and Plains states (see Figure 20.2).

Utilizing a conceptual approach developed by Babb (1979), we postulate that the structure of a local farm sector is influenced by international and national policies and events (USDA, ERS), as well as the attributes of the local area (Babb 1979; Carlin and Green 1988). Sommer and Hines (1988), for example, identified U.S. counties most affected by the swings in farm exports. Unraveling the complex relationships between national and international policies and the local community's farm structure is outside the bounds of this chapter. We are instead concerned with the local community or region's interaction with local farming structure.

The Structure of the Local Farming Sector

Formulating structural relationships in a way that emphasizes their great diversity around the nation opens the door to a much broader discussion of factors affecting local farm structure and the relationship of farm structure to the local economy. Prominent on most lists would be factors directly related to farming. These include potential enterprise combinations suitable for the area, availability of water, level of technology adopted by local producers, land characteristics, level of capital investment in the local farming plant, etc. These factors are distributed differently across the United States, and they influence the way the local farming sector evolves over time.

A local community's nonfarm economic activities affect local farm structure because they provide alternative uses of labor, land, and capital. As such, they establish the opportunity cost for farm resources. Included here are situations where farm resources are underemployed, and thus nonfarm economic activities are in fact complementary to local farming.

A wide variety of other variables including population size and settlement patterns, human capital, and public services influence local

FIGURE 20.2   U.S. Large Farm and Small Farm Counties, 1982

Large Farm Counties*

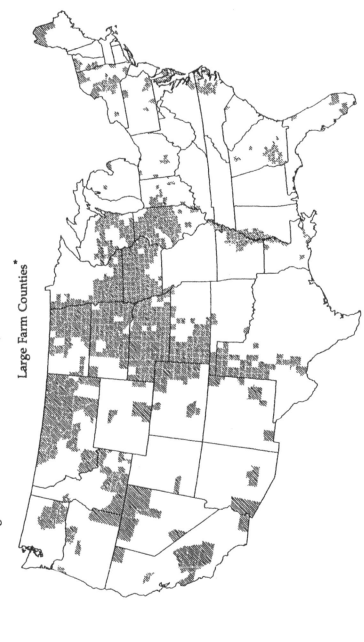

*Less than 59.3 percent of farms with 1982 gross farm sales of less than $40,000

(continued)

FIGURE 20.2 *(continued)*

Small Farm Counties*

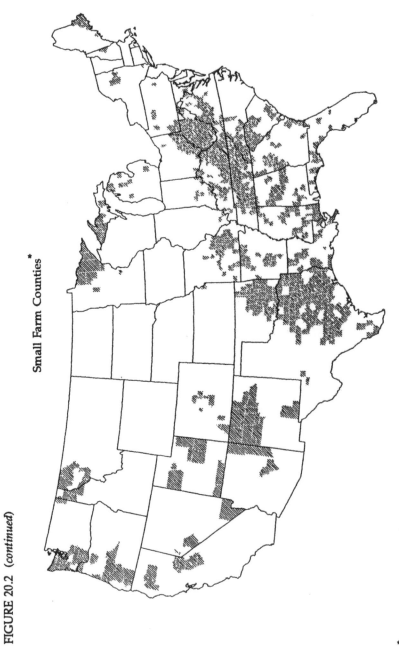

*88 percent or more of farms with 1982 gross farm sales of less than $40,000

farm structure. Human capital, which encompasses formal and informal education, health, and aesthetic and recreational experiences, affects the productivity of labor both on and off the farm. Population size and public sector relate to the patterns of demand for land and the level of private and public services a community can maintain. These variables, in concert, influence the structure of the local and, consequently, the national farming sector.

Likewise, rural community characteristics are but one of several factors affecting farming structure. The physical and social geography of the area, including soil type and topography, precipitation level and seasonal distribution, length of growing season, access to water, demographics of the farm operator population, Federal farm commodity programs, public and private investment in research and development of technology, and Federal and state farm credit and tax policies can all influence the structure of farming.

How Farm Structure Affects Communities

There is a rich body of literature pertaining to the effect of farm structure on the local rural community. Some studies have focused on a single community, others have attempted to look at several communities or regions. A common characteristic of these studies is that they assume a particular farm structure as "given" and then examine how that structure affects the local community. However, we would first note that structural change in farming is only one of several factors that determines the nature of rural communities. Proximity to larger urban centers, the adequacy of transportation and communication systems, the presence of mineral or forest resources, the recreational attributes of the area, a restructuring of the retail sector, government initiatives for community economic development, and the original settlement patterns and cultural beliefs (Solomon 1989) can all contribute to rural community characteristics and change.

Goldschmidt and Critiques

Goldschmidt's (1946) classic study of the effects of farm scale on community life was part of a series on how business enterprises affected the social, cultural, and economic environment. Published in 1946, it focused on two similar California towns, Arvin and Dinuba, and is generally considered the genesis of the area of research.

The towns selected were in the Central Valley of California, of similar size, and with similar total value of agricultural production by the surrounding farms. He considered their major difference to be in the size of farm, with farms near Arvin averaging 497 acres compared with 57 acres near Dinuba. However, in the latter community, three-fourths of the farms were fully owned by their operators, versus about one-third near Arvin.

The central theme from this work was that community vitality was enhanced in the area dominated by owner-operated family farms. Goldschmidt (1946) found that occupational grouping was the greatest contrast between the two communities, as 65 percent of Arvins' employed work force were farm laborers and 11 percent were farm operators, compared with 29 and 34 percent respectively in Dinuba. By his measures Dinuba also enjoyed a higher standard of living, the public service needs of the people were better answered, the schools were better, and the citizens had a higher level of participation in community institutions than in Arvin. Retail sales were about twice as great in Dinuba as in Arvin.

Goldschmidt (1946) concluded that because the large farms near Arvin were dependent on migrant wage labor, they fostered a skewed and segregated class structure. As a result, the working class in Arvin was poorer, less educated, and more alienated than the middle class small farmers from Dinuba (Salant and Munoz 1981).

Gilles and Dilecki (1988) note that at least 17 studies have examined the relationship between agriculture and socioeconomic well-being since 1972, most supporting Goldschmidt's (1946) thesis, but some with contradictory findings. They fault the Goldschmidt (1946) analysis for not making distinctions between structure of farming, farm size, land tenure, and farm labor systems, and for drawing conclusions about change from cross-sectional analysis.

Hayes and Olmstead (1984) examined Goldschmidt's data and method and concluded that because of metholodogical flaws his study offered little support for his conclusions. Goldschmidt (1946) had used four criteria to compare the two communities to establish that they were closely matched regions and differed importantly only in the size of farms that surrounded them. In examining the data, Hayes and Olmstead (1984) found instead that Arvin had experienced a one-third increase in population during a four year period shortly before the study. Dinuba had been platted and promoted by the railroad as a development scheme while Arvin had emerged as a community more than 15 years later. Arvin was near the center of a substantial oil-bearing region while Dinuba had no known petroleum deposits, and the farm costs of irrigation were twice as high in Arvin. The authors concluded that too many differences other than farm size were present to accept the Goldschmidt (1946) interpretation.

Several other studies including those by Heffernan and Lasley (1978) and Markousek (1979) also examine the effect of different farming structures on the nonfarm sectors, including the public sector. They all examine in some way the tradeoffs between farmers and the nonfarm sector under alternative farm structures. With some exceptions and regional differences, they in general suggest that the nonfarm sector of rural communities fare better under a farm economy dominated by small and medium sized farms.

Modelling Structural Impacts

Heady and Sonka (1974) examined the effect of alternative farm structures on several economic variables, including the secondary income effects on nonfarm sectors. They hypothesized that the atomistic nature of the farming sector implied that reductions in net revenue and farm income would have resulted from the structural change in American agriculture since the 1920s. To prevent inequities between farmers and consumers, Federal policy intervened with extensive farm programs, but little attention was given to other rural groups adversely affected. They concluded that the nonfarm sector in rural areas (the communities) bore the major costs of structural change in farming.

Heady and Sonka (1974) addressed the interrelationships between such communities and the number and size of farms using the Iowa State University national linear programming model with some 150 production areas and 31 consuming regions. The impacts of an agriculture composed of different sizes of farms on farm prices, location of production, farm income, number of farms, farm labor, consumer food costs, and the income generated in the rural nonfarm and agribusiness sectors were evaluated.

Four different farm size structures were used, one of which reflected the current farm size distribution, and served as the control. The other three represented a national agriculture composed of small, medium, or large farms. Differences in farm output among the farm size systems were carried forward in the model through the impacts on farm income, income from activity in the agribusiness sector, and income from sales of consumer goods. Compared with the control, a farm structure dominated by smaller farms resulted in 16.5 percent greater income generation in nonfarm businesses in rural communities (but with some regional differences), and also the greatest total net farm income. However, because of the large number of farms in this alternative, per farm net income was less than half that of the control. Under the small farm system, food costs were about four percent higher for consumers than in the control.

Their results emphasized the tradeoffs among producers, consumers, and rural communities among the alternative farm size structures. The large farm system, for example, would result in per farm income one-third higher than the control, four percent lower costs for consumers, but 16 percent less nonfarm income generated in the rural communities.

Henry et al. (1987) examined some effects of an agriculture composed of fewer medium-sized and thus more large farms on the nonfarm economy, adding another dimension to the Heady and Sonka (1974) analysis. Their analysis controlled for the total level of sales to users of farm products while the size distribution of farms was varied. To do this they merged a farm income and production expense account into the national input-output model and examined the direct and indirect changes in input use resulting from changes in the farm size distribution.

They concluded that within the nonfarm sectors, the locally oriented service and trade sectors would be required to produce more output and thus would benefit from a farm structure dominated by medium sized farms instead of larger farms. While this would be a cost to society, it would be a benefit to the rural communities where the additional demand for local goods and services would reside.

Community Analyses

Henderson, Tweeten, and Schriener (1989), in a study based on central place theory, examined how community retail businesses change as a result of changes in farm structure, and found that the effects on retail businesses vary by community size. In their study area, the farming dependent Oklahoma Panhandle, the smallest communities lost the most market share (caused by a shift in farm induced demand for goods and services) as the number of farms and crop acreage declined, and in spite of an increase in per farm income.

There was a smaller decline in the market share of mid-sized communities and the largest communities actually increased market share. The number of retail businesses in the smallest communities declined. There was also a shift in the regional employment structure with employment declines in building materials, grocery, gasoline, and miscellaneous retail businesses. On the other hand, employment increased in apparel, furniture, and restaurant businesses. The authors conclude that community hierarchies are not static and adjust to changes in farm structure. In general, farm structural change that features reduced farm numbers, etc, results in the growth of larger communities at the expense of nearby small communities.

Additional information about both the positive and negative effects of the farm sector on rural communities is contained in a report by Stone

(1987). He noted that the adoption in the past of machine and chemical technology by farmers had contributed to the trend toward fewer but larger farms, which in turn had resulted in gradually declining retail sales for some rural communities. That trend had been accelerated by the development of regional shopping centers, improvement in highways, and fuel efficient vehicles. The farm financial circumstances of the 1980s accelerated the rural to urban shopping trend. However, this was in fact a reversal of a 1970s trend in Iowa, in which small town resurgence was driven by increased retail sales of farm equipment, automobiles, building materials, and by other farm oriented businesses.

Rural lay persons' views of farm-community linkages may also be of interest. The perceptions and opinions of members of the county government, community officials, and farmers in a southern county regarding how farmers affect communities were reported by Moxley and Liles (1987). Among the positive responses were that farmers are retail customers, they operate an (export base) industry that brings outside money into the community, and they are a major property taxpayer. Taff (1989) also mentions the property tax, noting that it is the most important source of revenue for most local jurisdictions and is a major link between the farming sector and the local community.

How Communities Affect Farm Structure

The U.S. rural economic transformation that occurred during the last score years has resulted in a decline in farming as a source of rural employment and income. Ahearn, Bentley and Carlin (1988) note that by the early 1980s, farming accounted for less than 10 percent of total earnings in almost 60 percent of all nonmetro counties in the contiguous United States. As the economic influence of farming wanes in most rural areas, future structural changes in farming will have less effect on rural communities. When rural communities grow as a result of nonfarm influences, the local farming sector may be altered by the new economic and social environment. In these cases, the research question might be posed differently. How do changes in the community affect the local farm sector? Obviously, the answers differ depending on where the community is located.

Community Case Studies

There were a number of case studies conducted from the late 1950s through the 1970s which examined the effect of rural nonfarm employment growth on the local farm sector. Bertrand and Osborne

(1959) studied the effect of a wood products plant locating in a rural community in southeastern Louisiana. Agriculture in that area was characterized by small marginal farms, indicating the existence of underemployed farm labor resources. The researchers found that farm operators who were employed in the plant made little change in their farming operations.

Fuller (1960) studied the effect on farming in a five county area of a manufacturing plant locating in north central Pennsylvania. Fuller (1960) found that few of the sample farm families actually took jobs at the plant and, for those that did, there were negligible changes in their farming operations. Maitland and Friend (1961) reviewed the results of five studies of rural industrialization in Iowa, Utah, Mississippi, and Louisiana. All the areas were characterized as small, low income farming areas. In general, industrial employment was associated with a decline in the farm operator's contribution to farm labor and subsequent substitution of unpaid family labor.

Scott and Chen (1973) modeled the effects of industrialization (new steel rolling plant) on the farm sector in Putnam County, Illinois. Using a three-stage linear programming model involving six representative farm sizes, the authors concluded that small farmers could benefit because of their underemployed labor resources. They could decrease labor-intensive livestock enterprises, continue with crop production and take off-farm jobs. (The authors assumed that the prevailing nonfarm wage rate, the opportunity cost of farm labor, would increase.) Faced with higher labor costs, large farmers could also substitute less labor intensive enterprises. The net result would be higher and more evenly distributed income in the local sector and general economic stimulus in the community.

In general, the literature suggests that increased nonfarm employment opportunities in a rural community are related to positive increases in total family income for small farmers and are also associated with a change toward less intensive farming operations.

The perceptions and opinions of persons in the local county government, community officials, and farmers regarding how communities affect farmers may be of use. Moxley and Liles (1987) reported such views from a southern county. Among the positive responses were that the community provided retail services, off-farm job opportunities, seasonal hired labor for farmers, credit institutions, agricultural Extension Service, local grocery stores, local recreation services, local health services, good roads, rural water system, seed and fertilizer retailers, farm product buyers and processors, and feed mills. The community also provided the leadership in developing a Federal grant request for a livestock processing plant and in pursuing improvement in highways. Negative responses referred to competition for hired labor, and property taxes assessed by local governments.

National Studies

Brooks, Reimund and Peterson (1989), using an analysis of variance procedure with county data, found that the rate of change in farm structure variables over the decade of the 70s differed by the rate of total population growth in the region and the degree of urbanization in the county. They found that increases in the number of farms could not be attributed to regional increases in the total population, while decreases in land in farms was associated with urbanization of the county. Farmers in metro and adjacent counties shifted to less labor intensive types of farming compatible with off-farm employment. High population growth rates in the region were associated with increased likelihood that farms were operated by full-owners, perhaps because they were the smaller, part-time farms of urban workers.

Carlin and Green (1988), in a national study of local farm structure and community ties, calculated the proportion of farms in each county with gross sales of less than $40,000. They arrayed counties from lowest to highest using the proportion of farms in a county with gross farm sales of less than $40,000, and divided the array into quartiles. Mapping these counties reveals significant geographic groupings across the United States, with large-farm counties concentrated in the Midwest and small-farm counties concentrated in the South.

Relative to small-farm counties, large-farm counties have a relatively small population and employment base and farming is a larger component of the local economy. They are also characterized by population decline or slow population growth. Large-farm counties are less likely to be in or adjacent to a major metropolitan area, making commuting to work more difficult. A high proportion of the land area is in farming, testifying, in part, to the favorable physical geography of the region.

The results suggest that the structure of the local farming sector is influenced not only by conditions within that sector but also by conditions in the local nonfarm sector. Conscious decisions on the part of local community leaders to attract nonfarm employment to their communities are likely to alter the structure of the local farming sector.

Henderson and Brooks, in a follow up study to Carlin and Green (1988), suggest that the farmers in large-farm counties use a different farm management strategy than those in small farm counties. Farm operator households allocate their resources among alternative farm and non-farm activities in order to maximize family welfare. The physical and locational environment of large-farm counties correlates with management strategies that take advantage of economies of farm size, extensive crop production, part-ownership, and full-time farm employment. This reflects, in part, a lack of nearby nonfarm employment opportunities, thus expanding the farm business is the most practical way to increase family income. The physical and locational environment of small-farm counties favors strategies of diverting farm labor to full-time, off-farm employment and smaller, less capital intensive, fully owned farms. These alternative management strategies collectively are reflected in the structure of the local farming sector.

A farm structure analysis by Edwards, Smith and Peterson (1985) used a longitudinal Census of Agriculture file in a Markov Chain model and yielded very useful information that has altered some of our farm structure paradigms. For example, changes in farm size display a great deal of symmetry; for every farm that was likely to increase in size, there is one that is likely to decrease in size between census years. Previous research used synthetic models that usually specified that farms either grew or exited the system. While this research contributed significantly to our understanding of farm structural change, conventional analysis that uses a stable transition matrix to project future farm structure is troublesome because it is doubtful that the relations observed in the transition matrix constructed for a specific time period remain constant over time.

Smith (1988) explored ways to accommodate this concern in a study to predict the size distribution of dairy farms. Following the lead of Stavins and Stanton (1980) in their analysis of New York dairy farms, Smith (1988) used multinomial logit functions to develop nonstationary transition probabilities for the U.S. Markov Chain model based on the 1974-78 transition matrix. Exogenous variables that he included in his analysis were age of the existing operator population, extent of off-farm work by the existing operator population, change in farm product prices, change in farm asset prices, and change in nonfarm incomes.

While the independent variables considered had little or no explanatory power for many of the cells in the matrix, for mid-sized commercial farms, the proportion of operators age 65 and older in 1974 was positively associated with the probability of exit by 1978. When statistically significant, nonfarm income growth was positively related to the probability of declines in farm sales and negatively related to farm growth. The proportion of operators working off the farm 200 days or more was positively related to the probabilities of both exit and growth for small commercial farms. That is, the combination of full-time off-farm work and a farm of this size is not sustainable; operators tend either to leave farming completely or increase their farm size to improve total income. Smith (1988) demonstrated that including these variables in the model resulted in better projections of farm structure.

Peterson (1989), in a follow up study to Smith (1988), notes that the structural change in the size distribution of farms in the U.S. is related to a variety of economic-demographic forces which have impact on strength and survival of agricultural operations. His methodological study also involves the development of a technique that deals with the variability in Markov Chain matrices over time, allowing the measurement of changes in selected structural measures. His empirical results for change in U.S. agriculture during the mid-1970s indicated the importance of demographic variables in changes among size classes of farms. The percentages of farm operators over age 65 and under age 35 and the percentage working off-farm 200 days or more were the most important variables.

Ahearn, Bentley, and Carlin (1988) examined the relationship between individual farm financial stress and the extent to which farming dominates the local economy. They attached "farming dependent" or "farming important" county codes and individual "farming stress" codes to each of 12,428 farmer observations from the Farm Costs and Returns Surveys. Their results point to important relationships between the well-being of farm operator households and the characteristics of the counties in which they live. These relationships can be observed both through the farm household income statement and the farm balance sheet.

They hypothesized that off-farm incomes are important for sustaining many farm households. The larger and more diverse the nonfarm sector in the local economy, the more likely that farm household members will be able to obtain nonfarm employment to help maintain household income and the farm business. The more dependent a county's economy is on farming, the larger the average size farm and the less likely the operator is to work off-farm. Also, there are fewer alternative opportunities for the use of farmer's labor in such farming counties, thus there are incentives for farmers to expand their farms to achieve fuller employment and higher income.

Even though there are differences in the observed allocation of farm household labor among the county types, there were no major differences in total economic risk among the counties. That is, a farm household is as likely to report economic risk in a farming dependent county as in a nonfarming dependent county.

There are, however, substantial differences in the type of economic risk observed among the county types. In general, the more a county depends on farming, the more likely a farm operator household is to be in a financially risky position (i.e. have a relatively high debt-to-asset ratio) as opposed to lower income position. This suggests that community characteristics can affect a farm household's well-being through the balance sheet. Land values are affected by expectations about the ability of the land to generate income. Thus, in counties dominated by farming activities, land values will be sensitive to expectations about farm income. In counties dominated by nonfarm industries, competition from alternative users of farmland will ameliorate declines or actually increase farmland values, thus strengthening the equity position of farmers. Even though farmers, as a group, reduced their total debt burden during the 1980s, debt reduction could not keep pace with falling land prices. Thus, the sector's equity position deteriorated. It appears that farmland owners in farming dependent communities bore the brunt of asset value declines.

The lower incidence of low income in farming dependent counties was a bit puzzling. One explanation might lie in Government payments. Farms in farming dependent counties specialize in producing those crops included in Federal farm commodity programs. Sixteen percent of the farms and 23 percent of agricultural sales were in farming dependent areas, but they received one-third of direct Government payments in 1986. Government payments played a role in ameliorating economic stress in farming dependent areas; without direct Government payments economic stress would have surely been higher in farming dependent counties during the mid 1980s.

Deaton and Weber (1988) indicate that among the issues that emerge from the interrelationships between the agricultural economy and the community is the effect of expanding nonfarm employment opportunities on the farmer's perception of risk, with implications for the selection of farm product mix, the technology used in production, and the capital intensity in farming. Risk analysis has often been from the view of the farmer as an entrepreneur whose major focus is on markets, prices, credit, and technology. Risk analysis, however, should also recognize the allocation of farm household labor between farm and nonfarm employment. They suggest that risk averse farmers may be more likely to participate in nonfarm employment. Similarly, size economies in farming have been exploited. In regions where geography does not favor farming, the risk averse farmers may prefer full-time off-farm employment to farm expansion.

Conclusions

In this chapter we have discussed the synergistic, two-way relationship between farm structure and the local community. The linkages between farm structure and community characteristics operate in both directions. The more a local area depends on farming the more likely that changes in the fortunes of the farm sector will be felt in the local community. Farming communities are in essence a special case of the "one company town". The number of farming communities has declined substantially during the past 40 years. Most of the remaining farming communities are concentrated in the Plains states and the Western Corn Belt. This is the same area where farming is relatively large scale and where the production of commodities covered by Federal price and income support programs predominate. Often characterized as America's empty quarter, this region is sparsely settled and nonfarm employment opportunities are limited. Overall population and employment growth were stagnant during the 1980s and most counties actually lost jobs and people. Future farm structural change that features continued farm consolidation suggests that state and local government officials in this region will continue to struggle with managing overall community decline.

In most rural areas, farming has been a declining source of both employment and income and the chances of it becoming a major driving force for future rural economic growth are low at best. Farm employment has been declining even in relatively "good times" for farmers. While there are farm input and processing industries in local communities that "depend" on the well-being of the farm sector, much of the farm input and processing employment is metropolitan based. Those who advocate keeping the farm sector strong to "preserve rural America" should note that this argument applies to fewer and fewer places as the decades pass.

As communities diversify and grow, it becomes more likely that changes in the community will affect the structure of the local farming sector. We have noted the effect that nonfarm job opportunities have on the farm sector. But there are other issues proffered by basically nonfarm constituents that will continue to shape American agriculture. Both nationally and in local communities there is a pervasive environmental awareness that leads to several agricultural concerns. Whether valid or not, there are concerns about pesticide residue in food, animal stress in confinement livestock operations, and farmland protection and conservation. Half of all U.S. counties are reported to have potential for groundwater contamination from farm fertilizers and pesticides. While there are effective methods for detecting small quantities of contaminants, there is a lack of data on their effect on humans, leading to circumstances "full of suspicion but short on verification". Demand for state and Federal legislation to prevent contamination and addressing public concerns follow. New rural residents, living close by and observing current farming practices, are among the most vocal. Historically, adoption and use of farm technology has been voluntary but regulation of farming technology in the future could influence farming structure substantially. Analysis of how community characteristics affect farm household well-being and farm structure will be a useful approach for students of farm structure as they attempt to understand where the farm sector is heading.

References

Ahearn, Mary, and David Banker. 1988. "Urban Farming Has Financial Advantages." *Rural Development Perspectives.* Econ. Res. Serv., USDA 5(1): 19.

Ahearn, Mary, Susan Bentley, and Thomas Carlin. 1988. "Farming-Dependent Counties and the Financial Well-Being of Farm Operator Households." Econ. Res. Serv., USDA, AIB 544.

Babb, E. M. 1979. "Consequences of Structural Change in U.S. Agriculture." *Structural Issues of American Agriculture.* Econ. Res. Serv., USDA, AER No. 438.

Bender, Lloyd D., et al. 1985. "The Diverse Social and Economic Structure of Nonmetropolitan America." Econ. Res. Serv., USDA, RDRR No. 49.

Bertrand, A. L., and H. W. Osborne. 1959. "Impact of Industrialization on a Rural Community." *Journal of Farm Economics:* 1127-1134.

Brooks, Nora, Donn Reimund, and R. Neal Peterson. 1989. "Effects of Population Growth and County Type on Farm Structure, 1970-80." Staff Report No. AGES 89-37, Economic Research Service, U.S. Department of Agriculture.

Carlin, Thomas A., and Bernal L. Green. 1988. "Local Farm Structure and Community Ties." Econ. Res. Serv., USDA, RDRR No. 68.

Deaton, Brady J., and Bruce A. Weber. 1988. "The Economics of Rural Areas," Chapter 14 in R. J. Hildreth, et al. ed., *Agriculture and Rural Areas Approaching the Twenty-first Century: Challenges for Agricultural Economics.* Pp. 403-439. Ames, IA: Iowa State University Press.

Edwards, Clark, Matthew G. Smith, and R. Neal Peterson. 1985. "The Changing Distribution of Farms by Size: A Markov Analysis." *Agricultural Economics Research* 37(4), Econ. Res. Serv., USDA.

Francis, Wyn., and Mindy Petrulis. 1988. "Food Processing and Beverage Industries: Moving Toward Concentration." *National Food Review.* Econ. Res. Serv., USDA, 11(4): 23-27.

Fuller, T. E. 1960. "Description and Analysis of a Rural Area in North Central Pennsylvania Prior to the Establishment of an Industrial Plant." Agricultural Economics and Rural Sociology Report 26, Penn. Ag. Exp. Stat., Penn. State Univ.

Gilles, Jere Lee, and Michael Dalecki. 1988. "Rural Well-being and Agricultural Change in Two Farming Regions." *Rural Sociology* 53(1): 40-55.

Goldschmidt, Walter, R. 1946. "Small Business and the Community, A Study in Central Valley of California on Effects of Scale of Farm Operations." Report of the Special Committee to Study Problems of American Small Business, U.S. Senate, Washington DC.

Hayes, Michael N., and Alan L. Olmstead. 1984. "Farm Size and Community Quality: Arvin and Dinuba Revisited." *American Journal of Agricultural Economics* 66(4): 430-436.

Heady, Earl O., and Steven T. Sonka. 1974. "Farm Size, Rural Community Income, and Consumer Welfare." *American Journal of Agricultural Economics* 56(3): 534-542.

Heffernan, William D., and Paul Lasley. 1978. "Agricultural Structure and Interaction in the Local Community: A Case Study." *Rural Sociology:* 348-361.

Heimlich, Ralph E. 1989. "Metropolitan Agriculture-Farming in the City's Shadow." *APA Journal:* 457-466.

Henderson, David A., and Nora L. Brooks. "Bimodal Distribution in Agriculture: A Geographical Perspective." Econ. Res. Serv., USDA, RDRR, forthcoming.

Henderson, David, Luther Tweeten, and Dean Schreiner. 1989. "Community Ties to the Farm." *Rural Development Perspectives* 5(3): 31-35.

Henry, Mark S., et al. 1987. "Some Effects of Farm Size on the Nonfarm Economy." *North Central Journal of Agricultural Economics* 9(1): 1-11.

Holmes, Thomas, Elizabeth Nielsen, and Linda Lee. 1988. "Managing Groundwater Contamination in Rural Areas." *Rural Development Perspectives* 5(1): 35-40.

Kennedy, Donald. 1989. "Humans in the Chemical Decision Chain." *Choices.* Third Quarter 1989, pp. 4-7.

Maitland, Sheridan T., and Reed E. Friend. 1961. "Rural Industrialization, A Summary of Five Studies." AIB No. 252, Econ. Res. Serv., USDA.

Markousek, Gerald. 1979. "Farm Size and Rural Communities: Some Economic Relationships." *Southern Journal of Agricultural Economics*, Vol. 11.

Moxley, Robert L., and James Liles. 1987. "Agriculture and Locality Interrelationships: Perspectives of Local Officials and Farmers." Unpublished paper presented at the annual meetings of the Rural Sociological Society in Madison, WI (August 1987). North Carolina Agricultural Research Service Project NC13741.

Nielsen, Elizabeth G., and Linda K. Lee. 1987. "The Magnitude and Cost of Groundwater Contamination from Agricultural Chemicals: A National Perspective." Econ. Res. Serv., USDA, AER No. 576.

Peterson, R. Neal. 1989. "A Single Equation Approach to Estimating Nonstationary Markov Chain Matrices: The Case of U.S. Agriculture 1974-78." Unpublished report, ERS, USDA.

Salant, Priscilla, and Robert D. Munoz. 1981. "Rural Industrialization and Its Impact on the Agricultural Community: A Review of the Literature. Econ. and Stat. Serv., USDA, Staff Rpt. No. AGESS810316.

Scott, John T., Jr., and C. T. Chen. 1973. "Impact of Rural Industrialization on Farm Organization Within the Area Where Industry Locates." Working Paper Series RID 72.9, Cent. of App. Soc., Dept. of Rur. Soc., Univ. Soc., Dept. of Rur. Soc., WI.

Shaffer, Ron. 1989. *Community Economics: Economic Structure and Change in Smaller Communities.* Ames, IA: Iowa State University Press.

560

Shaffer, Ron, Priscilla Salant, and William Saupe. "Understanding the Synergistic Links Between Rural Communities and Farming." In *New Dimensions in Rural Policy: Building Upon Our Heritage*. Joint Committee Print, Joint Economic Committee, Congress of the United States, 99th Congress, 2nd Session, June 5, 1986, pp. 308-321.

Smith, Matthew G. 1988. "A Conditional Approach to Projecting Farm Structure." Econ. Res. Serv., USDA, Staff Rpt. No. AGES880208.

Solomon, Sonya. 1989. "What Makes Rural Communities Tick?" *Rural Development Perspectives* 5(2): 19-24.

Sommer, Judith E., and Fred K. Hines. 1988. "The U.S. Farm Sector: How Agricultural Exports Are Shaping Rural Economies in the 1980s." Econ. Res. Serv., USDA, AIB 541.

Stavins, R. N, and B. F. Stanton. 1980. "Using Markov Models to Predict the Size Distribution of Dairy Farms." New York State, 1968-1985. A.E. Res. 80-20, Ithaca, N.Y., Cornell Univ., Ag. Expt. Sta.

Stone, Kenneth E. 1987. "Impact of the Farm Financial Crisis on the Retail and Service Sectors of Rural Communities." Unpublished paper, Department of Economics, Iowa State University.

Taff, Steven J. 1989. "Farming, Farm Programs, and Local Economies." *Minnesota Agricultural Economist*, No. 659, pp. 4-5.

U.S. Department of Agriculture, Econ. Res. Serv. 1986. *Agricultural Finance: Outlook and Situation Report*. AFO-26.

Additional References

Farm structure-community interrelationships have been investigated by social scientists since at least the 1940s. A sizeable body of literature has emerged. Besides articles and reports, there are entire volumes that are relevant to the issues as well as annotated bibliographies. To facilitate a search by others we have referenced a selection of such sources here.

Leistritz, F. Larry, and Brenda L. Ekstrom. 1986. *Interdependencies of Agriculture and Rural Communities*. Garland Reference Library of Social Science, Vol. 383, Garland Publishing, Inc., New York and London.

Congressional Research Service, Library of Congress. "Agricultural Communities: The Interrelationship of Agriculture, Business, Industry and Government in the Rural Economy." Committee Print, 98th Congress, 1st Session, U.S. Government Printing Office, October 1983.

Joint Economic Committee, Congress of the United States. "New Dimensions in Rural Policy: Building Upon Our Heritage." Joint Committee Print, 99th Congress, 2d Session, U.S. Government Printing Office, June 1986.

Guither, Harold, and Harold G. Halcrow. 1988. *The American Farm Crisis, An Annotated Bibliography With Analytical Introductions*. Pierian Press, Ann Arbor, Michigan, Chapters 1 and 5.

21

Farm Structure and Stewardship of the Environment

Jay Dee Atwood and Arne Hallam

Introduction

Agricultural structure is linked to environmental outcomes through the impacts of the production technologies associated with the structure. Coincidentally, environmental (and agricultural) policies impact the structure of agriculture through their effects on the profitability of alternative production technologies. Some types of structure change the decision environment and affect the choice of production practices. Some aspects of structure are the result of the choice of practices. Therefore, a set of two-way causal relationships exist between the state of the environment, the chosen mix of environmental and agricultural policy, and the structure of agriculture. Or, where causal relationships may not exist, associations are at least apparent.

The current political climate for agriculture includes a mix of intertwined agricultural and environmental policies -- for example, the conservation compliance provisions of the 1985 Food Security Act (Glaser 1986). Recent agricultural income and commodity policies have been blamed for adverse impacts on the environment (American Farmland Trust 1990; Osteen 1985; and Reichelderfer and Phipps 1988). The treadmill theory of Cochrane (1979) provides a positive link between income and commodity policies, and farm size. Hence, it seems that direct links between farm structure and the environment might exist.

The 1980s agriculture finance crisis increased the rate of farm failures and the turnover of asset ownership. At the same time perceived agricultural damage to the environment has greatly increased (Phipps, Allen, and Caswell 1989). To some extent the political forces for environmental regulation have also endorsed the "small farm" aspect of agricultural structure (Center for Rural Affairs; The Land Stewardship Project). As farms grow they typically become subject to more stringent existing regulations, for instance, large feedlots must meet industrial waste management standards.

The desire for increased environmental regulation of agriculture is increasing both from within and from outside the sector (American Farmland Trust 1990; Batie 1988; Benbrook 1989; Reichelderfer and Phipps 1988). Batie (1988) explains that much of the agricultural policy agenda is now controlled by non-agricultural interests. Agriculture is increasingly perceived as a problem rather than a solution, particularly with regard to environmental problems. The rights traditionally associated with cultivating the soil are giving way to restraints. There is also significant interest in the concept of sustainable development and ecologically sound policy (Batie 1989).

Clearly, concern over the environment and over the health of the agricultural sector is leading to increased environmental regulation. Although many lobbyists infer a direct link, the cause and effect relationship between the structure of farming and perceived environmental damage is not readily apparent. In this chapter possible links between changes in agricultural structure and qualitative aspects of the environment are suggested. Though strong empirical data to support the proposed links are lacking, some reasoning and available studies are used to suggest relevant policy issues and some possible areas of further study.

A Conceptual Model of Policy, Farm Structure and Environmental Impacts

A conceptual model illustrating the two way relationships (or at least associations) between agricultural and environmental policy, farm structure, and environmental impacts is shown in Figure 21.1. Different structures of farming are associated with different choices of production methods. Various production methods result in more or less environmental damage. Policies addressing the perceived environmental damage impact the profitability of different production methods, and the overhead cost of farming. Different farm structures fare better or worse as overhead costs and relative production technology profitability changes. These links, or associations, as indicated by arrows in Figure 21.1, are carefully defined in this section of the chapter. The remainder of the chapter provides detail and reviews existing studies relative to this conceptual model.

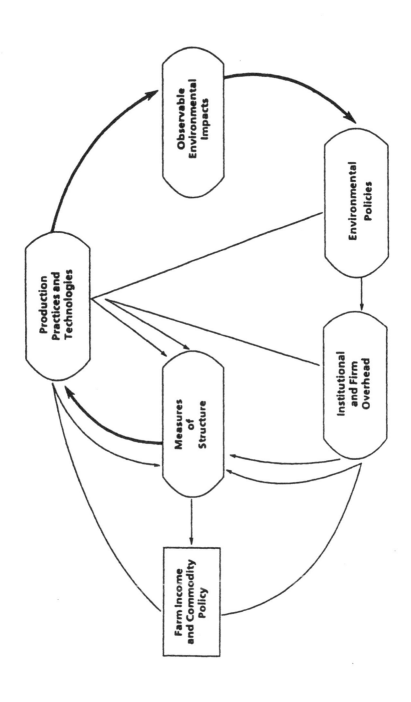

FIGURE 21.1 Relationships Between Policy, the Structure of Agriculture and Environmental Outcomes

The Link of Structure to the Choice of Practice

Different farm structures have access to different production technology choices, or else face differing relative profit levels for the same choices. For example, two farms of the same size and structure may choose to use either conservation or conventional tillage with different profit levels, while a large farm may have lower costs of conservation tillage than a smaller farm. Different practices may result in more or less environmental damage per unit of resource. The costs of the alternative production choices often fail to account for environmental externalities. Changing farm structure may lead to changes in comparative advantages across regions. Detailed regional analysis of input use levels and the resulting environmental damage by farm size is required for policy evaluations.

Production Practices and Environmental Outcomes

Alternative production practices result in differing environmental outcomes. Continuous mono-cropping culture instead of rotations results in more erosion and sediment generation, more susceptibility to pests, and greater rates of chemical application. Confinement feeding of livestock leads to a concentration of waste and results in a disposal problem. No-till cropping technology reduces erosion damage, requires more horsepower for planting, and may require more chemicals. Numerous other examples are also available.

The Impact of Environmental Outcomes
(Perceived or Real) on Policy

Public perception of environmental problems leads to formulation of environmental policy. As postulated in Figure 21.1, the public may observe some undesirable environmental impacts or risks which seem to be associated with agriculture. Research to determine which production practices are responsible for the environmental problem is conducted. Public policies are formulated and corrective programs initiated to mitigate the causes of the environmental problem. Research to determine less damaging agricultural practices may also be sponsored.

The Impact of Policy on Structure

Policies have indirect effects on structure in two ways. First, policies change the relative profitability of alternative technologies and some structures fare better when certain technologies are relatively more profitable. For example, integrated crop-livestock operations may fare better than intensive livestock-only operations if laws restrict off-site disposal of animal wastes or the building of sewage lagoons. Secondly, policy affects the firm overhead requirements, such as record keeping, and again some structures are better able to adopt. Often policy designed for one purpose will not only affect structure, but also have other direct negative impacts on the environment. For example, an erosion reduction policy may result in use of more chemicals.

Several studies on the factors leading to formulation of state level water pollution legislation have been done. Wise and Johnson (1990) found that for state legislators involved in formulation of new regulations, the "expert" advice or information put out by economists and other academicians have very little influence on the decision to sponsor legislation. Batie et al. (1989) found two factors that in combination lead to state level water regulations: the dependence of a large population on the water source, and available information on high use levels of chemicals by the area farmers.

Factors Affecting Choice of Production Technique. Policies, both agricultural and environmental, impact the relative profitability of production alternatives and some farm structures fare better under a given policy than do others. Examples are requirements that the method of manure application to cropland be injection or that a certain type of waste water treatment equipment be installed. In both these examples, there may be a returns to scale factor such that smaller farms leave the sector rather than comply.

Institutional and Farm Overhead Impact on Structure. Policies may also impact the institutional and financial overhead of the farm leading to changes in structure. An example of overhead expense is the requirement that pesticide usage be recorded and reported. The cost of meeting these requirements would likely vary by farm size. If small farms could not cope with the regulation and left the industry, farm structure, as measured by farm size, would increase.

Policy, structure, and the environment are all integrated into the economic and social system. Policies adopted to manage the environment affect technology as producers adjust to external requirements or costs. Changes in technology often lead to changes in farm structure as firms attempt in the long run to adjust the size of the operation to best maximize returns to resources. Policies adopted to favor one particular size class also affect the environment since different structures often employ different technologies with different environmental impacts. Thus the causal structure is direct, indirect and two-way.

Structural Change Indicators Relevant to the Environment

The structure of agriculture is usually related to size, but the units of measurement may be acres cultivated, value of produce sold, number of livestock unit produced, or number of input units employed. Value added has been suggested as the ideal measure, but data are generally unavailable (Reimund et al. 1987; Stanton, Chapter 2). Structural definitions sometimes consider other characteristics, such as educational levels, tenure, and owner occupations. Empirical testing is required before environmental links can be definitely stated. For the purpose of this chapter a number of size and structure indicators are utilized. These are:

1. Farm size (acres operated, gross sales, acres planted, etc.).
   a. Size of enterprises (equipment and investment) of the farm.
   b. Whether or not the farm is a contiguous parcel or not.
2. Education levels of the farm population.
   a. Operator and/or owner.
   b. Laborers.
3. Degree of labor intensity (versus other factors).
4. Degree of specialized management.
   a. Computerization.
   b. Use of integrated pest management (IPM).
5. Use of commercial or leased operations versus self.
6. The ownership of resources and the right to the stream of income, i.e., whether farming is by an owner/operator or by a tenant.
7. Aspects of ownership characteristics in the utility function.
   a. Family farm versus corporate (farm and non-farm corporate).
   b. Degree of prevalence of "hobby", "part-time" and "limited resource" farmers.
8. The changing set of "property rights" associated with farming.
9. Degree of vertical and/or horizontal integration.
10. Use or non-use of irrigation.

Common measures of *farm size* are acres operated and value of sales. Enterprise size and the associated machine use, field size and whether the farm is one contiguous parcel or not are also structural issues. The impact of farm size on technology choice and hence on environmental outcomes depends on treatment of individual resource units. As farms grow they may remove existing conservation measures and feedlots may exceed the waste assimilative capacity of the area; however, the larger farms may involve higher levels of education and may be more subject to existing legislation.

The educational level of the farm population would seem to have an impact on the operation of the farm relative to the environment. Owners and operators who understand both the on-site and off-site impacts of their practices may be more likely to be careful. Farm labor with higher levels of education perform their tasks more precisely and are likely to have fewer accidents. On the other hand, more education may imply a higher profit motive and a better ability to circumvent existing legislation, perhaps resulting in higher levels of environmental damage.

Labor intensity, particularly skilled or management labor, has decreased steadily relative to chemicals and machinery since 1920. As chemicals and machinery have become cheap relative to labor, farmers have increased their use, with some undesirable impacts on the environment. However, whether "large" farms differ relative to small "farms" in chemical use per unit of resource is an empirical issue not yet resolved (National Research Council 1989).

The degree of *specialized management* occurring in agriculture is increasing continually. The use of computers and management models such as integrated pest management (IPM) enable much better management of production externalities relative to the environment. The continuous monitoring and updating required for specialized management is clearly more possible with larger, more specialized farm structures.

Some *farm operations*, such as fertilizer application, may increase in relative proportions as farm structure changes. These operations may be more carefully completed by farm operators than when done by commercial off-farm people. On the other hand, a degree of specialization exists within the agri-business community, and for a given operation professionals may be more precise in applying inputs than the farm labor. Bender (1987) argues that larger equipment, which is more likely to occur in farm operations of larger farms, is beneficial to the environment. Large equipment enables timeliness for application of chemicals, and for avoiding field work when conditions are poor. Weight per foot of tilled soil will be less and fewer turns mean less compaction. Larger equipment is also generally more fuel efficient.

The proportion of farmers who are *renting versus those being owner/operators* is increasing and is an important aspect of structural change. The issue is whether having legal right to the value of the future stream of income from the resources leads to more careful management relative to the environment or not (Basu 1989). While owners will tend to internalize long run intertemporal externalities, long run leases may accomplish the same thing.

Ownership characteristics are aspects of structural change which likely have impact on the environment. The impact on the environment likely comes through the differing utility function arguments between those

568

who live in the rural areas and are associated with the land and those separated from the land but who derive profits from its use. It has been claimed that ownership of farmland by large non-farm corporations leads to increases in undesirable environmental outcomes (*The Land Stewardship Letter* and the *Center for Rural Affairs Newsletter*). No definitive studies exist.

How the increase in the mix of *"hobby"*, *"part-time" and "limited resource" farmers* impacts the environment is not clear (Gertel et al. 1985). These farmers may not have the appropriate education and or freedom to make the best management choices. On the other hand, these farmers are more likely to use pastoral type management techniques, which though perhaps not as productive, are less damaging. Being educated in other areas may lead to more of a conservation ethic for these farmers.

As farm structure changes, particularly as small production units are combined into large ones, the *property rights* associated with resource units employed by the farm change. For instance, local regulations relative to the environment usually become effective as the sales of the farm increase. Also, the availability of various farm subsidies depend on farm size and the subsidies distort resource use which then results in undesirable outcomes (Batie 1988; Benbrook 1989; Osteen 1985; and Reichelderfer and Phipps 1988).

Legal *property rights* are also changing as the list of allowable actions associated with property decrease. For example, the farmer may not allow runoff to escape his field or his feedlot and he may face restrictions on input usage. Agriculture is increasingly seen as a threat to the well-being of society which must be regulated by outside interests (Batie 1988). A study by Batie et al. (1989) found that state regulation of agricultural practices relative to groundwater were being developed in areas where a large population depended on groundwater for consumption, and where it was observed that farmers were using large quantities of chemicals.

The *degree of vertical and/or horizontal integration* existing in agriculture has clear impacts on the environment. An example is the increasing trend in specialization from mixed farming to either crop or livestock production. Also, there is an increase in the number of very large feedlots not producing their own feed on adjacent land to which waste can be applied. On the other hand these integrated farms are really "firms" and so are typically operated with a higher level of education and are subject to more of the industrial type regulations.

The adaptation and/or reliance on *irrigation* is also an important measure of the structure of agriculture (Reisner 1986). Irrigation often involves higher uses of chemical, fertilizers, and sometimes results in degraded quality of the soil and water resources. The existence of legal entitlements for irrigation water for agriculture which prohibit use of the water in higher valued uses also indirectly impose environmental costs as alternative resources must be used for other purposes, such as in the case of fossil fuel electrical generation when hydro power lacks water rights.

Production Practices Linking Farm Structure to the Environment

For any given structural change indicator such as farm size or labor intensity, various environmental outcomes might be correlated with levels of the indicator, with the choice of production practices being the link. Some of the production practices that might change as structure changes and which could give differing environmental outcomes are listed below this paragraph.

1. The degree to which a "farm" becomes subject to existing environmental regulation as its size and/or scope of activity changes.
2. The type of landscape alterations such as changes in fence rows, terraces, ponds, etc.
3. The adoption of Low Input Sustainable Agriculture (LISA).
   a. Chemical use.
   b. Soil stewardship.
   c. Purchased versus non-purchased inputs.
4. The timeliness of operations.
   a. Chemical management.
   b. Residue management.
   c. Product quality.
5. The choice of input formulation/carrier.
6. The use of IPM.
7. The choice of machine complement.

Existing environmental regulations on production activities generally have different provisions depending on the size of the firm. At a minimum it is typical for small farms to escape most of the existing water and air quality regulations. As farms grow in size (the current trend in structural change) more of the existing regulations become effective. At the same time, as regulations increase, not all firms will be able to comply and so the regulations force some farmers out, which in turn causes further structural change.

The *landscape* of U.S. agriculture has definitely changed, with drastic decreases in both natural and man-made conservation measures. Examples include terraces, fence-rows, filter strips, and wetlands. These landscape changes have caused well documented environmental problems. It is not completely clear how changes in these conservation measures are linked with the various measures of structural change.

Adoption of LISA production techniques could be considered as a separate measure of structural change, i.e., the degree of LISA adoption in the sector could be considered as a qualitative aspect of structure. A more interesting issue is whether current structural patterns will lead to a greater or lesser adoption of LISA. Empirical studies are needed on the use of chemicals, soil stewardship, and purchased versus produced fertilizer and other inputs.

A large part of the debate over the impact of agriculture on the environment evolves around the question of *timeliness of operations*. Better chemical and tillage management would go a long way towards reducing the negative environmental impacts. The link between timeliness and structure is an empirical issue since farmers with larger more efficient equipment may have proportionately larger enterprises.

The *formulation or carrier of chemicals and fertilizers* also influence the way agriculture impacts the environment. As structure changes the choice of the chemical and fertilizer carrier or formulation may change due to different application technology, timeliness, education, etc. With a different chemical formulation, persistence and movement in the environment may be affected.

The use of advanced *management techniques, such as IPM* or computer simulation, can also indirectly impact the environment since more accurate input use decisions can be made. The link between the use of these techniques and the structure of agriculture is not clear.

The choice of *machine complement* may be correlated with changes in agricultural structure and with changes in the environment. For example, if higher education leads to the choice of ridge till, or if high profits lead to a complete replacement of machinery, then practices change as well as do environmental outcomes.

Environmental Problems Associated with Agriculture

The environmental outcomes listed below are associated with agriculture and might be linked to structural change. Following this list of impacts, each will be discussed in detail.

1. On-site environmental impacts of agriculture.
    a. Local amenity values, such as odor, visual impact, etc.
    b. Productivity of the natural resources used for production.
    c. Quality rating of the wildlife habitat.

    d. Quality of the ground water for other uses.
    e. Operator safety and household hazards.
2. Off-site environmental impacts of agriculture.
    a. Air pollution (chemical and particulate).
    b. Surface water quality.
    c. Food safety.

Amenity value includes all those items detectable by the sensory functions, i.e., odor, visual, and sound. Livestock production activities as well as meat processing operations give off unpleasant odors. Many non-farm dwellers derive satisfaction from viewing a pastoral scene in the countryside, i.e., fence rows, ponds, pastures, etc. All agricultural operations involve sound and as the structure changes these sounds may be both larger and nearer to non-farm populations.

Nassauer (1989) conducted a survey of Minnesota rural dwellers to determine feelings about conservation and aesthetic values. Nassauer (1989) found that herbicides and conservation measures with high visual impact were applied beyond cost effective levels due to the perceived visual indicator of good stewardship of clean, well managed fields. On the other hand, conservation tillage and land idled in the Conservation Reserve programs was considered to give an unkempt appearance. Nassauer (1989) states that "the aesthetic value of good stewardship depends upon whether people can tell that they are looking at planned conservation practices."

Resource productivity is an intertemporal issue in that current negative impacts imply higher costs for the next generation, an example being erosion of topsoil. While soil productivity is an on-site problem, agriculture also impacts the resource costs of other industries through, for example, changes in the quality of water which other industries use for production processes. A good example is the cost of surface water sediment removal.

Wildlife habitat suitability can be characterized both by total organism populations, as well as by specie diversity. Total populations depend on crop cover quantity while specie diversity depends on the mix of vegetation available. Wildlife is also impacted by the effect of agriculture on nearby water resources.

On-site water quality concerns are reflected in the growing awareness that agricultural producers are dependent on the water that they are contaminating. Farmers and their livestock drink the water and water is also applied to crops where irrigation is prevalent. Irrigation water return flow typically has a higher concentration of salts than before use.

Operator safety may be impaired by environmental conditions such as dust impaired vision or by the mix of inputs (for example, chemicals) used for production. Agricultural labor must deal with more lethal and ever changing chemical mixes as well as larger and more complex equipment. Operator fatigue also becomes an issue where specialization is increasing. The air quality investment in livestock confinement buildings likely depends on whether it is the owner who will do the work in the building or whether work is assigned to hired labor.

Air pollution impacts may include changes in the chemical composition of the air, changes in physical qualities of the air such as temperature or moisture content, and changes in the particulate loading.

Surface water quality characteristics that may be impaired include the chemical loading and its effect on human health and industrial processes, sediment loadings which impact use of the water directly and which also fill reservoirs, temperature of the water, and the value of the water for recreational quality.

Food safety concerns evolve around chemical residues in the food chain, animal health, and nutrient makeup of the finished product (examples being fat or cholesterol levels).

Policy and Institutional Impacts
on Structure and Environment

In this section of the chapter the mechanisms by which policies impact both structure and production technology choice are reviewed. Other factors of the economy, labeled institutional here, also have an impact by the same mechanisms and are briefly discussed here.

1. Debt/equity ratios and profit margins.
2. Tax laws.
3. Technological change.
4. Food Grading and Marketing Orders.
5. Crop Insurance and Disaster Payments.

Debt/equity ratios and profit margins definitely impact the farming decisions, resulting in changes in environmental outcomes (Nielsen et al. 1989; Gustafson and Barry Chapter 15). However, the relationship between these factors and the structure of agriculture is not so clear. Farmers with good cash flow and/or wealth positions may feel more positive about the future and be willing to invest current income to increase future income. Conservation measures, or environmental stewardship in general, may be a consumer good and increase with better financial situations.

Tax laws greatly impact the management of the natural resources used in agriculture. Even with the tax code held constant as farm structure changes, the evolving farms become subject to other aspects of the code.

Technological change impacts the set of profitable production activities but the relationship between technological change and the structure of agriculture is not clear (Batte and Johnson Ch. 12), particularly with regard to the adoption of "sustainable" systems. The argument that no-chemical farming can be easily accomplished seems to imply a drastic change in the production function from that in existence at the time chemicals were sought after.

A comprehensive 1982 study (National Research Council) of existing conditions finds that technological change is having several adverse environmental effects: prevalence of larger, contiguous fields devoted to single crops is increasing; improved varieties of crops capable of growing on marginal soils are available; fertilizer and pesticide crop responses are greater with improved varieties and larger quantities of these inputs are applied; and expansion of drainage and irrigation development occurs at the expense of natural ecosystems. It seems clear that larger farms have the comparative advantage for larger, contiguous fields, for drainage of lowlands, and for irrigation development.

Federal *grading standards* or standards adapted under market orders encourage high chemical use for crops, and high concentrate rations for livestock (National Research Council 1989). Chemicals must be used on fruits and vegetables to meet cosmetic requirements that have little to do with nutritional value. Similarly, meat and dairy standards require a high fat content in the food product.

Crop insurance and disaster payment regulations currently require certification of use of "Best Management Practices" which typically require use of high levels of fertilizer and pesticides.

Studies on Structure and the Environment

The policy issue of soil erosion generated numerous studies on the structural impacts on conservation decisions in the 1980s. Many of these studies examined the off-site, externality aspects of the soil erosion problem. Many of the factors impacting farmer's decisions about soil conservation are likely to have the same type of influence on environmental stewardship as a whole. Since studies on general environmental stewardship are lacking, the relevant conservation studies are reviewed in this section.

Factors Impacting Conservation Attitudes and Decisions

Factors influencing government program participation, attitudes about conservation, and resulting farming decisions have been examined relative to farm structure in several studies. These studies have a narrow soil conservation focus but some implications can be generalized to stewardship of the environment as a whole.

A survey by Demissie (1989) finds that educational level is a major factor of non-participation in government programs by limited resource farmers in the South. Many of the farmers did not know about current commodity and conservation programs. The informational materials available and the delivery systems of the responsible agencies were inadequate for these farmers. For many farmers who had heard about the programs, the paperwork required for participation was considered too complex.

In their *Alternative Agriculture* report the National Research Council (1989, 9) makes the following statement: "Alternative farming practices typically require more information, trained labor, time, and management skills per unit of production than conventional farming." For alternative agriculture livestock and crop mixes and timeliness of operations are all more complex. A structure of agriculture with less education, less reliance on specialized management tools, and more hobby and part time farmers will have trouble moving to environmentally less damaging alternative agriculture techniques.

Lee (1980) found no significant differences nationally in erosion rates by landownership type after adjustment for inherent erodibility. Differences in erosion by ownership type in the Southeast could be attributed to the specialized crops grown by some large farms and by the small, undercapitalized nature of many of the region's farmers. Lee (1980) also found no significant difference between tenure groups or for net income classes at the national level. For the full owner operator class, erosion rates decreased as income increased but results were insignificant for other classes. Bromley (1980) questioned Lee's use of legal descriptions of farm ownership categories as proxies for different types of decision units. This criticism applies to other studies cited in this section also.

Lee and Stewart (1983) found that full owner operators and landowners have lower minimum tillage adoption rates than do other groups and that the corporate structure of non-family farmers has no impact on tillage choice. Lee and Stewart (1983) conclude that a small operating size poses a larger problem for conservation tillage adoption than does separation of ownership from operation.

Nowak and Korsching (1983) proposed a model showing factors leading to the adoption and maintenance of Best Management Practices and applied the model with Iowa data. They rejected the idea that large corporate farms "mine the soil" and propose that more complex forms of legal arrangements of ownership and operation coincide with higher levels of education, more availability of discretionary funds, and the ability to bear risk (or experiment on a small scale basis). Nowak and

Korsching (1983) found that Soil Conservation Service (SCS) cooperator farmers were likely to have owned the farm longer and also to have higher incomes than non-cooperators.

A more complete model of conservation behavior is proposed by Ervin and Ervin (1982). Four sets of factors (personal, institutional, physical, and economic) simultaneously impact and cause feedback between three stages of behavior (perception of an erosion problem, decision to use one or more conservation practices, and actual reduction in erosion). Ervin and Ervin (1982) applied their model to Monroe County, Missouri data and found the following results. Less experienced (or younger) farmers are more likely to accept the merits of a wider range of conservation practices. Younger farmers have higher educations, a better perception of the erosion problem, lower risk aversion/shorter planning horizons, and, on the negative side, a lack of adequate capital to engage in all the conservation practices they desire. Cash grain farmers use less conservation options but this is likely due to their crop specialization choice. Participation in SCS programs did not have a significant impact on either the number of practices used, or the resulting erosion outcome. Higher erodibility did not imply a higher degree of use of practices.

Concepts from social psychology and economics were merged to generate an indirect utility function, $U_m = U_m(y_m, P_m, A_m, F_m)$, by Lynne et al. (1988) where m is the level of conservation effort, y_m is income, P_m is the price of conservation practices, A_m reflects attitudes about farming and conservation, and F_m is farm features. The model was applied to data from three counties in the Florida panhandle and the following results were found. Renters use less conservation effort than owners, but this may be due to the owners having larger farms with different crops. Full owners expended the most conservation effort. Higher inherent erodibility and stronger views on the need to preserve the non-renewable resources and on the responsibility for externalities generated by one's own actions all lead to more conservation effort. Strong beliefs about the probability of technology offsetting resource problems led to less conservation effort. The discount rate and strong professional feelings about farming as a way of life had little impact on conservation decisions. Those willing to bear more risk expended more conservation effort. Over the income range of the data, effort increased with income at an increasing rate.

The national impact of cropland rental versus cropland ownership by the operators on soil conservation was studied by Bills (1985). Bills (1985) found that erosion rates of renters and owner operators were not significantly different when adjustments for land quality and other factors were made. The erosion potential of rented and owner-operated land is about equal. However, nearly two-thirds of leased land is used for erosive row crops compared to about half of the owner operated. Owner operators grow more non-erosive crops (hay and pasture) than do renters. Use of conservation practices (crop rotations, tillage practices, crop residue management, terraces, etc.) were nearly similar by both classes of farmers. Bills (1985) argues that some important soil-conserving management practices are neutral with regard to tenure because they are so cost effective in their own right.

Kraft et al. (1989) survey 264 southern Illinois farmers about their primary and secondary goals for the farming operation. The farmers were stratified into 20 categories based on size, age, type of enterprises, tenure and off-farm income. Soil conservation was chosen as the primary goal by only 1.8 percent of the sample. Of farmers with moderate to large size enterprises, aged early 50s, not working off the farm and owning most of their land, 13 percent stated conservation as their main goal. The only other group (2.6 percent) with conservation as the main goal were the very small farmers deriving more than 70 percent of their income from off-farm sources. Seven groups had some proportion listing conservation as their second most important goal, but five of these had less than 15 percent making that choice.

Land Improvement Investments and Structure

Investments in land improvements (conservation measures, drainage, and land clearing) have historically been used as a measure of good stewardship. Now the adverse environmental externalities of some of these practices are also recognized. However, some of the factors leading to these investments may also serve as proxies for causes of environmental stewardship. Recent national studies of factors impacting land improvement investment include Baron (1981); Gertel et al. (1985); and Nielsen et al. (1989). Some findings of these studies relevant to this chapter are summarized here.

Gertel et al. (1985) used 1975-77 data and studied the impact of farm ownership and operator tenure on land improvement capital expenditures. They found that the proposition that absentee owners invest less in land improvement is not supported. Retired non-operator owners made the smallest investments and there was no significant difference between retirees living in or out of the same county as their land. Non-family operators made improvement expenditures to a greater extent than did individuals or families. A higher proportion of farm operator owners made capital expenditures for improvement than did owners who operated none of their land and this result held for all classes of ownership, i.e., for individuals, families, partnerships, and corporations.

Nielsen et al. (1989) utilized 1980-86 national farm expenditure data to evaluate the impact of financial and government policy factors on land improvement investment. Operators investing in improvements in 1983 operated 1.5-2.0 times as much cropland as non-investors and also owned more acres. However, combinations of the three categories of land improvement on the same farm imply that much of the expenditure was on new, marginal cropland. It was also found that farmers with higher incomes were more likely to make improvements. An econometric investment model was estimated to link the expenditures to external factors. The estimated national level impacts are given in Table 21.1.

Relative Input Use Levels by Alternative Farm Structures

The National Research Council (1989, 12) found that expenses for machinery, pesticides, fertilizers, and interest on operating capital vary

TABLE 21.1   Factors Impacting Investments in Land Improvements, 1980-86

Factor	Change in the Factor	Change in Investment
Expected farm income	a $100 increase	a $1 increase
Expected real interest rate	1 percentage point up	$9 million decrease
Government idled acreage	a 1 acre increase	$2 increase
Ratio of land improvement costs to land value	a 1-unit increase	$73 million decrease
Land needing improvement	a 1 acre increase	$2-3 increase
Conservation subsidy	a $1 increase	$1.50

Source: Nielsen, Elizabeth G., John A. Miranowski, and Mitchell J. Morehart. *Investments in Soil Conservation and Land Improvements: Factors Explaining Farmer's Decisions.* Washington, DC: U.S. Dept. Agri., Econ. Res. Ser. Agri. Econ. Rep. No. 601, 1989.

more among farmers in a region than across regions; these cost components "account disproportionately for differences in per unit production costs." Environmental damage might also vary more within a region than across regions, and the variation may have to do with farm size. Interregional variance in costs and environmental outcomes may be a structural issue.

As an example of the kind of studies that can be constructed from available data, the relative expenditures by corn acres grown have been summarized at the national level (Ahearn et al. 1990). Ahearn et al.'s study is summarized in Table 21.2. Mid-sized farms (100-499 acres of corn) used less chemicals per acre than other sized farms. The mid-sized farms used more fertilizer than smaller farms but less than larger farms. The farms producing more than 500 acres of corn used more technical services, less custom operations, and more hired labor than other categories.

Distribution of Land Ownership

With different types of owners and operators exhibiting different levels of conservation stewardship, it is important to evaluate the distribution of farmland. The most recent comprehensive national survey of farm ownership was by Lewis (1980). Lee (1983) provides a summary of Lewis' data. Recent studies on land transfer and changes in distribution include Harrington and Carlin (1987); Reimund et al. (1987); and Wunderlich (1989).

The following structural characteristics of land ownership are from Lee (1983). High quality and erosion prone cropland are distributed among classes of landowners in approximately the same patterns as total cropland, except for a slight concentration of prime cropland in medium sized holdings. Owners of potential cropland (land currently in range or forest production) tend to have smaller holdings, lower net farm income and are less involved in agriculture than current cropland owners. Range and forestland tend to be held in large tracts (1000 or more acres) by absentee landowners except for 22 percent of the total which is held in tracts of less than 100 acres by non-farmers. Non-operator landlords own 31 percent of cropland and 78 percent of cropland owners reside in the same county as their landholdings. Owners of pastureland tend to be employed in non-farm occupations. Cropland is owned by farmers (48 percent), retired people (17 percent) and white collar workers (10 percent).

Harrington and Carlin (1987) examine the impacts of farm financial stress in the early 1980s on farmland ownership transfer and conclude that the impact on food supply is minimal. Most farmland changing hands remained in production without even a year's break. About 75

TABLE 21.2 Input Use (Cost) by Corn Production Acreage Class

		<25	Acres of Corn		
			25-99	100-499	500+
		Base	Value/ Base	Value/ Base	Value/ Base
Yield		91.00	1.16	1.33	1.41
Total variable	cost/acre	108.09	1.02	1.06	1.21
	cost/bu	1.19	0.87	0.80	0.87
Fertilizer	cost/acre	31.75	1.06	1.14	1.29
	cost/bu	0.35	0.91	0.86	0.91
Lime & gypsum	cost/acre	1.72	1.52	1.56	0.73
	cost/bu	0.02	1.00	1.00	0.50
Chemical	cost/acre	19.15	1.01	0.99	1.17
	cost/bu	0.21	0.86	0.76	0.86
Custom operation	cost/acre	15.31	0.71	0.39	0.22
	cost/bu	0.17	0.59	0.29	0.18
Elec., lube, fuel	cost/acre	10.69	0.92	0.72	0.71
	cost/bu	0.12	0.75	0.50	0.50

(continued)

TABLE 21.2 (continued)

			Acres of Corn		
		<25	25-99	100-499	500+
		Base	Value/ Base	Value/ Base	Value/ Base
Hired labor	cost/acre	1.79	1.94	3.98	7.01
	cost/bu	0.02	1.50	3.00	5.00
Technical ser.	cost/acre	0.07	2.71	3.43	6.00
	cost/bu	0.00	2.13	2.50	4.13
Irrigation	cost/acre	0.08	8.63	18.13	43.63
	cost/bu	0.00	7.22	13.33	30.33

Source: Ahearn, Mary, Gerald Whittaker, and Dargan Glaze. 1990. "Cost Distribution and Efficiency of Corn Production." Paper presented at the NC-181 Meeting, Albuquerque, NM, January 6, 1990. The paper was based on the 1987 Farm Costs and Returns Survey, USDA, ERS.

percent of the land was purchased by continuing farmers. Most of the remaining 25 percent was purchased by retired farmers who kept the land in production. The land which is marginal for cropping and that reverts to other uses as well as the land lost to higher valued real estate purposes are all minimal and occur in areas away from the greatest financial stress and ownership change.

Empirical studies have proven inconclusive as to the effects of farm structure on the environment. While the popular press has reported a variety of case studies showing a positive relationship between size and environmental degradation, the comprehensive studies reviewed here find no strong correlation between size and conservation practices. Clearly more research in this area is needed.

Conclusions

This chapter suggests some economic policy relevant questions about the structure of agriculture and stewardship of the environment. It is clear that one cannot understand the relationship between the structure of agriculture and the environment without an understanding of the impact of structure on choice of production practices. Likewise, complete information on the other links of the model shown in Figure 21.1 is scarce.

Insufficient Information is Available for Policy Formulation

The major conclusion of this chapter is that insufficient empirical studies exist to provide guidance for policy makers concerned about the impact of changing agricultural structure on the environment. The chapter has shown that the most important area in which information is lacking is on how alternative farm structures treat individual resource units. A summary report on the status of sustainable agriculture in the U.S. contains the following statement (National Research Council 1989):

> There is inadequate scientific knowledge of economic, environmental, and social costs and thresholds for pest damage, soil erosion, water contamination, and other environmental consequences of agricultural practices. Such knowledge is needed to inform farm managers [and policy makers] of the trade-offs between on farm and off farm consequences.

Existing U.S.D.A., Bureau of Census, state level, and private firm data could provide some answers, if access to unaggregated raw survey results on farm structural characteristics and input use were made available, and if sufficient funds for analytical work were provided.

Larger Farm Structures Have the Ability for Stewardship

As argued by several studies cited in the previous section larger and more complex forms of farm ownership and operation indicate a higher level of education and managerial skill. In addition, the various conservation studies cited are either neutral or positive in their assessment of the conservation attitudes of the larger farms. It was noted that these larger farms, with higher levels of skill, will be much more flexible in their choices and so can adopt to new technology faster. However, when the profit margin is strong, this ability to adopt also may lead these farms to the use of marginal lands and temporary removal of existing conservation measures. The empirical studies cited did not find significant differences in conservation attitudes and practices between these farms and the midsize family farms.

Since mid to large size farms utilize a relatively larger portion of the natural resources, the findings of this chapter imply that environmental policy incentives should be designed to take advantage of the flexibility and expertise of the larger farms. At the same time environmental policy (or commodity policy) must be structured so as to prevent rapid changes of land use at both extensive and intensive margins.

The Special Policy Needs of Smaller Farmers

The sources cited in this chapter indicate that smaller farmers may control resources with a relatively higher environmental damage risk as well as have less capability of practicing good stewardship. These farmers derive more of their income from off-farm sources which implies that financial incentives of current polices may not be so attractive. These farmers may operate at such small scales as to make the acquisition of new information and/or technology unfeasible. The fact that off-farm employment is a major source of income implies that timing, and overall labor and management decisions may be quite constrained for these farmers. In many cases the delivery mechanisms of current government programs are not compatible with the educational and or cultural capabilities of these farmers. Since the number of farmers in this group is large and they tend to have the more marginal resources, policies and programs to address their needs may be relatively more expensive than what is currently considered.

Economics and Multi-disciplinary Studies Are Needed

Analysis of farm structure and the environment can be categorized in the same manner as the model components shown in Figure 21.1.

Detailed studies are needed in five main areas. First, identifying the environmental impacts of agriculture. Secondly, identifying the links between the environmental impacts and production practices in use. Third, identifying the association between production practices/technologies and various measures of structure. Fourth, identification of the impacts of policies on farm structure and on the choice of production technologies. Fifth, identification of the factors leading to formulation of policy. Economists could contribute cost/benefit information to interdisciplinary studies in all five of these areas of investigation. The feedback effects of policy action must also be taken into account, such as the fact that an environmental outcome leads to a policy change which leads to a change in farming methods, which then changes structure. Once the five areas have been studied, comprehensive models accounting for interactions could be constructed for use in policy analysis.

References

Ahearn, Mary, and Dargon Glaze. 1990. "Cost Distribution and Efficiency of Corn Production." Proceedings, NC-181 meeting, Albuquerque, New Mexico, January 6, 1990.

American Farmland Trust. 1990. "Farm Bill 1990: Agenda for the Environment and Consumers." Washington, DC: American Farmland Trust.

Baron, Donald. 1981. *Landowner Characteristics and Investment in Soil Conservation.* Washington, DC: U.S. Dept. Agri. Econ. and Stat. Ser. Staff Rep. No. AGES810911.

Basu, Kaushik. 1989. "Technological Stagnation, Tenurial Laws, and Adverse Selection." *The American Economic Review* 79: 251-255.

Batie, Sandra S. 1988. "Agriculture as the Problem: New Agendas and New Opportunities." *Southern Journal of Agricultural Economics* 20: 1-11.

_____ . 1989. "Sustainable Development: Challenges to the Profession of Agricultural Economics." *American Journal of Agricultural Economics* 71: 1083-1101.

Batie, Sandra S., William E. Cox, and Penelope L. Diebel. 1989. *Managing Agricultural Contamination of Ground Water: State Strategies.* Washington, DC: National Governors Association.

Benbrook, Charles M. 1989. "Agriculture and Groundwater Quality: Policy Implications and Choices." Paper presented January 17, 1989, as part of the "Technical Session on Agriculture and Groundwater Quality," 1989 Meeting of the AAAS, San Francisco.

Bender, Jim. 1987. "Large Farm Equipment and Soil Erosion." *Journal of Soil and Water Conservation* 42: 169-170.

Bernat, G. Andrew, Jr. 1987. *Farmland Ownership and Leasing in the United States, 1982.* Washington, DC: U.S. Dept. Agri., Econ. Res. Ser. Staff Rep. No. AGES870225.

584

Bills, Nelson L. 1985. *Cropland Rental and Soil Conservation in the United States.* Washington, DC: U.S. Dept. Agri., Econ. Res. Ser. Agri. Econ. Rep. No. 529.

Boxley, Robert F. 1989. "Separating Land Ownership and Farm Operations." *The Farm and Food System in Transition,* Paper No. 55. Coop. Ext. Ser. and Dept. Agri. Econ., East Lansing: Michigan State University.

Bromley, Daniel. 1980. "The Impact of Landownership Factors on Soil Conservation: Discussion." *American Journal of Agricultural Economics* 62: 1089-1090.

Center for Rural Affairs Newsletter. Various issues. Center for Rural Affairs, Walthill, NE.

Cochrane, Willard W. 1979. *The Development of American Agriculture, A Historical Analysis.* Minneapolis: The University of Minnesota Press.

Daberkow, Stan G. 1988. "Low Input Agriculture: Trends, Goals and Prospects for Input Use." Invited paper, annual meeting of the American Agricultural Economics Association, Knoxville, Aug. 3, 1988.

Demissie, E. 1989. "Improving Government Farm Programs for Limited-Resource Farmers." *Journal of Soil and Water Conservation* 44: 384-387.

Ervin, Christine A., and David E. Ervin. 1982. "Factors Affecting the Use of Soil Conservation Practices: Hypotheses, Evidence, and Policy Implications." *Land Economics* 58: 277-292.

Gertel, Karl, Douglas G. Lewis, and Kenneth M. Miranda. 1985. *Investment in Land by Landowner Classes.* Washington, DC: U.S. Dept. Agri., Econ. Res. Ser. Staff Rep. No. AGES841029.

Giles, R. H., Jr. 1981. "Assessing Landowner Objectives for Wildlife." in R.T. Gumke, G. V. Burger, and J. R. March, eds., *Wildlife Management on Private Lands.* Pp. 112-129. Lacrosse, Wisconsin: Lacrosse Printing Co.

Glaser, Lewrene K. 1986. *Provisions of the Food Security Act of 1985.* Washington, DC: U.S. Dept. Agri., Econ. Res. Ser. Agri. Info. Bul. No. 498.

Gustafson, C. D., and P. J. Barry. "Structural Implications of Agricultural Finance." in *Determinants of Size and Structure in American Agriculture.* Ch. 15, forthcoming.

Hanchar, John J. 1989. *Irrigated Agriculture in the United States. State-Level Data.* Washington DC: U.S. Dept. Agri., Econ. Res. Ser. Stat. Bul. No. 767.

Harrington, David, and Thomas A. Carlin. 1987. *The U.S. Farm Sector: How is it Weathering the 1980s?* Washington DC: U.S. Dept. Agri. Econ. Res. Ser. Agri. Info. Bul. No. 506.

Johnson, James B., and Richard T. Clark. 1989. "Conservation Compliance: What Soil Loss with What Level of Farm Income Impact?" *Journal of Soil and Water Conservation* 44: 458-462.

Johnston, George M. 1989. "Wildlife and Agriculture: A Case of Interdependence." *The Farm and Food System in Transition.* Paper No. 35. Coop. Ext. Ser. and Dept. Agri. Econ., Michigan State University, East Lansing.

Kraft, Steven E., Paul L. Roth, and Angela C. Thielen. 1989. "Soil Conservation as a Goal Among Farmers: Results of a Survey and Cluster Analysis." *Journal of Soil and Water Conservation* 44: 487-490.

The Land Stewardship Letter. Various issues. Land Stewardship Project, Marine, MN.

Lee, Linda K. 1980. "The Impact of Landownership Factors on Soil Conservation." *American Journal of Agricultural Economics* 62: 1070-1076.

_____. 1983. *Linkages Between Landownership and Rural Land*. Washington, DC: U.S. Dept. Agri., Econ. Res. Ser. Agri. Info. Bul. No. 454.

Lee, Linda K., and William H. Stewart. 1983. "Landownership and the Adoption of Minimum Tillage." *American Journal of Agricultural Economics* 65: 256-264.

Lewis, Douglas, and Thomas McDonald. 1984. *Improving U.S. Farmland*. Washington, DC: U.S. Dept. Agri., Econ. Res. Ser. Agri. Info. Bul. No. 482.

Lewis, James A. 1980. *Landownership in the United States*. Washington, DC: U.S. Dept. Agri. Econ. Stat. Coop. Ser. Agri. Info. Bul. No. 435.

Lynne, Gary D., J. S. Shonkwiler, and Leandro R. Rola. 1988. "Attitudes and Farmer Conservation Behavior." *American Journal of Agricultural Economics* 70: 12-19.

National Research Council. 1982. *Impacts of Emerging Trends on Fish and Wildlife Habitat*. Washington, DC: National Academy Press.

_____. 1989. *Alternative Agriculture*. Washington, DC: National Academy Press.

Nassauer, Joan Iverson. 1989. "Agricultural Policy and Aesthetic Objectives." *Journal of Soil Water Conservation* 44: 384-387.

Nielsen, Elizabeth G., John A. Miranowski, and Mitchell J. Morehart. 1989. *Investments in Soil Conservation and Land Improvements: Factors Explaining Farmer's Decisions*. Washington, DC: U.S. Dept. Agri., Econ. Res. Ser. Agri. Econ. Rep. No. 601.

Nowak, Peter J., and Peter F. Korsching. 1983. "Social and Institutional Factors Affecting the Adoption and Maintenance of Agricultural BMPs." in Frank W. Schaller and George W. Bailey, ed., *Agricultural Management and Water Quality*. Ames, Iowa: Iowa State University Press.

Osteen, Craig D. 1985. *The Impacts of Farm Policies on Soil Erosion: A Problem Definition Paper*. Washington DC: U.S. Dept. Agri. Econ. Res. Ser. Staff Rep. No. AGES841109.

Phipps, T. T., K. Allen, and J. A. Caswell. 1989. "The Political Economics of California's Proposition 65." *American Journal of Agricultural Economics* 71: 1286-92.

Reichelderfer, Katherine, and Tim T. Phipps. 1988. "Agricultural Policy and Environmental Quality." A briefing book prepared for November 28, 1988, Briefing at Resources for the Future, Washington, DC.

Reimund, Donn A., Thomas A. Stucker, and Nora L. Brooks. 1987. *Large-Scale Farms in Perspective*. Washington, DC: U.S. Dept. Agri. Econ. Res. Ser. Agri. Info. Bul. No. 505.

Reisner, Mark. 1986. *Cadillac Desert*. New York: Viking Penguin, Inc.

Schertz, Lyle P., and Gene Wunderlich. 1982. "Structure of Farming and Landownership in the Future: Implications for Soil Conservation." in Harold G. Halcrow, Earl O. Heady, and Melvin Cotner, ed., *Soil Conservation Policies, Institutions, and Incentives*. Ankeny, Iowa: Soil Conservation Society of America.

Shoemaker, Robbin. 1986. *Effects of Changes in U.S. Agricultural Production on Demand for Farm Inputs*. Washington DC: U.S. Dept. Agri. Econ. Res. Ser. Tech. Bul. No. 1722.

586

Stanton, B. F. "Farm Structure: Concept and Definition." in *Determinants of Size and Structure in American Agriculture*. Chapter 2, forthcoming.

U.S. Department of Agriculture. 1989. *Economic Indicators of the Farm Sector. Costs of Production, 1987*. Washington DC: U.S. Dept. Agri. Econ. Res. Ser. ECIFS 7-3.

Wise, Sherry, and Stanley R. Johnson. 1990. "A Comparative Analysis of State Regulations for the Use of Agricultural Chemicals." in Nancy Boekstael and Richard Just, eds., *Commodity and Resource Policies in Agricultural Systems*. Berlin: Springer-Verlag.

Wunderlich, Gene. 1989. *Transfers of Rural Land*. Washington DC: U.S. Depart. Agri., Econ. Res. Ser. Agri. Info. Bul. No. 568.

22

Projections of Structural Change and the Future of American Agriculture

Kent D. Olson and B. F. Stanton

Projections of the structure of American agriculture have been made by several agricultural economists, particularly in the years since World War II. The methods used and their results are reviewed at the outset of this chapter. This is followed by a set of projections to the year 2000 based heavily on agricultural census data for recent periods with 1987 as the point of departure. Finally, the implications of these projections and alternative forces, which might speed up or slow down structural change, are considered. Comment is also made on what, if anything, can be said about the future of family farms in America.

Past Projections and Methods

Various methods have been used to study structural issues. The one point that ties these studies together is the expectation that past trends tell us something about the future. That is, the historical trend of declining number of farms is always projected to continue. What differentiates these studies is the choice of the future rate of decline, the change in that rate, and how to estimate that rate and change. The studies by Lin, Coffman, and Penn (1980), and the Office of Technology Assessment (U.S. Congress) are examples of these studies. A few studies have attempted to understand why past trends happened and what that means for the future. Reimund, Martin, and Moore (1981), and Tweeten

(1984) are examples of studies examining the underlying structural issues of change. In this section, a selection of these studies is reviewed for their methods, projections, and any criticisms or suggestions that may help guide future projections. Due to space limitations, the selection is small compared to the potential list of publications covering size and structure issues. The discussion is arranged by method.

Markov Process

Several studies have used the Markov process to describe the process of farms moving from one size class to another, remaining in the same class, or exiting from farming. The critical component of the Markov process is the transition matrix, P, which contains the probabilities of moving between size categories $(S_1, S_2, ... S_n)$:

$$
\begin{array}{c|cccc}
 & S_1 & S_2 & ... & S_n \\
\hline
S_1 & P_{11} & P_{12} & ... & P_{1n} \\
S_2 & P_{21} & P_{22} & ... & P_{2n} \\
\vdots & \vdots & \vdots & \vdots & \\
S_n & P_{n1} & P_{n2} & ... & P_{nn}
\end{array}
$$

In this notation the probability of moving from size S_i to S_j in the next period is P_{ij} where $\sum_{j=1}^{J} P_{ij} = 1.0$; and $P_{ij} \geq 0$ for all i and j. These restrictions say that all farms in one size will remain in that size class or will be in another class in the next period. Consequently, to apply this to farming, there has to be a class to allow a farm to exit between periods and another to allow a farm to enter the farming sector between periods. To project the distribution of farms across size classes in the tth period, X_t, the initial distribution, X_0, is multiplied by the transition matrix, P, raised to the tth power:

$$X_t = X_0 P^t.$$

Most applications of the Markov process are modifications of the pure process because of its restrictive assumptions and insufficient micro-level data. Anderson and Goodman (1957) show the maximum likelihood estimator of the stationary transition probability to be:

$$P_{ij} = \frac{\sum\limits_{i=1}^{T} m_{ijt}}{\sum\limits_{j=1}^{n} \sum\limits_{t=1}^{T} m_{ijt}}$$

where m_{ijt} is the number of farms moving from the ith size to the jth size in the tth period. That is, the transition probability is based on the actual frequency of farms moving from one size to another. The four critical assumptions about the size distribution of farms are:

· Farms can be grouped into size-classes according to some criteria;
· The evolution of a farm through these classes can be regarded as a stochastic process;
· The probability of a farm moving from one class to another is a function only of the basic stochastic process; and
· The transition probabilities remain constant over time (Stavins and Stanton 1980).

In an early study, Krenz (1964) projected the trend in the number of farms in North Dakota. He used census data from 1935 through 1960 to estimate the Markov transition matrix. To make his estimate, Krenz (1964) had to make several restrictive assumptions due to the lack of data on individual farms: (1) operators will expand, if possible, (2) farmers who do expand are those initially larger than the average in that size category, (3) increases are gradual, and (4) decreases are not likely to occur. These assumptions were translated into these conditions: the largest farms will remain in farming, increases in sizes will be to the next class only, and farms will either grow, remain the same size, or exit; they will not shrink.

In early studies which used sales to define size, the probability of moving (or not moving) from one class to another was estimated using nominal census data. This ignored inflation which pushed farms into larger classes without an actual, real increase in size. This was not a large concern when inflation was low. Between the 1969 and 1974 census, however, there was nearly an 80 percent increase in prices received by farmers. To account for this inflation, Lin, Coffman, and Penn (1980) adjusted the 1974 distribution of farms to 1969 prices. They assumed a uniform distribution of farms within the unadjusted classes, recalculated the class boundaries to account for inflation, and redistributed the farms by linear interpolation. By using this method, they felt that only "real" changes between size classes would be left. The transition matrix estimated using 1964 and 1969 census data was very similar to the transition matrix using estimated 1969 and adjusted-1974 census data; so they assumed the transition probabilities were stable. When sales classes were used, the projection was first made with the "deflated" transition probabilities. The resulting distribution was readjusted by assuming a future inflation rate and a uniform distribution of farms within each class. With a price inflation of 7.5 percent per year, Lin, Coffman, and Penn (1980, 43-44) projected the number of farms to decline to 2.2 million in 1990 and 1.86 million in 2000 (Table 22.1). Using a lower inflation rate, they projected 1.85 million farms in 2000. Using acreage classes, the number of farms was projected to decline to 2.1 million in 1990 and 1.7 million in 2000.

Stavins and Stanton (1980) improved the Markov process in their study of the New York dairy industry. First, they followed size changes over time in a sample of dairy farms. This sample data allowed them to obtain the transition probabilities without using the restrictions required by census data. Second, they regressed the estimated probabilities on a structural variable (i.e., the milk-feed price ratio) and not time. Third, by using a structural variable rather than time, they could project the structural variable and a new set of probabilities under different scenarios of the future. As a final step, they took the projections based on the sample and extrapolated the results to the entire New York dairy industry.

In a more recent analysis of the size and structure issues at the national level, the Office of Technology Assessment (U.S. Congress, OTA 1986) used a Markov process to project farm distributions to 2000 (Table 22.2). Using data from 1969 through 1982, OTA said "farm numbers are likely to decline from 2.2 million in 1982 to 1.8 million in 1990 and 1.2 million in 2000" (p. 96). The 1987 Census of Agriculture counted 2.09 million farms; based on a linear extrapolation between 1982 and 1990 this is more than the OTA projections expected. OTA also concludes that the future distribution of farms will be a bimodal or bipolar distribution; that is, there will be a large proportion of small and part-time farms, an increasing proportion of large farms, but a decreasing proportion of mid-sized farms (p. 96). Another measure of structural change is the concentration of sales in the larger farms; the OTA study projects that approximately 50,000 of the largest farms in 2000 will account for 75 percent of the agricultural production (p. 9). OTA projected similar trends for individual commodities and regions with some differences in the details.

TABLE 22.1 Projections of Farm Numbers to the Year 2000 by Acreage Classes

| | | | | Size by Acreage | | | |
Method	1-99	100-219	220-499	500-999	1000-1999	2000 & up	All Farms
				(1,000 farms)			
Most likely:	827	302	264	183	102	71	1,750
Trend Extrapolation:	751	300	286	205	108	61	1,711
Negative Exponential:	320	313	501	430	224	37	1,826
Markov Process:	865	290	230	153	92	77	1,705
Age Cohort:*	934	301	220	156	98	54	1,772

*For the Age Cohort analysis, the projection reported in this table is for the year 2004.

Source: Lin, Coffman, and Penn.

TABLE 22.2 Projections of Farm Numbers to the Year 2000 by Sales Class

	From: Up to:	0-2.5	2.5-10	10-20	20-40	40-100	100-200	200-500	500 & up	All Farms
Method					*(1,000 farms)*					
Most likely:		604	186	100	88	214	161	183	217	1,750
Trend Extrapolation:		456	201	164	443	539	188	81	27	2,109
Negative Exponential:		13	42	53	102	271	354	606	417	1,857
Markov Process:		640	180	108	88	262	168	190	226	1,862
Age Cohort:*		655	219	100	100	190	—>	600	<—	1,864
OTA:		———>		638	—>	363	75	125	50	1,250

*For the Age Cohort analysis, the projection reported in this table is for the year 2004.

Sources: Lin, Coffman, and Penn for the first 5 projections and U.S. Congress for the OTA projections.

Statistical Models

Two of the four projection methods used by Lin, Coffman, and Penn (1980) were the statistical models: trend extrapolation and negative exponential functions. For *trend extrapolation*, they evaluated four functional forms:

- linear: $N = b_0 + b_1T$
- polynomial: $N = b_0 + b_1T + b_2T^2$
- semilog: $\ln N = b_0 + b_1T$
- log-linear: $\ln N = b_0 + b_1\ln T$

where N = the number of farms and T is the time variable (1959 = 1, 1960 = 2, ...). The linear and polynomial forms were eliminated due to projecting the disappearance of a size or a reversal of past trends. The choice between the semi-log and log-linear forms was based on goodness-of-fit and the reasonableness of the projection. A function was chosen for each size class and projected. The total of the projection was the sum over all size classes.

The resulting projections for the year 2000 were for 1.7 million farms using the acreage distribution and 2.1 million farms using the sales class distribution (Tables 22.1 and 22.2). They noted that the projections by sales class were larger than the projections by acreage and reasoned that this was due to a statistical misreading of the direction in the $20,000 to $39,999 class between 1969 and 1974.

For another projection, Lin, Coffman, and Penn (1980) used *negative exponential functions* to represent the distribution of farms over size classes. For the acreage distribution, they estimated this function to be:

$$\ln y = 2.0011 - 0.4160 \, X_i / \bar{X}$$

where y is the percentage of farms lying above a size limit, X_i; X_i is the lower class limit in acres; and $\bar{X}$ is the average farm size in acres. With this function and projections of total land and average farm size, the number of farms, and then the size distribution, can be projected. This projection also requires the distribution over size classes to be stable over time. Lin, Coffman, and Penn's (1980) test for structural change could not reject the hypothesis of no structural change (p. 28). Using the negative exponential functions and the acreage classes, farm numbers were projected to decline from 2.9 million in 1974 to 1.8 million in 2000 (Table 22.1). Except for a slowing in the rate of decline after 1980, the negative exponential function projected a pattern of decline similar to historical trends (p. 29). Projections for size classes on the basis of sales were made, but the results departed from current trends in several important ways and were rejected (p. 31 and 32).

Simulation Models

Schatzer et al. (1983, 7) used a simulation model to estimate the potential impacts of changes in average farm size. Their analysis showed that society's goals and policies will affect whether average U.S. farm size will grow. They state that if the goal is to increase farm income, current trends should continue. If the goal is a larger supply of cheaper food, however, they find that larger farms will be needed, but at a large cost to U.S. farmers.

Teigen (1988) developed an artificial model to analyze the linkages between farm policy instruments, the introduction and adoption of technology, and the structure of agriculture. After the introduction of a new technology, Teigen (1988) estimated that government programs of acreage diversion, price supports, and parity income slowed the change to a new equilibrium and slightly increased the number of farms compared to no government intervention. Instituting a new government program of exit annuities paid to farmers who leave farming and not to farmers who remain in farming speeded adjustment to the new equilibrium.

Age Cohort Analysis

The fourth projection method used by Lin, Coffman, and Penn (1980) was age cohort[1] analysis. This method uses population dynamics and the life cycle of farmers. Projections are made by assuming historical patterns of changes in the number of farmers by age cohort will continue into the future. As an age cohort grows older, the number of farmers in that group is first expected to increase as parents are replaced. Historical data show this happens up to age 35. After that age, the number of farmers in each age cohort decreases--that is, the number of exits is greater than the number of new entrants. Using census data, Lin, Coffman, and Penn (1980) calculated that by the time the 35-44 year-old cohort aged 10 years, only 98 percent of the group were farming. After another 10 years, only 77 percent of the 98 percent were farming. In another 10 years when this cohort was 65-74 years old, only 65 percent of the 77 percent were farming. Due to data limitations, all farmers were assumed to retire at age 75. Thus, as farmers are projected to quit farming, there are fewer farmers to be replaced by children. Entry rates of the youngest cohort were based on replacing parents who were up to a 40-year age difference. Historically, this replacement rate was less than 1.0, which further decreases the number of farmers.

To project the number of farms using age cohort analysis, they start with the actual number of farms in the 1974 census and adjusts the cohorts into the future by these historical changes. Thus, the total number of farms is projected to be 1.61 million in 2004 (p. 49). Using these procedures on each size class measured by acreage also resulted in a projection of 1.61 million farms in 2004; using sales classes resulted in a projection of 1.65 million farms (Table 22.1 and 22.2). When the sales were adjusted to reflect 1964 price levels, the projected number of farms was reduced to 1.5 million in 2004.

Programming Models

Sonka and Heady (1974) used a linear programming model of U.S. agriculture to estimate the impact of four farm size scenarios: small, medium, large and mixed sizes. They evaluated the effects of these scenarios on: prices, location, income, labor, food costs, and income generated in the rural nonfarm and agribusiness sectors. They decided that none of the four scenarios were clearly superior. Their analysis showed that small farms would provide greater total net farm income, more agricultural and off-farm employment, and more income generated in off-farm sectors. Large farms were shown to have larger net income per farm, lower commodity prices, lower consumer prices, and some incentives for exports.

Linked Census Data

Structural change in Canada was analyzed by Ehrensaft et al. (1984) using linked census data from 1966 to 1981. Unlike U.S. census data, the Canadian data was "linked," that is, individual farms could be identified in each census. Thus, they were able to follow farms from census to census. They could tell whether each farm stayed the same size, grew or decreased in size, or exited from farming. They could see new entrants and then follow their future paths. Ehrensaft et al. (1984) divided the farms into size classes, not by sales volume but by percentiles formed by ranking all farms in each year. They measured size change by changes in a farm's percentile ranking. Thus, they avoided the problem of deciding whether the growth in size was due to inflation or real growth.

Because they had linked data, Ehrensaft et al. (1984) found a much higher rate of exit and entry than expected by considering just the relatively small net change in farm numbers. Indeed, after reviewing the length of life for new entrants, they commented, "life in the farm sector ... appears to be distinctly Hobbesian: nasty, brutish, and short" (p. 824). The turnover rate was much higher for the smaller classes than for the larger classes. In the midst of this turnover, the upward mobility was "stately rather than bustling" (p. 826). They found very few farms decreasing in size; most farms were either staying the same size or exiting, with a few growing. All rates (exit, entry, and growth) decreased in the larger size classes. Farms which started in the smaller sizes rarely grew into the larger classes. "The most viable path to the upper ranks of the size scale over the 1966-81 period, then, was to start out among the biggest third of the 1956 farm operators in the first place" (p. 827). They found that entry rates varied by commodity and did not conform to preset ideas that supply management hindered entry. When compared to the whole of the U.S. (including the large-farm areas in the Sunbelt), the Canadian changes were slower than the U.S. experience. But they found that the experience in the northern U.S. states was similar to the Canadian experience of change.

Compilation of Methods

Upon comparing their four methods, Lin, Coffman, and Penn (1980, 54) note that all "the projections point to a continuous decline in farm numbers, to about 1.75 million farms by 2000". In testing the projections from 1974 using a Thiel-U test statistic, the Markov processes performed the best. Their most likely projections of farm numbers and size distributions were developed in two steps. First, the total number of farms was estimated based on the acreage distribution using the trend extrapolation and the Markov process. Second, projections by acreage class were computed by multiplying the most likely total number of farms by a synthesized distribution that they obtained from trend extrapolation and Markov process projections (Table 22.1). Also, as part of the second step, projections by sales class were computed by multiplying the most likely total number of farms by a synthesized distribution of farm numbers obtained from Markov process and age cohort analysis (Table 22.2).[2]

After comparing their four methods, Lin, Coffman, and Penn (1980, 10) concluded "farm numbers are likely to decline from 2.87 million in 1974 to 2.32 million in 1985, 2.09 million in 1990, 1.89 million in 1995, and 1.75 million in 2000". Using the new definition of a farm that requires minimum sales of $1,000, they stated that "farm numbers are likely to decline from the 2.37 million in 1978 to 2.05 million in 1985, 1.85 million in 1990, 1.66 million in 1995, and 1.54 million in 2000" (p. 10). (The 1987

Agricultural Census counted 2.09 million farms so they have overestimated the rate of decline.)

Using both acreage and sales class distributions, they projected the trends towards both larger and smaller farms would continue. However, most of the increase in large farms as measured by sales class would be due to "the expected rise in the index of prices received by farmers rather than a rise in the real output per farm" (p. iii). Related to the growth in size is the concentration of farmland ownership, farm production, and farm wealth. They predicted that the percent of sales from farms with sales of $100,000 and up would increase from 53.7 percent in 1974 to 95.8 percent in 2000; the percent share of the largest 50,000 farms was predicted to increase from 31 percent to 63 percent (p. 13). They also expected these trends to continue: (1) farm operators renting more of their farmland and (2) the increasing importance of contractual arrangements between farmers and food processors. Their estimates of changes in financial structure may have been biased by the upturn in asset values in the 1970s but not affected by the yet unseen declines in the 1980s.

Forces Affecting Structural Change

Reimund, Martin, and Moore (1981) evaluated the conditions, forces, and processes of structural change in the broiler, fed cattle, and processing vegetable subsectors since World War II. Out of this analysis, they concluded that forces or factors outside farming trigger structural change to exploit new or changed conditions, and then to manage new risks. Their list of external forces or factors included: new mechanical, biological, or organizational technology; shifting market forces and demand; and new Government policies and programs. These observations were developed into what they call a preliminary model of the agricultural change process. They divided this process into four parts or stages. First, new technology is adopted by the innovators who are often new entrants. The innovators include input suppliers, processors, and distributors, as well as farmers. Second, there is a shift in location of production to new areas whose resources are more amenable to the new technologies. Third, growth and development occurs with increases in total production, firm size, specialization, and concentration. Fourth, the industry adjusts to new risks through the increased use of forward sales, production contracts, and other forms of increased coordination between suppliers, producers, and processors. More of the control over the product shifts from the farmer to stages closer to the consumer. The subsector becomes more industrialized.

On the basis of their work, Reimund, Martin, and Moore (1981, 65) concluded "it would be difficult if not impossible to control structural change in agriculture solely through manipulation of existing policy variables" because policy variables exert their influence primarily through interaction with technological development and market forces. They do add that "policies could, however, provide a basis for influencing the structural basis through their impacts on such structural dimensions as adoption of technology, geographic relocation, and producer risks" (p.65).

In another comprehensive study, Tweeten (1984) examined the causes of structural change and made several conclusions. Technological change has resulted in the need to increase farm size to maintain a given labor-management return (p. 23). National economic growth increases the needed size of farms to keep up with nonfarm earnings (p. 25). Farmers' off-farm income had the opposite effect on farm size; because, as off-farm income increases, farm size could actually decrease (p. 25). Concerning the level and variability of prices, "no compelling case can be made for instability providing an overwhelming competitive edge for any particular organization [i.e., size] of farms" (p. 28). However, middle-sized farms may be disadvantaged by having neither the substantial off-farm income of smaller farms nor the sources of capital and risk management strategies of larger farms. The "concentrated agribusiness structure may contribute to the trend toward fewer, larger farms" (p. 31) as a response of countervailing market power. However, the effect is not well understood and "probably much overshadowed by other forces" (p. 31). Different farm commodity programs have opposite effects and "have been neutral in their overall impact" on the whole farming industry but not necessarily for farms producing program commodities (p. 33). Favorable monetary-fiscal policy (which promotes steady economic progress without sizeable inflation) may decrease the number of farms while unfavorable policy (which reduces aggregate demand and creates high real interest rates, inflation, and instability) may help current farmers but hurt the entry of beginning farmers and speed the separation of land ownership and operation (p. 33-34). Even though FmHA does loan to larger farms, their lending does not encourage the trend toward fewer, larger farms and may help increase the number of rural residences (p. 36). The effect of federal tax policies has been detrimental to the family farm, but their impact on the trend toward larger, fewer farms is overshadowed by other forces (p. 72). Public research and extension has slowed the trend toward larger farms by giving smaller farms increased access to new technology (which is usually scale neutral from public institutions) and to market information and analysis which larger farms could purchase.

Current Projections of Structural Change

The structure of American agriculture in future decades will necessarily be influenced by its current structure and the way it has evolved through time. One approach to making projections of likely change to the year 2000 is to extend trends based on changes over recent time spans. An examination of changes in the distributions of farm numbers by standard size classes using census data makes use of a widely recognized data base.

One of the problems when comparing changes in size distributions based on sales through time is that both price levels and technologies change. Both changes push farms into larger size classes even if there is no "real" change in size. For example, the size class, $10,000-20,000 of sales, in 1969 is roughly equivalent to $20,000-40,000 of sales in 1978 because prices approximately doubled in that period (Table 22.3). Thus the size of farms appears to have grown even though the farms may have not changed physically. Stanton (this volume) discusses the impact of inflation in more detail. Also, any adjustment for inflation by itself does not take into account how much more output one worker produced because of changes in technology. Technological changes are important to consider because they affect what size of farm one person or family can manage or, in other terms, the minimum size that can support one person or family. This change can affect the public's perception of whether changes in size are important or not.

For an initial comparison, unadjusted data for the three most recent census periods, when price changes were relatively small, are summarized in Table 22.3 along with the 1969 size distribution adjusted to 1978 prices. Observations with nominal sales of $500 or less in 1969 were omitted because of the change in the definition of a farm. Between 1969 and 1987, there was a reduction of 297,000 farms or 12.5 percent of the 1969 total. Fifty percent of the total change between 1969 and 1987 occurred between the 1982 and 1987 censuses. While changes in farm numbers between 1969 and 1987 are important, they are small relative to the changes in the previous 20 years. From 1950 to 1969, farm numbers in the United States had been cut in half, from 5.4 to 2.7 million, with no change in the definition of a farm.

Other important points can be made when looking at the recent changes between 1969 and 1987.

1. The group of residential and part-time farms with sales of less than $20,000 accounted for about 60 percent of the total number in each census period, when 1969 is adjusted to a 1978 base.

TABLE 22.3 Farm Numbers by Sales Class (Census, United States, 1969, 1978, 1982, 1987)

Sales Class	1969 in 1978 Prices		1978		1982		1987	
Index of prices received by farmers, 1977=100	(59) 118		115		133		126	
				Number of Farms				
Under $5,000	686,176		761,234		814,535		753,214	
$ 5,000 - 19,999	748,347		613,303		540,809		525,566	
Subtotal	1,434,523	(60.2%)	1,374,537	(60.9%)	1,355,344	(60.5%)	1,278,788	(61.3%)
$ 20,000 - 39,999	395,472	(16.6%)	299,175	(13.3%)	248,825	(11.1%)	225,671	(10.8%)
40,000 - 99,999	396,697	(16.6%)	360,093	(16.0%)	332,751	(14.9%)	287,587	(13.8%)
$100,000 - 249,999	103,990		165,493		215,912		202,550	
250,000 - 499,999	40,460		38,202		58,668		61,148	
500,000 and over	11,535		17,973		27,800		32,023	
Subtotal	155,985	(6.6%)	221,668	(9.8%)	302,380	(13.5%)	295,721	(14.2%)
Total	2,384,788†		2,255,473*		2,239,300*		2,087,759	

*Totals adjusted downward by unclassified abnormal farms.
†Reduced from 2,730,250 to account for all farms with sales of $500 or less in 1969 because of definition change.

2. Reductions in numbers occurred mainly among the part-time and small commercial farms with sales of $20,000-39,999 and $40,000-99,999.
3. The number of farms with sales of $100,000 or more increased with the rates of increase less between 1982 and 1987 than in earlier years.

Adjustments of 1978 Census Data to a 1987 Base

While the changes in prices received by farmers between 1978 and 1987 are modest compared to the doubling of prices between 1969 and 1978, it is important to recognize these differences as well and their effects on the distributions. Farm prices rose modestly between 1978 and 1982, by 15.65 percent; they fell on average between 1982 and 1987 by 5.3 percent. The net change over the nine years was 9.1 percent. To examine the change in the size distribution between 1978 and 1987, adjustments were made in the 1978 distribution to reflect 1987 prices. That is, the 1978 class intervals were increased as if 1987 prices had been effective and the number of farms in each class was adjusted assuming a uniform distribution within the unadjusted class (Table 22.4). Numbers in each size class were rounded to the nearest thousand to facilitate comparisons.

Reductions in numbers between the adjusted-1978 and 1987 censuses were largest in the part-time and small commercial farm groups with sales of $20,000-39,999 and $40,000-99,999. The smallest size class (sales of less than $20,000) includes 60 percent of farm numbers; had the smallest percentage reduction; but accounted for about one-third of the absolute decrease in the number of farms. In contrast, numbers increased in each of the three largest classes; half of these were in the $100,000-249,999 size group. Thus, even after adjusting for inflation, we see a decrease in the number of smaller farms and an increase in the number of larger farms. The much-discussed trend towards a bipolar distribution can be seen in the smaller percentage decline in the smallest size class, the large increase in the largest size classes, and the relatively large decreases in the middle although the absolute numbers show the middle classes continuing to dominate the commercial farms.

Value of Sales by Size Class

While it is both interesting and instructive to examine changes in the numbers of farms, it is also important to recognize what is implied by changes in each size class's share of total farm production. The changes in the unadjusted shares of total value of production that have occurred between each of the three most recent census periods is more striking than the changes in numbers (Table 22.5). In 1978, output from all farms

TABLE 22.4 Number of Farms by Sales Class (1987 Base, United States, 1978 and 1987 Census)

Sales Class	Actual 1978	Net Adjustments for Prices 9.1%	1978 Census on 1987 Base	Actual 1987	Percentage Change
			Number of Farms		
Residential and Part-time:					
Under $20,000	1,375,000	-27,000	1,348,000	1,279,000	-5.4
$ 20,000 - 39,999	299,000	-3,000	296,000	226,000	-31.0
Small Commercial:					
$ 40,000 - 99,999	360,000	-5,000	355,000	288,000	-23.3
Large Commercial:					
$100,000 - 249,999	165,000	17,000	182,000	202,000	+11.0
250,000 - 499,999	38,000	13,000	51,000	61,000	+19.6
500,000 and over	18,000	+5,000	23,000	32,000	+39.1
Total	2,255,000*		2,255,000	2,088,000*	-9.1

*Reduced by number of unclassified "abnormal" farms.

TABLE 22.5 Total Value of Sales Unadjusted, by Size Class (Census, United States, 1978, 1982, 1987)

Sales Class	1978		1982		1987	
			Millions			
Under $20,000	$ 8,181		$ 7,260		$ 6,967	
$ 20,000 - 39,999	8,599		7,142		6,448	
Subtotal	$ 16,780	(15.7%)	$ 14,402	(10.9%)	$ 13,415	(9.9%)
$ 40,000 - 99,999	$ 22,869	(21.4%)	$ 21,642	(16.4%)	$ 18,764	(13.8%)
$100,000 - 249,999	$ 24,772		$32,930		$31,178	
250,000 - 499,999	12,848		19,851		20,740	
500,000 and over	29,559		42,764		51,952	
Subtotal	$ 67,179	(62.9%)	$95,545	(72.7%)	$103,870	(76.3%)
Total	$106,828*		$131,589*		$136,049	

*Reduced by aggregate sales from "abnormal" farms.

with sales of less than $40,000 accounted for a larger share of the total than that produced by farms with sales of $250,000-499,999: 15.7 percent compared to 12.0 percent. The three classes for the largest farms made up 62.9 percent of the total in 1978; those with sales of $500,000 or more made up 27.7 percent. In contrast, by 1987 the smallest farms (less than $40,000 of sales) accounted for about 10 percent of the total. The three largest classes were now 76.3 percent and those with sales of $500,000 or more, 38.2 percent. Thus, a larger share of a larger total was produced by the largest farms in 1987. The small commercial farms (sales of $40,000-99,999), still a very important part of the numbers of units where farm income was the primary source of family income, had decreased to only 13.8 percent of the total.

Projections of Farm Numbers Based on Trend 1978-1987

In this set of projections to the year 2000, recent percentage changes in trend for individual size classes were used after correction for price level changes rather than Markov processes or a more complete simulation model. Here, we assume the rate of change over recent years will be repeated from 1987 to 2000. This assumption implies a linear rate of change. Hence, if the rate of change is indeed slowing down, the linear rate will overestimate both decreases and increases. The potential overestimation means these projections may be overstating the decline in farm numbers. To compensate for this potential overestimation, a series of different scenarios for these projections, which reflect the effects of a continuation of different percentage changes within size classes, were considered more appropriate to reflect future rates of change in prices and technology than a less heuristic methodology (Just and Rausser 1989). These are discussed later in this chapter.

The first set of projections was based on the percentage changes, which occurred in each of the six size classes, for the period 1978-87, after correction of the 1978 data to a 1987 base (Table 22.4). Using the adjusted-1978 data as a base, the average annual percentage change between 1978 and 1987 was extended to 2000 for each class and the total. When these percentages were used to make a first estimate of farm numbers, the sum over the six classes was greater than the projected total. To correct this imbalance the numbers in each class were adjusted to reflect the projected total. This adjustment was done simply by calculating each class's percentage of the class sum in the first estimate and using that percentage to allocate the projected total. The projected percentage change between 1987 and 2000 was recalculated for each class using these adjusted projections. Since this projection is an approximation to facilitate discussion, the final percentages are set as integers (with some small adjustments so that the class sum equals the total) and farm numbers are rounded to thousands.

Using the steps just described and the 1978 to 1987 trend, the total number of farms is projected to decline by 11 percent to 1.9 million farms in 2000 (Table 22.6). The loss in farm numbers in the three smallest size classes is continued. In the group of farms with sales less than $20,000 (1987 prices), the reduction is 125,000 or 10 percent of the 1987 number. Most of this decrease is most likely for those units with sales between $10,000 and $20,000 based on historical experience. Similarly, there is an important reduction of 81,000 part-time farms with sales between $20,000 and $40,000. Some of these go out of business while others advance into the next largest category. Further, there is a large decrease of 30 percent or 86,000 farms in the small commercial category. This total is more than the total increase for the next three size categories. Thus, this projection implies that an important part of the exits in the small commercial category are net entrants into larger classes.

The three larger size groups increase by 14, 27, and 55 percent, respectively. In total, these increases are modest; 62,000 farms are added to the totals. The net decrease in numbers considering all six classes is 230,000 farms; 11 percent of the total. In reviewing these results, it is well to remember that from 1978 to 1982 farm numbers in the U.S. decreased by only 17,000. Most of the fallout in numbers occurred between 1982 and 1987. Thus, this projection includes both a period when expectations about the future were good and farm land prices were rising, as well as the tough years when land prices fell dramatically.

Projections of Farm Numbers Based on 1982-1987 Trend

Another projection uses the rate of change in the most recent five-year census period: 1982-1987. This is the period when the farm debt crisis was at its peak and when many considered the rate of exodus from commercial farming to be higher than normal. One then could consider projections on this kind of basis to emphasize a continuing rapid rate of exodus and change for the future. For this projection, the same procedures are used as for the 1978-87 trend except the base year is 1982.

Based on the 1982-87 trend, the total number of farms is projected to decline by 18 percent to 1.7 million farms in 2000 (Table 22.7). Four of the size intervals show decreases in numbers. There is a small increase of 5,000 farms in the next to largest class and an increase of 12,000 in the largest. The net decrease in farm numbers projected is 376,000. Again, the bulk of the decrease is observed in the two smallest classes: 253,000 or approximately two-thirds of the numbers. But there is a decrease of 104,000 in farms selling $40,000 to $99,999 and 36,000 in the $100,000 to

TABLE 22.6   Projected* Number of Farms by Sales Class (1987 Prices, United States, 1987 and 2000)

Sales Class	Actual Distribution 1987	Trend Percentage Change (1978-87)	Projection for 2000
		Number of Farms	
Residential and part-time:			
Under $20,000	1,279,000	-10%	1,154,000
$20,000 - 39,999	226,000	-36%	145,000
Small commercial:			
$40,000 - 99,999	288,000	-30%	202,000
Large commercial:			
$100,000 - 249,999	202,000	+14%	230,000
250,000 - 499,999	61,000	+27%	77,000
500,000 and over	32,000	+55%	50,000
Total	2,088,000	-11%	1,858,000

*Projection based on annual rates of change between 1978 and 1987.

TABLE 22.7 Projected* Number of Farms by Sales Class (1987 Prices, United States, 1987 and 2000)

Sales Class	Actual Distribution 1987	Trend Percentage Change (1982-87)	Projection for 2000
	Number of Farms		
Residential and part-time:			
Under $20,000	1,279,000	-15%	1,083,000
$20,000 - 39,999	226,000	-25%	169,000
Small commercial:			
$40,000 - 99,999	288,000	-36%	184,000
Large commercial:			
$100,000 - 249,999	202,000	-18%	166,000
250,000 - 499,999	61,000	+9%	66,000
500,000 and over	32,000	+38%	44,000
Total	2,088,000	-18%	1,712,000

*Projection based on annual rates of change between 1982 and 1987.

$250,000 size class. Since the increase in the two largest classes is much less than the decreases in other classes, this projection implies a much higher rate of net exits from farming. This projection is, remember, based on the trend during the period of financial crisis and would be expected to show a higher rate.

Projections of Farm Numbers
Compared to OTA Projections

A summary of these two trend-based projections are compared with OTA's 1986 estimates using Markov processes (Table 22.8). The OTA projections did not adjust for changes in prices over this period; hence, the numbers assume similar rates of changes in prices in succeeding years. The projections made in Tables 22.6 and 22.7 are based on 1987 prices. One of the difficulties in making these comparisons is the slightly different class intervals used by OTA and the need to arbitrarily divide those projections into the current sales categories established by the Office of Management and Budget (OMB) for such distributions.

The most striking difference among the projections is that reported for farm numbers with sales of less than $20,000. Actual reductions in numbers between 1978 and 1987 were less than 100,000 as reported by the census (Table 22.3). It is unlikely that the exodus projected in residential and small, part-time units will be as large as that suggested in the OTA report. Tweeten (1984) points out that, even though small farms headed by full-time operators have declined in numbers, the number of part-time operators on the small farms has increased substantially. This would explain why the rate of exodus has slowed. The difference in the projections for this class alone accounts for over half of the total change in farm numbers in the two projections made in this chapter.

A comparison of the projections for the two classes with sales between $20,000 and $100,000 is interesting as well. The combined totals of the two classes are similar in each of the three projections, falling in a narrow range, from 347,000 to 363,000 farms. All three projections suggest a decrease of more than 150,000 farms from the 1987 level, continuing the reductions that occurred in the 1970s and 1980s.

The two projections in this chapter are similar to those by Lin, Coffman, and Penn (1980) in total but not in the distribution across sizes (see Table 22.2). Lin, Coffman, and Penn (1980) project many more farms moving into the largest size classes. This difference is due to the data base used to make the projections. The rate of change has slowed in recent years compared to earlier years.

TABLE 22.8 Alternative Projections of Size Distribution of Farms (United States, Year 2000)

Size Class	1987 Census	Trend 1978-87 Change	Trend 1982-87 Change	OTA Projection 1982 Base
Under $20,000	1,279,000	1,154,000	1,083,000	638,000
$20,000 - 39,999	226,000	145,000	169,000	363,000
40,000 - 99,999	288,000	202,000	184,000	
$100,000 - 249,999	202,000	230,000	166,000	100,000*
250,000 - 499,999	61,000	77,000	66,000	100,000*
500,000 and over	32,000	50,000	44,000	50,000
Total	2,088,000	1,858,000	1,712,000	1,251,000

*OTA used $100,000-199,999 and $200,000-499,999 as the size classifications with 75,000 and 125,000 farms, respectively; proportionately these were reallocated to 100,000 farms in each of the Census classes.

Comparisons for Farms with $40,000 of Sales or More

Much of the interest in changes in the structure of agriculture is centered on what is happening in the commercial sector ($40,000 of sales or more), where more than 90 percent of total sales were obtained in 1987 (Table 22.5). One of the areas of greatest interest has been rates of change in the numbers of "small, family" farms, often associated with those selling between $40,000 and $100,000 annually. In 1987, this class accounted for nearly half of all commercial farms, while in 1978, corrected to a 1987 price base, it amounted to 58 percent of the commercial total (Table 22.4).

The projections based on the 1978-87 trend have the smallest deviations from the 1987 census distributions (Table 22.9). A small change (i.e., 4 percent) in the total is projected. Much of the decrease in the smallest commercial class is seen moving into the next largest class, $100,000-249,999. There are important increases in each of the larger size categories. The number of farms with sales of $1,000,000 and over is projected to increase to 17,000 by 2000. This is a relatively small number of farms, but this class accounted for 30.9 percent of all sales in 1987--the first year this class was reported separately (Table 22.10).

The projections based on trends between 1982-87 show a larger decrease in the number of commercial farms, 21 percent. The larger size classes increase from 1987, but not as much as projected with 1978-87 trends. Most of the decreases in the two smaller sizes are not due to the farms moving into the larger sizes.

The 1978-87 projections are more nearly like those developed by OTA than those based on 1982-87. The numbers in the smallest commercial class and the two largest classes are essentially the same. OTA projects a much larger proportion moving into the $250,000-499,999 class than does either the trend between 1978-82 or 1982-87. This difference can be explained in large part by the fact that the OTA projections did not account for price changes during this period. All these projections, however, do point to the increasing importance of larger farms and a decrease in the relative importance of the middle size classes.

Projection Assuming More Rapid Reductions

Another projection was developed on the assumption that very rapid reductions in farm numbers might come about as a result of changes in technology, relatively low prices for agricultural products, and agricultural supply outrunning effective demand. The distribution of expected increases and decreases for each size class reflects changes observed in the recent past and are projected using the same procedures as before.

TABLE 22.9 Alternative Projections of Size Distribution, Commercial Farms (1987 Base, United States, Year 2000)

Size Class 1987 Prices	1987 Census Base	1978-87 Trend Base	1982-87 Trend Base	Adjusted OTA Projection 1982	25% Reduction in Numbers
			Number of Farms		
$ 40,000 - 99,999	288,000	202,000	184,000	203,000	167,000
100,000 - 249,999	202,000	230,000	166,000	100,000	160,000
250,000 - 499,999	61,000	77,000	66,000	100,000	66,000
500,000 - 999,999	21,000	33,000	29,000	33,000	29,000
$1,000,000 and over	11,000	17,000	15,000	17,000	15,000
Total	583,000	559,000	460,000	453,000	437,000

TABLE 22.10   Total Value of Sales and Proportions of Total by Size Class (1987 Base, Projections for United States, Year 2000 (1987 Prices))

Size Class (using 1987 Prices) and Class Averages	Actual 1987 Census Base	Projection Based on 1978-87 Trend	Projection Based on 1982-87 Trend	Adjusted* OTA Projection 1982	25% Reduction in Numbers
			Percent of Total Sales		
$ 40,000 - 99,999 (65,000)	15.3	8.5	9.2	9.2	8.4
100,000 - 249,999 (155,000)	25.4	23.0	19.7	10.9	19.3
250,000 - 499,999 (340,000)	16.9	16.9	17.2	23.8	17.4
500,000 - 999,999 (675,000)	11.5	14.4	15.0	15.6	15.2
1,000,000 and over (3,400,000)	30.9	37.3	39.0	40.5	39.6
Total sales, billions	$122.7	$155.0	$130.7	$142.8	$128.7
Total farms	583,000	559,000	460,000	453,000	437,000

* Adjustments made to split classes proportionately from OTA distributions.

A reduction of 25 percent in total commercial operations in a span of 13 years is much greater than the annual rates experienced between 1982 and 1987. This could be looked at as a "worst case scenario" of rapid exodus from agriculture. About 163,000 farms are forced out of the two smallest classes so that 17,000 can be added to numbers in the three largest size classes. The three largest classes have the same number of farms as those based on the 1982-87 trend. The brunt of this rapid loss is taken by the smallest farms.

Total Value of Sales
Associated with Different Projections

One additional way to examine what is implied by different projections is to study total sales for each class. The assumption of average sales for each class is basic to this process. Because the projections use Census data, the total sales per class was calculated from the average sales per farm for each class in 1987 (Table 22.10). The percentage distribution of total sales from all farms with sales of $40,000 and over in 1987 is provided as a basis for comparison.

For farms with sales of $40,000-99,999, each of the projections indicates about 8 or 9 percent of total sales will come from this group − down from 15 percent in 1987. In all cases at least 35 percent of "commercial" farm numbers remain in this class. There is much greater variation in the percentages attributed to farms with $100,000-249,999 of sales. In all cases, more of total sales comes from this group than from the smallest commercial class. In all of the projections, 65 to 75 percent of the total number of commercial farms are included in these two classes, but they account for only 20 to 35 percent of total sales.

The midpoint of the three largest classes, particularly that for farms with $1,000,000 or more of sales, has an important effect on the percentages. The 1987 census had an average of $3.4 million per farm in the largest class. No doubt, those units moving into this class from the $500,000-999,999 class would initially help to move that average down if the number moving in were large. On the other hand, the continuing units also have the capacity and management to continue to increase sales. The projections all suggest increasing proportions of total sales to come from this relatively small number of farms, from 15,000 to 17,000 in number.

The amount of historical data available concerning this class of largest farms and its changes through time is meager at best. The census provided summary data nationally for the first time in 1987. Also for the first time, USDA has published information on this class in its 1990 annual estimates of economic indicators, but only at a national level (USDA). It is an important class in terms of the future structure of

614

American agriculture. Some special tabulations from past censuses on an aggregate basis would be useful in trying to improve estimates of change in this increasingly important component of U.S. agricultural production.

Farms with $500,000 or More of Sales by State, 1987

Information about the distribution of farms with sales of $500,000 or more by state was shown in the 1987 census but not for farms with sales of $1,000,000 or more. These large farms are widely distributed across the United States; however, 36 percent of the farms and 49 percent of total sales are in five states (Table 22.11). If one divides the number of these farms in each state into the total value of sales, it is clear that an important number of farms with $1,000,000 or more of sales also are located in Colorado and Arizona as well from among the next five states. The top ten states include 51 percent of the farms and nearly 64 percent of sales. In contrast, the 35 states with the smaller numbers of large farms make up 35 percent of the total number and 27 percent of sales for this size class. The largest farms in the United States are in the irrigated West, the Great Plains, the Western Corn Belt and parts of the South. An important part of future increases in the largest farms are likely to occur in these locations, although the adoption of new technology may well change the incidence of change in some agricultural sectors.

Influence of Major Forces
on Projections of Structural Change

Most quantitative projections of farm numbers and total output by size classes assume the trends of the recent past will continue into the future. Most also recognize that this simple assumption is not likely to hold true. Thus, some additional qualitative statements may be helpful in interpreting the direction of change that is most likely when past trends are interrupted by major forces affecting production agriculture. These include: (1) economic growth or recession in the nation's economy; (2) changes in agricultural policy that increase or decrease the amount of government intervention either in the United States or internationally; (3) a speed-up or slow-down in the availability of new agricultural technology and its associated potential for increasing or decreasing productivity; (4) public concerns and regulation associated with conservation of natural resources, the environment and food safety; and (5) international turmoil or disturbance associated with war, political instability, or major food shortages. An historical view of forces affecting farm structure can be seen in the early chapters by Stanton, Rasmussen, and Hornbaker (this volume).

TABLE 22.11 Farms With $500,000 or More of Sales (States by Rank, Census, 1987)

State	Number of Farms		Value of Sales	Percent of Total
			Millions	
1. California	5,641		$10,313.8	19.9
2. Texas	2,142		5,573.4	10.7
3. Kansas	957		3,533.8	6.8
4. Florida	1,455		3,214.3	6.2
5. Nebraska	1,279		2,829.2	5.4
Subtotal	11,474	(35.8%)	$25,464.5	49.0
6. Colorado	688		$ 1,932.3	3.7
7. Iowa	1,630		1,643.4	3.2
8. Washington	962		1,460.7	2.8
9. North Carolina	1,084		1,287.8	2.5
10. Arizona	555		1,271.0	2.4
Subtotal, 10	16,393	(51.2%)	$33,059.7	63.6
11. Idaho	662		$ 1,077.1	2.1
12. Arkansas	903		1,005.0	1.9
13. Georgia	935		993.0	1.9
14. Illinois	1,059		978.0	1.9
15. Minnesota	860		952.1	1.8
Subtotal, 15	20,812	(65.0%)	$38,064.9	73.2
Other 35 states	11,211	(35.0%)	13,887.4	26.8
Total, United States	32,023		$51,952.3	100.0

Economic Growth or Recession

During much of the 1970s and 1980s, the national economy has grown rather steadily at modest rates, between one and five percent of GNP per year. There have been relatively short periods when growth was negative for a few quarters but the turn around was achieved quite rapidly and the economy continued to expand. The health of the general economy will affect structural change in agriculture because agriculture's health is directly affected by what occurs in the rest of the economy (Schultz 1945). When the rest of the economy is growing rapidly, opportunities for employment outside agriculture are strong; expectations for the future are good; and productivity in the input supply and processing sectors is rising. Rates of structural adjustment within agriculture are more likely to keep pace with the general economy. Specifically, marginal producers can find attractive employment outside agriculture--as they did in large numbers during the growth period of the 1950s and 1960s. Thus, if the general economy is growing faster than in recent years, some acceleration toward fewer, larger farms is likely.

In contrast, during periods of recession, structural change should be expected to slow. Alternative employment opportunities are less available. Productivity gains in the industries serving agriculture are less available. Aggregate demand will be down; new investment in agriculture will slow. One should expect that recession in the general economy will be reflected in the farm economy as well, and the struggle to survive will be central to farm businesses of all sizes. Hence, the trend towards fewer, larger farms would slow because fewer farmers will quit farming and fewer farmers will be able and willing to expand their current size.

The effect of the general economy's health on agriculture will be especially evident if the period of growth or downturn is prolonged. If the economy continues to have short cycles of growth and downturn--as in the 1970s and 1980s, structural change can be expected to approximate more closely the trends established in these two decades and which were used to make the projections in the previous section.

Increases or Decreases
in Government Intervention in Agriculture

Efforts to identify the long-term effects of government intervention on structural change in agriculture has been investigated by several agricultural economists (Gardner 1978; Sumner 1985; Tweeten 1984, this volume); Helmers (this volume)). Different effects are found depending on the programs in force and the commodity groups to which they are applied. Caution must be exerted in making generalizations because we cannot know what would have happened in the same commodity groups in the absence of these programs. Nevertheless, the authors suggest that, in general, mandatory supply control programs involving production agriculture tend to slow down the rate of structural change. Hence, a move away from government programs and controls toward free market conditions will speed up structural change, if other conditions in the economy encourage it or at least do not discourage it.

The atmosphere and public attitudes toward government intervention are important both here and overseas. The GATT negotiations on agricultural issues reflect a strong presence of protectionism in Japan, the European Communities (EC), and other developed countries. The late 1980s have been marked with several new, domestic quota programs in both North America and Western Europe. Associated with them are restrictions to exit and entry, restrictions to movement of production across political boundaries, and new costs of production associated with the capital value of the production quotas themselves (Nott and Doyle 1988; Schaub 1990).

If the nation moves increasingly toward a free market environment for storable commodities with essentially no restrictions on production, areas planted, and little or no price support activity, it will have to occur in an international environment where similar steps are being taken by other major exporting countries. It is unlikely that the United States will take such action alone without important steps to open markets in the EC and Japan. In such a free-market setting, based on the work cited earlier, structural change can be expected to move forward somewhat more rapidly than recent trends. Likewise, if protectionism grows internationally and barriers to agricultural trade continue to rise, then a slow-down in changes in farm size for units producing storable commodities seems likely to follow.

Availability of New Technology

The greatest impetus for structural change in production agriculture during the 20th century has been the availability and subsequent adoption of new technology (Kislev and Peterson 1982). Almost all of the projections for continued structural change assume that a steady stream of new, cost-reducing technologies will flow rather steadily into American and world production systems. Batte and Johnson (this volume) discuss emerging technologies and the impact on agriculture. If the promise of important new advances associated with applications of biotechnology are realized, then increases in productivity and cost reductions per unit of output can be expected to occur for many commodities at various locations around the world. A smaller number of larger, more specialized producing units is a likely consequence in this country. Past trends should continue or accelerate for farms with sales of $40,000 or more.

If the promised advances from biotechnology are found to be more elusive and advances from new technology begin to slow perceptibly, a slow down in the rate of structural change can be expected as well. A combination of public concerns about agriculture's impacts on the environment, required reductions in the use of pesticides and fertilizers, and fewer yield increasing technologies might alter both the rate and nature of structural change. Farming might require more management time per unit of output which would decrease the competitiveness of both larger and smaller, part-time farms. The competitiveness of small, full-time family farms might increase. Lack of cost-reducing technology in this environment could slow expected change.

The discussion of new technology should not be limited to physical technology. Advances in management technology can be just as important. The chapters by Barry, Robison, and Janssen (this volume) point out several impacts. Technology changes in related sectors as discussed by Barry and Hudson (this volume) can also affect the farming sector. As with biotechnology, the speed at which other technology changes will affect the speed of structural change on farms.

Public Concerns About the Environment

Public awareness and concerns about the environment in which we live have increased (Atwood and Hallam, this volume). Key concerns center on: (1) contaminants in both surface and ground water supplies; (2) conservation of our scarce natural resources, especially land and forests; and (3) air pollution, acid rain and global warming (Batie 1989). Almost coincidentally has come greater demands for food safety. While it is easy to recognize these public concerns, it is less easy to assess their future effects on numbers and sizes of farms.

Increased state and federal regulation of the use of pesticides and fertilizers seems inevitable particularly where agricultural production occurs near large numbers of people or the resources (e.g., water supply) which they use. Higher standards of management will be required and, in some cases, more labor per unit of output may be needed to comply with the new regulations. Under such circumstances, national standards (not individual state standards) must be established to ensure fair competition within the country. International competitiveness will also be a concern. The net effects on farm structure will depend on the extent and nature of regulation and farmers' abilities to adapt to this production setting. Requirements for increased technical management and capital investment to meet national standards gives advantage to large specialized operations. The only exception to this advantage will be requirements placed only on farms larger than a minimum size. In these cases, the smaller operations will not incur the costs of meeting the requirements and will retain some competitiveness at the expense of not meeting environmental goals.

Low input, sustainable agriculture (LISA) gives the illusion that small, family farms will have an increased advantage if appropriate technology and the associated practices are developed. Much of the impact of LISA depends on the management systems and markets that evolve with these efforts. While sustainability is a worthy objective, the question remains about what size and type of production units will be best prepared to accomplish this goal. It may well differ by commodity, natural resource situation, and the markets available. Improvements in information management technologies and knowledge of environmental impacts may well remove the advantages of smaller operations that, now, may be able to monitor their physical environment better than very large operations. High capital and management requirements will inevitably favor larger, more specialized operations despite the intent of many who see this as a possible return towards "family farming."

International Turmoil or World Food Shortages

In a period when excess capacity and surplus grain production has been a primary concern of public policy for agriculture in North America, the possibility of food shortages and political instability may seem an unlikely consideration. Yet the balance between shortage and surplus is surprisingly small given world populations and the potential for disaster arising from weather or political upheaval. In a disaster setting, food policy becomes an imperative; price variability would increase; and expectations of increased profitability from agricultural production would rise.

In the short run, instability and higher grain prices would slow down exits from farming. Land would be seen as a safe harbor for investment. Longer run adjustments are much more difficult to assess. Historically, period of war and turmoil have led to important adjustments and changes in society. The after effects of the two World Wars in the twentieth century are clear reminders of this reality. Perhaps the most important point to recognize is that United States agriculture is an integral part of the world economy. What happens outside our borders will affect our economy and its evolving agricultural system. The structure of agriculture cannot escape such effects.

The Survival of the Family Farm -- A Disappearing Issue

The survival of the family farm has been addressed either directly or indirectly in much of the literature concerning farm size and the structure of American agriculture (e.g., Brewster 1979; Economics and Statistics Service 1979; Bergland 1981; Belden et al. 1980). The concept of the family farm is still used by groups trying to shape policy (e.g., Nodland 1990). However, the authors of this chapter increasingly believe the family farm is becoming a disappearing issue for political, social, and economic reasons. This is not to say that the emotion that surrounds the concept of the family farm has disappeared; it hasn't. Neither is this to say that many farms are not under economic stress; many are. Nor is it to say that the decline in the number of medium-sized farms did not and will not happen; it did and will. Indeed, the authors believe that the medium-sized farms which account for most family farms will continue to be a large and important part of U.S. agriculture. Rather, the authors believe the effectiveness of the family farm ideal in garnering interest and political support is diminishing. Let us explain the reasons we believe this to be happening.

Politically and socially, the concept of the family farm has been powerful because of the public's perception of the family farm, our past as a nation of farmers, and our own close ties to farmers. This power of the family farm can be seen in the statement regularly examined and reaffirmed in recent years by the Congress of the United States in its omnibus agricultural legislation:

> Congress hereby specifically reaffirms the historical policy of the United States to foster and encourage the family farm system of agriculture in this country. Congress firmly believes that the maintenance of the family farm system of agriculture is essential to the well-being of the Nation and the competitive production of adequate supplies of food and fiber. Congress further believes that any significant expansion of nonfamily owned large-scale corporate farming enterprises will be detrimental to the national welfare. (95th Congress)

Today, there is increasing recognition both within and outside agriculture that the mythology surrounding the family farm has lost some of its allure. The image of the hard-working, God-fearing, independent farmer taking risks against weather, disease, and uncertain prices is now being replaced with an image of a person on the dole from the taxpayer relying on guaranteed target prices, deficiency payments and disaster payments for income rather than the sweat of his/her brow and an uncertain market (e.g., *The New York Times,* June 19 and July 23, 1990; Berg 1990; Clift 1990; *Newsweek* 1990). While the popular image of farmers is still generally positive, more people recognize that the continuum of farmers is as wide and diverse as in many other industries.

More people today are likely to recognize the lack of connection between reality and idealism in Vogeler's (1981) concluding statements:

> Dialectically, family farmers can turn the very instruments of their oppression and extinction into their liberation. The *myth* of the family farm can be destroyed through self-education and group action. Through the democratic process, progressive farmer-labor alliances can turn the ideals of the family farm into reality and turn agribusiness into a vanishing species.... Working together, consumers and farmers can have both lower food costs and higher agricultural prices if they share even a fraction of the immense profits made by the food industry. Such sharing would lead to greater economic justice and greater equality of income distribution. (pp. 295-6)

While Vogeler's (1981) ideas may sound wonderful to a segment of society, the realities of forming these alliances, of making the connections to share profits and lower food costs are much more difficult than implied. Consumers in New York City, Los Angeles, and other large and small communities have shown by their purchasing, that they are not ready for these changes. In contrast, Paarlberg's (1980) statement on the family farm issue are more credible, not only to economists but to increasing numbers of the public.

> Agriculture need not be, nor is it likely to become, monolithic. We are a pluralistic nation, socially, politically, and economically. That the trend has been in the direction of large-scale units does not mean it will automatically extend itself until it embraces all of agriculture. Nor does it mean that large-scale farming should be abolished.... Why try to obliterate all these differences and homogenize this heritage? Perhaps our present mix of large farms, small farms, and part-time farms has considerable justification. Those who believe in market competition should also believe in the appropriateness of competing institutional forms. (pp. 202)

The disappearance of the family farm as an issue also can be traced to the official farm definition. As long as our definition of a farm includes any place from which $1,000 or more of agricultural products were produced and sold or normally would have been sold, farm numbers will be large and diverse in terms of size. The minimum of $1,000 is very small in today's agriculture. Thus, most "official" farms will be owned and operated by people who get the bulk of family income from employment outside agriculture. While 90 percent of national agricultural output may come from a slowly declining number of farms, diversity in their size and organization is the likely rule. Such a judgment is based on both historical trends and expectations about returns from investment in agriculture compared to alternative investments.

One other reason that the survival of the family farm is a diminishing issue is the difficulty we have in finding a generally accepted, quantifiable definition of a family farm. Without a quantifiable, acceptable definition, Congress is unable to pass any legislation specifically targeted to only the full-time family farm. Tweeten's (1984) definition of a family farm is as good as any:

> ... a crop and/or livestock producing unit in which the operator and his/her family:
>
> - control most of the decisions;
> - supply most of the labor;
> - derive most of their income from farming;
> - receive family income and rates of returns on resources comparable to those in the nonfarm sector.

Even this definition was not immune from criticism. A farmer on the committee sponsoring the report took issue with this definition arguing that the labor requirement was not realistic. Others might find the need to obtain equal returns on resources to other sectors difficult. Deciding who meets and does not meet this or any other worthy definition is complex. Is a large farm with $1,000,000 of annual sales not a family farm if it is owned and operated by one farm family or some kind of family partnership? Establishing the border lines of any definition soon becomes academic rather than useful. Hence, a general qualitative definition meets whatever needs remain.

Another reason that the family farm and farming in general is losing political support and interest is the fact that fewer people have a direct, or even a remote connection to farming. Most families have never lived on a farm. In those families who once were farmers, many left the farm two or more generations ago. City cousins do not have the same view of farming today as they did a generation ago. This lack of connection causes consumers and voters to see farmers as another group competing for government help and not as a group that should be given priority over themselves.

One final reason why the survival of the family farm can be viewed as a non-issue is economic. Many family farms continue to survive and compete effectively. If one excludes all those units where the family gets most of their living from non-farm sources, the great majority of the remaining farms have been family enterprises throughout the 20th century. While numbers of "commercial" or "full-time" farms declined markedly particularly between 1950 and 1970, those who remained were predominately family farms by almost any definition you might choose. Now, in the last decade of the century, more and more of total production is coming from the 10,000 largest farms. Are they more efficient and profitable per unit of output than their smaller competitors? The answer provided in other chapters in this book is, No. Most of the economies of size in production for most agricultural enterprises are captured by a large, well-managed family business that has exercised its available options. In many cases, these family units dominate production for crops like corn, wheat, soybeans, and horticultural crops in most parts of the country. The same is true for cow-calf operations and dairying. Historical trends and projections of the future show family farms still competing and surviving in the future.

Many argue that the patterns of agriculture found in the irrigated west for crops and livestock will come to dominate the rest of the country in the twenty-first century. If that should occur because of advances in technology or pecuniary economies, the underlying issue should then be about size of business and the potential for large units to exercise some kind of monopoly power, not about who can qualify as a family farmer.

To all intents and purposes, family farms are going to survive in production agriculture at least into the beginnings of the 21st century. The practical debate needs to be about the structure of agriculture and whether or not public policy in fact will have much, if any, significant effect on that structure in the years ahead. Debate also needs to be about the quality of our food supply, of rural communities, and of our use of the environment (as pointed out in other chapters). Survival of family farms is simply not central to these issues, because they will survive.

Concluding Comments

The process of making specific projections of farm numbers by size classes is more useful for thinking about the forces that have led to changes in the recent past and those that are likely to prevail than for the projected numbers themselves. A few central points stand out from this exercise:

1. Historically, about 60 percent of all the producing units defined as farms sell less than $20,000 of agricultural product (1987 prices). Unless the definition of a farm is changed, and that is unlikely, this proportion is likely to continue or might even grow

a little larger. It is here that the trend-based analysis developed in this report is most divergent from the OTA projections.

2. Part-time farms with sales of $20,000-39,999, have declined in importance in terms of both numbers and total value of sales in the past 20 years. It is likely that this trend will continue under nearly any scenario projected, particularly if environmental regulation and requirements increase in the 1990s.

3. The much talked about decline in the number of small, family farms, where farming provides the major source of family income (sales of $40,000-99,999), is likely to continue as it has in the past 20 years. The importance of this group as a proportion of the total number of "commercial" farms has also decreased somewhat. *Nevertheless, in any serious projection for the year 2000, this group will continue to be the most important class in terms of total numbers among all farms with sales of $40,000 or more.* The rapid disappearance of this group and the demise of small family farms does not emerge from any of the projections considered likely by 2000.

4. In nearly all of the projections, the number of farms with sales of $100,000-249,999 is substantially smaller than those with $40,000-99,999 but larger than the next three size classes. While it is possible that the number of these farms in the year 2000 will be larger than 1987 as smaller family farms expand modestly, most projections suggest somewhat smaller numbers.

5. The rate at which there are increases in each of the three largest size classes based on sales is most difficult to project given the limited historical bases, especially for farms with sales of $1,000,000 or more. This largest size category deserves more attention than it has received. All of the projections indicate that a small absolute increase in the number of these farms is associated with a substantial increase in the proportion of total sales arising from this size group. The impact of these farms on other commercial farms in the area, on input supply and on markets for agricultural output deserves increased attention.

Notes

Each author contributed equally to the analysis and the chapter; senior authorship is not assigned.

1. An age cohort are all the people born in the same decade.

2. See Lin, Coffman, and Penn's (1980) Tables 5 and 6 for a more detailed reporting of these projections (p. 11).

References

Anderson, T. W., and A. Goodman. 1957. "Statistical Inference about Markov Chains." *The Annals of Mathematical Statistics* 28: 89-110.

Batie, Sandra. 1989. "Sustainable Development: Challenges to the Profession of Agricultural Economics." *American Journal of Agricultural Economics* 71: 1083-1101.

Belden, J., D. E. Brewster, J. C. Doyle, B. B. King, and P. Stolfa. 1980. "A Dialogue on the Structure of American Agriculture: Summary of Regional Meetings." United States Department of Agriculture.

Berg, S. 1990. "City Folk's Tough Questions Shake up Farm Bill Debate." *Star Tribune*, Minneapolis, MN, July 6, 1990, p. 1A, 12A-13A.

Bergland, B. S. 1981. "A Time to Choose: Summary Report on the Structure of Agriculture." United States Department of Agriculture.

Brewster, D. 1979. "The Family Farm: A Changing Concept," in *Structure Issues of American Agriculture.* Agricultural Economics Report 438, Economics, Statistics, and Cooperative Service, U.S. Department of Agriculture.

Carter, H. O., and W. E. Johnston. 1980. "Farm-Size Relationships, with an Emphasis on California," in C. F. Nuckton, ed., Department of Agricultural Economics, Giannini Foundation Project Report, California Agricultural Experiment Station, Division of Agricultural Sciences, University of California, Davis.

Clift, E. 1990. "The Inside Guerrilla: Dick Armey fights farm supporters in Congress." *Newsweek,* July 16, 1990, p. 22.

Economics and Statistics Service. 1979. "Status of the Family Farm." Second Annual Report to the Congress, Agricultural Economics Report 434, United States Department of Agriculture.

Economics, Statistics, and Cooperatives Service (ESCS). 1979. "Structure Issues of American Agriculture." Agricultural Economic Report 438, U.S. Department of Agriculture.

Ehrensaft, P., P. LaRamee, R. D. Bollman, and F. H. Buttel. 1984. "The Microdynamics of Farm Structural Change in North America: The Canadian Experience and Canada-U.S.A. Comparisons." *American Journal of Agricultural Economics* 66: 823-828.

Gardner, Bruce L. 1978. "Public Policy and the Control of Agricultural Production." *American Journal of Agricultural Economics* 60: 836-43.

Hayenga, M., V. J. Rhodes, J. A. Brandt, and R. E. Deiter. 1985. *The U.S. Pork Sector: Changing Structure and Organization.* Ames: Iowa State University Press.

Just, Richard, and Gordon Rausser. 1989. "An Assessment of the Agricultural Economics Profession." *American Journal of Agricultural Economics* 71: 1177-1190.

Kislev, Yoav, and Willis Peterson. 1982. "Price, Technology and Farm Size." *Journal of Political Economy* 90: 578-595.

Krenz, Ronald D. 1964. "Projection of Farm Numbers for North Dakota With Markov Chains." *Agricultural Economics Research* 16(3): 77-83.

Lin, William, George Coffman, and J. B. Penn. 1980. "U.S. Farm Numbers, Sizes, and Related Structural Dimensions: Projections to Year 2000." Technical Bulletin No. 1625, National Economics Division, Economics, Statistics, and Cooperatives Service; U.S. Department of Agriculture.

Moore, C. V., D. L. Wilson, and T. C. Hatch. 1982. "Structure and Performance of Western Irrigated Agriculture With Special Reference to the Acreage Limitation Policy of the U.S. Department of Interior." Bulletin 1905, Information Series 82-2, Giannini Foundation of Agricultural Economics, Division of Agricultural Sciences, University of California.

Newsweek. 1990. "Do American Farmers Need New Handouts." July 16, 1990, p. 22.

The New York Times. 1990. "Get Fat Farmers Off Welfare." Editorial, June 19, 1990, p. A14.

The New York Times. 1990. "Fat Farmers at the Public Trough." Editorial, July 23, 1990, p. A12.

Nodland, R. 1990. "Make Farm Programs Work as They Should." Letter to the Editor, The New York Times, July 10, 1990, p. A14.

Nott, S. B., and C. J. Doyle. 1988. "Milk Quotas in the European Community." Michigan State University, S.E. Rpt. 519.

95th Congress. 1977. Food and Agriculture Act of 1977. House of Representatives, Report 95-599, September, 1977.

Paarlberg, Don. 1980. Farm and Food Policy: Issues of the 1980s. Lincoln: University of Nebraska Press.

Reimund, Donn A., J. Rod Martin, and Charles V. Moore. 1981. "Structural Change in Agriculture: The Experience for Broilers, Fed Cattle, and Processing Vegetables." Technical Bulletin No. 1648, ESS, USDA.

Schatzer, R. J., R. K. Roberts, and E. O. Heady. 1983. "A Simulation of Alternative Future in U.S. Farm Size." North Central Journal of Agricultural Economics 5(1): 1-7.

Schaub, James. 1990. "The Peanut Program and Its Effects." National Food Review 13: 1.

Schultz, T. W. 1945. Agriculture in an Unstable Economy. New York: McGraw-Hill Book Company.

Sonka, S. T., and E. O. Heady. 1974. "American Farm-Size Structure in Relation to Income and Employment Opportunities of Farms, Rural Communities, and Other Sectors." CARD Report 48, Dept. of Economics, Iowa State University, Ames, IA.

Stavins, R. N., and B. F. Stanton. 1980. "Using Markov Models to Predict the Size Distribution of Dairy Farms, New York State, 1968-1985." Cornell University Agricultural Experiment Station, A. E. Res. 80-20.

Sumner, Daniel A. 1985. "Farm Programs and Structure Issues." U.S. Agricultural Policy: The 1985 Farm Legislation. Amer. Ent. Inst., 283-329.

Teigen, Lloyd D. 1988. "Agricultural Policy, Technology Adoption, and Farm Structure." Staff Report No. AGES880810, Agriculture and Trade Analysis Division, Economic Research Service, U.S. Department of Agriculture, Washington, DC.

Tweeten, Luther. 1984. *Causes and Consequences of Structural Change in the Farming Industry.* NPA Report 207, Food and Agriculture Committee, Washington, DC; National Planning Association.

U.S. Congress. 1986. "Technology, Public Policy, and the Changing Structure of American Agriculture." OTA-F-285, Office of Technology Assessment, Washington, DC: U.S. Government Printing Office.

U.S. Department of Agriculture. 1991. *Economic Indicators of the Farm Sector: National Financial Summary, 1990.* Economic Research Service, ECIFS 10-1, Washington, DC.

Vogeler, Ingolf. 1981. *The Myth of the Family Farm: Agribusiness Dominance of U.S. Agriculture.* Westview Press.